Spirits of Life and Perception

Studien und Texte zur Geistesgeschichte des Mittelalters

Begründet von

Josef Koch

Weitergeführt von

Paul Wilpert, Albert Zimmermann und
Jan A. Aertsen

Herausgegeben von

Andreas Speer

In Zusammenarbeit mit

Wouter Goris, Kent Emery, Jr.
und Georgi Kapriev

BAND 136

The titles published in this series are listed at *brill.com/stgm*

Spirits of Life and Perception

Albert the Great's Early and Mature Psychophysiology in Light of His Arabic Sources

By

Michele Meroni

BRILL

LEIDEN | BOSTON

The Library of Congress Cataloging-in-Publication Data is available online at http://catalog.loc.gov
LC record available at http://lccn.loc.gov/2024020939

Typeface for the Latin, Greek, and Cyrillic scripts: "Brill". See and download: brill.com/brill-typeface.

ISSN 0169-8028
ISBN 978-90-04-70473-2 (hardback)
ISBN 978-90-04-70674-3 (e-book)
DOI 10.1163/9789004706743

This book is printed on acid-free paper and produced in a sustainable manner.

To my parents, Franco and Raffaella

Do they never reflect on the camels, and how they were created?
The heaven, how it was raised on high?
The mountains, how they were set down?
The earth, how it was made flat?

The Holy Qur'ān, 88: 17–20

•••

Most of the more complex emotions are common to the higher animals and ourselves.
Everyone has seen how jealous a dog is of his master's affection, if lavished on any other creature; and I have observed the same fact with monkeys.
This shows that animals not only love, but have the desire to be loved.

CHARLES DARWIN, The Descent of Man

Contents

Acknowledgements

This book represents the reworking of my doctoral dissertation ("Spirits of Life and Perception. Greek and Arabic Legacies in Albert the Great's Early Theology and Natural Philosophy"), defended in July 2021 amidst the COVID-19 pandemic to achieve the Ph.D. title at the Ludwig-Maximilians Universität München, in joint supervision (cotutelle) with the University of Milan. Ironically enough, the primary subject of both dissertation and book is *pneuma*, a notion intimately linked to breathing, from the Antiquity to the Renaissance until the dawn of modern science and medicine. The hardship that I, as every single scholar of all academic fields, had to go through during the pandemic years makes my gratitude to my doctoral supervisors, Professor Luca Bianchi, Prof. Peter Adamson, and Prof. Luigi Campi, grow even stronger. I am deeply thankful to Professor Luca Bianchi for having warmly welcomed my research proposal when I was still a newbie in the history of medieval philosophy with a strongly Classicistic background, for supervising my research with the utmost attentive scrutiny and the sincere and honest critical attitude vital for the scientific growth of a young scholar. I owe my second supervisor, my German *Doktorvater* Peter Adamson, a broader vision on the history of philosophy—not without any gaps, for my competences are limited to the history of medieval philosophy—and a constant encouragement to dig always deeper into the most materialistic explanation of natural phenomena. I am also indebted to Professor Luigi Campi for his meticulous supervision of the most minute formal details of my work, and his constant encouragement throughout my doctoral years. Given that this book does not exactly range within his research interest, his contribution is even more commendable. I cannot thank enough my home department of philosophy "Piero Martinetti" (Università degli Studi di Milano) for providing me with a supportive working environment and giving me the chance to grow both scientifically and as a person. It was there I was able to meet and collaborate with excellent scholars in the history of medieval and renaissance philosophy such as Diana Di Segni, Andrea Fiamma and Pierdavide Accendere. My sincere gratitude goes out also to my doctoral colleagues Ivan Malara, Sara Pavan, David Dal Bianco, and Jessica Murano for being there all (or most of) the steps of this long journey, for their constant and friendly help, support, and encouragement.

On the other hand, this book simply could not have existed without the time spent at LMU-Munich's School of Ancient Philosophy (MuSAPh) and the team of the ERC Project "Animals in the Islamic World" led by Peter Adamson. Among many people, I wish to thank Tommaso Alpina for being the most determining scientific influence in my study of Arabic sciences and philosophy, for guiding me through the interpretation of the most complex Avicennian pages and constantly helping me discuss the thorniest issues in this book. I am also sincerely grateful to Rotraud Hansberger for warmly welcoming me at MuSAPh and constantly helping me develop and refine my skills in philosophy in the Islamic World and philosophical Arabic. I also thank my colleagues Sarah de Mendonça-Virgi, Alex Lamprakis, Max Bergamo, and Caterina Pavoni for helping me settle in and teach me how to make the most out of my time at MuSAPh. I also wish to express my gratitude to Henryk Anzulewicz (Albertus-Magnus-Institut, Bonn) for his generous and punctual advice that helped me correct the flaws of the dissertation which led to this book. Since my undergraduate studies, he has been a constant presence in my development of an independent approach to Albert's works, based on hard textual evidence as much as possible. At the same time, I am deeply thankful to Silvia Donati (Albertus-Magnus-Institut, Bonn) for her precious indications and discussions on the core issues of the thought of such a complex yet fascinating thinker like Albert. I am also grateful to the teaching staff of the Abjadiyya Institute (Fes, Morocco), especially Issam Marjani and Fatima Benhabyb, for their generous and helpful advice in improving my Arabic. I also thank Katja Krause (MPIWG-Berlin) for her insightful comments on my dissertation, Marc-Aeilko Aris (Albertus-Magnus-Institut) for his stimulating observations during the viva of my doctoral dissertation, and Carson Bay for the careful and patient revision of my English. I also thank the two anonymous reviewers for their stimulating and constructive comments and the board of STGM for their insightful helpfulness. The rest of formal and content errors are solely a responsibility of mine.

This book is the result of a long journey started in 2014–2015, during my last academic year of undergraduate studies. It owes much of its rationale to the impulse of Paolo Branca, who inspired and encouraged me to pursue the study of the cross-cultural influences between the Islamic World and our Western world, stressing the points of contact and constructive influence rather than insisting on and emphasizing the natural differences among different cultures. The interest in these studies grew during my Erasmus at Freiburg, thanks to the advice of Nadja Germann, Ulrich Rebstock and Catherine König-Pralong, and has been further developed throughout my doctoral studies thanks to all the people I have met, among whom I wish to deeply thank (besides my official supervisors and those listed above) Amos Bertolacci, Gabriella Zuccolin, Anna Tropia, Marilena Panarelli, Mario Loconsole, and Evelina Miteva. I thank my closest friends, Mirko, Stefano, Giacomo, Marta, for sticking with me through thick and thin and supporting me every step of the way. As for my other friends who have been supportive of my scholarly endeavors, scattered through Milan,

Genova, Turin and other Italian cities, their list would be too long here not to be an anonymous and dispersive one. They know how crucial their staunch support has been to me through these years and I will not thank them enough for that. Last, I want to thank my parents Franco and Raffaella for their unconditioned love and support through this journey, and for teaching me since my childhood to value each and every culture, looking at different languages, religions or cultural products such as philosophy simply as different yet valuable manifestations of humanity in all its possible colors. I dedicate this book to them.

Abbreviations

1 Arabic Sources and Latin Translations

The Arabo-Latin translations of Arabic works constitute the key part of this work and will be cited by putting the Latin as the primary text of reference and the Arabic in square brackets. Quotes from the Latin translation will follow the model of referencing provided by Theodor's Köhler book "Homo animal nobilissimum",[1] and the manner of quotation of the Arabic sources will be uniformed to it. The bracketed reference to the Arabic page number will not signal book and chapter of the work, except when the textual divisions between Arabic and Lating differs, as in the two examples:

Avicenna, Abbreviatio de animalibus, IX, c. 5, ed. Venezia 1508, fol. 43rb [ed. Cairo 1970, 174.1–2].
Avicenna, Abbreviatio de animalibus, IV c. 3, ed. Venezia 1508, fol. 33va [IV c. 2, ed. Cairo 1970, 64.6–7]

When we shall refer only to the Arabic text or to untranslated Arabic works, we shall use their Arabic name alone and translate the passage into English:

Avicenna, Ḥayawān, I c. 1, ed. Cairo 1970, 1.10–13: "Let us now etc. [...]"

To achieve consistency in quotations throughout the whole book, if the Arabic title is not present in the book title, this title will be specified in the following list of editions according to the alphabetical order of authors.

1.1 *Algazel (Abū Hāmid al-Ġazālī)*

Algazel (Abū Hāmid al-Ġazālī): Maqāṣid al-falāsifa: fi-l-manṭiq wa-l-ḥikma al-ilāhiyya wa-l-ḥikma al-ṭabīʿiyya, Cairo 1936.
Algazel (Abū Hāmid al-Ġazālī): Metaphysica, ed. Joseph. T. Muckle, *Algazel's Metaphysics: A Mediaeval Translation*, Toronto 1933/1967.
Algazel (Abū Hāmid al-Ġazālī): Metaphysica [IV.1–5], ed. Eva St. Clair, "Algazel on the Soul: A Critical Edition," *Traditio* 60 (2005): 47–84.[2]

1 Homo animal nobilissimum. *Konturen des spezifisch Menschlichen in der naturphilosophischen Aristoteleskommentierung des dreizehnten Jahrhunderts*, 3 voll. (Studien und Texten zur Geistesgeschichte des Mittelalters 114.2/1), Leiden-Boston 2014, vol. 2/1.

2 When the two editions will overlap, we shall use St. Clair's more recent one. References to

1.2 *Al-Kindī*

Al-Kindī: Liber de somno et visione quem edidit Jacobus Alchindus, magister vero Gerardus Cremonensis transtulit ex arabico in latinum, ed. Albino Nagy, in *Die philosophischen Abhandlungen des Ja'qūb ben Isḥāq al-Kindī*, (Beiträge zur Geschichte der Philosophie des Mittelalters 5/2), Münster 1897, 12–27

Al-Kindī: Risāla fī Māhiyyat al-nawm wa-l-ru'yā, in: Rasā'il falsafiyya, ed. Muḥammad 'Abdu Abū Rīda, Cairo/Frankfurt a.M., 1950–1953/1999, 293–311.

1.3 *Averroes (Ibn Rušd)*

Averroes (Ibn Rušd): Al-Kulliyyāt fī al-tibb, ed. Mohamed 'Ā. al-Ǧābirī, Beirut 1999.

Averroes (Ibn Rušd): Colliget (Aristotelis opera cum Averrois commentariis, Suppl. I), Venice 1562/Frankfurt a.M. 1962, fols. 1–172.

Averroes (Ibn Rušd): Commentarium magnum in Aristotelis de anima libros, ed. Frederick St. Crawford (Corpus commentariorum Averrois in Aristotelem, Versionum Latinarum VI/1), Cambridge (Mass.) 1953.

Averroes (Ibn Rušd): Compendium libri Aristotelis de memoria et reminiscentia, ed. Henricus Blumberg, Averrois Cordubensis, Compendia librorum Aristotelis qui Parva Naturalia vocantur (Corpus commentariorum Averrois in Aristotelem: Versionum Arabicorum VII), Cambridge (Mass.) 1972, 36–51.

Averroes (Ibn Rušd): Compendium libri Aristotelis de memoria et reminiscentia, ed. Aemilia L. Shields/Henricus Blumberg, Averrois Cordubensis Compendia librorum Aristotelis qui Parva naturalia vocantur (Corpus Commentariorum Averrois in Aristotelem, Versionum Latinarum VII), Cambridge (Mass.) 1949, 47–72 [henceforth: Averroes, De memoria et reminiscentia].

Averroes (Ibn Rušd): Compendium libri Aristotelis de sensu et sensato, ed. Henricus Blumberg, Averrois Cordubensis, Compendia librorum Aristotelis qui Parva Naturalia vocantur (Corpus commentariorum Averrois in Aristotelem: Versionum Arabicorum VII), Cambridge (Mass.) 1972, 1–35.

Averroes (Ibn Rušd): Compendium libri Aristotelis de sensu et sensato, ed. Aemilia L. Shields/Henricus Blumberg, Averrois Cordubensis Compendia librorum Aristotelis qui Parva naturalia vocantur (Corpus Commentariorum Averrois in Aristotelem, Versionum Latinarum VII), Cambridge (Mass.) 1949, 3–44 [henceforth: Averroes, De sensu et sensato].

Averroes (Ibn Rušd): In Aristotelis Metaphysicorum libros XIII commentarius (Aristotelis Opera cum Averrois Commentariis, vol. 8), ed. Venezia 1562/Frankfurt a.M. 1962, fols. 1ra–355vb.

St. Clair's text instead of Muckle's will be highlighted in the footnotes as following: Algazel, Metaphysica, pars, tract. IV, c. 1, ed. St. Clair 2005, 235.

Averroes (Ibn Rušd), In Aristotelis Meteorologicorum libros IV commentarius (Aristotelis Opera cum Averrois Commentariis, vol. 5), ed. Venezia 1562/Frankfrurt a.M. 1967, vol. 5, fols. 400ra–487vb.

Averroes (Ibn Rušd), Tafsīr mā ba'da at-ṭabī'a. *Texte arabe inédit établi par Maurice Bouyges*. 2nd edition, Beyrouth 1973.

1.4 *Avicenna (Ibn Sīnā)*

Avicenna (Ibn Sīnā): Abbreviatio de animalibus (Ed. Venezia 1508/Frankfurt a.M. 1961, 29–64).

Avicenna (Ibn Sīnā): Al-Šifā', al-Ṭabī'iyyāt VIII: al-Ḥayawān, ed. 'Abd al-Ḥalīm Muntaṣir Sa'īd Zāyid, 'Abdallāh Ismā'īl, Cairo 1970.

Avicenna (Ibn Sīnā): Al-Šifā', al-Ṭabī'iyyāt, Ma'ādin wa-Āṯār 'ulwiyya, ed. 'Abd al-Ḥalīm Muntaṣir Sa'īd Zāyid, 'Abdallāh Ismā'īl, Cairo 1965.

Avicenna (Ibn Sīnā): Al-Šifā', al-Ṭabī'iyyāt, al-Samā' al-ṭābī'ī, ed. Sa'īd Zāyid. With a preface by Ibrāhīm Madkūr, Cairo, 1983.

Avicenna (Ibn Sīnā): Canon (Ed. Venezia 1507/Hildesheim 1964).

Avicenna (Ibn Sīnā): De anima, ed. Fazlur Rahman, *Avicenna's* De anima (*Arabic text*): *being the Psychological Part of* Kitāb al-Shifā', London 1959.

Avicenna (Ibn Sīnā): De diluviis, ed. Manuel Alonso Alonso, "Homenaje a Avicena en su milenario. Las traducciones de Juan Gonzáles de Burgos y Salomón," *Al-Andalus* 14 (1949): 306–308.

Avicenna (Ibn Sīnā): De medicinis cordialibus, ed. Simone Van Riet, Liber de anima, Avicenna Latinus: Liber de anima seu sextus de naturalibus, IV–V, Louvain–Leiden 1968, 187–210.

Avicenna (Ibn Sīnā): *Les plantes; texte etab. et éd. par Abd el-Halim Montasir, Said Zayed, Abdullah Ismail; introd. by Ibrahim Madkour*, Cairo 1965 [henceforth: Avicenna, Nabāt].

Avicenna (Ibn Sīnā): Liber de anima, ed. Simone Van Riet, Avicenna Latinus: Liber de anima seu sextus de naturalibus, I–II–III, Louvain–Leiden 1972; IV–V, Louvain–Leiden 1968.

Avicenna (Ibn Sīnā): Liber de philosophia prima, ed. Simone Van Riet, Avicenna Latinus: Liber de philosophia prima sive scientia divina, V–X, Louvain–Leiden 1980.

Avicenna (Ibn Sīnā): Sufficientia, ed. Simone Van Riet, Avicenna Latinus: Liber primus naturalium, Louvain-la-Neuve–Leiden 1992.

Avicenna (Ibn Sīnā): al-Qānūn fī al-ṭibb, ed. Institute of History of Medicine and Medical Research, vols. I, III.1, and IV, New Delhi 1981–1996.

Avicenna (Ibn Sīnā): Risāla fī-'adwiya al-qalbiyya, ed. Muhammad Zuhayr al-Bābā, Min mu'allafāt Ibn Sīnā al-ṭibbiyya, Aleppo 1404/1984, 221–294.

Avicenna (Ibn Sīnā): *The Metaphysics of The Healing* (al-Šifā': al-Ilāhiyyāt), *a parallel English-Arabic Text; translated, introduced and annotated by Michael E. Marmura*, Provo 2005.

1.5 *Costa Ben Luca (Qusṭā ibn Lūqā al-Baʿlabakkī)*, De differentia spiritus et animae

Costa Ben Luca (Qusṭā ibn Lūqā al-Baʿlabakkī): *Épître sur la différence entre l'esprit et l'âme. Éd. et trad. par Gérard Troupeau avec la collaboration de Joseph Dagher. Hommage d'André Miquel.* Paris: Librairie orientaliste Paul Geuthner, 2011, iii–xxxi. [henceforth: Costa Ben Luca, Risālat al-faṣl bayna al-rūḥ wa-l-nafs].

Costa Ben Luca (Qusṭā ibn Lūqā al-Baʿlabakkī): De differentia spiritus et animae, ed. Judith Carol Wilcox, *On the Transmission and Influence of Qusta Ibn Luqa's "On the Difference Between Spirit and Soul"*, Unpublished Ph.D. Thesis, New York 1985, 143–191.[3]

Costa Ben Luca (Qusṭā ibn Lūqā al-Baʿlabakkī): *Le livre des caractéres*, ed. Paul Sbath, "Le livre des caractères de Qosta ibn Louqa (1941) grand savant et célèbre médecin du IXe siècle," *Bulletin de l'Institut d'Égypte* 23 (1941), 103–169 [henceforth: Costa Ben Luca, Kitāb min ʿilal iḫtilāf al-nās fī aḫlāqihim].

1.6 *Haly Abbas (ʿAlī ibn ʿAbbās al-Maǧūsī)/Constantine the African*, Pantegni

Haly Abbas (ʿAlī ibn ʿAbbās al-Maǧūsī): Kāmil al-ṣināʿa al-ṭibbiyya, ed. Khālid Harbī, Alexandria 2018.

Constantinus Africanus: Pantegni [Theorica IV.19] ed. Charles Burnett, "The Chapter on the Spirits in the *Pantegni* of Constantine the African," in *Constantine the African and ʿAlī Ibn al-ʿAbbās al-Maǧūsī. The* Pantegni *and Related Texts*, ed. Charles Burnett and Danielle Jacquart, (Studies in Ancient Medicine) 10, Leiden-New York-Köln 1994, 99–120.

---

3 Note that, given the alternated disposition of the two texts in Wilcox's thesis, John's version appears in pages with odd numbers, while the anonymous in pages with even numbers. Because of this peculiarity, quotes from successive pages will go from odd to odd or even to even—and not consecutive—numbers: e.g.: *De differentia*, 257.10–259.31. A second edition of Costa ben Luca's text: *Excerpta e libro Alfredi Anglici De motu cordis item Costa-Ben-Lucae De differentia animae et spiritus liber translatus a Johanne Hispalensi*, edited by Carl Sigmund Barach, (Bibliotheca philosophorum mediae aetatis) 2 (Innsbruck: Verlag der Wagnerschen Universitaets-Buchhandlung, 1878), 117–139, is based only on three manuscripts, later than those deployed by Wilcox and much less adherent to the Arabic text established by Troupeau. Since Wilcox's edition relies on a much broader, older and solid series of manuscript testimonies than Barach's edition, her edition is the natural candidate to be used in a scientific work.

2 Albert, the *editio Coloniensis* and the *Aristoteles Latinus*

In this book, we shall follow the notation of Albert's *editio Coloniensis* concerning the presence of words borrowed from the Latin versions of Aristotle. The notation of Albert's works will be the following:

Albertus Magnus, Metaphysica VII tr.2 c. 5 (Ed. Colon. XVI/1, 345.49–51): "*Sanitas autem est ratio* factiva, quae est *in anima* medici, *et* est species *in scientia* factiva, non speculativa."

To assess the relevance of Aristotelian doctrines in this book, we shall refer to the most relevant passages of the *Aristoteles Latinus* to put the reader in a better condition to understand the material condition according to which Albert became acquainted with Aristotle's doctrines. To do so, we shall rely on use the following editions of the *Aristoteles Latinus*:

Aristotle, De anima [Translatio vetus], ed. Decorte-Brams, Iacob Veneticus translator Aristotelis. De anima, textus editionis quae paratur a +J. Decorte et J. Brams [A.L. XII.1]. Text available online at the ALD: https:/clt.brepolis.net/ald [Henceforth: Aristoteles Latinus, De anima [A.L. XII.1], ed. Decorte-Brams (online)].

Aristotle, De generatione animalium [Translatio Scoti], ed. Aafke M.I. van Oppenraaij, *Aristotle*, De animalibus. *Michael Scot's Arabic-Latin Translation*, Part Three: Books XV–XIX: Generation of Animals (Aristoteles Semitico-Latinus 5/2), Leiden-New York-Köln 1992 [Henceforth: Aristoteles Latinus, De generatione animalium (Aristotle Semitico-Latinus 5/2), ed. van Oppenraij 1992.]

Aristotle, De longitudine et brevitate vitae [Translatio vetus], ed. Manuel Alonso, in *Pedro Hispano: Obras filosóficas*, III (Instituto de Filosofía "Luis Vives", Serie A Núm. 4), Madrid 1952, 403–411.

Aristotle, De memoria et reminiscentia [Translatio vetus]. Textus praevius a S. Donati paratus [A.L. XIV.1]. Text available online at the at the ALD only: https:/clt.brepolis.net/ald [henceforth: Aristoteles Latinus, De memoria et reminiscentia [A.L. XIV.1], ed. Donati (online)].

Aristotle, De sensu et sensato [Translatio vetus]. Textus qui a L. Peeters paratur [A.L. XIII.1]. Text available online at the ALD only: https:/clt.brepolis.net/ald [henceforth: Aristoteles Latinus, De sensu et sensato [A.L. XIII.1], ed Peeters (online)].

Aristotle, Aristotelis De somno et vigilia liber adiectis veteribus translationibus et Theodori Metochitae commentario, ed. Hendryk J. Drossaart Lulofs, Leiden 1943 [henceforth: Aristoteles Latinus, De somno, ed. Drossaart Lulofs 1943].

Aristotle, De sompno et vigilia II–III [Vetus translatio–Nova translatio], ed. Hendrik J. Drossaart Lulofs, Aristotelis De insomniis et De divinatione per somnum. *A new*

Edition of the Greek Text with the Latin Translation, Index verborum (Philosophia antiqua 2), Leiden 1947. Text available online at the ALD: https://clt.brepolis.net/ald [henceforth: Aristoteles Latinus, De insomniis/De divinatione per somnum, ed. Drossaart Lulofs 1947].

3 Abbreviations and Secondary Literature

Titles of secondary literature will come in a complete form in the first occurrence, then will be abbreviated in the following appearances in the same manner it is shortened in Köhler's book.

Introduction

Albert the Great (1200–1280) stands out among medieval philosophers and theologians. Without having practiced medicine a single day in his life, he enjoyed a solid reputation as a medical authority for decades, if not centuries. Tellingly, Peter of Abano (1250–1316), professor of medicine at Padua and author of the famous *Conciliator differentiarum medicorum et precipue philosophorum*, recommended the reading of Albert's 26-book *De animalibus* to medical students.[1] Moreover, Albert's very detailed acquaintance with medical teachings often led Renaissance authors, who often refrained from naming medical authorities, to make exceptions for him.[2] His aura of medical authority and deep knowledge of the human body, mixed with his alleged knowledge (or, according to some, even practice) of magic, led to the formation of sometimes curious legends about him. One of the most outstanding is that of the android. The story goes that his pupil Aquinas, having knocked at his master's door, was greeted with a "Salve" by what he thought it was his master. Upon entering the room, however, Aquinas realized that what was standing in front of him was actually a man-made android, which he struck with his cane and destroyed.[3]

This Frankenstein-esque legend implies that Albert, along with his controversial relationship with magic, indeed knew the structure and function of the human body down to the smallest detail. This would seem to answer in part the most intuitive question arising from the fact of Albert's physiological popularity: why would men of medicine read the works of someone who was, in the first place, a theologian, and who had never seen a patient, let alone ever practiced a single phlebotomy or surgery?

Among Scholastic authors, it is well-recognized that Albert played a prominent role in the transmission of Graeco-Arabic wisdom. Having had a marked interest in the world of living beings, he felt the need to draw on medico-philosophical texts—mostly from the Islamic World—to update and complement Aristotelian psychophysiology. He fulfilled this task, especially in his mature Aristotelian paraphrases, by describing in detail human and animal

1 Nancy Siraisi, "The Medical Learning of Albertus Magnus," in *Albertus Magnus and the Sciences. Commemorative Essays*, ed. James. A. Weisheipl (Studies and Texts 49), Toronto 1980, 380 n. 2.

2 Siraisi, "The Medical Learning," 381.

3 Willehad P. Eckert, "Albert-Legenden," in *Albert der Grosse: seine Zeit, sein Werk, seine Wirkung*, ed. Albert Zimmermann and prepared for print by Gudrun Vuillemin-Diem (Miscellanea Medievalia 14), Berlin 1981, 19–20.

 | DOI:10.1163/9789004706743_002

anatomy and physiology in what became perhaps the most famous work on zoology in the Middle Ages, his *De animalibus*. In this work, his accurate exegesis of Aristotle's own views on generation, often read through the lenses of Avicennian doctrines, justified, for instance, the misattribution to him of the spurious yet famous *De secretis mulierum*.[4]

It is certain that Albert played a major role in the reception of Aristotelian zoology and psychophysiology at the beginning of the first half of the 13th century. Albert's works on natural philosophy not only displayed a marked interest in the use of medical learning but also insisted on what Albert called "experience", although Albert's concept of experience proves a slippery one. Yet despite his strong interest in scientific medicine, Albert can in no way be considered a scientist; much less can he be compared to the fathers of modern science such as Galileo when it comes to scientific methodology. Some of his theories are quite bizarre, as his famous report about the miscarriage of more than 100 fetuses at a time clearly suggests, or they conceal the outright appropriation of passages from other sources when speaking of "personal experience". It is hardly a coincidence that Marcantonio Zimara, the famous Paduan philosopher, chief editor of many Averroean texts and a deep connoisseur of Albert's works—which, for other things, he deeply valued—degraded Albert as "*rusticus mendax*", especially in the field of natural philosophy.[5]

Yet, this *rusticus mendax* played a key role in the early reception of medico-philosophical works from the Islamic world and built a study of the natural world which has not yet been fully explored, despite its massive historical popularity and influence. Animals, plants, minerals, humans: there was hardly a subject of the living world which escaped Albert's eager curiosity and his program of scientific and philosophical formation of a new Dominican intellectual élite in Cologne. Until now, much emphasis has been placed upon how Albert integrated large chunks of medical teachings—especially Avicenna's *Canon*—into works such as his zoology. State-of-the-art studies have shown how the influence of Avicennian zoology and medicine were crucial for the construction of Albert's philosophy of the natural world of sublunar beings, present to some extent even in his early works.

Until now, these considerations have been confined to a very few articles: it is thus a good time for scholarship on Albert to bring the relationship between Albert's early and mature psychophysiology under closer and more compre-

---

4 Siraisi, "The Medical Learning," 381.

5 Luca Bianchi, "*Rusticus mendax*: Marcantonio Zimara e la fortuna di Alberto Magno nel Rinascimento italiano," in Luca Bianchi, Studi sull'aristotelismo del Rinascimento, (Subsidia Mediaevalia Patavina 5), Padova 2003, 209–223.

hensive scrutiny. As a matter of fact, if we turn back the clock to Albert's early theological works such as the *De homine* (1241/1242) and the *Commentary on the Sentences* (1242–1249),[6] we soon realize that we can gain a broader and clearer picture of his psychophysiology than one might expect. Contrasting Albert's *De homine* with those works—such as the *De animalibus*—that earned him fame as a medical authority reveal unexpected surprises, which we will uncover in this book. This comparison must not be concerned merely with the primary object of medico-philosophical texts, viz. human beings, nor only his mature *De animalibus*, though. A careful study of Albert's use of medical teachings needs to frame the epistemic domain within which medical theories can be employed and evaluate the possible developments of medical doctrines for the understanding of the world of minerals, plants and animals. This attitude alone, aptly termed "universal physiology" by Joan Cadden,[7] brings into the bigger picture of this kind of research Albert's *Parva Naturalia*, his botany (*De vegetabilibus*) and—to some extent—his mineralogy (*Mineralia*).

Following this line of reasoning, this book will try to investigate the story of Albert's reception and development of a secular concept, spirit (Gr. *pneuma*, Ar. *rūḥ*, Lat. *spiritus*), in his study of humans, animals and—surprisingly—plants. This tiny fluid, resulting from the process either of breathing or of digestion, was charged with a key role in Albert's early and mature psychophysiology, which deserves a substantial inquiry of its own. This book will consider the functions of spirit as an instrument of the soul, or vehicle of its faculties, in two respects: life-related operations (life, nutrition, growth, generation) and phenomena (length and shortness of life) and perception-related activities (sense perception and cognitive activities), hence the title "Spirits of life and perception". Among his sources, certainly the influence of medico-philosophical texts from the Islamic World are prevalent. For this reason, Albert's "Arabic"—denoting their language of composition—sources will be given primary attention.

Spirit is defined by Albert in different manners: as instrument of the soul (*instrumentum animae*), proper vehicle of life (*vehiculum vitae proprium*), vehi-

6 On the chronology of Albert's works, see: *Albertus Magnus und sein System der Wissenschaften. Schlüsseltexte in Übersetzung. Lateinisch–Deutsch*. Intr., transl. and prepared for print by Hannes Möhle/Henryk Anzulewicz/Maria Burger/Silvia Donati/Ruth Meyer/Martin Bredenbeck/Susana Bullido del Barrio, Münster 2011, 28–31.

7 Joan Cadden, "Albertus Magnus' Universal Physiology: the Example of Nutrition," in *Albertus Magnus and the Sciences. Commemorative Essays*, ed. James A. Weisheipl (Studies and Texts 49), Toronto 1980, 231–248.

cle of the (soul) faculties (*vehiculum virtutum*) and subtle body. On the one hand, because of its nature as instrument of the soul and substrate for the inherence of vital heat, spirit plays the role of intermediary agent through which the soul imparts life to the body and exercises the activities specific to each living being. At the same time, spirit as a "subtle body" (*corpus subtile*) is subject to the laws of the world of coming-to-be and perishing and, in Galenic fashion, makes the capacities of the soul (also, life) dependent on the mixtures of the body. Amidst the body's intermixed make-up, spirit was known to be the element most akin to the soul. Therefore, its material properties made it into the material substance most apt to be the receptacle through which life (*vehiculum vitae*) and soul powers (*vehiculum virtutum*) were spread into the body. Like any material substance, spirit can be affected by the accidents to which matter is prone, and thus can fail in achieving its end or destination, just like a broken vehicle. Nonetheless, a third and final aspect that will be investigated here is the connection in Albert's thought between the material nature of spirit and God's providential design, often summed up by the famous axiom *opus naturae est opus intelligentiae*. If spirit is the vehicle through which life comes about, marking the moment when life begins, then how is it that this tiny and bubbly, yet crucial and life-giving substance, connects to the providential order of the universe and to its First Cause (God)?

Chapter 1 investigates Albert's understanding of the relationship between natural philosophy and medicine. A comparison between the criteria for the use of medical doctrines in Albert's earlier works of theology and his mature works on natural philosophy aims at clarifying the epistemic framework within which he includes medico-philosophical doctrines such as spirit itself. Chapter 2 represents the first step in the study of the causal relationship between spirit and life-related phenomena. Moving from the definition of life, this chapter sheds light on the correlation between Albert's emanative psychology and his theory of spirit, investigating the connection of this material substance with the most basic, necessary and sufficient conditions for life in living beings. Chapter 3 examines the role of spirit in generation, highlighting its function as connecting point between the world of heavenly influences and the sublunar world. The causal relevance of spirit in generation appears to be twofold: *qua* carrier of the *virtus formativa*, it is instrument of the formal cause (the father's soul) and vehicle of the cosmic heat, crucial in the peculiar case of spontaneous generation. Given the extensive treatment given by Albert to generation, this chapter will be split into two logical parts: human generation (3) and, on the other hand, animal, plant and spontaneous generation (4). Chapter 5 closes the section on "Spirits of Life"' and investigates the indirect causal connection between spirit *qua* carrier of vital heat and the material cause of longevity, radical moisture. Chapter 6 takes on the role of spirit in sense perception and the functions of common sense related to sense perception, shedding new light on the role of spirit in distal (sight, hearing, smell) and contact (touch and taste) senses and on the hierarchy of senses. Chapter 7 and 8 analyze the role of spirit in the so-called internally apprehensive powers (*potentiae apprehensivae deintus*), viz. common sense, imagination, estimation and memory. In particular, each of the two chapters analyzes a key argument: on the one hand, the cognitive divide between animals and humans in instinctual reactions and memory (chapter 7) and, on the other hand, the psychophysiology of sleep, divinatory and God-given dreams and prophecy (chapter 8). This study is conducted considering the different epistemic domains (philosophy and theology) where the psychophysiology of sleep and dreaming is tackled. For reasons of length and clarity, the relationship between spirit and motion and emotions (*passiones*) had been left out of this study, though it remains a desideratum for future research on Albert's psychophysiology. A final aspect that could complement the inquiry presented in this book is represented by the relationship between spirit, radical moisture and growth, only touched upon in chapters 3, 4, and 5 but which would require a study of its own. A last and possible step towards a qualitatively better knowledge of Albert's spirit theory would be represented by a detailed study of the relationship between spirit theory and Albert's doctrine on humoral complexion, especially as respects as the role of blood.

A close look at the role of spirit in the world of natural causes presents a case study that can flesh out the methodological and doctrinal peculiarities of Albert's early and mature psychophysiology. In terms of methodology, one aim of this book is to investigate the place reserved by Albert for medical doctrines in his psychophysiology and upon what epistemic parameters medicine is not allowed to infringe. This goal will be clearly exemplified by two main topics, viz. the psychophysiology of generation, with its debates on the role of females and on the origin of veins, and the complex case of God-given, divinatory dreams and prophecy. To do so, we shall need to contextualize the use of these explanations within the broader background of Albert's early and mature understanding of the relationship between philosophy and theology. Moreover, the primary purpose of this book will be to investigate the causal role played by spirit in life- and perception-related phenomena. By drawing on the generalizations which can be inferred from a detailed study of Albert's psychophysiology, this book will bring to the fore three key hitherto neglected aspects of Albert's spirit theory: (1) the strict causal correlation of the physiological processes in living beings, (2) the relationship between Albert's understanding of psychosomatic unity of living beings and the causal role of spirit in psychophysiological processes and (3) the place of spirit in a world ruled by divine providence, an idea substantiated in the axiom *opus naturae est opus intelligentiae*. As we shall set out to demonstrate, all of these three key aspects are inherited and reworked from Albert's Arabic medico-philosophical sources.

CHAPTER 1

A Matter of Principles: The Epistemic Status of Medicine in Albert's Early Theology and Aristotelian Paraphrases

Throughout the history of philosophy, the topic of the disputes between philosophers and physicians has been a recurrent and famous one, at least as far as Medieval philosophy goes. In the Islamic World, many philosophers—such as Avicenna, al-Rāzī, Averroes and Maimonides—were also physicians who engaged themselves with practice on field, or at least wrote on medical subjects, as in the case of al-Kindī, al-Farābī. In the Latin Middle Ages many theologians were also famous for being highly competent in matters of medicine: Giles of Rome wrote the famous *De formatione corporis humani in utero*, Albert the Great would become himself one of the most revered medical authorities without practicing medicine but having integrated and reshaped large portions of medical doctrines into his works of natural philosophy. The disagreement between physicians and philosophers in the Latin Middle Ages, best represented at an early stage by Peter of Abano's *Conciliator*, was a well-known topic of interest for many different reasons. During the formation of the universities, disputes between philosophers and physicians touched upon the theme of autonomy of medicine from philosophy, the (presumed) lack of adequate demonstrative cogency in medical arguments and, more in general, the compatibility between the epistemic principles of medicine and natural philosophy, strongly permeated by Aristotelian hylomorphism. Under this respect, the discussion of the Galenic-Aristotelian quarrels on the role of females in generation, the origin of veins, and the entrance of masculine semen into the embryo, represented the perfect exemplification of conflict between physicians and natural philosophers.

Natural philosophy appeared, on the one hand, as a science based on the epistemic system set out by Aristotle's *Posterior Analytics*, founded upon a rigidly deductive structure deriving necessary conclusions from self-evident premises. On the other hand, medicine suffered from problems of definition, viz. whether it had to be considered a 'science', an 'art' or just an instance of practical wisdom. Indeed, its aim was practical (the attainment of health) and essentially tied to the particular cases with which it was confronted. On the other hand, its definition in the Aristotelian corpus and in the subsequent tradition was quite wavering: at times it was defined as an "art" (as in Aristotle's *Metaphysics*), viz. a practical skill with no apparent theoretical framework,

 | DOI:10.1163/9789004706743_003

sometimes as a "science", divided into a theoretical part and a practical part. This latter division of medicine into theoretical and practical was influenced, during the Middle Ages, by authors of the Islamic World.[1] At any rate, the aim of a good physician was to heal the patient in the manner that was best suited to the nature of each individual, thus adopting a perspective in clear opposition to the universalizing purpose of philosophy (of nature). Still, as was often noted, a good physician also needed a firm training in the theoretical knowledge of human anatomy and the laws of nature to which the human body was subject. On the other hand, a philosopher of nature who wanted to take his exegesis of Aristotelian philosophy of nature down to the most minute physiological details had indeed to confront himself with medical doctrines, as Aristotle's physiology was largely outdated in many fields. The theoretical part of medicine significantly overlapped with the area of natural philosophy called psychophysiology, which examined the body-soul relationship in life-related processes. This overlap was particularly evident in Aristotelian works such as the *Parva Naturalia* and the zoological corpus. Given this overlap of subject matter and of the different epistemic aims (practical vs. theoretical) of medicine and natural philosophy, what was one to do in the case of conflicting doctrines? How could Galen's critique of Aristotle's theories on the role of females in generation, or on the origin of veins, e.g., be judged? Are these critiques epistemically sound or, given the lack of demonstrative cogency in medical argumentation, by nature tied to particulars, must be rejected as devoid of insufficient to hold true necessarily and always? In short, the question that this chapter will address revolves around the role of medico-philosophical doctrines in works of natural philosophy and their scientific, i.e. philosophic, underpinning.

In Medieval debates on the relationship between natural philosophy and medicine, one of the elements of this answer often lied in the interpretation of the subject of medicine, the body inasmuch as it can be healed (*corpus sanabile*). To understand the relationship between natural philosophy and medicine at the time of Albert, it is therefore of paramount importance to understand the different meanings of health and their influence on Albert's approach. Having elucidated this first crucial aspect of the topic, we will then have the conceptual tools to assess Albert's understanding of the relationship

1 On these terminological issues and struggles, see: Danielle Jacquart, "La médecine entre théorique et pratique: retour sur quelque définitions originelles," in *Handlung und Wissenschaft—Action and Science. Die Epistemologie der praktischen Wissenschaften im 13. und 14. Jahrhundert—The Epistemology of the Practical Sciences in the 13th and 14th Centuries*, ed. Matthias Lutz-Bachmann/Alexander Fidora (Wissenskultur und Gesellschaftlicher Wandel 29), Berlin 2008, 34–42.

between natural philosophy and medicine throughout his early theological and mature works. To do this, two more key factors will have to be considered: the influence of Albert's main medico-philosophical source (Avicenna) and Albert's mature and peculiar understanding of natural science. In particular, Avicenna's works constituted a first and crucial contribution to the history of medico-philosophical debates in the Latin West, fitting much of Galen's medical teachings in a staunchly Aristotelian system of natural science and being crucial for the formation of generation of physicians in the Latin Middle Ages.[2]

1 What Is Health and What Is Healing? A Conundrum in Medieval Discussions

During the 13th century, medicine had become a form of scientific knowledge with a proper course of study at the universities, codified by curricula and mandatory stages under the supervision of a senior physician. In fact, together with Law and Theology, medicine represented one of the three main faculties of European universities. Access to these studies, though, was conditioned by completing the mandatory course of studies at the arts faculties, where students focused on the arts of the trivium, quadrivium and natural philosophy, based largely on the Aristotelian *corpus vetus*. The subject of medicine, as physicians understood it, involved the *corpus sanabile*. In Medieval debates, it was this formal qualification of the subject matter that led to different conclusions concerning the relationship between natural philosophy and medicine, thus raising a second question about the mutual compatibility between the epistemic principles of medicine and those of natural philosophy.

As far as the human body (*corpus*) itself represented a partial conceptual overlap between the object of medicine and that of natural philosophy, it was the interpretation of the concept of "treatable" (*sanabile*) which became the subject of different interpretations and doctrines. To find an answer to this epistemic question, many Medieval philosophers referred to the Aristotelian theory of subalternation (*subalternatio*) of sciences. In their scholarly contributions, Reynolds and Zuccolin have investigated precisely the meaning of the term "sanabile" (Zuccolin) and its implication (Reynolds, Zuccolin) when it comes to the determination of the epistemic relationship between medicine and natural philosophy. In the eyes of a medieval philosopher, a science could

2 Danielle Jacquart, François Micheau, *La medecine arabe et l'Occident médiéval* (Collection Islam-Occident 7), Paris 1990.

be subordinated to another if its object fell within or was subalternated to the latter, given that the principles of the latter were derived from those established in the former.[3] In this context, the key role was played by the concept of healing and health. While it is intuitive that healing represents, in the end, a process causally produced by a natural potency, the individuation of the primary agent responsible for the process, nature or the art of medicine, is much less uncontroversial. By assuming that healing is a natural process triggered by an artificial practice and that the human body is a natural body, then one can conclude that medicine is actually a part of natural philosophy. By contrast, if healing is considered a process principled by an art whose principles are derived from natural philosophy, then medicine appears to be subalternated to natural philosophy and not a part of it. This latter position is, famously, Aquinas'. This conundrum about the relationship between medicine and natural philosophy is further complicated by questions of practical employment of medical doctrines, viz. the part of medicine which could be included into works of natural philosophy and the compatibility of Galenic anatomy and its theoretical foundations with the principles of natural philosophy, strongly permeated by Aristotelian hylomorphism.

While these questions concern health from a purely scientific perspective, there is still a second way to consider health that was quite popular in the Middle Ages. Chiara Crisciani has shown how, traditionally, disease was considered to be among the *tria mala*—together with moral weakness and physical vulnerability—which had entered the world in the wake of the Original Sin.[4] Diseases were viewed as a representation of the normal human condition, somehow a somatization of the postlapsarian state.[5] Thus, health primarily meant "spiritual health", which was often juxtaposed to the physical health derived from medical practice to shed light on theological doctrines via com-

3 Gabriella Zuccolin, "Il corpo sanabile e l'embrione. Il *De animalibus* come punto di intersezione tra filosofia naturale e medicina," in *La Medicina nel Basso Medioevo. Tradizioni e conflitti. Atti del 55. Convegno storico internazionale, Todi, 14–16 ottobre 2018* (Atti dei convegni del Centro italiano di studi sul basso Medioevo, Nuova Serie 32) Spoleto 2019, 119 n. 10; Philip L. Reynolds, *Food and the Body. Some Peculiar Questions in High Medieval Theology* (Studien und Texte zur Geistesgeschichte des Mittelalters 69), Leiden 1999, 137.

4 Chiara Crisciani, "Medicine and Philosophy in the Middle Ages. Sisters, Companions, Rivals," in *Medical Ethics. Premodern Negotiations between Medicine and Philosophy*, ed. Mariacarla Gadebusch Bondio, Stuttgart 2014, 30.

5 In her article, Crisciani argues that sickness acquires two further meanings. On the one hand, sickness acquires a parenetic value, prompting the ill patient to heal from his or her spiritual sicknesses as well. On the other hand, sickness makes the patient, to some extent, an image of the suffering Christ: Crisciani, "Medicine and Philosophy," 30.

parison. As we shall see, this view appears in Albert's early works and holds a key to understanding the early Albert's attitude towards medicine as the principle of the natural process of healing and, on the other hand, as a corpus of scientific writings with their own theoretical doctrines. Albert also followed one of the most commonplace comparisons at play in the early Middle Ages concerned the role of remedies and medicines: sacraments were considered 'medicines of the soul'. By the same token, while physical health had its own expert, the physician, spiritual health had its most prominent physician in Christ. During the administration of sacrament, his role was taken on by priests and clergymen. These comparisons between spiritual and physical health were commonplace in key texts commented upon by Albert during his activity as bachelor. This is the case with Peter Lombard's *Sentences*, whose fourth book was entirely devoted to the doctrine of sacraments.[6] The juxtaposition between spiritual and physical health, typical of earlier texts, offers us an insight with which to begin the examination of the early Albert's position on the relationship between natural philosophy and medicine.

2 Healing the Soul, Healing the Body: Medicine and Natural Philosophy in Albert's Early Theology

Albert's early theological works can be divided into two major groups: on the one hand, the *De sacramentis, De incarnatione, De resurrectione, De quattuor coaequaevis, De homine* and *De bono* (1242–1245). On the other hand, his *Commentary on the Sentences*, redacted between 1243 and 1249.[7] These works were written, according to current chronology, before 1245, but they were also subject, as demonstrated by Caterina Rigo, to later revisions, which were concluded by the year 1250.[8] In these works, Albert retained much of the traditional tension between physical and spiritual health, comparing the purgative agency of the 'sacred medicines' (sacraments) to the homonymous function of Purgatory.[9] Some of these passages, much as they retain this traditional tension

6 Reynolds, *Food and the Body*, 129.

7 For a periodization of Albert's works, see above: "Introduction", 3, n. 6.

8 Caterina Rigo, "Zur Redaktionsfrage der Frühschriften des Albertus Magnus," in *Albertus Magnus und die Anfänge der Aristoteles-Rezeption im lateinischen Mittelalter: vom Richardus Rufus bis zu Franciscus de Mayronis*, ed. Ludger Honnefelder/Rega Wood/Mechthild Dreyer/Marc-Aeilko Aris (Subsidia Albertina 1), Münster 2005, 325–374.

9 On Albert's doctrine of sacraments, see: Franz-Josef Nocke, *Sakrament und personaler Vollzug bei Albertus Magnus* (Beiträge zur Geschichte der Philosophie und Theologie des Mittelalters 41/4), Münster 1967, 15–41. According to Albert, a further type of sacred medicine is repre-

between two different kinds of health and the role of sacraments as spiritual remedies, offer us a key to understanding Albert's mature position on the relationship between natural philosophy and medicine. In particular, a passage from Albert's *Commentary on the Sentences* comparing the causal agency of Divine Grace at the moment of Baptism and during the sacrament of Confirmation seems particularly enlightening:

> The work of medicine is like a servant of nature, and nature, strengthened by medicine, works by itself; so likewise in spiritual [matters]. A sacrament, moreover, is like medicine; for this reason, its operation concerns the spiritual being as if [it was] serving [it]. Thus, the work of the strength-giving Sacrament (sc. Confirmation) will concern the matter of preexisting virtue and grace without being essentially distinct [from them].[10]

According to Albert, Confirmation (*operatio sacramenti praestantis robur*) restores the strength of pre-existing Divine Grace received at the moment of Baptism (*circa materiam virtutis et gratiae praeexistentis*); likewise, the medicament strengthens the nature of the living being (*natura confortata per medicinam*).[11] In both cases, Grace received in Confirmation and medical remedies (most likely, drugs) are only auxiliary in this process, since the ultimate primary causes are Divine Grace received in Baptism and nature. In Albert's view, nature and Divine Grace share a common goal: health, though from different perspectives.[12] While nature is undeniably deemed the primary agential cause in healing (*natura … operatur per se*), it remains unclear how the operation of

---

sented by the Revealed Law of Moses and Christ, both meant to preserve the health of God's people through history: see: Albertus Magnus, De bono tr. 5 q. 2 a. 2 (Ed. Colon. XXVIII, 286.15–23).

10 Albertus Magnus, Super IV Sententiarum, (dist. 1–22), ed. Auguste Borgnet, Paris 1893 (Ed. Paris. XXIX, 161a): "Item, medicinae operatio est sicut naturae ministra, et natura confortata per medicinam operatur per se: ergo similiter est in spiritualibus: sacramentum autem est sicut medicina: ergo operatio sua est circa esse spirituale sicut ministra: ergo operatio sacramenti praestantis robur, erit circa materiam virtutis et gratiae praeexistentis, et non distincta per se." Translation is mine.

11 Note that here the formulation is potentially ambiguous, as *medicina* can refer both to the art of medicine and the medicament itself. Given the primary practical focus of the passage and its comparison to the sacrament, the most fitting comparison is that of a drug or medicament. However, the theoretical implications of this line of reasoning which foresee nature as the primary agent in healing will appear more clearly in Albert's later *Metaphysics*.

12 On the polarity between health and Salvation, see: Crisciani, "Medicine and Philosophy," 29–30.

nature and of the pristine Divine Grace relate to each other and, more generally, whether these processes of physical and spiritual healing are voluntary, necessary etc. In a passage again from the 4th book of his *Commentary on the Sentences*, Albert seems to give an answer to the question on the role of nature in healing. Discussing the difference between spiritual and physical diseases, he argues that Sacraments work if only and if there is a voluntary assent of the subject, while nature operates through necessity.[13] According to Albert, nature is the agent which produces health necessarily in a body, once the right therapeutical decisions have been made. While it can be agreed that nature may be the primary and necessary agent through which health comes about in a human body, there are two further questions: the first concerns the limits of the effectiveness of medical practice, the second the meaning of healing in terms of physiological processes.

The latter question seems to be answered, although in general terms, in a passage comparing the purging role of medicines (drugs) from physical diseases to the function of Purgatory. Just as Purgatory expiates our faults and frees us from our previous sins before entering Paradise and achieving Salvation, so the purging action of drugs removes the corrupted humors which stand in the way of achieving a correct physical state of health.[14] Despite this relatively clear analogy, the concrete physiological meaning of *inductio sanitatis* in the passage is left, once again, unexplained. The manner through which corrupted humors are expelled, though, can be inferred from another passage of the *Commentary on the Sentences* where Albert explains that the medical remedy is the ultimate trigger of a series of involuntary reactions (in other words, there is no placebo effect) dependent on primary qualities and, therefore, involuntary (*nec voluntarie nec per intellectum*). By contrast, sacraments require the cooperation of human will.[15]

13 Albertus Magnus, Super IV Sententiarum (dist. 1–22), dist. 8 art. 5, ed. Auguste Borgnet, Paris 1894 (Ed. Paris. XXIX, 182a).

14 Albertus Magnus, Super IV Sententiarum (dist. 1–22), dist. 21 art. 4, ed. Auguste Borgnet Paris 1894 (Ed. Paris. XXIX, 866b): "Adhuc, in medicinalibus ante inductionem sanitatis, necesse est purgari naturam ab humoribus corruptis: sed in ethicis vitia humoribus proportionantur: ergo etiam in illis necesse est praecedere purgationes ad sanitatem perfectam: sed sanitas perfecta est secundum gloriam tantum, quae excludit culpam et poenam: ergo necesse est post mortem purgationem fieri ante gloriae inductionem."

15 Albertus Magnus, Super III Sententiarum, dist. 1, art. 2, ed. Auguste Borgnet, Paris 1894 (Ed. Paris. XXVIII, 7b): "Dicendum, quod secus est in medicina spiritus, et in medicina corporis: quia corporis medicina agit cum natura et juvat illam, et hoc per actiones primarum qualitatum, quae non agunt voluntarie nec per intellectum: et ideo melius est illi morbo occurrere in principio, quamdiu natura fortis est, antequam succumbat. Sed medicina spiritualis non agit nisi cum nostra voluntate ad minus consentiente [...]."

According to Albert, medicaments do not produce health, but facilitates nature's production of health (*medicina corporis agit cum natura et iuvat illam*) by removing impediments to it, such as corrupted humors. The triggering of this natural process of removing corrupted humors is in turn based on the interaction between primary qualities, which in turn build up humors. While we may try and put together a picture of Albert's conception of healing and health, it is more complicated to understand the epistemic constraints according to which the notion that nature "necessarily" (*per necessitatem*) heals can hold true.

In Albert's early works, any consideration of the practical limits of medicine and healing appears in singular discussion on the effectiveness of sacraments. While limited, some insights from here will resurface in his mature works, especially his commentary on Aristotle's *Topics*. In the first place, Albert argues that a first limit of medical practice consists in the non-responsivity of patients to treatment, comparable to unrepentant heretical behavior. To achieve health, physicians might want to induce sickness as a means of remedy, just like clergymen excommunicate unrepented sinners as an *extrema ratio* to bring them back to the Christian community.[16] A second limit of medical practice may be represented, as highlighted in Albert's explanation of confession, by the joint presence of manifold diseases, which make healing difficult. This may get worse if the physician is not an expert in all possible diseases and overlooks the presence of one of them. For this reason, the spiritual physician (the priest) should know all possible moral diseases if he is to cure them.[17] In his mature commentary on the *Topics*, Albert analyzed the limits of medical science, comparing them to the activities of attorneys and dialecticians. According to Albert, there are three factors hampering the effectiveness of medicine: *malitia infirmitatis*, *fallacia experimenti* and *inoboedientia aegroti*.

These factors are quite ambiguous as respects their lexical formulation, apart from the hardly ambiguous *malitia infirmitatis*, i.e. the severeness of the disease. This is because the second factor, *fallacia experimenti*, can have a manifold meaning: it can relate either to an error in the diagnostic process, or in the treatment of the patient by the physician, or to a side effect of the drug or of the treatment itself. The third limit, *inoboedientia aegroti*, can also have a twofold meaning, since *inoboedientia* can either be willing (refusal to take a prescribed medicine) or involuntary (unresponsiveness of a patient to

16 Albertus Magnus, De sacramentis, tr. 8, q. 8, a. 1 (Ed. Colon XXVI, 152.11–19); De sacramentis, tr. 8, q. 8, a. 3 (Ed. Colon. XXVI, 153.18–25).

17 Albertus Magnus, Super IV Sententiarum (dist. 1–22), dist. 16, art. 36, ed. Auguste Borgnet, Paris 1894 (Ed. Paris. XXIX, 622b).

treatment).[18] Strictly speaking, Albert's considerations on the practical limits of medicine found in the *Topics* are a bit richer than those found in his earlier ruminations, although, as we can see, they are quite in keeping with them on some points (unresponsiveness of the patient, difficulties in diagnosing or treating the illness).

These brief passages on the practical limits of medicine represent a first clue as to how Albert's early works offer a key to understanding his mature positions. In the case of the relationship between medicine and natural philosophy, these clues provide increasing yet not conclusive clarity. If we consider Albert's take on this topic, healing is a primarily natural process, triggered by nature and produced by the means of the removal of corrupted humors from the body, most often because of the non-voluntary interaction between primary qualities. As we have seen, the different doctrinal position on the process of healing directly affected the position of Medieval authors: if the subject of medicine is the body insofar as it can be healed (*corpus sanabile*) and healing is a natural process, then medicine should fall within the remit of natural philosophy as a part of it. By contrast, if healing is considered a natural process caused by a science whose principles are derived from natural philosophy, then medicine is subalternated to natural philosophy. While the ideas of Albert that we have garnered so far seem to point toward the first solution, there is no explicit statement concerning the relationship between the two sciences. In his *Commentary on the Sentences*, Albert famously argued that

> Augustine is to be preferred rather than the philosophers in case of disagreement in matters of faith. But if the discussion concerns medicine, I would rather believe Galen or Hippocrates, and if it concerns things of nature, Aristotle or anyone else experienced in natural things.[19]

18 Albertus Magnus, Topica, I tr. 1 c. 5, ed. Auguste Borgnet, Paris 1890 (Ed. Paris II, 247b–248a): "Neque enim rhetoricus advocatus omnino et universaliter persuadebit, impedimento triplici impeditus: malitia causae, perversitate judicis, et debilitate allegationis suae. Nec medicus similiter omnino sanabit propter triplex impedimentum, scilicet fallacia experimenti, malitia infirmitatis, et inobedientia aegroti: et ideo non semper consequuntur finem in alio. Et similiter impeditur dialecticus ex debilitate probabilis quod adducit, et ex protervitate respondentis, et difficultate problematis circa quod disputat, ita quod non semper consequitur finem in alio."

19 Albertus Magnus, Super II Sententiarum, dist. 13, art. 2, ed. Auguste Borgnet, Paris 1894 (Ed. Paris. XXVII, 247a): "[…] sciendum, quod Augustino in his quae sunt de fide et moribus plusquam Philosophis credendum est, si dissentiunt. Sed si de medicina loqueretur, plus ego crederem Galeno, vel Hipocrati: et si de naturis rerum loquatur, credo Aristoteli plus vel alii experto in rerum naturis." This translation by Nancy Siraisi in: Siraisi, The Medical Learning, 382.

According to Albert, medicine would seem to be an autonomous science independent from natural philosophy and with authorities of its own, namely Galen and Hippocrates. Still, this passage does not say anything about the epistemic relationship between medicine and natural philosophy other than that medicine has their own authorities, distinguished from Aristotle. In Albert's early theological works, the epistemic compatibility of Galenic and Aristotelian doctrines is not addressed *ex professo* at an abstract level. However, some keys to this question emerge from Albert's early use of the Avicennian *corpus* of zoology (more limitedly, medicine), which deals with concrete cases of medico-philosophical controversies such as generation and origin of veins.

3 Avicenna: Between the First Teacher and the Physician

In the *De homine*, Albert drew extensively on Avicenna's *De animalibus* as concerns the anatomical structures of living beings, especially humans. His descriptions of the workings of cranial nerves, the brain ventricles and the role of medico-philosophical notions such as humors and spirits owe much to Avicennian zoology. More to the point of this chapter, Albert himself confronted serious issues of epistemic compatibility between medical and philosophical doctrines, as highlighted by the case of generation. As many previous studies have shown, in doing so Albert adhered to the position of Avicenna in his zoology, the *Book of Animals* (*De animalibus*/ *Kitāb al-Ḥayawān*), which we should now briefly recall.[20]

Composed around 1027/1028, the *Ḥayawān* represents the only extant Avicennian work on zoology. While its structure faithfully follows the Aristotelian template, Avicenna's work does not display any distinction between the *Histo-*

20 On the influence of Avicennian zoology in Albert's theory of generation, see: Gabriella Zuccolin, "Seed Matters. Albert the Great on Human Generation as a Disciplinary Conflict," *Quaestio. Yearbook of the History of Metaphysics. Special Issue Natural Philosophy in Albert the Great—A Dialogue of Disciplines*, ed. Mario Loconsole/Evelina Miteva/Marilena Panarelli, 23 (2023), 55–84; Zuccolin, "Il corpo sanabile e l'embrione," 113–134; Evelina Miteva, "*Iam patet igitur veritas eius quae dixit Aristoteles, et causa deceptionis Gallieni.* Philosophers vs. Medics in Albertus Magnus Account on Human Body," in *Irrtum-Error-Erreur*, ed. Maxime Mauriége/Andreas Speer (Miscellanea Medievalia 40) Berlin 2018, 107–122; Luciano Cova, "Seme e generazione umana nelle opere teologiche di Alberto Magno," in Summa doctrina et certa experientia. *Studi su medicina e filosofia per Chiara Crisciani*, ed. Gabriella Zuccolin (Micrologus Library 79), Firenze 2017, 237–256; Luke Demaitre–Anthony A Travill, "Human Embryology and Development in the Works of Albertus Magnus," in *Albertus Magnus and the Sciences. Commemorative Essays*, ed. James A. Weisheipl (Studies and Texts 49), Toronto 1980, 405–440.

ria Animalium (I–X), *De partibus animalium* (XI–XIV), and *De generatione animalium* (XV–XVIII) like its Arabic *Vorlage*, "The book of animals" (*De animalibus/ Kitāb al-Ḥayawān*). This is not the only difference between Aristotelian and Avicennian zoology, though. The main differences lie in the content of these works. In the *Book of animals*, large sections from book XI–XIV (Aristotle's *De partibus animalium*) and book IX on generation were excerpted from Avicenna's masterpiece, the *Canon of medicine* (*Canon medicinae/ Qānūn fi al-ṭibb*), composed earlier, sometime around 1015. This fact alone gives these books, as noted by the pioneering study of Basil Musallam on the relationship between Avicenna's *Canon* and Avicenna's zoology, a clearly human-oriented perspective on generation, anatomy of nerves, muscles, bones, and the like.[21]

Moreover, the inclusion of large sections from the *Canon* reifies the insertion of medical learning into natural philosophy. To some extent, Avicenna's extensive use of medical teachings even contradicts his statements in the prologue of the *Book of Animals*, extant only in Arabic and recently analyzed by Tommaso Alpina, in his study on the Avicennian discussion on the origin of veins whose exposition we shall rely on. Alpina shows that, in the opening of his work (extant only in Arabic), Avicenna argued that he would deal with anatomy "in a single place" (*fī mawḍiʿ wāḥid*, i.e. the *Canon*), that he would summarize what Aristotle had been prolix about (in matters of physiology) and that he would insert the necessary parts of "theoretical science" (*kalām naẓarī*) whenever needed.[22] These words, though, clash with Avicenna's exegetical practice: medical teachings are, *de facto*, part of his zoology,[23] as Albert himself, though deprived of the precious prologue of the *De animalibus*, could have observed. Both Avicenna and Albert were well aware, though, that the inclusion of medical doctrines needed to be subject to appropriate epistemological restrictions.

---

21 On the relationship between Avicenna's *Ḥayawān* and *Qānūn*, see: Basil Musallam, "AVICENNA X. Medicine and Biology," *Encyclopaedia Iranica*, III/1, 94–99, available online at http://www.iranicaonline.org/articles/avicenna-x (accessed on April 22, 2019).

22 Avicenna, Ḥayawān, I, c. 1, ed. Cairo 1970, 1.10–13: "Let us now talk about animals by following in all this book the first teaching as a model (*muḥtaḏīna fī ǧamīʿ hāḏā l-kitāb ḥaḏwa l-taʿlīm al-awwal*), except in the case of the anatomy of the organs of the human being (*illā fī tašrīḥ aʿḍāʾ al-insān*)—actually, we prefer to put together anatomy (*al-tašrīḥ*) and [its] usefulness (*al-manfaʿa*) in one single place (*fī mawḍiʿ wāḥid*)—and in the case of few [other] things. Then we shall cut off in terms of information that which he (sc. Aristotle) was prolix about. We shall mention of the theoretical discourse (*min al-kalām al-naẓarī*) what is appropriate for our opinion and our collection of these sections (*bi-raʾyinā wa-ǧamʿinā li-hāḏihi l-funūn*)." This translation by Alpina in: Tommaso Alpina, "Exercising Impartiality to Favor Aristotle: Avicenna and 'the accomplished anatomists' (*aṣḥāb al-tašrīḥ al-muḥaṣṣilūna*)," *Arabic Sciences and Philosophy* 32 (2022/2), 142 n. 15.

23 More on this in: "Alpina, Exercising Impartiality," 138–145.

A clear example of these epistemological concerns is found in the opening chapters of the *Canon*, especially the chapter on the subjects of medicine (*De subiectiis medicinae*/*Fī mawḍuʿāt al-ṭibb*). Dimitri Gutas has shown how in those lines, crucial for the history of medicine, Avicenna argued that the establishment of the existence, essence and numbers of humors, spirits, elements, bodily dispositions and the like could not be discussed by physicians. Given their lack of suitable conceptual tools to discuss these concepts, physicians had to trust natural philosophers with the handling of such matters.[24] In Avicenna's view, this entailed the subordination of medicine, classed as a "derivative natural science" (*ḥikma farʿiyya ṭabīʿiyya*), to natural philosophy.[25]

Now, this epistemic subordination is clearly observable in the epistemic issues of generation—dealt with in the *Canon* and copied in the zoology—and the origin of veins, discussed in Avicenna's zoology (III, 1). While these issues will be clarified in detail in the third chapter, it is still relevant to provide some information about the truth value which Avicenna assigns to Galenic doctrines that are patently in conflict either with the crucial laws of natural philosophy or with his own views on the soul. In Avicenna's view, Galen's arguments against Aristotle do not imply a sheer correction of flaws in Aristotelian physiology—something to which Avicenna, following Galen, subscribes—but rather posed a serious threat to the fundamental rules of Aristotelian natural philosophy such as the contemporary presence of two formal, agential causes (as in the case of feminine semen) or undermined Aristotle's cardiocentric conception of formative power, whose astrologized interpretation is a key in Avicenna's philosophy of nature and biology.[26] This methodological framework is key to understand

24 Avicenna, Canon I fen 1 doctr. 1, ed. Venezia 1507, fol. 1va [ed. New Delhi 1981–1996, vol. 1, 36.10–14]. On this passage, see: Dimitri Gutas, "Medical Theory and Scientific Method in the Age of Avicenna," in *Before and After Avicenna. Proceedings of the First Conference of the Avicenna Study Group*, ed. David C. Reisman, with the assistance of Ahmed H. Al-Rahim (Islamic Philosophy, Theology and Sciences. Texts and Studies 52), Leiden 2003, 150–151.

25 Peter E. Pormann, "Avicenna on Medical Practice, Epistemology, and the Physiology of the Inner Senses," in *Interpreting Avicenna. Critical Essays*, ed. Peter Adamson, Cambridge 2013, 93–94.

26 On semen, see more generally: Remke Kruk, "Ibn Sīnā *On Animals*. Between the First Teacher and the Physician," in *Avicenna and his Heritage: Acts of the International Colloquium, Leuven-Louvain-la-Neuve, September 8-September 11, 1999*, ed. Jules Janssens/Daniel De Smet (Ancient and Medieval Philosophy [Series 1] 28), Leuven 2002, 325–341. This debate will be laid out more thoroughly in chapter 3, with richer bibliographical support. A similar yet more summarized discussion on the Avicennian doctrine on the origin of the veins, based on Alpina's contribution as well, in: Michele Meroni, "*Natura confortata per medicinam operatur per se*. The Role of Medicine in Albert's Early Theology and Aristotelian Paraphrases," *Quaestio. Yearbook of the History of Metaphysics* (*Special

Albert's position in his early works imbued with anatomy and physiology, most of which derived from Avicenna's *De animalibus*.

4 The *De homine* and the First Entry of Avicenna's *De animalibus*

Translated into Latin by Michael Scot in the early 1230s,[27] Avicenna's *Abbreviatio de animalibus* (its Latin title) was extensively used by Albert in the *De homine*, as we shall see throughout his study, to discuss generation, the anatomy of sense perception and cognitive activities, the role of brain and of the nerves in humans and, partially, in animals. Therefore, Albert's *De homine* (ca. 1242) represents one of the earliest cases in the Medieval Latin reception of Avicennian zoology: in this book, we shall expand on the extremely concise yet fundamental observation by Romana Martorelli Vico, according to whom Albert's *De homine* represents the first example of the Latin reception of this Avicennian work.[28] While the issue of generation—and the role of *spiritus* in it—will be dealt with specifically in the third chapter, it is vital to remember the roots of Albert's approach to Aristotle's and Galen's conflicting views on generation. In the *De homine*, Albert adhered to the Avicennian view that the mother possessed a semen—called so only by analogy—whose potency to act on the determination of the sex and shape of the embryo was inversely proportional to the formative action of the masculine semen. In other words, while the action of the masculine semen was called 'formative', that of the feminine semen 'informative.'[29] Luciano Cova has shown how (like Avicenna) Albert played down the role of menstrual blood in generation, considering it a sheer nutrimental substance for the embryo during pregnancy.[30]

In this manner, Albert firmly rejected the coexistence of two formal causes (male and feminine semen) implied by an adoption of Galen's standpoint in works of natural philosophy. Staunchly faithful to the epistemic parameters of

---

Issue Natural Philosophy in Albert the Great—A Dialogue of Disciplines, ed. Mario Loconsole/Evelina Miteva/Marilena Panarelli), 23 (2023), 116–124.

27 For a general overview of the translated parts of the *Šifāʾ* and their periodization, see: Amos Bertolacci-Tommaso Alpina, Tommaso, "Introduction," *Documenti e Studi sulla Tradizione Filosofica Medievale* 28 (2017), vii–xviii.

28 Romana Martorelli Vico, *Medicina e filosofia: per una storia dell'embriologia medievale nel XIII e XIV Secolo* (Hippocratica Civitas 4), Milano 2002, 29.

29 On Albert's rejection of the active, formal role attributed to the feminine semen, see: Albertus Magnus, De homine (Ed. Colon. XXVII/2, 129.13–15). More on this in chapter 3.

30 On the role of menstrual blood as food for the embryo, see: Avicenna, Abbreviatio de animalibus IX c. 5, ed. Venezia 1508, fol. 43rb [ed. Cairo 1970, 174.1–2]; Albertus Magnus, De homine (Ed. Colon. XXVII/2, 126.45–57); Cova, Seme e generazione umana, 251.

the Aristotelian science of nature and to Aristotelian cardiocentrism, Albert not only rejected Galen's position on the feminine role in generation but denied also that veins—as argued by Galen—could be originated by the liver. Unlike his mature zoology, where the issue of the origin of veins would be dealt with *ex professo* in both his commentary on *De animalibus* and in his *Quaestiones super de animalibus*, in his early theology Albert straightforwardly assumed Aristotle's embryological cardiocentrism, thus leaving little doubt as to the origin of veins themselves. Albert's early cardiocentrism has been showcased in a recent study by Katja Krause we shall rely on for this section in chapter 3.[31] We will see in the third chapter how his mature refutation of Galen's doctrine on the origin of veins in the *De animalibus* owes much to Averroes' *De corde*, known to him as early as the *De homine*.[32] On the other hand, we will see how Avicenna's zoology is crucial to Albert's take on the issue of the origin of veins in the *Quaestiones super de animalibus*. Be that as it may, Albert's criticism of Galen's view is due to the fact that, as he would argue later in his *De spiritu*, his putting the liver—also the brain—on an equal footing with the heart thanks to its formative power would have jeopardized the substantial unity of the human soul.[33] For him, the soul consists in one single substance and its different faculties, as in Avicenna's case, flow from it. At a physiological level, the heart is the source of this life-giving flow.[34] Consequently, arguing that the liver was endowed with a formative power analogous to that of the heart would have disrupted the psychological structure of the living being as conceived by Albert. For this reason, Galen's position needed to be rejected. In the third chapter we shall observe his argumentative strategy in relation to his sources (Avicenna's *De animalibus*, Averroes' *De corde*).

31 Albertus Magnus, De homine (Ed. Colon. XXVII/2, 49.37–41): "[…] sicut dicit Aristoteles in XVI *De animalibus*, primo generatur cor et ex corde producuntur omnia membra animalis […]." Katja Krause, "Albert the Great on Animal and Human Origin in his Early Works," *Lo Sguardo* (*Rivista di filosofia*) 18 (2015/2), 226.

32 Albertus Magnus, De homine (Ed. Colon. XXVII/2, 244.23–27): "Ad hoc autem quod obicitur in contrarium, respondendum est secundum Avicennam in libro suo De animalibus et secundum Averroem in capitulo *De corde* […]."

33 Albertus Magnus, De spiritu et respiratione I tr. 2 c. 1, ed. Auguste Borgnet, Paris 1894 (Ed. Paris. IX, 231b–232a): "Cum igitur, ut inquiunt, cor, cerebrum, et hepar, loco, subjecto, et numero differant: necesse est quod motores eorum loco, subjecto, et numero differant, et inde tres in homine animos, et similiter in animalium corporibus esse concesserunt vel contenderunt."

34 Albertus Magnus, De homine (Ed. Colon. XXVII/2, 52.43–49): "[…] anima secundum se est actus totius et secundum suas partes est actus partium. Unde pars animae est in organo corporis, et ita actus organi specialis erit actus partis animae. Secunda probatur per suppositionem naturalium et medicorum, scilicet quod a corde per arterias est influxus vitae et spiritus vitalis in totum corpus."

If we leave aside for a moment the controversies on the generation and origin of veins, we can observe as a well-established fact in scholarship that the early theological works by Albert did value Galenic medical learning in his study of the anatomy of sense perception, cognitive activities and sleep. In the *De homine*, large sections of the *quaestiones*, especially those devoted to the organs of sense perception and cognitive activities (*de organo visus/ de organo auditus* etc.), contained rather accurate quotes from Avicenna's *De animalibus*, Costa Ben Luca's *De differentia spiritus et animae*, and Nemesius' *De natura hominis*. The picture which emerges from the *De homine* is that of an author who dedicates an extensive treatment to human physiology, much more detailed than his contemporaries, as shown by the study of Steneck.[35] To achieve this level of detail in his psychophysiology, Albert proved himself to be very open to the use of medical teachings, provided that their doctrines did not represent a vital threat to the fundamental laws of natural philosophy nor go against Albert's most important psychological tenets. In that case, they needed to be amended in a manner that fit the epistemic parameters of natural philosophy (generation and feminine semen) or rejected if they disrupted the psychological unity of the living being (origin of veins) and Aristotelian cardiocentrism. This is the case, at least, according to Albert's interpretation of Aristotle's notoriously difficult position on the unicity or plurality of forms.

All in all, Albert's position on the key issues of the generation and origin of veins—which we will examine in greater detail in the third chapter—betrays a strong and consistent Avicennian methodological influence, which merges with the influence of Averroes' *De corde*. While, as we have observed, medicine may not formally be said to be part of natural philosophy in Albert's early works, great amounts of medical learning are *de facto* injected into his early *De homine* via Avicenna's *De animalibus* and, to a much lesser extent, the *Canon*, and other sources such as Costa ben Luca and Nemesius. In Albert's mature works, as already noted by Nancy Siraisi, the integration of medical learning was mostly derived from Avicenna's *Canon*, plus a number of other sources already known by him at the time of his studies in Paris.[36] While the Avicennian epistemic parameters for the inclusion of medical learning into works of natural philoso-

35 Nicholas H. Steneck, "Albert on the Psychology of Sense Perception," in *Albertus Magnus and the Sciences, Commemorative Essays* (Studies and Texts 49), ed. James A. Weisheipl, Toronto 1980, 275: "Few Latin writers in the thirteenth and fourteenth centuries paid more than routine attention to this aspect of sense perception. [...] Throughout his writings it is clear that he is localizing the senses in an anatomically real body and not simply repeating descriptions handed down from an earlier tradition."

36 Siraisi, "The Medical Learning," 379–404.

phy represent a common doctrinal ground between Albert's early and mature approaches to medicine, Albert's mature understanding of the scope of natural sciences and, therefore, their relationship to medicine needs a more detailed inquiry, for it presents relevant elements of discrepancy with Avicenna.

5 Medicine in the Aristotelian Paraphrases: Theoretical Underpinnings and Historical Context

In his mature works, Albert famously set out to make the entirety of the Aristotelian *corpus* accessible to his Domenican pupils (*ut Aristotelem competenter intelliger possint*) in Cologne. In his view, this enterprise would entail not only a work of commentary on Aristotelian, pseudo-Aristotelian (*Liber de causis*, *De plantis*, *De causis proprietatum elementorum*) and non-Aristotelian (Boethius' *De divisione*, Pseudo-Gilbertus Porretanus' *De sex principiis*, Costa Ben Luca's *De differentia spiritus et animae*) texts. It would also encompass the composition of works which filled conceptual voids left by Aristotle, such as the *De nutrimento* and the *De intellectu et intelligibili*, or replaced genuine but unavailable Aristotelian works such as the *De motibus animalium*—meant to replace Aristotle's *De motu animalium*, later paraphrased in the *De principiis motus processivi*—and the *De aetate sive de iuventute et senectute*, meant to replace Aristotle's homonymous work.[37]

For his part, Albert pledged to write digressions whenever needed to clarify obscure points in Aristotle's text, thus making his work of paraphrase a singular case among medieval scholars. Within this framework, his understanding of the domain of natural sciences represents a unique case study, too. In his

37 On the content and methodology for the composition of Albert's *Parva Naturalia*, see: Pieter de Leemans, "*Secundum viam naturae et doctrinae*, Lire le *De motu animalium* et les *Parva Naturalia* d'Aristote au Moyen Âge," in *Les* Parva Naturalia *d'Aristote. Fortune antique et médiévale*, ed. Cristophe Grellard/Pierre-Marie Morel, Paris 2010, 201–206. On Albert's conception of natural sciences, see: Silvia Donati, "Alberts des Großen Konzept der *scientiae naturales*: Zur Konstitution einer peripatetischen Enzyklopädie der Naturwissenschaften," in *Albertus Magnus und der Ursprung der Universitätsidee: die Begegnung der Wissenschaftskulturen im 13. Jahrhundert und die Entdeckung des Konzepts der Bildung durch Wissenschaft*, ed. Ludger Honnefelder, Berlin 2011, 354–381. On the more general methodology adopted by Albert in the construction of his scientific system, see: Katja Krause–Henryk Anzulewicz, "Albert the Great's *interpretatio*. Converting Libraries into a Scientific System," in *Premodern Translation: Comparative Approaches to Cross-Cultural Transformations*, ed. Sonja Brentjes/ Alexander Fidora (Contact and Transmission; Intercultural Encounters from Late Antiquity to the Early Modern World 2), Turnhout 2021, 89–132.

catalogue of sciences set out in the opening of the *Physics*, Albert clearly distinguishes between physics or science of nature, mathematics and metaphysics. Within his classification of natural sciences, zoology, botany and mineralogy, consider three different kinds of mixed bodies, from the simplest (minerals) to the most complex (humans and animals).[38] These sciences are conceived of in a strongly unitary manner and closely connected in terms of epistemic parameters of natural philosophy, necessary to reorder and reorganize the philosophically unsatisfying pseudo-Aristotelian *De plantis* and the scattered material available to Albert on mineralogy. As shown in the contributions of Amalia Cerrito and Gila Wöllmer to botany and Mario Loconsole (previously, Riddle and Mulholland) to mineralogy, Albert shaped *de facto* two sciences on the model of Aristotelian zoology, reorganizing previous and often obscure texts (the *De plantis*) or construing a science based on the scattered remarks, for instance, of Avicenna's *De conglutinatione lapidum*.[39]

With zoology, the case was somewhat different, as Albert had Michael Scot's Arabo-Latin translation of Aristotelian zoology on his desk and Scot's translation of the *Abbreviatio de animalibus* by Avicenna. These three sciences, though, share a methodological peculiarity: while they should be, according to Aristotelian epistemology, organized as a study of minerals, plants and animals according to part and functions, their structure seems explicitly to contravene Aristotle's own epistemic principles. As a matter of fact, all of these works present what in the past have been labelled as 'alphabetic books'.[40] These A–Z lists of minerals, plants and animals do contravene Aristotle's own principles in the beginning of the *De partibus animalium*—and Albert shows himself to be aware of that (*licet enim hic modus non omnino sit philosophicus*). In a recent contribution, Katja Krause has convincingly demonstrated, though, that Albert

---

38 Minerals are defined *commixta*, plants *complexionata* while animals and humans are defined *organizata* in: Albertus Magnus, Physica II tr. 2 c. 1 (Ed. Colon. IV/1, 97.27–35).

39 Amalia Cerrito, "Botany as Science and exegetical Tool in Albert the Great," *Aisthesis* 11 (2018/1) 97–107; Gilla Wöllmer, "Albert the Great and His Botany," in *A Companion to Albert the Great: Theology, Philosophy and the Sciences*, ed. Irven Resnick (Brill's Companions to Christian Tradition 38), Leiden 2013, 221–267. On minerals, see: Mario Loconsole, "A Theory on the Formation of Minerals. Albert the Great and the Constitution of Scientific Mineralogy," *Quaestio* 20 (2021) 369–387; John M. Riddle-James A. Mulholland, "Albert on Stones and Minerals," in *Albertus Magnus and the Sciences. Commemorative Essays*, ed. James A. Weisheipl (Studies and Texts 49), Toronto 1980, 203–234.

40 Miguel de Asùa, "Mineral, Plants and Animals from A to Z. The Inventory of the Natural World in Albert the Great's *philosophia naturalis*," in *Albertus Magnus. Zum Gedenken nach 800 Jahren. Neue Zugänge, Aspekte und Perspektiven*, ed. Walter Senner/Henryk Anzulewicz (Quellen und Forschungen zur Geschichte des Dominikanerordens, Neue Folge 10), Berlin 2001, 389–400.

regarded these books as an integral part of his zoology.[41] This is because he considered books XXII–XXVI from his *De animalibus* as a study of mixed bodies according to their species (*specialiter*) and not as a merely redundant study of different species.[42] If Krause's interpretation is correct—and there is solid evidence to believe it is—then, by the same token, this line of reasoning should apply to *De mineralibus* II, 2 (a list of stones with their properties) and *De vegetabilibus* VI, 3 (a list of herbs and their pharmacological properties). In Albert's eyes, the study of minerals, plants and animals according to their species represents therefore an integral part of mineralogy, botany and zoology and not a purely accidental part of it, built only to satisfy the curiosity of his students. It will not have escaped those attentive of zoology that Albert's *De animalibus* is a 26-book long enterprise. While 19 books can be traced back to the Aristotelian original and the last five are represented by the study of animals according to their species, books 20 and 21 are as alien to Aristotelian and Avicennian zoology as books 22–26.

According to Albert himself, the first represents a study of animal bodies on the whole,[43] while the second lists animals according to their ascendent degrees of cognitive perfection, a clear example of *scala naturae*.[44] As will be by now crystal-clear thanks to Krause's contribution, the mature Albert's

41 Albertus Magnus, De animalibus XXIII tr. 1 c. 1 n. 1, in: Albertus Magnus, ed. Hermann Stadler (Beiträge zur Geschichte der Philosophie des Mittelalters 15–16), Münster 1916–1920, 2 voll. (ed. Stadler, II, 1430.6–9); Albertus Magnus, De animalibus XXIV c. 1 n. 1 (ed. Stadler, II, 1515.10–15).

42 Albertus Magnus, De animalibus, XXIII tr. 1 c. 1 n. 1 (ed. Stadler, II, 1349.7–15): "Nos tamen adhuc sub alterius libri principio specialiter secundum ordinem nostri alfabeti quasdam sub nominibus propriis ponemus naturas. Quamvis enim hunc modum non proprium phylosophiae supra esse dixerimus eo quod in eo saepe eadem reiterare oportet, tamen quia sapientibus et insipientibus nos esse recognoscimus debitores, et ea quae particulariter de particularibus narrantur, rusticam melius instruant contionem, talem in fine nostro libro tractatum apponemus, incipientes ab hiis quae singulis literis in principio praenotantur animalibus." Katja Krause, "Medieval Philosophy of Nature popularized? Albert the Great's *De animalibus*," in *The Popularization of Philosophy in Medieval Islam, Judaism, and Christianity*, ed. Marieke Abram/Steven Harvey/Lukas Muehlethaler (Philosophy in the Abrahamic Traditions of the Middle Ages 3), Turnhout 2022, 185–201. For further insights into the epistemic structure of the *De animalibus*, see also: Katja Krause, "Grenzen der Philosophie. Albert der Großen Kommentar zu *De animalibus* und die Medizin," *Documenti e studi sulla tradizione filosofica medievale* 30 (2019) 286–289.

43 Albertus Magnus, De animalibus XX tr. 1 (ed. Stadler, II, 1273.1): "Incipit liber vicesimus animalium qui est de natura corporum animalium."

44 Albertus Magnus, De animalibus XXI tr. 1 (ed. Stadler, II, 1321.1–4): "Incipit liber animalium vicesimus primus qui est de perfectis et imperfectis animalibus et causa perfectionis et imperfectionis."

conception of natural science, in particular zoology, cannot be reduced to Avicenna's and Aristotle's. This premise is absolutely crucial when we speak about Albert's mature understanding of the relationship between natural philosophy and medicine, which is addressed twice in Albert's philosophical works, namely the opening lines of the *De sensu et sensato* and of the *De animalibus*.

6 Medicine in the *Parva naturalia*: An Open Question

The *De sensu et sensato* belongs to the body of the *Parva Naturalia* which, according to Albert, represent the second and supplementary part of psychology (*scientia de anima*). The subject of the group of works classed under the name *Parva Naturalia*, a name first used by Giles of Rome, was the operations and the activities of the soul (*scientia de operibus animae*).[45] In his preface to the research program of the soul's activities in the body, Aristotle had listed four pairs of opposites (waking and sleeping, youth and old age, inhalation and exhalation, life and death) as subject to inquiry, which lend themselves well to some epistemic consideration on the relationship between medicine and natural philosophy.[46] In his paraphrases of the opening lines of the *De sensu et sensato*, Albert writes:

> And therefore many of the **natural philosophers** and the more skilled among the **doctors** who *use **their art most philosophically*** end up at the same point (*terminantur adinvicem*). For the **natural philosophers, going** from first principles to their consequences and from **universals** to particulars, end their **inquiries into** living things *by reaching matters which have to do with medicine*, namely the particular causes of health and sickness. But those **doctors** who use **their** art ***according to the criteria of natural philosophy*** (*physice*) ascend from particular diseases to general signs, causes, and **common accidents**; for they cure the disease **only by removing** its causes and draw out the causes of health.[47]

---

45 By contrast, the subject of the *De anima* is the science of the soul in itself and its powers: Albertus Magnus, Physica I tr. 1 c. 4 (Ed. Colon. IV/1, 7.12–58).

46 Other passages in the Aristotelian works dealing with the relationship between the two sciences are *De respiratione*, 480b22–31 (unavailable to Albert) and *Nicomachean Ethics*, 1102a18–26. On Aristotle's position, see: Philip van der Eijk, *Medicine and Philosophy in Classical Antiquity: Doctors and Philosophers on Nature, Soul, Health and Disease*, Cambridge 2005, 192–201.

47 Albertus Magnus, De sensu et sensato I tr. 1 c. 1 (Ed. Colon. VII/2, 20.47–57): "Et ideo *plurimi physicorum et* peritiores *medicorum, qui magis philosophice arte utuntur*, terminantur ad

Judging from these lines, Albert seems to be quite open about the relationship between the two sciences, at least when medicine is considered as an autonomous body of texts and doctrines.[48] Luca Bianchi has recently shown how the wording of Albert's statement closely resembles the commentary on the *De sensu et sensato* attributed to Adam of Buckfield and Roger Bacon's *Opus maius* and *De erroribus medicorum*.[49] In his *Opus maius*, Roger Bacon stresses that knowledge of universal causes is needed for the sake of healing both spiritual and physical illnesses. Recalling explicitly Aristotle's *De sensu et sensato*, Bacon claimed that the principles of medicine begin where those of natural philosophy have their end (*ubi terminant principia philosophiae naturalis incipiunt principia medicinae*).[50] Bacon held the same position in the *De erroribus medicorum*, where he harshly criticized those physicians who despised the rel-

---

invicem. Physici enim venientes a primis in posteriora, et ab universalibus ad particularia terminant considerationes suas de animatis *ad ea quae* sunt *medicinae*, quae sunt causae particulares sanitatis et aegritudinis. Medici autem *physice* arte utentes a particularibus aegritudinibus in signa et causas et accidentia communia ascendunt, eo quod morbum non sanant nisi remota causa morbi, et inducta causa sanitatis." The translation is based on: Siraisi, "The Medical Learning," 382. I have modified the translation following the text of the recent edition by Silvia Donati: differences in translation based on the different textual data are marked in bold. I have also opted to translate the term *physici* as "natural philosophers," while the term is left untranslated by Siraisi.

48 Albert's terminology when referring to medicine is quite wavering at times. When considering its role as a scientific system of doctrines and endowed with a practical purpose, he tends to call medicine *scientia practica*: see: Albertus Magnus, Metaphysica VII tr. 2 c. 5 (Ed. Colon. XVI/2, 345.49–58): "*Sanitas autem est ratio* factiva, quae est *in anima medici, et* est species *in scientia* factiva non speculativa." By contrast, when medicine is likened to other arts which require technical craftsmanship, Albert defines it *ars mechanica*: Albertus Magnus, Metaphysica I tr. 1 c. 10 (Ed. Colon. XVI/1, 15.40–48): "Differentia autem artium est in modo illius admirationis; quoniam cum *plures repertae* per inventionem sint *artes*, quaedam quidem *ad necessaria* sive *utilia* vitae civilis, sicut mechanicae vocatae, quae sunt panifica, medicina [...]" Note that medicine is termed *ars* (*peritiores eorum medicorum qui magis philosophice arte utuntur*) in the above-mentioned passage from the *De sensu* too.

49 Luca Bianchi, "*Ubi desinit physicus, ibi medicus incipit*," in Summa doctrina et certa experientia. *Studi su medicina e filosofia per Chiara Crisciani*, ed. Gabriella Zuccolin (Micrologus Library 79), Firenze 2017, 5–28.

50 Roger Bacon, Opus Maius, pars 1 c. 9, ed. John H. Bridges, *The* 'Opus majus' *of Roger Bacon*, I–II und Suppl., Oxford 1879–1900/Frankfurt a.M. 1964, vol. 1, 18: "Necesse vero est causas universales has praemitti, ad hoc ut error vitetur et veritas relucescat. Nam in morbo spirituali evenit sicut in morbo corporali; medici per signa causas proprias et particulares morbi cognoscunt; sed tam haec quam illas praecedit causarum universalium notitia, quas actionibus naturalium medicus habet scire; quia dicit philosophus in libro de Sensu et Sensato, *ubi terminantur principia naturalis philosophiae incipiunt principia medicinae*."

evance of sense experience, knowledge of languages and sciences like natural philosophy,[51] which seems a preparatory science for the study of medicine.[52] Aside from this more specific consideration on Bacon's position, the passages from the *De erroribus medicorum* and the commentary on the *De sensu et sensato* evince a position very close to that found in the second recension of the commentary on *De sensu et sensato*, ascribed to Adam of Buckfield.[53]

Adam, Master of Arts at Oxford in the 1240s and one of the earliest Latin authors to comment extensively on the Aristotelian corpus (*De anima, Parva Naturalia, Metaphysics*), commented on the Aristotelian passage with language that is different only terminologically from Bacon's. According to Adam, it is the principles of medical inquiry in an abstract sense, and not of the physician himself, that end up at the same point as natural medicine (*incipit consideratio medici ubi terminatur consideratio phisici*).[54] Adam's position seems, in the end, quite in keeping with those of Roger Bacon and, partially, of Albert. In short, Albert's reading of this Aristotelian passage is analogous to some illustrious contemporary authors who commented on Aristotle (Adam) or used the Aristotelian text as a reference for the discussion about the relationship-between natural philosophy and medicine (Bacon).

Much different, by contrast, is Aquinas' reading of this passage. Albert's pupil comments this passage by clearly arguing for the subordination of medicine to natural philosophy. Aquinas proposes two reasons why it should be so, i.e. a demonstrative argument (*ratio*) and a sign (*signum*). The argument (*ratio*) runs as follows: health and illness can only describe a living being, hence living

51 Roger Bacon, De erroribus medicorum, ed. Andrew G. Little/Edward Withington, in: Roger Bacon, De retardatione accidentium senectutis cum aliis opusculis de rebus medicinalibus (Opera hactenus inedita Rogeri Baconi IX), Oxford 1928, 158: "Ultima autem causa manifestans ignorantiam vulgi medicorum est quod ignorant naturalem philosophiam. *Aristoteles enim dicit* quod *ubi terminatur philosophia naturalis, ibi incipit medicina*, et naturalis philosophus habet dare principia ultima sanitatis et infirmitatis, ut dicit."

52 Bianchi, "*Ubi desinit physicus*," 22 n. 41.

53 Bianchi, "*Ubi desinit physicus*," 25.

54 Adam of Buckfield, De sensu et sensato (London, The Wellcome Historical Medical Library, 3, fol. 3): "Consequenter removet dubitationem. Quereret enim aliquis quare non facit mentionem de hec contrarietate: sanitas—egritudo, et excusat se dicens quod physicus considert principia sanitatis et egritudinis. Sunt enim ille proprietates in habentibus vitam. Unde eadem sunt prima principia ipsarum, scilicet et mortis et vite, et ideo inspectant ad naturalem philosophum. Ipse enim considerat prima principia medicine et *terminat suam considerationem* ad propria principia ipsius. Sed medici incipiunt a propriis principiis medicine sive sanitatis et egritudinis. Et ita *incipit consideratio medici ubi terminatur consideratio phisici*, et hoc est: 'verum phisici'." Quoted according to Bianchi, "*Ubi desinit physicus*," 26.

beings must be the subject of such states. Since the principles which explain illness and health are those which explain the subject, for those states belong to the subject, it follows that it is for the natural philosopher to consider the nature of those states from a more general and theoretical perspective.[55] The 'sign' relies on the fact that for Aristotle himself natural philosophers who inquire into the particular causes of life and physicians who devote themselves to the search for more general causes of illness and health end up (*finiunt*) at the same point. Aquinas adds to this the idea that, even though health can be caused sometimes by nature and sometimes by art, art produces health only when it serves nature and its principles, just as the steersman must trust himself to the astrologer.[56] As Mark Jordan has brilliantly highlighted analyzing these passages, the difference between Albert's and Aquinas' approaches lies in the tendency of the latter to stress more the teleology of nature and to avoid an excessive focus on the details of material bodies Albert was so fond of.[57]

As we have shown, Albert's reading of the opening lines from the *De sensu et sensato* give us a rather open and not clearly defined position on the relationship between medicine and natural philosophy in the *Parva Naturalia*. While this position is similar to Bacon's and Buckfield's and very much different from Aquinas', it still leaves an open question about the concrete use of medical learning in the *Parva Naturalia* themselves. As noted by Miguel de Asùa, the fact that Albert mentions an open position on the relationship between the

55 Thomas Aquinas, Sentencia libri de sensu et sensato, Proh. (Ed. Leon. XLV/2), 8–9; Reynolds, Food and the Body, 136–139.

56 Thomas Aquinas, Sentencia libri de sensu et sensato, Proh. (Ed. Leon. XLV/2), 9.

57 Mark Jordan, "Medicine and Natural Philosophy in Aquinas," in *Thomas von Aquin. Werk und Wirkung im Licht Neuerer Forschung*, ed. Albert Zimmermann (Miscellanea Mediaevalia 19), Berlin 1988, 246: "On the one hand, Thomas depends on Albert for much technical knowledge in medicine and natural philosophy. On the other hand, Thomas seems deliberately to exclude much that Albert has to offer. If this exclusion is not to be dismissed as a matter of intellectual taste, then it must be considered as reflective revision of Albert's concerns and procedures. In reducing the great mass of detail offered through Albert, Thomas is clarifying the hierarchical logic of lower studies. Thomas' parsimony in the matter of medical sources is his strengthening of the philosophical and theological teleology of Aristotelian speculation about nature." Aquinas' position may well have been influenced not only by Avicenna but also by Dominicus Gundissalinus: Alexander Fidora, "Zum epistemologischen Status der Medizin in der *Summa Avicennae* und bei Thomas von Aquin," in *Handlung und Wissenschaft—Action and Science. Die Epistemologie der praktischen Wissenschaften im 13. und 14. Jahrhundert—The Epistemology of the Practical Sciences in the 13th and 14th Centuries*, ed. Matthias Lutz-Bachmann/Alexander Fidora (Wissenskultur und Gesellschaftlicher Wandel 29), Berlin 2008, 97–105.

two sciences signals Albert's openness to employing both bodies of knowledge in these works.[58] The question, though, is how.

To answer this question, few examples can be brought to the fore. In the first place, Albert's *De morte et vita*, a paraphrase of Aristotle's *De longitudine et brevitate*, bases the paraphrase the Aristotelian text on the key medical concept of radical moisture (*humidum radicale*).[59] Moreover, Albert draws again and extensively on Avicennian medicine; more specifically he employed the chapter about the complexion of ages from Avicenna's *Canon* in writing the *De aetate sive de iuventute et senectute*, meant to replace the unavailable Aristotelian *De iuventute et senectute*.[60] We will see these aspects of Albert's *De morte et vita* and *De iuventute et senectute* more closely in the fifth chapter of this book.

Additionally, Albert also included in the *Parva Naturalia* the *De spiritu et respiratione*, loosely based on Costa Ben Luca's *De differentia spiritus et animae*. This work was meant to substitute Aristotle's own and lost work on respiration. As we shall detail more at length in this book, though, the *De differentia*, much as it was relevant for Albert's spirit theory, was filled by conceptual tensions and inconsistencies. To build his own theory of spirit, Albert relied primarily

---

58 Miguel de Asùa, "War and Peace," 277: "The inclusion of the rationale of the precise relationships between *physici* and *medici* at the beginning of the *Parva naturalia* amounts to an implicit declaration of his intention of using the two bodies of knowledge in the same plane, for, as Albert claims, both of them '*terminantur ad invicem*'."

59 Miguel de Asùa, "War and Peace," 284–285; Chiara Crisciani-Giovanna Ferrari, "Introduzione," in Arnaldus de Villanova, Tractatus de humido radicali, ed. Michael R. McVaugh, with an introduction and commentary by Chiara Crisciani and Giovanna Ferrari (Opera Medica Omnia, V.2), Barcelona 2010, 385–396; Chiara Crisciani, "Aspetti del dibattito sull'umido radicale nella cultura del tardo medioevo (sec. XIII–XIV)," in *Arxiu de textos catalans antics* 23/24 (2005), 350–354.

60 Note that Avicenna had also dealt with the complexion of the four ages in his *De animalibus* (ed. Venezia 1508, XII, c. 3, fols. 45^{rb-vb} [ed. Cairo 1970, XII, c. 4, 200–204]), taken almost verbatim from the corresponding chapter of the *Canon* (ed. Venezia 1507, I fen 1 doctr. 3 c. 3, fols. 3ra–4va [ed. New Delhi 1981–1996, vol. 1, 44–47]). Albert must have read also the *Canon* and not only the *De animalibus*, because he mentions the so-called *istimbrae* (ar. *istimrāʾ*), a form of spontaneous appetite towards food, especially vehement in the first age. This appetite is mentioned only in the *Canon* (ed. Venezia 1507, I fen I doctr. 3 c. 2, fol. 3vb [ed. New Delhi 1981–1996, vol. 1, 46.10, 46.12]) and found in: Albertus Magnus, De aetate sive de iuventute et senectute tr. 1 c. 4, ed. Auguste Borgnet, Paris 1890 (Ed. Paris. IX, 312a): "Istimbra autem est appetitus cibi naturalis cum bona digestione." More on this in chapter 4. Clearly, this claim could be better substantiated or, by contrast, revised, upon the appearance of a critical edition of Avicenna's *De animalibus* (in preparation by Aafke van Oppenraaij) and of Albert's *De iuventute et senectute* from the Albertus-Magnus-Institut.

on Avicenna and integrated the doctrinal teachings from Costa ben Luca with other sources, such as Nemesius' *De natura hominis* or Alfred of Shareshell's *De motu cordis*, all of which are heavily indebted to the Galenic medical tradition.

From these examples, it should be clear that Albert makes substantial use of medical learning in his corpus of *Parva Naturalia* both to explain Aristotelian works and to make up for the lost ones. Therefore, there is more to a purely theoretical motivation for the inclusion of medical doctrines into works of natural philosophy. The rationale behind Albert's use of Avicennian medicine and zoology or Costa ben Luca's *De differentia* are therefore both the combination of theoretical and historical ones, driven by the need to fill in and supplement certain gaps in the Aristotelian corpus Albert set out to comment on. In short, this inclusion is historically influenced by Albert's need to make his Dominican brothers able to understand Aristotle appropriately (*competenter intelligere possent*). Still, this inclusion of medical learning is conditioned by epistemic criteria: as Albert maintains in the *De spiritu*, he will deal with phenomena such as respiration or the six *non-naturalia* (food, drinks, sleep, exercises, etc.) not in the way physicians do.[61] The bottom line, if we consider Albert's attitude towards medicine in the *Parva Naturalia*, is that medical teachings constitutes a fundamental complement to natural philosophy on the condition that such doctrines do not discuss particulars which properly fall to physicians and can be brought into agreement or are consistent with the principles of an Aristotelian philosophy of nature.

Even though Albert touches from time to time upon the opinion of Galen (*tangemus etiam quandoque opinionem Galeni*) as necessary,[62] this fact alone does not entail that medicine-related topics will be handled in the same manner as in medicine but, rather, according to the standards of natural philosophy (*quantum confert scientiae naturali*).[63] This expression must be referred both to respect for the fundamental laws of natural philosophy (Aristotelian hylomorphism) and the epistemic suitability of the inquiry being conducted. This

61 Albertus Magnus, De spiritu et respiratione, II tr. 2 c. 3, ed. Auguste Borgnet, Paris 1894 (Ed. Paris. IX, 249b): "Sunt autem multae differentiae anhelitus, sicut et pulsus, de quibus non per physica, sed per medicinalia particularia oportet expediri." A concrete example is represented by the so-called *non-naturalia*: Albertus Magnus, De morte et vita tr. 2 c. 6, ed. Auguste Borgnet, Paris 1890 (Ed. Paris. IX, 361b): "Causae autem horum omnium peccatorum in calido et humido sunt valde multiplices [...] sed de his in medicina intenditur, et non spectat ad scientiam universalem naturalem."

62 Albertus Magnus, De somno et vigilia I tr. 1 c. 1 ed. Auguste Borgnet, Paris 1890 (Ed. Paris. IX, 123a).

63 Albertus Magnus, De morte et vita tr. 2 c. 1, ed. Auguste Borgnet, Paris 1890 (Ed. Paris. IX, 351a).

means, for Albert, as it had for Aristotle and Avicenna, that proper knowledge of the causes of the human body according to their universal nature is the proper domain of the natural philosopher and not of the doctor (*scire enim causas corporis humani physici est, non medici*).[64] The main difference is that Albert's concept of "what is suitable in natural philosophy" is notably different from Aristotle's and Avicenna's, especially if we consider zoology and botany, where causal explanation of the living world revolves not only around the study of parts and functions but also includes the study of individual species.

7 Medicine as a Part of Natural Philosophy (?) in Albert's Zoology

Albert's zoology represents perhaps one of the most successful examples of the integration of medical doctrines into natural philosophy, as testified by its good reputation in the medical field. In Albert's zoology, the presence of medical learning is pervasive, and, as noted by the pioneering study of Nancy Siraisi, is mostly influenced by Avicenna's *Canon* which has replaced the less extensive *De animalibus*.[65] In addition to the Avicennian descriptions of the organs of sense perception and cognition (brain, nerves, bones and sinews), and the psychophysiology of generation, Albert also includes in the first book of his zoology large chunks from the *Canon* on physiognomy.[66] In short, medical learning does appear to play a substantial role in Albert's zoology. As we have shown in the previous sections, though, a proper evaluation of Albert's position on the relationship between medicine and natural philosophy needs to consider both abstract statements on the matter and concrete examples in the use of medical learning. While we have until now focused on concrete examples in the use of medical learning (*Parva Naturalia*), we can turn to Albert's abstract statements on the relationship between natural philosophy and medicine in the *De animalibus* to complete the picture:

> Thus does a physician (who is a particular kind of natural philosopher) work when he defines medicine in terms of sensible matter, saying that it is the science of the healthy, the sick, and those who are neither, because

64 Albertus Magnus, De anima I tr. 1. c. 7 (Ed. Colon. VII/1, 14.76–85).

65 Siraisi, "The Medical Learning," 393.

66 Heinrich Schipperges, "Das medizinische Denken bei Albertus Magnus," in *Albertus Magnus. Doctor Universalis 1280/1980*, ed. Gerbert Meyer, OP/Albert Zimmermann (Walberberger Studien, Philosophische Reihe 6), Mainz 1980, 283–286.

> it is the science by which the sensible dispositions of the human body are known with respect to those things by which it is healed and those by which it is made ill.[67]

If we follow Albert's line of reasoning, we can see how, if the object of medicine is the subject of a primarily natural process (healing) and the consideration of the object determines the relationship between the two sciences, then medicine is part of natural philosophy. This is why a physician can be considered a kind of natural philosopher (*particularis est physicus*). Gabriella Zuccolin has explained that this is precisely the opposite of Aquinas' position, expressed in the *Super Boetium de trinitate*, which insists on the subordination of medicine to natural philosophy since the natural process occurs because of artificial processes, whose principles are derived from the laws of natural philosophy.[68] Among Albert's contemporaries, Peter of Spain (1205–1277) also considered medicine to be subordinated to natural philosophy and based most of his solutions to medico-philosophical controversies on Avicenna.[69] In his commentary on the *De animalibus*, Peter distinguished the criteria according to which medicine and philosophy operate: while the former operates according to particulars, does not go beyond sense data and deals with the contingency of bodily operations, the latter approaches them from a perspective based on

67 Albertus Magnus, De animalibus XI tr. 1 c. 2 n. 15 (ed. Stadler, I, 766.11–15): "Sic enim facit medicus, qui particularis est physicus, quando diffinit medicinam per materiam sensibilem dicens quod est scientia sanorum et aegrorum et neutrorum, eo quod ipsa est scientia qua humani corporis dispositiones sensibiles cognoscuntur secundum ea quibus sanatur, et secundum ea quibus infirmatur." English Translation by Kenneth F. Kitchell Jr. and Irven Resnick in Albertus Magnus, *On Animals: a Medieval* Summa Zoologica, *transl. and annotated by Kenneth F. Kitchell and Irven Resnick* (Foundations of Natural History), 2 voll., Baltimore 2020, vol. 2, 862.

68 Zuccolin, "Il corpo sanabile e l'embrione," 119 n. 10; Thomas Aquinas, Super Boetium De trinitate, ed. Fratres Predicatores (Pierre M.J. Gils), Roma–Paris 1992 (Ed. Leon. L) q. 5, art. 1, ad. 5, 140–141: "Medicina ergo non ponitur sub phisica ut pars: subjectum enim medicinae non est pars subjecti scientiae naturalis secundum illam rationem qua est subjectum medicine: quamuis enim corpus sanabile sit corpus naturale, non tamen et subiectum medicine prout est sanabile a natura, sed prout est sanabile ab arte. Set quia in sanitatione que fit etiam per artem, ars est ministra nature, quia ex aliqua naturali virtute sanitas perficitur auxilio artis, inde est quod propter quid de operatione artis oportet accipere ex proprietatibus rerum naturalium, et propter hoc medicina subalternatur physice."

69 Miguel de Asùa, "Medicine and Philosophy in Peter of Spain's Commentary on the *De animalibus*," in *Aristotle's Animals in the Middle Ages and the Renaissance*, ed. Carlos Steel/Guy Guldentops/Pieter Beullens (Mediaevalia Lovaniensia [series 1] 27), Leuven 1999, 209.

concepts such as soul, which are required to transcend the realm of sense data and produce a universal account of the world of living beings.[70]

Returning to a closer scrutiny of Albert's text, there is more to the meaning of *particularis* than is necessarily embedded in the Latin word. By *particularis physicus*, Albert has in mind a natural philosopher who deals with the human body insofar as it can be healed and according to its particular causes and accidents; by definition, study according to particulars is ill-suited to the epistemic parameters of natural philosophy and its search for universal causes. As we have argued above, though, Albert's consideration the subject of medicine as a part of natural philosophy is a statement that does not come out of the blue. Rather, it entails a paramount reflection on the relation between the concept of healing and its subsequent relation to the object of medicine (*corpus sanabile*), which we find laid out in Albert's commentary on the *Metaphysics* and which shares doctrinal commonalities with Albert's early *Commentary on the Sentences*.

In book 7 of his *Metaphysics*, Albert takes on the relationship between art and nature when discussing their relationship with artificial products. If an architect is that who builds a house, what is the primary agential cause involved in the process? If a farmer makes his land fertile, what is the primary agential cause involved in the process of the soil becoming fertile? To answer these questions, Albert distinguishes between (1) objects produced by nature and art (*quae sunt ab arte et natura*) and (2) objects produced by art alone (*ab arte solum*). By the latter category, Albert means entirely artificial products like a house; by the former, he understands products of an art-nature cooperation such as health, agriculture (fruits, etc.) and the possible outcomes of the astronomer's action on elemental mixtures.[71] In these sciences, Albert takes the power of artificial actions to be merely auxiliary with respect to natural processes. According to him, the function of art is purely the removal of obstacles hampering nature's course.[72] In the end, Albert concludes that nature is

70 De Asùa, "Medicine and Philosophy," 204–207.

71 Albertus Magnus, Metaphysica VII tr. 2 c. 10 (Ed. Colon. XVI/2, 352.7–16): "Quia vero quaedam sunt quae fiunt simul a natura et arte, et in his non videtur conveniens a convenienti fieri, eo quod non potest idem secundum idem duobus differentibus convenire, ideo *dubitabit aliquis, quare alia sunt* simul ab *arte et ab automato* sive per se existente generante, *sicut sanitas* et agricultura et multa quae fieri per caelestem effectum procurat astronomus; de quo dicit Ptolemaeus, quod sapiens iuvat caelestem effectum, sicut iuvatur terra aratione et seminatione. […]" In this passage astronomy must be understood, following Ptolemy, as a scientist devoted to the second part of astronomy, i.e. the science dealing with heavenly influences, which Albert refers to often as "astrology".

72 Albertus Magnus, Metaphysica VII tr. 2 c. 10 (Ed. Colon. XVI/2, 352.54–58): "In omnibus

the primary agential cause and art only an auxiliary cause in those processes involving the cooperation of art and nature (*in illis quae sunt ab arte et natura, natura est causa et intellectus apprehensor et minister*).[73] Healing, for its part, is considered a primarily natural process consisting in the removal of physiological obstacles standing in the way of nature, similar to what he had argued in his early *Commentary on the Sentences*.

It should be noted that these ideas are clearly consistent with the early Albert's position on the nature of the process of healing: nature is considered the primary causal agent of this process, while medical art is considered just an aid to natural processes. Moreover, healing is understood, at a physiological level, as a removal of physiological obstacles, such as corrupted humors. While Albert's emphasis on the role of non-voluntary reactions triggered by the interaction of primary qualities and due to medicaments is not spelled out here (unlike in his early works), Albert does examine cases of medicine for different animals (e.g. falcons) elsewhere in his zoology.[74] Moreover, in his *De vegetabilibus* (VI, 2), he examines the humoral qualities of some herbs and their healing properties.[75] All in all, Albert's position on the inclusion of medicine into natural philosophy either at a theoretical or a practical (use of medical doctrines) level seems quite consistent throughout his work, if not for his peculiar understanding of the scope of certain parts of natural science (mineralogy, botany, zoology) displayed in the Aristotelian paraphrases.

Conclusions

Albert's early theological works (*De homine, Commentary on the Sentences*) constitute a valuable source for understanding his mature approach to the epistemic role of medical learning. In particular, Albert's remarks on the role of nature and art in the healing process, presented in his *Commentary on the Sentences* and often embedded within the comparisons between sacred (sacraments) and physical medicines (drugs or the like), enable us better to grasp his

---

huiusmodi virtus incipiens motum est naturae, et artis est removere prohibens, dummodo illa ars habeat partem eius quod fit vel aliquid quod consequitur id quod fit, sicut est calidum respectu sanitatis […]"

73 Albertus Magnus, Metaphysica VII tr. 2 c. 10 (Ed. Colon. XVI/2, 353.89–91).

74 Robin S. Oggins, "Albert on Falcons and Hawks," in *Albertus Magnus and the Sciences. Commemorative Essays*, ed. James A. Weisheipl (Studies and Texts 49), Toronto 1980, 441–462.

75 Albertus Magnus, De vegetabilibus, *Buch VI, Traktat 2: lateinisch-deutsch. Transl. and comm. by Klaus Biewer. With an introd. by Rudolf Schmitz* (Quellen und Studien zur Geschichte der Pharmazie 62), Stuttgart 1992.

mature position in the *Metaphysics* and his inclusion of medicine within natural philosophy in the *De animalibus*. Both *Commentary on the Sentences* and *Metaphysics* insist on a shared doctrinal core: (1) nature is the primary causal agent in healing; (2) accordingly, art and its products are only auxiliary agents in the achievement of health. Moreover, (3) health is defined physiologically as the removal of obstacles, such as corrupted humors, which get in the way of achieving a healthy condition. As we have shown, the first two arguments (1–2), which lead Albert to consider the *corpus sanabile* to be subject to a purely natural process of healing, accord conceptually with the inclusion (rather than its subordination) of the medicine within natural philosophy in the *De animalibus*.

But there is more to these purely theoretical motivations for the inclusion of medical doctrines into works of natural philosophy or parts of a *Summa* devoted to the human soul. If we consider Albert's production, there are some historic motivations too. In Albert's Aristotelian paraphrases, the need to fill in the gaps left by unavailable Aristotelian works further encouraged Albert to make a substantial use of medical works. The void left by the unavailable *De iuventute et senectute* or the *De respiratione* was filled by paraphrastic reworkings of Avicenna's chapter on the complexion of ages in the *Canon* and Costa ben Luca's *De differentia*. In this manner, Albert was able to complete his systematic paraphrasis of what he considered the "Aristotelian corpus", welding it together with medical doctrines as long as those doctrines were compatible with the principles of natural philosophy. Even though lacking the programmatic purpose of his Aristotelian paraphrases, Albert's early theology (*De homine*) has also historic reasons behind its use of medical learning. In the first place, it was common practice of theologians to make use of medical doctrines at the university of Paris at that time. Moreover, the use of Avicennian zoology and Costa ben Luca's detailed anatomy (along with other sources) allowed Albert to produce an anatomically accurate depiction of the soul activities in the human body if compared to his contemporaries, who focused more on the psychological aspects of the psychophysiology of soul faculties in treatises on the human soul. Given the detailed analysis of each and every faculty of the soul produced in the *De homine*, one can well see how the use of medical doctrines could play well into Albert's early philosophical psychology.

So far, we have observed that the theoretical reasons behind the inclusion of medicine within natural philosophy may shine through Albert's *De animalibus* and *Metaphysics* and possess clear historical reasons of their own. All of Albert's corpus, from his *De homine* on, betrays *de facto* a detailed yet epistemologically selective use of medical learning. Strongly influenced by Avicenna (*De animalibus* and, later on, the *Canon*) and to a lesser extent by Averroes' *De corde*,

Albert's take on the value and use of medical learning shares with Avicenna epistemic concerns towards the Galenic arguments on the role of females in generation and on the origin of veins. Galen's theories were rejected by Avicenna because of two main reasons: on the one hand, to avoid the infringement of the hylomorphic pattern of natural causation if male and feminine semen were to form two concurrent formal causes. On the other hand, Albert rejects the origin of veins from the liver to prevent the disruption of the unitary psychological structure of the living being grounded in Aristotelian cardiocentrism and in the Avicennian-Boethian physiological structure of Albert's faculty theory we shall see throughout this book. In the third chapter, though, we will try to isolate and discuss an aporia in both Avicenna's and Albert's embryological account, viz. the entrance of masculine semen into the material composition of the offspring, which caused the Aristotelian hylomorphic scheme of generation some serious trouble.

Albert's *De homine* reflects perfectly the epistemic concern of Avicennian zoology, whose detailed anatomy and physiology is combined with other relevant medico-philosophical sources such as Costa Ben Luca, Nemesius and Alfred of Shareshell. The result is an accurate picture of the anatomy and physiology of organs of sense perception and cognitive activity, such as the brain and the nerves. On the other hand, Albert firmly denies any active causal agency to feminine semen, while not refraining from discussing the topic of sexual differentiation. On the origin of veins, Albert remains steadfastly faithful to Aristotelian cardiocentrism, although not really addressing the Galenic-Aristotelian quarrel *ex professo*. By contrast, all of these three epistemically sensitive main themes (feminine semen, sexual differentiation, origin of veins) would be tackled more in detail in his zoology, as we will see in the third chapter.

Albert selectively included large chunks of medico-philosophical doctrines consistently throughout his writing to produce a more accurate depiction of the psychophysiology of living beings. This approach, common to his *De homine* and *De animalibus*, applies to his *Parva Naturalia* as well and is driven, besides theoretical reasons, by concrete historical motivations. In these works, although the opening lines of the *De sensu et sensato* betray an approach formally different from the *De animalibus*, for medicine and natural philosophy are said to "meet halfway" (*terminantur ad invicem*), Albert maintains a substantially consistent approach to the epistemological role of medical learning. He injects a substantial amount of medical learning into his *Parva Naturalia*, either to paraphrase and explain the Aristotelian texts (*De morte et vita*), to rephrase and reshape non-Aristotelian texts transmitted with the Aristotelian corpus (*De spiritu*, paraphrase of the *De differentia*) or even to fashion new texts, such as the *De iuventute et senectute*, meant to supply the lost Aristotelian *De iuventute*. According to Albert, the inclusion of medicine depends more in general on three main crucial factors: first, a more abstract consideration of the subject of medicine *qua* being primarily healed by nature (hence, the subject of natural philosophy), the abidance of medical doctrines to epistemological rules of natural philosophy and the use by natural philosophy of medical doctrines according to the standards of natural philosophy (*sicut confert scientia naturali*). This inclusive but selective attitude towards medicine, typical of the *De homine* and developed originally in the Aristotelian paraphrases, is simply impossible to imagine without Albert's most influential medico-philosophical source, Avicenna.

As we have emphasized, though, Albert's mature concept of natural sciences differs notably from Aristotle and Avicenna. By including books aimed at the study of minerals, plants and animals according to their species, Albert transgresses the usual boundaries of Aristotelian natural science. The upshot is a substantial difference compared to the Avicennian model in the scope of zoology and botany and, in turn, in the use of medical learning in works of natural philosophy. The generally positive—though epistemologically selective—attitude towards medicine puts us in a much better position to understand the place and development of a key psychophysiological notion such as *spiritus* throughout Albert's intellectual career. In the first part of this book (*Spirits of life*), we shall focus on the causal relationship between spirit and life in three cases: the definition of life and the most basic conditions for its subsistence, the origin (generation) and duration (length and shortness) of life.[76]

76 A summarized version of this chapter in: Meroni, *Natura confortata per medicinam*, 109–136.

PART 1

Spirits of Life

CHAPTER 2

Vita abstrahitur a vegetatione: Spirit and Albert's Emanative Psychology of Life

The concept of spirit has a complex history straddling medicine, natural philosophy and theology. On the one hand, it appears fragmentarily in the Aristotelian *corpus*, especially in the zoology and to a smaller extent in the *Parva Naturalia*. On the other hand, it represents a cornerstone of the physiology in the Galenic medical tradition. Heir to a long-durée medical tradition and transmitted by the Arabo-Latin translations of medico-philosophical works, the concept of *pneuma* as the "instrument of the soul", "medium" or "vehicle"[1] of the soul faculties played well into the hands of those who, like Albert, were seeking to build a comprehensive account of the psychophysiology of life-related operations. In both Aristotle and the Galenic medical tradition, the word spirit (Gr. *pneuma*/Ar. *rūḥ*/Lat. *spiritus*) is etymologically and primarily related to respiration. According to Aristotle and Avicenna, lungless animals such as insects and smaller watery animals possess a material substance functionally analogous to spirit, the so-called connate *pneuma*. As we shall see, spirit and connate spirit are the cornerstone upon which Albert the Great will build part of his psychophysiology of soul activities.

In this chapter, the main focus will be placed on the relationship between the causal role of spirit and the most basic activities of the soul, viz. respiration (or the motion of connate *pneuma* in lungless animals), pulse and nutrition. To understand this relation, a first step is to understand the concept of life and its

1 On spirit as "medium", see: Costa ben Luca, De differentia spiritus et animae, ed. Wilcox 1985, 187.528–530 [ed. Troupeau 2011, xxix.8–9]: "Anima quoque movet corpus et prestat ei vitam atque sensum mediatore spiritu (*bi tawassuṭ al-rūḥ*), et spiritus operatur hoc absque aliquo mediatore." Aristotle often defines spirit as a vehicle (ὄχημα) in his *De motu animalium* (c. 10, 703a4–703a27). A similar formulation of spirit as "vehicle" (*maṭiyya, markab*) in Avicenna and al-Ġazālī: Avicenna, Liber de anima III c. 7, ed. Van Riet 1972, 257.50 [ed. Rahman 1959, 144.2]; Liber de anima V c. 8, ed. Van Riet 1968, 175.49–51 [ed. Rahman 1959, 263.9–10]; Canon, I fen 1 doctr. VI c. 1, ed. Venezia 1507, fol. 23[rb] [ed. New Delhi 1981–1996, vol. 1, 123.1]; Algazel, Metaphysica pars 2 tr. 5 c. 4, ed. Muckle 1933/1967, 188.19–22 [ed. Cairo 1936, 66.6–8]: "Spiritus vero est corpus subtile (*ǧism laṭīf*) compositum ex vaporibus humorum, cuius sedes est cor, et ipse est vehiculum virtutum vitalium, et animalium (*wa-huwa markab al-quwà al-nafsāniyya wa-ḥayawāniyya*), per quem virtutes sensibiles et motive transeunt ad sua instrumenta (*wa-tattaṣilu bihā al-quwà al-ḥassāsa wa-l-mutaḥarrika ilà ālātihā*)."

 DOI:10.1163/9789004706743_004

definition in Albert and in his main sources, Aristotle and Avicenna. A proper explanation of the role of spirit in the psychophysiology of life necessitates a close connection between the fundamental life-related activities (pulse, respiration, nutrition), Albert's theory of spirit and his emanative conception of life as an act of the soul.[2] A comparison between Albert's early and mature interpretation of the same concept (spirit) will uncover doctrinal influences and differences between Albert and his sources and among the early and mature Albert, as it were. Given the polysemic nature of the term *spiritus*, though, a short semantic disambiguation and historic contextualization for the concept of spirit in Albert's works seems in order here.

1 Medieval Conceptions of Spirit: Between Language of Life and Doctrine of Intermediaries

In his key 1984 article, "Medical spirits and the medieval language of life," James Bono highlighted how the concept of spirit represented for medieval thinkers (among which Albert) one of the cornerstones in the formation of a language which blended Aristotelian psychology—often read through the lenses of Avicenna—and medical doctrines in theological treatises on the human soul.[3] In her book on the body-soul relationship at Paris until 1250, Magdalena Bieniak has stressed how the concept of spirit was used in the early 13th century in theological discussions on the body-soul union and, in particular, the so-called doctrine of intermediaries. This doctrine, first introduced into the discussions of the Theology faculty at Paris by Philip the Chancellor's († 1236) *Summa de bono* and perhaps inspired by the anonymous *De potentiis animae et objectis*,[4] was supposed to explain how a simple, incorruptible and incorporeal entity such as the soul could operate its activities in the composite, material and corruptible human body.[5] Albert, who was a bachelor student of theology (*baccalaureus*) at Paris during the years 1241–1245, was acquainted with these discussions. In his pioneering study, though, Chenu showed how, the term "spirit" (*spiritus*) at the beginning of the 13th century could take on quite a few theological and psychological meanings (spirit as "soul" or "imagination",

2 Thérèse Bonin, "The Emanative Psychology of Albertus Magnus," *Topoi* 19 (2000), 45–57.
3 James Bono, "Medical Spirits and the Medieval Language of Life," *Traditio* 40 (1984), 91–130.
4 Magdalena Bieniak, *The Soul-Body Problem at Paris, ca. 1200–1250: Hugh of St. Cher and His Contemporaries* (Ancient and Medieval Philosophy [Series 1] 42) Leuven 2010, 129.
5 Bieniak, "The Soul-Body Problem," 119–124.

as in Augustine) and not just the medico-philosophical one.[6] As a child of his time, Albert the Great felt the need to distinguish and clarify them.

The most organic example of the semantic disambiguation of the term *spiritus* can be found in Albert's *De spiritu et respiratione* (1256/1257), a work inspired to Costa Ben Luca's *De differentia spiritus et animae*.[7] Here, Albert singles out three different concepts of spirit:

1. Spirit as "soul", a position attributed both to "Platonic" philosophers and theologians;
2. Spirit as subtle body (*corpus subtile*) according to its medico-philosophical construal;
3. Spirit as "imagination", an heritage of the Augustinian tradition (Augustine's *De genesi ad litteram*, ps.-Augustine's *De spiritu et anima*).[8]

The coexistence of these three possible meanings of spirit mirror quite well the terminological and semantical struggle Albert had to put up with whenever mentioning *spiritus* in front of an audience or readership of any kind. Bono has shown that Albert imposed strict methodological caveats on language to avoid consequent confusion: by spirit, in works of natural philosophy, was only meant the medical understanding of the term as subtle body, the instrument of the soul and vehicle of its faculties.[9] A similar yet not identical strategy of linguistic disambiguation can be found in his early *De homine* (1241–1242): in this work, Albert distinguished sharply between a philosophical (corporeal) and a theological (incorporeal) meaning for the term spirit too: this time, the term *spiritus* in an incorporeal sense only meant "soul" and not—at least not explicitly—the Augustinian "imagination". This clearcut distinction was achieved by quoting the definition of spirit from Costa Ben Luca's *De differentia spiritus et animae* in opposition to a theological meaning of spirit as "soul" (*aut accipitur spiritus modo theologico pro eo quod est anima vel eiusdem*

6 Marie-Dominique Chenu, "*Spiritus*: Le vocabulaire de l'âme au XIIe siècle," *Revue des Sciences philosophiques et théologiques* 41 (1957/2), 217–232.

7 Albertus Magnus, Physica I tr. 1 c. 4 (Ed. Colon. IV/1 7.54–56).

8 Albertus Magnus, De spiritu et respiratione II tr. 2 c. 4, ed. Auguste Borgnet, Paris 1894 (Ed. Paris. IX, 250b): "Relinquitur autem ex dictis, quod nomen spiritus aequivoce dicitur de anima et spiritu corporeo: anima enim apud Platonicos dicitur spiritus [...]. Spiritus autem, de quo diximus, dicitur essentialiter spiritus, eo quod est substantia spiritualis subtilis et tenuis. Si quis autem dicat animam proprie et essentialiter dici spiritum, hoc non erit verum secundum quod Philosophi utuntur hoc nomine, spiritus: sed hoc erit secundum scientiam divinam: divini enim in multis significationibus hoc nomine spiritus utuntur: aliquando quidem pro animae superiori parte: aliquando pro imaginativa, eo quod dicunt quod spiritus est vis animae quaedam mente inferior, in qua imagines rerum imprimuntur: sed hoc modo non sumus hic locuti de spiritu, eo quod in hoc opere non nisi physica tractare suscepimus."

9 Bono, "Medical Spirits," 119.

naturae cum ipsa).[10] The fact alone that Albert felt the need to clarify the term *spiritus* at least twice in his works speaks volumes in terms of the conceptual tensions which, without prior clarification, this term could generate. This is an approach to the concept of spirit that was not entirely new, though, for authors of the likes of Alain of Lille (1128–1203) had already used the strategy of a careful distinction between a metaphysical and a physiological sense of spirit prior to Albert's early theological works.[11]

The use of the medico-philosophical concept of spirit in theological treatises on the human soul is no innovation on Albert's side, but rather reflects the discussions sparked from the above-mentioned doctrine of intermediaries by Philip the Chancellor. The starting point of Philip's doctrine was to consider body and soul as two dualistically separate substances, standing as complete ontological opposites: on the one hand, the simple, incorporeal and incorruptible soul; on the other hand, the composite, corporeal and corruptible body. His doctrine of intermediaries is meant to account for the operation of the soul in the body. At the very basic level of matter (body), Philip argues, there are the four elements, whose chief primary quality is represented by vital heat, crucial to make the body alive. Next there is the spirit in the medico-philosophical sense, which gives the heat of life to the body.[12] Then come the

10 Albertus Magnus, De homine (Ed. Colon. XXVII/2, 154.57–67): "Item, si esset substantia composita, aut esset spiritualis, ita quod esset spiritus, aut corporalis sive corpus. Si primo modo, aut accipietur spiritus pro corpore prout accipiunt eum philosophi, sicut habetur in libro Constabuli, quod 'spiritus est quoddam corpus subtile, quod in humano corpore oritur ex toto corde et fertur in venis pulsativis ad vivificandum corpus, et operatur vitam et anhelitum et pulsum; et oritur ex cerebro et fertur in nervis et operatur sensum et motum,' aut accipitur spiritus modo theologico pro eo quod est anima vel eiusdem naturae cum ipsa." See: Costa ben Luca, De differentia spiritus et animae, ed. Wilcox 1985, 145.33–38 [ed. Troupeau 2011, V.2–3].

11 Alanus de Insulis, De fide catholica contra haereticos sui temporis, praesertim Albigenses liber quatuor, PL 210, Paris 1855, col. 329D–330A: "Est namque in homine duplex spiritus, spiritus rationalis et incorporeus, qui non perit cum corpore; et alius qui dicitur physicus sive naturalis, quo mediante anima rationalis unitur corpori, et hic spiritus est subtilior aere, et etiam igne, quo mediante fit sensus et imaginatio: et ille perit cum corpore. Talis spiritus naturalis est in corpore bruti animantis, et illud vegetat, et perit pereunte corpore." On the concept of spirit during the 12th century and its medical, philosophical and theological implications, see: Anne Marie Bautier, "'*Spiritus*' dans les textes antérieurs à 1200. Itinéraire lexicographique médiolatin: du souffle vital à l'au delà maléfique," in Spiritus. *IV Colloquio Internazionale del Lessico Intellettuale Europeo. Roma, 7–9 gennaio 1983*, ed. Marta Fattori/Massimo Bianchi, Rome 1984, 113–132.

12 Philippus Cancellarius, Summa de bono, De bono nature pars IV q. 8, ed. Nikolaus Wicki (Corpus Philosophorum Medii Aevi. Opera Philosophica Mediae Aetatis Selecta 2, 2 voll.) Bern 1985, vol. 1, 286.170–287.174: "Exiguntur ergo ad coniunctionem anime rationalis cum

vegetative and sensitive souls, the remote and proximate media between the rational soul and the body. The characteristic of Philip's discussion on the doctrine of intermediaries is that each of these media has something in common with the one that is inferior to it as well as the one that is superior to it, and it is those commonalities that ultimately enable the body to be joined to the soul.

The doctrine of intermediaries (sensitive soul, vegetative soul, spirits, elements) appears not only in Philip and the *De potentiis* (his likely source), but in many other authors and texts such as Hugh of St-Cher, an anonymous question on the immortality of the soul, the *Summa Duacensis*,[13] and John of La Rochelle's *Summa de anima*,[14] which in turn influences the *Summa Halensis*.[15] Other authors such as William of Auvergne and Roland of Cremona made use of the concept of spirit according to its medico-philosophical understanding but without reference to Philip's doctrine of intermediaries. While Roland spoke more generically of corporeal dispositions necessary for the soul to operate in the body,[16] William limited himself to a few remarks on the mediating

corpore anima sensibilis et anima vegetabilis et iterum spiritus, qui est a natura corporis superioris, et calor elementalis. Sciendum autem quod spiritus triplex est. Est enim quidam animalis, et hic est in cerebro; et est spiritus vitalis, et hic est in corde; et iterum spiritus naturalis, et hic est in epate." On Philip's doctrine, see: Bieniak, "The Soul-Body Problem," 119–124.

13 Bieniak, "The Soul-Body Problem," 123–124.

14 Johannes de Rupella, Summa de anima c. 39, ed. Jacques G. Bougerol, Jean de la Rochelle, Summa de anima (Textes philosophiques du Moyen Âge 19), Paris 1995, 120.25–123.56.

15 Summa fratris Alexandri (Summa Halensis), II pars 1 inq. 4 tr. 1 s. 1 q. 3, tit. 2 c. 1 a. 3, (Ed. Quaracchi II, 421): "Anima rationalis est substantia incorporea, motiva, cognoscitiva, separabilis sive non dependens a corpore; corpus vero, cui unitur tamquam organo sive perfectibili, est corpus compositum ex elementis, complexionatum. Ex parte vero incorporei est substantia incorporea, motiva, non cognoscitiva, inseparabilis a corpore, scilicet vegetabilis, et substantia incorporea, motiva, cognoscitiva, inseparabilis, scilicet sensibilis: et ita duo sunt media. Ex altera vero parte est corpus compositum ex elementis, non complexionatum, et corpus non compositum ex elementis: et ita duo sunt media. Et ita quatuor erunt media inter animam ratiorialem et suum corpus, duo ex una parte et duo ex alia: spiritus enim est corpus non compositum ex elementis, humor vero est corpus compositum ex elementis, sed non complexionatum."

16 Rolandus Cremonensis, Summae magistri Rolandi Cremonensis, O.P. Liber secundus, CXVII n. 1, ed. Alessio Cortesi/Umberto Midali (Quaderni della Cortesiana 10), Bergamo 2017, 167: "Vincenus (sc. Avicenna) dixit quod spiritus sunt vehicula virtutum. Quasi idem dixit. Et nos loquemur de istis spiritibus quando nos prosequemur de viribus" On the role of spirit in sensation and motion, see also: Rolandus Cremonensis, Summa II CXVIII n. 14 (ed. Cortesi–Midali 2017, 170). On Roland, see also: Bieniak, "The Soul-Body problem," 131–136.

role of spirit in sensation and motion, deferring a more detailed approach of the topic to physicians.[17]

In keeping with the previous tradition, Albert tackled the role of spirit in the *De homine* in a *quaestio* which shares the identical subject (*De unione animae et corporis*) with the one in which Philip the Chancellor presents his doctrine of intermediaries. First, Albert recalls the traditional opposition between the soul as a simple, incorporeal and incorruptible substance (*simplex est et incorporea et incorruptibilis*) and the body as corruptible (*quantum grossum corruptibile*). Albert considers the soul from a threefold point of view: as the body's form, efficient cause and principle of the soul activities. According to Albert, the soul is joined to the body immediately as form to matter (*patet quod anima sine medio est in corpore*) and, unlike his predecessors and contemporaries, he does not consider the vegetative and the sensitive soul as mediums between the rational soul and the body. In his view, both vegetative and sensitive powers of the soul flow, in accordance with Avicennian psychology, from the soul's essence. Accordingly, they are not *media* between the rational soul and the body in terms of existence but only of activity.[18] Per this perspective, Albert's position, even though maintaining the nature of the three spirit as mediums for the soul's operation in the body,[19] notably differs from the previous tradition in terms of psychological structure of the living being. Unlike Philip and a major part of the tradition influenced by this reading of Aristotelian psychology, which lends

---

17 William of Auvergne, Tractatus de anima c. 2 p. 14 (Opera omnia II, Suppl., Paris 1674, 84b): "[…] et eosdem modo se habet de spiritibus, quos in corporibus animalium medici ponunt esse: per eos namque imperant motum nervis, et chordis, aliisque ligaturis membrorum."

18 Albertus Magnus, De homine (Ed. Colon. XXVII/2, 572.22–31): "Ad id quod contra obicitur, dicendum quod anima vegetabilis et sensibilis in homine non sunt diversae in substantia ab anima rationali, sed sunt aliae potentiae, et propter hoc non sunt mediae, ut per quas anima rationalis sit in corpore, sed ut per quas operetur. Ad aliud dicendum quod anima sensibilis et vegetabilis corrumpuntur in suis operibus, eo quod sunt potentiae affixae organis, et non corrumpuntur per se secundum quod sunt in homine. Et hoc supra probatum est in quaestione de immortalitate animae." On the early Albert's faculty theory and its debt to Avicenna (and Boethius), see: Dag N. Hasse, "The Early Albertus Magnus and his Arabic Sources on the Theory of the Soul," *Vivarium* 46 (2008) 243–244.

19 Albertus Magnus, De homine (Ed. Colon. XXVII/2, 571.58–572.7): "Consideratur etiam in suis operationibus, et sic indiget medio duplici, scilicet generali et speciali. Generale medium est spiritus, qui secundum medicos corpus medium est inter aerem et ignem, et est triplex, scilicet naturalis et vitalis et animalis. Naturalis autem est instrumentum trium virium animae vegetabilis; vitalis vero est ille, per quem anima a corde vitam et pulsum operatur in toto corpore; animalis vero deservit praecipue virtutibus existentibus in capite, sive sint apprehensivae, sive motivae."

itself to plurality of forms thesis,[20] Albert goes in a much different direction by adhering to the Avicennian psychology rooted in the Neoplatonic theory of *fluxus* and trying to emphasize the unity of the substantial form.[21]

Even though Albert considers certain faculties to be subservient (*deserviunt*) to the higher ones, such as nutrition to sensation and phantasy to intellection, his understanding of their relation to the rational soul is not that of mediums which exist for the sake of the conjunction of the rational soul with the body but of powers flowing from the essence of the rational soul which need vital, natural and animal spirits as appropriate tools and instruments to act in the body. To stress this teleological function of spirits, Albert often reiterates a cosmical analogy between the nature of spirits and celestial lights *qua* vehicles of certain virtues (*quoad similitutinem utilitatis et finis*) that is based on a reworking of a passage from Isaac Israeli's *De definicionibus*.[22] This passage will be turned into the recurrent analogy that "spirit moves in the body like lights in the world" (*spiritus moventur in corpore sicut lumina in mundo*) and will become a staple throughout his works, attributed either to Isaac,[23] Costa Ben Luca,[24] or left anonymous.[25]

It is precisely in the context of one of these comparisons between celestial lights and spirits that Albert presents, in his early *De resurrectione*, a slightly different view on the nature of spirit. When commenting on a passage from

20 On the plurality of substances in Philip the Chancellor, see: Bieniak, "The Soul-Body Problem," 120–122.

21 Bonin, "The Emanative Psychology," 51.

22 Isaac ben Salomon Israeli, Liber de definicionibus, ed. Joseph Muckle, in: Archives d'Histoire Doctrinale et Littéraire du Moyen Âge 11 (1937/1938), 319.19–22: "Causa autem quare non est possibile ut spiritus et sanguis incedant in canali una et curso uno est quod spiritus subtilitate sui currit in corpore subito sicut currit splendor candelae in domo subito absque spacio temporis. Sed sanguis grossicie sua et gravitate sui motus non currit nisi paulatim paulatim sicut currit aqua in poris terrae et canalibus eius, etsi sit sanguis grossior aqua et tardius incedens. Si ergo spiritus et sanguis incederent in canali una et curso uno, vinceret sanguis spiritum et prohiberet ipsum ne curreret in corpore subito, et esset in hoc corrupcio ordinis naturalis."

23 Albertus Magnus, De quattuor coaequaevis tr. 4 q. 53, ed. Auguste Borgnet, Paris 1895 (Ed. Paris. XXXIV, 592b); De homine (Ed. Colon. XXVII/2, 155.52–59; 328.66–329.1); De somno et vigilia I tr. 1 c. 5, ed. Auguste Borgnet, Paris 1890 (Ed. Paris. IX, 130a); De spiritu et respiratione, I tr. 1 c. 3, ed. Auguste Borgnet, Paris 1894 (Ed. Paris. IX, 217b).

24 Albertus Magnus, De causis et processus universitatis a prima causa II tr. 2 c. 37 (Ed. Colon. XVII/2, 130.17–20); Albertus Magnus, Summa theologiae sive de mirabili scientia dei, pars I, tr. 7, q. 31, c. 3 (Ed. Colon. XXXIV/1, 245.16–22).

25 Albertus Magnus, De homine (Ed. Colon. XXVII/2, 572.7–15); De anima II tr. 3 c. 11 (Ed. Colon. VII/1, 115.56–60); De animalibus, XX tr. 1 c. 7 n. 37 (ed. Stadler, II, 1292.16–24). On Albert's use of this cosmic analogy, see: Bono, "Medical Spirits," 120–121.

the prophet Ezekiel 28:9 dealing with the resurrection of bodies (*Et insufflat super interfectos et reviviscunt*), Albert argues that the resurrection of the dead will occur through the blowing of the unperishable (*incorruptibiles*) and quintessential (*sunt de natura quinti corporis*) spirits. These spirits, which move in the human body like celestial lights in the sublunar world, do not perish after the death of the human body; rather, they are always ready to bring the dead back to life at God's order.[26]

The passage from the *De resurrectione* does not have parallels in other works by Albert, though: in the *De quattuor coaequaevis* the same comparison between the motion of spirit and celestial lights appears already stripped of any consideration on the quintessential nature of spirit.[27] The other passages devoted to spirit in Albert's early theological works share this medico-philosophical and non-quintessential view on spirit. The passage from the *De resurrectione* appears to be therefore an isolated one and not indicative of the early Albert's consideration of the ontological nature of spirit.

From these first pages it should be clear why spirit in the medico-philosophical sense of subtle body understood as the "proper vehicle of life" (*vehiculum autem vitae proprium est spiritus*)[28] is a natural candidate to represent one of if not *the* cornerstone of Albert's psychophysiology. Crucial in 13th-century discussions on the body-soul relationship, this concept was framed by Albert within a psychological structure that preferred the Neoplatonic and Avicennian theory of *fluxus* as a voucher for the unity of the substantial form over Philip's traditional doctrine of intermediaries, prone to criticisms regarding the plurality of substances. To provide an accurate depiction of Albert's own views on the causal relationship between spirit and life, clearly characterized by a Neoplatonic and Avicennian model of psychology, this chapter will consider:

---

26 Albertus Magnus, De resurrectione tr. 1 q. 5 (Ed. Colon. XXVI, 248.22–32): "Ad ultimum dicendum secundum Isaac in libro *De diffinitionibus*, quod in fabricatione humani corporis conveniunt spiritus, qui moventur in corpore ad modum luminis in mundo. Unde illi spiritus, inquantum sunt luminosi, sunt de natura quinti corporis et ideo incorruptibiles et in dissolutione corporis humani revertuntur ad circulos. Et isti sunt qui a quattuor ventis caeli et a quattuor plagis mundi per duplicem motum superiorum distinctorum veniunt, et sunt vivificantes mortuos, quia illi sunt spiritus vitales."

27 Albertus Magnus, De quattuor coaequaevis tr. 4 q. 53, ed. Auguste Borgnet, Paris 1895 (Ed. Paris. XXXIV, 592b): "Sed tunc spiritus est multipliciter dictum. Proprie enim dicitur spiritus corpus subtile, luminosum, medium inter corpus et animam, principium et instrumentum vitae, qui discurrit in corpore secundum Isaac, ut lumen in mundo. Et hic dividitur in spiritum animalem et vitalem et naturalem, quorum primus habet situm in cerebro, secundus in corde, et tertius in hepate."

28 Albertus Magnus, De spiritu et respiratione I tr. 1 c. 1, ed. Auguste Borgnet, Paris 1894 (Ed. Paris. IX, 214a).

the body-soul relationship, the faculty theory and the relation of the soul's faculties to spirit, and the causal role of spirit in the necessary and sufficient conditions for life in living beings (respiration, heartbeat and nutrition).

If "living", as Aristotle says, "is spoken of in many ways"[29] in plants, animals and humans, and spirit, according to Albert, is the proper vehicle of life, the study of the relationship between spirit and life offers the intriguing opportunity to inquire into the manifold variety of life on earth: its generation, its subsistence and its duration. To understand how life is generated (chapters 3 and 4) and its length (5), the first step to take is defining life in relation to spirit, and to understand Albert's interpretation of life as an act of the soul (*actus animae secundum se*) in the light of his philosophical and medical sources. The beginning of this philosophical journey cannot start but with the Philosopher par excellence, Aristotle.

2 Keeping It Cool: Aristotle on Soul, *pneuma*, and Life

According to Aristotle, life is "being among living beings" or, as Albert would argue, *viventibus est esse*.[30] The soul, cause and principle of life,[31] is defined in two ways: substance as the form of a natural body[32] and as "the first actuality of a natural body which has life in potentiality".[33] To describe the body-soul relationship, Aristotle employed the famous and controversial metaphor of the steersman and the ship:[34] the soul relates to the body as the former does to the latter. In the Aristotelian and Neoplatonic traditions, the interpretations given to the term 'actuality' and its problematic association with the concept of 'form' would represent the key to the development of new psychological models aimed at granting immortality to the human soul. There is good reason to think that Aristotle denied any possible survival of the human soul after the body's death, an answer unacceptable both to Neoplatonists and adherents to monotheistic religions. In Aristotle's psychology, the soul is substance *qua* form of the animated body. Moreover, the soul is not just the formal cause of the living being, cause and principle of life and cause of organic cohesion against the

29 Aristotle, De anima II 2 (413a21–25).
30 Aristotle, De anima II 4 (415b12).
31 Aristotle, De anima II 1 (415b13–14).
32 Aristotle, De anima II 1 (412a12–21). On Aristotle's hylomorphism, see at least: Richard Sorabji, "Body and Soul in Aristotle," in *Articles on Aristotle* (IV). *Psychology and Aesthetics*, ed. Jonathan Barnes/Malcolm Schofield/Richard Sorabji, London 1979, 42–64.
33 Aristotle, De anima II 1 (412a27–28).
34 Aristotle, De anima II 1 (413a8–9).

mutual and corrupting interaction of elements. The soul is also the final cause of the living being, that for the sake of which the entire physiological structure of the living being, with the heart in its (purported) center, is arranged.

According to Aristotle, the soul becomes the actuality of the body endowed with organs once these organs reach a degree of suitable configuration to allow the operations of the soul to take place. The soul becomes the actuality of the body when the body becomes suitably organized, a process which takes a different amount of time, as we shall see more in detail in the next chapter, in humans, oviparous animals (20 days), and plants. According to Aristotle, the soul *qua* cause and principle of life and its activities exists in the body not only as an actuality whenever its operations are carried out (as during the activity of sensation) but also as an actuality that subsists regardless of the use of these capacities (as during sleep). The former kind is qualified by Aristotle as a 'second actuality' and does not represent the true essence of the soul. Rather the soul is classified as a first actuality, which can be fully actualized through its various capacities, unless death supervenes.[35]

The operations of the soul's faculties are divided by Aristotle into those which are inseparable from the body and those which are separable. Among the inseparable activities are the nutritive and the sensitive activities of the soul, while the intellect seems to enjoy a peculiar, hybrid status. On the one hand, Aristotle makes clear that it is not possible to have intellectual activity without images (φαντάσματα); on the other hand, Aristotle seems to suggest the incorruptible, separable and eternal nature of the intellect.[36] These controversial passages on the intellect would then pave the way for exegesis—such as that by John Philoponos—aimed at arguing for the immortality of the human soul, rooted in the nature of the intellectual faculty. This aim could have been achieved only at the cost of dissociating the Aristotelian concepts of form and actuality, more often defined—after the mediation of Arabic philosophers—as "perfection" (ἐντελέχεια).[37] According to Aristotle, all living beings are given a certain set of faculties which allow them to live in accordance both with their being (εἶναι) and their "well-being" (εὖ εἶναι). The connection between faculties and soul proper in Aristotle, though, is anything but uncontroversial. The issue of whether Aristotle is or is not an advocate for the pluralism of forms is a very thorny issue which need not be debated here. Suffice it here to say that the

35 Aristotle, De anima II 1 (412a22–27).

36 Aristotle, De anima II 1 (413a3–5); II 2 (413b24–27).

37 On the Arabic interpretations of this passage, see: Tommaso Alpina, *Subject, Definition, Activity. Framing Avicenna's Science of the Soul* (Scientia Graeco Arabica 28), Berlin 2021, 101–102 and bibliography therein.

interpretation we are assuming is that given by Albert, namely that Aristotle maintains that there is one single soul which acts as a form to all living beings and whose actions are seated in the heart, according to Aristotle's cardiocentrism. But this was not, as we have seen so far, the mainstream view among his contemporaries in Paris.

In humans, though, it would be better to speak of 'perfection' than 'form' (according to Albert, that is, not Aristotle). The soul's faculties, usually defined according to their object by Aristotle, are divided into nutritive, sensitive, locomotive, appetitive and intellectual. If life, in abstract and psychological terms, is the act or actuality of the soul's powers according to the capacities of each species, then life equates to the activities of the nutritive soul in plants, the activity of the nutritive soul and senses (at least, of touch) in animals, and the activity of nutrition, sense perception and intellect in humans. Since these faculties are powers of an immaterial soul, they are themselves immaterial; at the same time, though, they need suitable material instruments to carry out their operations and grant subsistence of life.

Vital heat is the first physical notion that comes to the mind of any Aristotelian scholar. It plays a crucial role since living beings move from life in potentiality to life in actuality, not only when their organs are suitably configured, but also when they are endowed with a sufficient amount of vital heat. Without vital heat, a living body with a suitable organic structure would still be just a corpse. Yet it is not exactly easy to fathom how vital heat, according to Aristotle, becomes diffused throughout the body. In his book on this topic, Gad Freudenthal has proposed a strong reading of Aristotle's fragmentary concept of *pneuma* to solve the puzzle on the motion and diffusion of vital heat through the body which shall constitute the backbone of our sections on Aristotle's doctrine of *pneuma*.[38] *Pneuma*, defined by Aristotle as "hot air",[39] is the candidate proposed by Freudenthal as the instrument of the soul for vehiculating heat in the body. Throughout this book, we shall follow mostly Freudenthal's inter-

38 Gad Freudenthal, *Aristotle's Theory of Material Substance: Heat and Pneuma, Form and Soul* (Clarendon Paperbacks), Oxford 1998, 137–144. See also: Christof Rapp, "Introduction part I," 20–28, in Aristotle's De motu animalium: Symposium Aristotelicum, ed. Christof Rapp/Oliver Primavesi. *With a new critical edition of the Greek text by Oliver Primavesi and an English transl. by Benjamin Morison*, Oxford 2020; Martha C. Nussbaum, "The *sumphyton pneuma* and the *De motu animalium*'s account of soul and body," 143–164, in *Aristotle's* De motu animalium. *Text with translation, commentary, and interpretive essays by Martha Craven Nussbaum*, Princeton 1978. For a less stronger reading of the concept of *pneuma* than Freudenthal's, see at least: Richard A.H. King, *Aristotle on Life and Death*, London 2001, 113–129. On the *status quaestionis* on *pneuma*, see: Silvia Berryman. "Aristotle on *Pneuma* and Animal Self-Motion," *Oxford Studies in Ancient Philosophy* 23 (2002) 85–97.

39 Aristotle, De generatione animalium II 2 (736a1).

pretation, though raising possible issues (e.g.: the role of blood and *pneuma* in sensation) on the role of *pneuma* in the different physiological processes. According to Aristotle's definition in the *De motu animalium* of *pneuma* as the "vehicle" (ὄχημα) of the soul's faculties, this material substance appears to be primarily seated in the heart,[40] the operational center of the soul in the body. In the Aristotelian *corpus*, references to *pneuma* are scattered through the zoological works (*De motu animalium*, *De animalibus*, especially *De generatione animalium*) and the *Parva Naturalia* (*De somno*, *De iuventute et senectute*, *De inspiratione et expiratione*). The main problem affecting the Aristotelian account of the role of *pneuma* in the definition of life is, simply put, its fragmentation.

Still, there are some useful indications that may point toward life-giving functions of *pneuma*. According to Aristotle, spirit is the substrate of vital heat travelling in the blood. This is the case, for instance, with sense perception. In some scattered remarks in the *De generatione animalium* and *De partibus animalium*, he argues that spirit reaches, together with the blood, the sense organs of sight, hearing and smell through subtle pores.[41] Given that blood is itself non-conducive to sensation, at least according to the interpretation of Freudenthal,[42] then it is likely that *pneuma* itself is the material substance conducive to sensation. Together with sensation and motion—as in the *De motu animalium*—spirit is connected to vital heat in a famous passage on human generation (*De generatione animalium* II, 1) which will be discussed in greater detail in chapter 3. Given the number of psychic activities (generation, sensation, motion) with which spirit seems to be charged, given the cardiocentric nature of *pneuma* stated in the *De motu animalium* and given the cardiocentric nature of the soul operations according to Aristotle, it is possible to conclude according to Freudenthal that *pneuma* is involved in nutrition as well.[43] This

40 Aristotle, De motu animalium 10 (703b14–16).

41 Aristotle, De generatione animalium II 6 (743b37–744b); De partibus animalium II 10 (656b17); Freudenthal, *Aristotle's Theory of Material Substance*, 132–133.

42 Aristotle, Historia Animalium III 19 (520b12–17); De partibus animalium II 10 (656b19–22); De partibus animalium III 4 (666a16–18). More on the interpretations of Freudenthal and Gregoric on the role of blood in chapter 5.

43 Freudenthal, *Aristotle's Theory of Material Substance*, 140: "If the heart controls all soul-functions by the means of the connate *pneuma*—why should the function of the nutritive soul be an exception? To be sure, Aristotle does not say it explicitly. But the logic of this programme would seem to point in this direction." Although the word "programme" related to Aristotle's theory of *pneuma* represents quite bold claim, the rationale behind Freudenthal interpretation of the concept of *pneuma* still stands as tenable. On the doubtful existence of such a "programme", see the cautious remarks in: Sophia Connell, *Aristotle on Female Animals: A Study of the Generation of Animals*, Cambridge 2016, 217 n. 105.

conclusion would actually square with Aristotle's remarks in the *De generatione animalium* that "nature makes the majority of her productions by means of *pneuma* used as her instrument".[44] Among her productions feature nutrition and respiration, but there is now a question concerning the most basic condition for the subsistence of life. Can it be identified with one among breathing, heartbeat or nutrition or, in case not, how are we to understand their mutual relation?

According to Aristotle, respiration must not be thought to be a function of nutrition. In his article on the teleology of respiration according to Aristotles, Lennox has shown how respiration is conceived by Aristotle not for the sake of nutrition but rather to allow the maintainance of balance in the vital heat which, in turn, allows for the correct performance of nutrition.[45] This role is all the more important inasmuch the action of vital heat produces, by acting on the nutritive substances, blood or a substance analogous to it. Blood plays a key role in the Aristotelian account of nutrition since it represents not only the physiological base of nutrition *qua* product of digestion but also, strange as it may sound, the base for the physiological foundation of living beings. Even bones, according to Aristotle, result from the concoction of blood.

Returning to the relationship between *pneuma*, respiration, pulse and nutrition, we find Aristotle's answer in two texts, viz. the *De respiratione* and *De partibus animalium* (III, 6). According to the Stagirite, without respiration the heat inside the body would destroy all the nutritive substances within it, like the fire of a greater candle would extinguishing that of a smaller one by burning up its material substrate. This is the comparison Aristotle draws when explaining why suffocation is lethal in breathing animals.[46] According to him, inspiration is designed by nature to draw in air to cool down the vital heat in the living being, while expiration is designed to waste away excessive heat through exhalation. In humans and breathing animals, the quantitative need for cooling substance is directly proportional, according to Aristotle, to the amount of blood. Lennox has shown that the more blood animals possess, the more external air they need

44 Aristotle, De generatione animalium V 8 (789b9–10). This translation by Freudenthal in: Freudenthal, *Aristotle's Theory of Material Substance*, 111.

45 Aristotle, De iuventute et senectute 14 (474a25–28): "We have said previously that living and the state of the soul is by aid of a certain warmth; for concoction, through which nutrition comes about in animals, occurs neither in the absence of soul nor in the absence of warmth; for everything works by means of fire." This translation by Lennox in: James Lennox, "Why Animals Must Keep Their Cool: Aristotle on the Need for Respiration (and Other Forms of Cooling)," in *Heat, Pneuma and Soul in Ancient Philosophy and Science*, ed. Hynek Bartoš/Colin Guthrie-King, Cambridge 2020, 225.

46 Aristotle, De iuventute et senectute 5 (469b21–470a4).

to cool down the vital heat. In animals with smaller amounts of blood, lungs are smaller and so is the amount of vital heat.[47] A notable exception is represented by fish, which do not cool down by the means of air drawn in through the lungs but, rather, by water filtered through the gills.[48] In both cases, respiration through lungs or gills seems to be directed at a clear purpose: tempering the vital heat in the heart. Respiration has the clear purpose of avoiding the spread of vital heat in the body, operated by *pneuma*, from becoming self-destructive to the being itself.

Lennox also explains that, according to Aristotle, respiration is not needed in bloodless animals such as insects, since the cooling function of outer air and, to some extent, of food taken in, is considered sufficient to cool down their body. If we consider the necessary and sufficient conditions to define life, in bloodless animals and plants there can be no talk of respiration but rather of 'cooling functions', which are fundamental to keeping a thermic balance in the living being.[49] These cooling functions are crucial for attaining a correct balance in the heat inhering in the *pneuma*, which is in turn spread into the body and is fundamental to nutrition and, thus, life.

Plants, by contrast, never appear to possess *pneuma* or anything analogous to it, according to Aristotle: rather, they possess vital heat as the instrument of the vegetative soul for their operations. Luciana Repici has higlighted how, according to the Aristotelian criteria of *scala naturae*, plants are considered 'upside down men' by Aristotle, since upright position is directly proportional to heat, concoction of the blood and intelligence. Plants, on their account, the least intelligent among living beings, possess the least concocted inner fluids (sap and not blood) and have an organ functionally akin to the mouth, viz. roots, placed down in the earth and not in the upper part of their body like the head in humans and animals.[50] Moreover, plants do not possess a heart (or an analogue to it), but only a central part which seems apt to be the central seat of the soul;[51] secondarily, their vital heat does not entirely come from the inside but is to some extent supplied from the surrounding air.[52] In the case of plants, digestion does not occur inside but outside the plant, i.e. in earth, which operates analogously to the stomach.[53] For this reason, *pneuma* clearly seems to be

47 Lennox, "Why Animals Must Keep Their Cool," 228–231.
48 Lennox, "Why Animals Must Keep Their Cool," 224.
49 Lennox, "Why Animals Must Keep Their Cool," 220.
50 Luciana Repici, *Uomini capovolti. Le piante nel pensiero dei greci*, Bari 2000, 13–17.
51 Aristotle, De iuventute et senectute 23 (478b32–479a1).
52 Aristotle, De iuventute et senectute 6 (470a20–b27).
53 Aristotle, De partibus animalium II 3 (650a20–23).

out of the picture. In terms of causal explanation, the only physiological cause at play in plant nutrition, i.e. the most basic material substance involved for the sake of life's subsistence in plants, is vital heat.[54] On the other hand, though, some scholars like Gad Freudenthal have emphasized how *pneuma* could even be viewed as the result of the action of vital heat on moisture in the so-called process of "pneumatization" taking place in the heart.[55] According to this kind of strong reading of the concept, it would seem there is room for some speculation. On his account, though, Aristotle flatly rules out the existence of *pneuma* and plants, even though both vital heat and moisture are material principles of life found in plants too. The rejection of the existence of *pneuma* in plants is shared by the pseudo-Aristotelian *De plantis*, which at the same time emphasizes the smoky nature of the material byproduct of the action of heat over moisture.[56]

All in all, in the Aristotelian corpus the respiration or cooling of the living being plays a fundamental role in balancing the vital heat needed for digestion: in the case of blooded animals, this cooling function is achieved thanks to intrinsic physiological processes (respiration), while lungless animals and plants achieve the same cooling process by the means of surrounding air and ingested food. There can be no digestion and survival of the living being without respiration or cooling, while there may be still breathing even when the digestive process is failing. Breathing, together with heartbeat, is among the last living functions to stop during the process of death.

In Aristotelian physiology the heart plays the key role of principle and source of the life-giving heat spread over the whole body. This heat inheres primarily

---

54 Giuli Korobili, "Aristotle on the Role of Heat in Plant Life," in *Nutrition and Nutritive Soul in Aristotle and Aristotelianism*, ed. Roberto Lo Presti/Giouli Korobili (Topics in Ancient Philosophy 9), Berlin 2020, 153–169. See: Nicolaus Damascenus, De plantis I c. 2 n. 32, in: Five translations, edited and introduced by Hendrik Joan Drossaart Lulofs and E.L.J. Poortman (Verhandelingen der Koninklijke nederlandse akademie van wetenschappen, Afd. letterkunde. Nieuwe reeks 139), Amsterdam 1989, 521: "Et non habet planta spiritum, licet Anaxagoras dixerit illas habere spiritum. Et iam invenimus multa animalia quae non habent spiramen."

55 Freudenthal, *Aristotle's Theory of Material Substance*, 120–121.

56 Nicolaus Damascenus, De plantis II c. 1 n. 148 (ed. Drossaart Lulofs/Poortman 1989, 540): "Ut in balneo calor humorem attrahit et in vaporem convertit qui elevatur, et quando superfluus fuerit vertetur in guttas, similiter quoque in animali et planta superfluitates ascendant ab inferioribus ad superiora et descendunt a superioribus ad inferiora in actionibus suis." These passages from the *De plantis* are based on the article by: Marilena Panarelli, "*Spiritus* and plant formation. Albert the Great's Restatement of the Pseudo-Aristotelian *De plantis*," in *Rencontres de Philosophie Médiévale* (forthcoming). I would like to thank the author for sharing her contribution with me in advance.

in the *pneuma*, which travels throughout the whole body as "hot air". While these remarks may resonate with later Galenic accounts of *pneuma*, it is necessary to make a couple of remarks to avoid jumping to improper conclusions. According to the Stagirite, the heart is a three-chambered organ,[57] which is itself—surprisingly—a blood vessel and is the ultimate source of all blood vessels, veins included. Unlike in later Galenic medicine, blood vessels are thought to exist for the sake of containing blood and avoiding its spread into the whole body and the destruction of organs, not for the sake of pumping blood and its life-giving powers into the whole body.[58] For this reason, it would be improper to speak of pulse for Aristotle in the same way we speak of it in Galen and in the medical tradition. It is much more appropriate to speak of 'beating' and 'palpitation', as Aristotle himself did in his *De respiratione*. While palpitation is defined as "the recoil of the heart against the compression due to cold", pulsation is "volatilization of the heated fluid".[59] Pulsation, which has nothing to do with the Galenic concept of it, represents a sympathetic vibration of the heart and the surrounding veins. This motion is not caused by the usual systolic (contractive) and diastolic (expanding) motions of the heart as Galen understood it; rather, this motion occurs because heat expands the fluid (blood) which eventually reaches the heart. Unlike in Galen, there is no clear connection between heartbeat, pulse and respiration, so it is difficult to understand how pulse would relate to respiration or, as in lungless animals and plants, the cooling functions of the living being.

All in all, the role played by *pneuma* in Aristotle's definition of life as "being among living beings" seems to be that of an instrumental cause of life with respect to the true formal cause of life, the soul, because of two main functions. On the one hand, *pneuma* maintains a correct thermic balance by spreading an amount of vital heat, mitigated by outer air in the case of breathing animals, into the whole body. This balance enables, in turn, a correct functioning of the activities of the nutritive soul. In non-breathing animals endowed with connate *pneuma* and plants these cooling functions are supplied by the outer air. Insofar as it is "hot air" and substrate of the material inherence of vital heat, *pneuma* is crucial for spreading the vital heat in the body, nutritive activities and, in turn, the subsistence of life. From a physiological point of view, the cooling functions of a living body seem even more basic to the existence of life, for without them

57 Aristotle, De partibus animalium III 4 (666b22–24); De somno 3 (458a16–19).

58 Aristotle, De partibus animalium III 5 (667b13–668b33).

59 Aristotle, De iuventute et senectute 26 (20) (480a13–15). Translation by Benjamin Jowett in: Aristotle, *The Complete Works of Aristotle: the Revised Oxford Translation*, ed. Jonathan Barnes, Princeton 1984, vol. 2.1, 22.

the heat employed in nutritive processes would end up destroying the living being itself. In plants, the diffusion of vital heat seems to follow different paths for the transmission of vital heat in the whole body of the plant, for *pneuma* is not involved in the process, but plans still require a movement of life-giving vital heat throughout the whole of the plant. A similar position is maintained in the pseudo-Aristotelian *De plantis*.

Unfortunately, part of these Aristotelian doctrines on the role of *pneuma* in breath were lost in translation and unavailable to Albert. Aristotle's entire *De respiratione* and *De iuventute et senectute* were translated from Greek into Latin by James of Venice, but their transmission amounts to an extremely small number of manuscripts (4), as discussed by Pieter de Leemans.[60] Therefore, they were unavailable to Albert. An alternative source for the transmission of the Aristotelian doctrine on breathing was represented by the so-called *Auctoritates Aristotelis*, a mnemonic tools for students at the Arts faculties of Paris which displays few sentences from the *De iuventute et senectute* and the *De respiratione*.[61] The periodization of this work, though, is posterior to Albert's period of intellectual activity and no traces of quotes from this text can be identified in Albert's works. Be that as it may, the most likely vehicle for Albert's acquaintance with Aristotle's theory of *pneuma* was Michael Scot's Arabo-Latin version of Aristotle's *De animalibus*. In particular, the chapter on the lungs (*De partibus animalium* III 6)[62] presented him with a less extensive picture that was nevertheless conceptually consistent with the *Parva Naturalia*: (1) lung represented an instrument of cooling needed by hotter animals, which usually possess more blood; (2) cooling agency is external in blooded animals while produced by external air in bloodless animals; (3) the abundance of blood is directly proportional to the abundance of heat and thus to the need of breathing as a cooling mechanism; (4) the cooling of fish, who do not breathe, takes place through water and not air.

---

60 Pieter De Leemans, *Secundum viam naturae et doctrinae*, 203.

61 Johannes de Fonte, Auctoritates Aristotelis, Senecae, Boethii, Platonis, Appuleii Africani, Porphyrii et Gilberti Porretani, ed. Jacqueline Hamesse, *Les* Auctoritates Aristotelis: *Un florilège medieval. Étude historique et édition critique* (Philosophes médiévaux 17) Louvain–Paris 1974, nn. 206–207, nn. 123–137. On respiration, see 206, n. 133: "Respiratio fit in animali ad refrigerandum cor et loca vicina cordi, unde quae magis calida sunt magis indigent respiratione."

62 Note that Albert was acquainted with Aristotelian zoology from the very outset of his works: Henryk Anzulewicz, "Die Aristotelische Biologie in den Frühwerken des Albertus Magnus," in *Aristotle's Animals in the Middle Ages and the Renaissance*, ed. Carlos Steel/Guy Guldentops/Pieter Beullens, (Mediaevalia Lovaniensia [Series 1] 27), Leuven 1999, 159–188.

The main point that gets lost in the process of translation is the Aristotelian distinction between 'beating' and 'palpitation'. As shall be seen, Albert's position on the topic will rather reflect that of his medical sources. A second text which mentioned the pushing and pulling of *pneuma* in the veins, viz. the *De motu animalium*, will be known by Albert only at a later point in his life.[63] Like many of his contemporaries, Albert would then have to turn to sources of Galenic medico-philosophical doctrines on spirit such as Avicenna, Costa Ben Luca, Nemesius and the *De motu cordis* to understand the process of pulse and the relation of heartbeat to respiration. More generally, Albert's use of these authors can be illustrated, as pointed out in the introduction, by his use of the concept of *spiritus* as a medium between body and soul in theological discussions on the operations of the human soul.

3 Vital Spirits, Vital Faculty, and Life: Costa Ben Luca, Nemesius, and the Galenic Medical Tradition

In his works, Galen discussed in detail both the usefulness of respiration (*De respiratione*) and the role of pulse (*De usu pulsuum*) in connection with the role of spirit, divided into vital (ζωτικόν) and animal/psychic (ψυχικόν) spirits. A third kind of *pneuma*, viz. natural (φυσικόν) spirit, responsible for the activities of the nutritive faculty, was introduced by Galen's successors.[64] According to Galen, vital spirit in humans is produced by the interaction of air drawn in via the lungs, vital heat seated in the heart and blood.[65] The purpose of respira-

63 In his Aristotelian paraphrases, Albert wrote in the first instance the *De motibus animalium* to make up for the loss of Aristotle's *De motu animalium*. Later on, he paraphrased in his *De principiis motus processivi* a remarkable textual version of Aristotle's work: De Leemans, "*Secundum viam naturae et doctrinae*," 201–206.

64 Owsei Temkin, "On Galen's Pneumatology," *Gesnerus*, 8 (1951), 182, mentions and discusses the only passage in Galen's works (*De methodo medendi*) on "natural spirit" (πνεῦμα φυσικόν).

65 Bono, "Medical Spirits," 92: "Pneuma […] derives chiefly from the air taken into the body (through inspiration) and from the blood (though this source is an aspect of Galen's pneumatology which tends to be ignored)." This mechanism is highlighted in: Avicenna, De medicinis cordialibus, I 8, ed. Van Riet 1968 205.92–227.26 (ed. Bābā 1404/1984, 238–240). The Avicennian text referred to as *De medicinis cordialibus* is the fragment copied at the end of book four from his *De anima*: Tommaso Alpina, "Al-Ǧūzǧānī's Insertion of On Cardiac Remedies in Avicenna's Book of the Soul: the Latin Translation as a Clue to his Editorial Activity on the Book of the Cure?", *Documenti e studi sulla tradizione filosofica medievale* 28 (2017) 365–400. Note that Avicenna connects the humoral complexion of blood to that of spirit in his account of emotions, which will not be dealt with in this book.

tion consisted in both cooling down and nourishing the heart's vital heat. In the end, the production of vital spirit, spread into the whole body through arteries, would have thus granted a correct thermic balance of paramount importance for nutrition to occur.[66]

This Galenic doctrine of the vital faculty landed on Albert's desk via texts such as Avicenna and Costa Ben Luca's *De differentia spiritus et animae*. Among the other medical sources besides Avicenna, the *De differentia* is the only one which explains in detail how respiration is meant for the sake of the cooling down and 'respiration' of the heart and how spirit results from the interaction between vital heat and breathed-in air. The only drawback of the Latin translation of Costa ben Luca's text is that it confuses the left ventricle (where the aorta departs) with the right one where the vena cava inferior comes in, thus shifting the mechanism of respiration from the left to the right portion of the heart.[67] Costa ben Luca does not develop the connection among spirit, breathing, vital heat and nutrition, even though it can be inferred that a situation of natural regularity in pulse can be correlated to a balanced diffusion of vital heat

---

It is nonetheless worth mentioning that Albert's *De motibus animalium* (Albertus Magnus, De motibus animalium, 1 tr. 2 capp. 6–7, ed. Auguste Borgnet, Paris 1890 [Ed. Paris. IX, 278b–281b]) draws heavily on Avicenna's *De medicinis cordialibus*, especially as respects as the connection between spirit and passions (*passiones*). However, these generic claims need be substantiated through more specific research assessing the impact of Avicenna's *De medicinis cordialibus* on Albert's psychophysiology of emotions (*passiones*).

66 Leonard J. Wilson, "Erasistratus, Galen and the Pneuma," *Bulletine of the History of Medicine* 33 (1959/4), 303.

67 Costa Ben Luca, De differentia spiritus et animae, ed. Wilcox 1985, 145.46–147.64 [ed. Troupeau 2011, v.10–vi.7]: "Et ex ventriculo dextro (Ar. *min taǧwif allaḏī fī ǧānib al-aysar*, "from the ventricle that [is] in the left side") procedunt due vene, una quarum vadit ad pulmonem, et fit per eam anelitus cordis. Cor enim colligitur atque extenditur, et per extensionem eius atque collectionem fit pulsus totius corporis, et ideo pulsus indicat esse cordis, id est eius passiones proprias tam equales quam inequales atque diversas, que fiunt causa diversi impedimentis cordis quod accidit ei a semet ipso vel a quibusdam membris sibi vicinis. Cor ergo cum extenditur attrahit per predictam venam ex pulmone partem aeris qui ingreditur ad pulmonem per anelitum ad refrigerandum calorem naturalem qui est in eo ut sit hoc nutrimentum vel sustentatio eiusdem spiritus qui continetur in eis ventriculis. Et cum colligitur, cor pulsat per ipsam venam ad pulmonem quicquid generatur in eo vaporum fumosorum expellit eos ab ignito calore qui est in eo, et pulmo expellit eos a corpore, et hec vena vocatur pulsus [Ar. *šaryān warīdī*, "venous artery"]." On the periodization (1123–1142) and a first assessment of Qusṭā's Latin translation by Johannes Hispalensis, see: Charles Burnett, ""Magister Johannes Hispalensis et Limiensis" and Qusṭā ibn Lūqā's *De differentia spiritus et animae*: a Portuguese Contribution to the Arts Curriculum?," in *Arabic into Latin in the Middle Ages: The translators and their Intellectual and Social Context*, ed. Charles Burnett (Variorum Collected Studied Series 939), London 2009, 221–267.

in the body via vital spirits which, in turn, is crucial for the correct functioning of the nutritive organs (stomach, liver, spleen). This void in Costa ben Luca's account, rather than being the author's own fault, reflects a conceptual gap between the faculty theory of the Aristotelian tradition and the Galenic physiology of life-related operations. To Galen, vital spirit does not have, as in Costa ben Luca's own account of faculty theory, a matching philosophical faculty, in Aristotelian terms, so it is hard to place, in the end, the Galenic vital faculty, responsible for life, pulsation and breathing, into the Aristotelian framework of faculty theory.[68] This is a problem that is common to Nemesius,[69] Costa ben Luca, Alfred (who depends on Costa ben Luca and Isaac Israeli),[70] and, later, Avicenna. All the other sources who mention the existence of vital spir-

68 This problem is highlighted by: Bono, "Medical Spirits," 95. On Costa's faculty theory, see: Costa ben Luca, De differentia spiritus et animae, ed. Wilcox 1985, 183.494–498; 183.501–185.510 [ed. Troupeau 2011, xxvii.2–5; xxvii.8–13]: " [...] et dicamus quod prime virtutes anime que sunt ei in fortitudine quasi genera sunt tres: prima scilicet vegetativa, secunda sensibilis, tercia rationalis, et he virtutes vocantur ad placitum anime. [...] Anima autem vegetabilis que in aliis libris vocatur 'anima sementis', est nobis communis cum arboribus et herbis atque animalibus; et anima sensibilis, que alibi vocatur 'vitalis', nobis et animalibus est communis; anima vero rationalis proprie inest homini. Opus autem vegetabilis anime est nutrire et reficere, et hoc facit per quatuor virtutes que vocantur naturales, que sunt attractiva, retentiva, digestiva, expulsiva."

69 The translation by Burgundio, unlike the later by Alfanus, does mention the vital faculty: Nemesius, De natura hominis c. 23 (De pulsibus), ed. Gérard Verbeke/Josep R. Moncho (Corpus Latinum Commentariorum in Aristotelem Graecorum. Suppl. 1), Leiden 1975, 107.20–23: "Pulsualis autem motus vocatur quidem et zotica (id est vitalis) virtus. Habens autem cor principium et maxime sinistrum eius ventriculum spiritualem vocatum, innatum et vitalem calorem dispertitur omni particulae corporis per arterias, sicut et hepar per venas cibum." On Albert's knowledge of both translations, see: Ana Palanciuc, "Nemesius d'Émése," in *Dictionnaire des Philosophes Antiques*, ed. Richard Goulet, Paris 2005, vol. 4, 648–650.

70 Alfredus Anglicus, *Des Alfred von Sareshel* (Alfredus Anglicus) *Schrift* De motu cordis, ed. Clemens Baeumker (Beiträge zur Geschichte der Philosophie des Mittelalters 23.1–2), Münster 1923, c. 10 n. 9, 42.4–5: "Tres autem in animali corpore spiritus esse medicorum scola testatur [...]"; De motu cordis 10 n §10 (ed. Baeumker 1923, 42.24–43.1–2): "Duo igitur tantum sunt spiritus. Horum haec prima distantia est, quod hic vitam operatur, ille sensum et motum." De motu cordis 10 n. 7 (ed. Baeumker 1923, 40.17–22): "Praecipui quoque phisici et philosophi duos tantum esse commemorant ut Aristoteles, Plato, Ysaac, Costa quoque Lucae filius in libro quem De differentia animae et spiritus edidit duos tantum esse ostendit, vitalem scilicet et animalem, et a vitali animalem originem sumere docet." On Alfred's psychology and pneumatology, see: Tilman Struve, "Die Anthropologie des Alfredus Anglicus in ihrer Stellung zwischen Platonismus und Aristotelismus," in *Staat und Gesellschaft im Mittelalter*, ed. Thieleman Struve (Historische Forschungen 80), Berlin 2004, 47. Unfortunately, Struve leaves Qusṭā out of the picture, while the *De differentia* represents a source of Alfred's pneumatology, as Alfred himself states. On Alfred's teleology

its such as the Isagoge,[71] the *Pantegni*,[72] Isaac Israeli's *De definicionibus* (likely dependent on Costa ben Luca as well),[73] the *Anatomia Richardi* and the work of magister Nicolaus Ursus of Salerno[74] do not even present the conceptual tension between medical and philosophical faculties. Rather, they only point out the cooling role of breathing in humans and the role of vital spirit as spreader of vital heat in the body.

In addition to the teleology of respiration, the texts from the medical tradition supply Albert with an aspect of respiration that Aristotle had not dealt with in detail, namely the connection between respiration and heart motion. According to Galen, the drawing in of air through the lungs corresponded to cardiac expansion (diastole), while cardiac contraction (systole) had a twofold purpose. On the one hand, pulse was meant to waste away the obnoxious substances (vapors, mostly) produced upon the formation of spirit. The heart, having drawn in air from the lungs, caused the motion of blood and spirit thanks to its systolic and diastolic motion: therefore, the regularity of pulsation could measure the healthiness of the cardiac condition. This is a feature of his account that comes mostly through Costa ben Luca's *De differentia spiritus et animae* and Alfred's *De motu cordis*. The other texts are quite limited in terms of the teleology of respiration, something Albert would find more extensively treated in Aristotelian zoology (*De partibus animalium* III, 6).

Unlike modern of explanation of the breathing process, Galen considered the attraction and expulsion of air to take place through a single blood vessel,

of respiration, see: De motu cordis c. 6 n. 4 (ed. Baeumker 1923, 22.17–23); c. 11 n. 16 (ed. Baeumker 1923, 52.4–15).

71 Johannitius, Isagoge ad Techne Galieni, ed. Gregor Maurach, *Sudhoffs Archiv* 62 (1978/2), n. 17, 155.

72 Constantinus, Pantegni [Theorica IV.19], ed. Burnett 1994, 114.

73 Isaac Israeli, Liber De definicionibus (ed. Muckle 1937/8, 318.18–23): "Alius sermo de spiritu vitali. Convenerunt antiqui quod spiritus est corpus subtile procedens a corde in arteriis ad totum corpus et facit ipsum habere vitam et anhelitum per pulmonem et pulsum per arterias et ascendit sursum ad cerebrum et procedit ab eo in nervis ad totum corpus et facit ipsum habere sensum et motum." Costa ben Luca, De differentia spiritus et animae, ed. Wilcox 1985, 145.33–38 [ed. Troupeau 2011, V.1–3]: "Spiritus est quoddam corpus subtile quod in humano corpore oritur ex corde et fertur in assurianet, idest in venis pulsus ad vivificandum corpus, operaturque vitam et anelitum atque pulsum; et similiter oritur ex cerebro in nervis et operatur sensum atque motum." On Isaac's likely dependence on Costa, see: Alexander Altmann, Samuel M. Stern, *Isaac Israeli: A Neoplatonic Philosopher of the Early Tenth Century*, Oxford 1958, 50.

74 Richardus (Anglicus), *Die Anatomie des* Richardus (ed. Victor Tarrasch, Berlin 1898, 17–18); Nicolaus Physicus, Anathomia, ed. Franz Redeker, *Die* "Anatomia magistri Nicolai phisici" *und ihr Verhältnis zur* Anatomia Cophonis und Richardi (Diss.), Leipzig 1917, 44–46, 63–64.

the 'veinous artery'. This blood vessel was so called because its role amounts to that of an artery, i.e. to convey blood and spirit, while retaining the morphological appearance of a vein. This vein, filled with more spirit than blood; reached the left ventricle of the heart; by contrast, the right ventricle, which was not connected to the veinous artery, contained more blood than spirit. The two ventricles (not three, as in Aristotle) were separated, according to Galen, by a membrane or septum, which filtered through blood and spirit from the right to the left ventricle. The cardiac septum represented the first passage through which blood, produced in the liver and vehiculated to the right ventricle of the heart, was able to enter the arterial system. The second doorway from the veinous to the arterial system was represented by the thick net of blood vessels in the lungs which granted the flow of blood from the arterial vein into the lungs but also into the veinous artery as well. The third passage through which veins, defined as non-pulsatile vessels, convey blood to arteries is represented by the small vesicles connecting them, called anasthomoses.[75]

According to Galen, the pushing of *pneuma* and blood from the heart into the arteries generates a vibration of the so-called "tunics" (χειτῶνες), viz. the different layers of the arterial tissue. This vibration of the different tunics produces, in turn, pulsation (or pulse). This doctrine, though, appears to be simplified in many texts from the Galenic medical traditions or which draw on Galen.[76] If we consider those examined here—except for Avicenna—there are two main simplifications which can be observed. On the one hand, pulse is viewed as the causal product of the pushing of spirit into the arteries and not, as Galen thought, of the vibration of the different tunics correlated to the pushing of pneuma in arteries. On the other hand, spirit is viewed as the byproduct of air and vital heat, often overlooking the key role played by blood and the complexion (*complexio/mizāǧ*) of blood.

The different views on the role of the vital spirit as the material instrument of the soul to carry out heartbeat, pulsation and respiration appear through some of the texts which Albert would employ in his early production. Some of them, such as Costa ben Luca's *De differentia spiritus et animae*, Nemesius' *De natura hominis*, Alfred of Shareshell's *De motu cordis* and Isaac's *De definicionibus* are faithful to the Galenic bipartition of spirits. Others, such as the *Pantegni*, the *Isagoge Johannitii* and other minor works such as the *Anatomia Richardi* and the work by the master Nicolaus, all display the traditional tripartition of spir-

75 For a more detailed account, see: Wilson, "Erasistratus, Galen and the Pneuma," 303–310.

76 On pulse in Galen: Wilson, "Erasistratus, Galen and the Pneuma," 303–305.

its. While this point on the classification of spirit appears *prima facie* a moot point, it is most certainly not if we want to assess the causal role of spirit in the definition of life and its necessary and sufficient conditions, not least considering the role of the nutritive faculty. Following Costa ben Luca, Nemesius and Alfred, it would seem that the nutrition-related processes are not carried out by an instrument like spirit; at this point, the only conjectural and viable option left is to assign the role of instrument of nutrition-related activities to vital heat.

The situation, by contrast, is different in sources such as the *Pantegni*, *Isagoge Johannitii* and others which include a taxonomy of three spirits. If natural spirits are included in the equation of the necessary and sufficient conditions for life, spirit plays an instrumental role both concerning the conditions which are primarily associated with life (heartbeat, pulsation, respiration) and those which grant life its subsistence (nutrition). This general physiological structure is shared by Albert's most influential source, Avicenna. In Albert's eyes, Avicenna's most remarkable psychophysiological contribution resides in his having blended an enlarged version of the Galenic concept of spirit (introduced to explain animal physiology as well) with a psychology capable of granting the human rational soul an after-death survival. The above-mentioned medical sources, much as they were detailed in their physiological account, neither produced an account of medical spirits which spilled over into the realm of zoology (for they were mostly medical texts) nor framed spirit in a complex metaphysical structure of the cosmos in the same way Avicenna did.

4 Neoplatonic Psychology Meets Galenic Medicine: Avicenna on Spirit, Soul, and Life

In the first chapter, we have placed much of the focus on how Avicenna shaped Albert's view on the relationship between the principles of natural philosophy and medicine and we have examined the relationship between natural philosophy and medicine themselves in both authors. According to Avicenna, medicine must be subordinated to natural philosophy as a "derivative natural science" (*ḥikma farʿiyya ṭabīʿiyya*). Thus, shared subjects such as elements, humors and spirits must be discussed by natural philosophers alone, since they possess suitable conceptual tools (demonstrative argumentations) to do so. Consistently with his approach, Avicenna gives his definition and discusses the causal role of spirit in the last chapter of his psychology (*De anima*, V, 8), where he discusses the role of the instruments of the soul, among which spirit itself:

> [...] the primary vehicle of psychic faculties in the body (*virtutum animalium corporalium vehiculum/ al-quwà al-nafsāniyya al-badaniyya maṭiyyatuhā al-ūlà*) is a subtle body (*corpus subtile/ ǧism laṭīf*) that passes through the outlets (*diffusum in concavitatibus / nāfiḏ fī l-manāfiḏ*), and which is spiritual (*spiritualis, rūḥānī*), and this body is the pneuma (*spiritus / rūḥ*).[77]

While in this passage spirit is defined as a single entity and may seem a single material substance, it is a well-known fact that Avicenna adopted the classic post-Galenic division of spirits into natural, vital and animal. Moreover, an attentive reader of psychology may not have certainly missed the zoocentric nature of the passage, Spirit, according to Avicenna, is not found in plants. If spirit is the vehicle for the soul embodied activities in humans and animals but not in plants, one may tempted to question the belonging of plants to the roster of living, ensouled beings. To fully understand the psychophysiological role of spirit in Avicenna's doctrine on life, attention must be turned to the Avicennian doctrine on the nature of the soul. After this preliminary exam, we will be in a much better position to assess the causal link between spirit, Avicenna's faculty theory and the most basic functions of life such as nutrition, breathing, pulsation and heartbeat.

4.1 *Vehicle of Which Life and of Which Faculties? Spirit and Soul in Avicenna's Psychology*

As often happens in the history of philosophy, a good approach in the study of a text is represented by reading its prologue first. In his study on the preface to Avicenna's *De anima*, Alpina has stressed the essentialist perspective adopted by Avicenna in his science of the soul: knowledge of something is taken to be primarily knowledge of its causes and principles. Since the soul is primarily associated with life as a cause of sensation and voluntary motion,[78] psychology comes to be associated primarily with the study of sensation and motion. This idea reflects an intrinsically zoological approach typical of Avicenna's psychology, rooted on two facts: in the first place, the primary connection between sensation and (voluntary) locomotion is typical of Aristotelian

77 This translation by Alpina in: Tommaso Alpina, "Is Nutrition a Sufficient Condition for Life? Avicenna's Position between Natural Philosophy and Medicine," in *Nutrition and Nutritive Soul in Aristotle and Aristotelianism*, ed. Roberto Lo Presti/Giouli Korobili (Topics in Ancient Philosophy 9), Berlin 2020, 239.

78 Avicenna, Liber de anima I (Prologue), ed. Van Riet 1972, 13.59–62 [ed. Rahman 1959, 3.9–11]. More on this passage in: Alpina, "Is Nutrition a Sufficient Condition for Life?," 236.

psychology as well.[79] Second, Aristotelian *pneuma* is a zoological concept as Avicennian spirit, as it is not found in plants. In the second place, Avicenna is convinced that the quiddity of the things which are remote from us, such as plants, is more difficult to grasp than those of animals and humans.[80] This difficulty is essentially grounded on the nature of the science of the soul, for it is based on the role of the human soul as both part of the investigation of psychology (subject-matter) and, at the same time, that which investigates the nature of the soul in living beings.[81] Based on our experience, the essence and the activities of plants are much more difficult to be detected, if compared to animals. This should not lead us astray into thinking that plants are excluded by Avicenna from the roster of living beings.[82]

In his recent contribution on the Avicennian concept of life, Alpina has rightfully called the reader's attention to the fact that the Avicennian concept of life is indeed primarily associated with sensation and voluntary locomotion. Therefore, it is of little surprise that the instrument of soul and life *qua* primarily connected to sensation and motion is represented by spirit, although this fact alone does not exclude plants from ensouled, living beings.[83] In Avicenna's "Book of Plants" (*Kitāb al-Nabāt*), the disagreement about the scope of living beings is attributed mostly to verbal and linguistic incomprehensions. To solve them, Avicenna points out the need to operate a careful distinction between the concept of "animal" (*ḥayawān*) and that of "living being" (*ḥayy*).[84] While the former class comprehends living beings endowed with sensation and voluntary motion (hence, animals), the second class encompasses all living beings, plants included. This is because life comes about and subsists thanks to heat and moistness, which exist also in plants.[85]

79 Alpina, "Is Nutrition a Sufficient Condition for Life?," 239.

80 Alpina, "Is Nutrition a Sufficient Condition for Life?," 243.

81 Alpina, *Subject, Definition, Activity*, 67.

82 Alpina, "Is Nutrition a Sufficient Condition for Life?," 246–252.

83 Alpina, "Is Nutrition a Sufficient Condition for Life?," 250.

84 Avicenna, Nabāt, ed. Cairo 1965, 1, 4.1–4: "The term animal seems coined for what has sensation and voluntary motion. Therefore, it seems that plants cannot be called *animal* at all. A group of people (*qawm*) has made this kind of distinction between living (*al-ḥayy*) and animal (*al-ḥayawān*). This distinction between the signification of the expression *having life* (sc. living) and [the signification of the term] animal (*bayna mafhūm lafẓa ḏī l-ḥayāwa-lafẓa l-ḥayawān*) is a difference that linguists (*asḥāb al-luġāt*) are not aware of." This translation by Alpina in: Alpina, "Is Nutrition a Sufficient Condition for Life?," 250–251.

85 Alpina, "Is Nutrition a Sufficient Condition for Life?," 251. Avicenna, Nabāt, ed. Cairo 1965, 1, 7.12–20: "Hence, vegetative life, and in general nutritive [life] (*fa-iḏan al-ḥayā' al-nabātiyya wa-bi-l-ǧumla al-ġiḏā'iyya*) depends on moistness and heat, the death, which is the opposite, occurs only for the annihilation of the mater of moistness and the extinc-

Following in the footsteps of previous scholarship, it should by now be clear that Avicenna, although developing the notion of spirit in a manner clearly influenced by Galen and the Galenic medical tradition (as highlighted by the key study of Danielle Jacquart),[86] and maintaining its same Aristotelian, zoological scope, never excludes plants from his psychological enquiry. Like Aristotle, Avicenna considered plants subjects of his psychological inquiry *qua* living and ensouled beings, but he did not attribute spirit to their internal features. Unlike the Stagirite, though, Avicenna operated a strong distinction between the quiddity (*quidditas*/*māhiyya*) of human, animal and plant souls.

4.2 *The Soul of, the Soul in Itself: Spirit in Human and Animal Psychology of Life*

Among the crucial tenets of Avicennian psychology stands most certainly the qualification of the soul as the first perfection of the body. Many studies, from the pioneering work of Dimitri Gutas to the latest by Tommaso Alpina, not to mention Robert Wisnowsky's crucial volume *Avicenna's Metaphysics in context*, have shown the Late Antique background of the Avicennian concept of perfection was strongly permeated by the reading of John Philoponos.[87] According to Avicenna, the soul must be understood as the 'perfection' of the body since the term 'perfection' encompasses both the cases of separable, viz. human and celestial, and non-separable souls, viz. plant and animals. Avicenna's exegesis of the concept of *perfectio*/*kamāl* was strongly influenced by the Philoponian exegesis of the famous yet controversial Aristotelian metaphor according to which the soul is the actuality of the body as the steersman is the actuality of a ship. While the point of the Aristotelian metaphor was—perhaps—to stress

---

tion of heat. For this life belongs to a moist and hot body [(...)] Then, when the matter of moistness ceases [to exist] and the heat connected with it is extinguished through the process of nutrition (*ʿalà sabīl al-taġaḏḏī*), as was said in other places (*wa-ʿalà naḥw mā qīla fī mawāḍiʿ uḫrà*, sc. in *Nafs*, I, 5 and II, 1) and as we have explained thoroughly in our big book on the discipline of medicine (*wa-ʿalà mā basaṭnahu kull al-basṭ fī kitābinā al-kabīr fī ṣināʿat al-ṭibb*, sc. *Qanūn*), the substance having this [kind of] life necessarily corrupts. Then, the transformation of a complexion similar to this into coldness and dryness is annihilation." This translation by Alpina in: Alpina, "Is Nutrition a Sufficient Condition for Life?," 251.

86 Danielle Jacquart, "La notion philosophico-médicale de *spiritus* dans l'*Avicenna Latinus*," in *Body and Spirit in the Middle Ages. Literature, Philosophy, Medicine*, ed. Gaia Gubbini, Berlin 2020, 13–33.

87 Tommaso Alpina, "The *Soul of*, the *Soul in Itself*, and the '*Flying Man*' Experiment," *Arabic Sciences and Philosophy* 28 (2018) 187–224; Dimitri Gutas, "Marginal Glosses on *De anima* and the Greek Commentatorial Tradition," in *Philosophy, Science and Exegesis in Greek, Arabic and Latin Commentaries*, 2 vols., ed. Peter. Adamson/Hans Baltussen, London 2004 (vol. 2), 83; Robert Wisnovsky, *Avicenna's Metaphysics in Context*, London 2003, 113–141.

the nature of the soul as moving and efficient cause as respects the body, Philoponos took this metaphor to mean that the human soul, while representing the body's actuality, is separable from it as the steersman is from a ship.[88] For Avicenna, the steersman metaphor serves to stress the separability—and hence immortality—of the human rational soul. On the other hand, though, Avicenna maintained the classical Aristotelian view that the soul is a substance *qua* form in animals and plants (*De anima* I, 3).[89]

The demonstration that the term 'perfection' is more indicative of the function of the soul in humans and celestial bodies does not exhaust the task with which Avicenna is concerned,[90] namely to demonstrate that the human soul does not subsist as form in matter (*De anima* V, 2). By contrast, the famous "Flying Man" argument was often misread as a demonstration (which is not). Rather, it is a pointer (*tanbīḥ*) towards the fact that the soul is immaterial.[91] The nature of the "Flying Man" argument as a pointer in no way implies a difference in content with the following Avicennian doctrines of the soul: rather than being a different argument, it is a different way to present the same argument, viz. that the principle of life (the soul) is immaterial. The implications of Avicenna's demonstration of the existence of the soul in the body not as a form in matter would be quite painful for Medieval Latin readers, though since Avicenna argues for the annihilation of the lower soul faculties (vegetative and sensitive) after the death of the body.[92] This doctrine would be rejected, for instance, by Albert the Great. According to Avicenna, the soul is indeed the principle of issuance or emanation of its faculties, which appear to be ontologically distinguished from it. On the other hand, though, it is difficult to capture exactly by way of definition what Avicenna thought of the essence of the soul and, as a consequence, of its definition. If the soul cannot be defined properly neither as referring to its faculties (for there is no identity between soul essence and soul faculties) and its being 'perfection' (hence, referring to its essence) according to the Aristotelian definition is not really an exhaustive definition for the term 'soul',[93] how can the Avicennian soul be defined? A fitting answer is found in Tommaso Alpina's recent monography on Avicennian psychology:

88 More on this in: Alpina, *Subject, Definition, Activity*, 101–103.

89 Alpina, *Subject, Definition, Activity*, 74–76.

90 Avicenna, Liber de anima I c. 1, ed. Van Riet 1968, 20.44–21.48 [ed. Rahman 1959, 7.8–10].

91 For a thorough analysis of the Flying-Man experiment, see: Alpina, "The *Soul of*, The *Soul in itself*," 202–208, 209.

92 Hasse, *Avicenna's* De anima, 42–47.

93 Alpina, "The *Soul of*, The *Soul in Itself*," 196: "The major obstacle to interpret this not as Avicenna's own definition of the soul is the use of the verb *ḥadda* (sc. to define) [...] However,

> This could then be considered Avicenna's own definition of the soul: the soul is not a substance because it is a perfection; rather, the soul is a perfection as substance. Its being perfection as substance results in its being the causal principle of activities. In doing so, he also succeeds in unifying the notion of *kamāl*: when applied to the soul, it always refers to a substance, regardless of its being an instance of separable or inseparable perfection or, what amounts to the same thing, of a self-subsistent substance or a substantial form.[94]

There are two different instances of soul investigated in Avicenna's *De anima*: the first is the soul as a relational entity in (i.e. the soul of) animals and plants, in which the Aristotelian equation between soul, form and substance fully applies. This part of Avicennian psychology could be rightfully called "general psychology" (*psychologia generalis*), as it belongs to the genus of ensouled, living beings. By contrast, there are sections of Avicenna's *De anima*—e.g. V, 2—explicitly and specifically devoted to the study of the human soul in itself and its fate (*psychologia specialis*). These souls are ontologically different, for the souls of humans are separable and immortal unlike others, but both kinds of soul share a common ontological structure in terms of faculty theory.

Avicenna, like Aristotle, maintained that the soul is endowed with a series of faculties defined by their specific objects. Unlike Aristotle, though, Avicenna introduced a rather problematic distinction between the soul's essence and the soul faculties, which he describes according to the Neoplatonic concept of "flow" (*fluxus*/*fayḍ*). These faculties flow from the essence of the soul and are received in their bodily locations when the organic structure of the living being has reached a degree of "appropriate preparation" (*praeparatio specifica*/*istiʿdād ḫāṣṣ*).[95] In this process, the role of spirit is crucial inasmuch as it represents the vehicle through which the soul performs its activities in the body, except for intellectual activities. The work where Avicenna deals per-

---

examination of the critical apparatus of the editions of Avicenna's text, and inspection of a part of the extant manuscripts preserving it, reveal that Avicenna's words might have been different: *naǧidu* (*we find*) is attested as a variant for *naḥuddu*, which is confirmed by the Latin translation of the *Nafs* (*invenimus*). Therefore, Avicenna might have meant to say that 'the soul that we find' and not 'that we define' is the first perfection of a natural, organic body, having the capacity of performing the activities of life, without committing himself to endorsing the Aristotelian definition of the soul [...]."

94 Alpina, *Subject, Definition, Activity*, 117.

95 Olga Lizzini, "Flusso, preparazione appropriata e *inchoatio formae*: brevi osservazioni su Avicenna e Alberto Magno," in *Medioevo e filosofia: per Alfonso Maierù*, ed. Massimiliano Lenzi/Cesare A. Musatti/Luisa Valente (I Libri di Viella 150), Roma 2013, 129–150.

haps more in detail with the elemental composition of spirit is "On cardiac remedies" (*De medicinis cordialibus*/*Fī l-ʿadwiya qalbiyya*). In this work, whose sections appear both in the *Canon* and copied in an attachment to book IV of Avicenna's *De anima* by Avicenna's disciple Abū ʿUbayd al-Ǧuzǧānī (†1070),[96] Jacquart has shown that Avicenna considers spirit to be the receptacle of the God-ruled emanative process,[97] a mixed substance endowed with a complexional form that is the only one suitable to receive psychic powers.[98] This view is consistent with Avicenna's study of the different faculties (*virtutes*/*quwà*) carried out in the famous chapter (I, I, VI) of his *Canon* of medicine.

4.3 *Spirit as a Juncture between Divine Providence and Animal Physiology*

Created by God in the left ventricle of the heart, spirit is the most noble among bodily substances, such that it resembles heavenly spheres and is a luminous substance.[99] Once it has reached the suitable degree of appropriate preparation, spirit acquires its life-giving power by being receptive to the emanation of the soul powers from the soul's essence.[100] On the one hand, spirit seems

96 Alpina, "Al-Ǧūzǧānī's Insertion," 365–400.

97 Jacquart, "La notion philosophico-médicale," 19, 25–31.

98 Avicenna, De medicinis cordialibus tr. 1 c. 1 in: Avicenna, Canon, ed. Venezia 1507, fol. 544ra (ed. New Delhi 1981–1996, vol. 1, 222.2–10): "Creauit deus ex concauitatibus cordis sinistram vt esset armarium spiritus et minera generationis illius. Creauitque spiritum latorem virtutum anime in membra animalium tendentium mediante ipso. Creauit itaque primum appendiculum virium anime spiritum et manationem secundam in membris corporis mediante ipso. Produxit etiam spiritum ex subtilitate humorum et igneitate quemadmodum produxit corpus ex grossitie ipsorum et terrestreitate. Etenim comparatio spiritus ad humorum subtilitatem est sicut comparatio corporis ad ipsos humores: quemadmodum enim ex humoribus membra constituuntur (*tataǧawharu*) propter commixtionem ipsorum adinuicem producentem ad formam aliquam compositionis (*ṣūra wāḥida mizāǧiyya*) qua commixtum disponitur ad recipiendum habitum non conuenientem ei ratione simplicium simili modo constituitur spiritus ex subtilitate humorum commixtione ipsorum quattuor ad formam complexionis qua spiritus preparatur ad receptionem virium anime quas ex simplicibus non acquirit. principium autem eius est ex manatione diuina sine auaritia et diminutione." Unfortunately, this section was not translated in the fragment of the *De medicinis cordialibus* attached to the *De anima* and available to Albert, as it was later translated by Arnaud of Villanova. Anyway, we have chosen to report this passage and another (see below, n. 100) from the untranslated chapter 1, for it captures very well the general attitude of the whole Avicennian work which blended Galenic medicine with Neoplatonic emanationism and which can be retrieved in the below listed passages of the *De medicinis cordialibus*, all found in the fragment attached to Avicenna's *De anima* and, unlike this one, available to Albert.

99 Avicenna, De medicinis cordialibus, ed. Van Riet 1968, 190.31–44 [c. 2, ed. Bābā 1404/1984, 225.12–226.7].

100 Avicenna, Canon I fen 1 doctr. 6 c. 4, ed. Venezia 1507, fol. 24[rb] [ed. New Delhi 1981–1996,

to be the subtlest byproduct of the interaction of humors. Its material subtlety makes it, among the bodily substances, the closest to the immaterial nature of the soul. This is the reason why spirit was regarded as the vehicle of the soul's faculties. Since its nature is the closest to immateriality among all bodily substances, this makes spirit apt to receive and spread the soul's powers into the whole body. On the other hand, the consequence of the material nature of spirit is its proneness, like any other material substance, to the laws of generation and corruption which rule the natural world.

All in all, spirit represents a key juncture between the Neoplatonic and emanative metaphysical framework ruled by divine providence and the Galenic anatomical system based on the concepts of humors and spirits. In this sense, spirit can be understood as single physical entity and as the instrument (*instrumentum*/*āl*) of the soul's acting in the body from the point of view of its being a tool of the soul. When considered in its activities, it is threefold (vital, natural, animal). Still, these considerations that place the spirit within Avicenna's broader psychological and metaphysical framework leave us to examine the relationship among breathing, heartbeat and nutrition and spirit to explain the necessary and sufficient causes for life in humans and animals.

4.4 *Vital Spirits, Natural Spirits, and the Most Basic Conditions of Life*

Avicenna's concept of spirit is, like Aristotelian *pneuma*, a zoological one. The conceptual framework within which the Avicennian theory of spirit is worked out, though, is basically indebted to Galenic medicine, especially his doctrine of humors, and its reception in the Islamic World. To understand Avicenna's take on the concept of life and respiration, there is therefore need to consider two main works, which in turn presuppose two different notions of life: the *Canon* of medicine, which takes into account human life, and the zoology of his "Book on animals" (*De animalibus*/*Kitāb al-Ḥayawān*), as clearly pinpointed in Alpina's recent article on the Avicennian concept of life.[101] To support any kind of conclusions regarding psychophysiology, though, the faculty theory expounded in Avicenna's psychology will prove crucial also for highlighting the compatibility among the medical, the zoological and the psychological notions of life. Let us start from Avicenna's medical masterpiece, the *Canon*.

---

vol. 1, 126.25–26]: "Et hic spiritus cum secundum complexionem prouenit quam debet habere ad hanc virtutem recipiendam preparatur. Deinde omnia alia preparat membra ad recipiendum omnes alias virtutes animales: et etiam reliquas." See also: Avicenna, De medicinis cordialibus, ed. Van Riet 1968, 191.53–66 [c. 3, ed. Bāba 1404/1984, 226.14–227.9] where spirit is defined "substantia lucida".

101 Tommaso Alpina, "Is Nutrition a Sufficient Condition for Life?," 221–258.

In this work, the passage where Avicenna deals with faculty theory and its relation to life is situated in a very well-known chapter (I, I, 6) on the soul's faculties or virtues (*virtutes/quwà*). Here, Avicenna divides the faculties into: (1) the general genera of the faculties (*de generibus virtutum secundum sermonem universalem/ fī aǧnās al-quwà bi-qawl kullīy*), (2) the natural serving faculties (*de virtutibus naturalibus quibus ministratur/ fī quwà al-ṭabīʿiyya al-maḫdūma*), (3) the natural served faculties (*de virtutibus naturalibus ministrantibus/ al-quwà al-ṭabīʿiyya al-ḫādima*), (4) the vital faculties (*de virtutibus animalibus/ al-quwwa al-ḥayawāniyya*),[102] and (5) the psychic, perceptive faculties (*de virtutibus animalibus comprehendentibus/ al-quwà al-ḥayawāniyya al-mudrika*). In this chapter, we shall consider the first four sections only.

Alpina has shown that Avicenna draws on the classical medical classification of faculties, leaving out any discussion on the soul faculties and deferring it to natural philosophy (see for instance *De anima*, I, 5 and its list of faculties). In the *Canon*, the terminology employed by Avicenna is quite close to classic sources of medical literature such as Johannitius' *Isagoge*, the *Pantegni* and Costa ben Luca's *De differentia*. According to Avicenna, the serving faculties (nutrition, growth and generation) are dependent on the subservient faculties such as the attractive (*attractiva/ǧāḏiba*), retentive (*retentiva/ māsika*), digestive (*digestiva/ hāḍima*) and expulsive (*expulsiva/ dāfiʿa*), which in turn depend on the four primary qualities. In the second section, Avicenna details how these faculties operate, serving the faculties of nutrition, growth and generation, whose actions are expounded in the third section. As has been remarked in the case of the works from the medical tradition, and most notably the *De differentia*, the problem consists in the relationship between the vital faculty and its related spirit (*comparatus/mansūb*)[103] and the general framework of Aristotelian faculty theory, which does not foresee any vital faculty, since breathing, heartbeat and pulse do not have, strictly speaking, a proper object like the other faculties.

In the *Canon*, Avicenna distinguishes two parts of the vegetative faculty: the nutritive (*nutritiva / ġādiyya*) faculty, aimed at the survival of the single living being (*ad hoc ut individuum remaneat / li-baqāʾ al-šaḫṣ*), which is able to

102 In this respect, the Latin translation does not fully convey the connection of *ḥayawāniyya* to life (*ḥayā*), rendered usually with *vitales*. Rather, it gives a sense closer to psychic mental activities, also designated by the *virtutes animales comprehendentes*, expounded in the final part of the chapter on the soul powers in the *Canon*.

103 Avicenna, Canon I fen 1 doctr. 6 c. 4, ed. Venezia 1507, fol. 24rb [ed. New Delhi 1981–1996, vol. 1, 126.21].

carry out the transformation of food into blood and bodily members thanks to its transformative faculty (*virtus immutativa / al-quwwa al-muġayyira*). This faculty groups under its name the growing faculty as well, which is served by the nutritive faculty.[104] The second part of the nutritive faculty aims at the preservation of the species (*ad hoc ut remaneat species / li-baqāʾ al-nawʿ*) and is constituted by two parts as well, the generative (*generativa / muwallida*) and the formative (*formativa / muṣawwira*). The complex examination into the causes of digestive failure carried out here and elsewhere—especially in the study of the concept of radical moisture—in the *Canon* deserves to be treated organically together with the study of the passage of time, youth and old age, the subject of chapter 5. What is of interest here is mainly the concept of life expounded in the *Canon* and its relationship to Avicenna's psychology, zoology, and theory of spirit.

In the *Canon*, Avicenna deals with the vital faculty and its relation to life from an explicitly medical point of view, which presupposes a narrower notion of life, closely based on sensation and motion. Unsurprisingly, the link between life, sensation and motion had appeared, under Aristotle's influence, in the opening of Avicenna's *De anima* and is reiterated in the *Canon*'s definition of the spirit as the "vehicle" (*vehiculum*/*markab*) of sensation and motion.[105] When Avicenna discusses the role of the vital faculty with respect to life, especially what the faculty is which is preparatory for the reception of life in a living being, his position seems to waver. At first, he associates life exclusively or primarily with the vital faculty,[106] then asserts that actually both vital and natural faculties are necessary and sufficient conditions for life; otherwise, without the activity of the nutritive faculty, the living being perishes.[107]

Right after having made this claim, Avicenna nuances his position. He denies the preparatory role of vital and natural faculties with respect to life. Otherwise, one could infer—even if Avicenna does not say so explicitly—that plants also, which possess natural faculties, could be prepared to receive sensation and locomotion. Avicenna argues that the vital faculty—and with it, vital spirit—is the first thing which comes to be in the cardiac *pneuma*. When spirit comes into being from the gametes as the first, most basic and preparatory

104 More on the relationship between growth and nutrition in chapter 4.

105 Avicenna, Canon I fen 1 c. 6 summa 1, ed. Venezia 1507, fol. 25[rb] [ed. New Delhi 1981–1996, vol. 1, 123.1].

106 Alpina, "Is Nutrition a Sufficient Condition for Life?," 242.

107 Avicenna, Canon I fen 1 doctr. 6 c. 4, ed. Venezia 1507, fol. 24[va] [ed. New Delhi 1981–1996, vol. 1, 127.4–5]: "Et preparans non est virtus nutriendi tantum (*wa-laysa hāḏā al-muʿidd huwa quwwa al-taġḏiya waḥdahā*) ita ut dicamus quod quamdiu virtus nutriendi affuerit viuum erit: et cum destruetur erit mortuum."

function of life, it necessitates the action of the natural faculty to preserve the living being—in this case, a human.[108] This line of thought is clearly shared by Avicenna in his *De anima*. In I, 5, he replicates the identical psychological structure of the *Canon*, dividing the nutritive soul into the classical servient and subservient faculties rooted in the action of primary qualities. In his psychology, most notably in the chapter dedicated to the nutritive faculties (II, 1), Avicenna assumes that subsistence of life in living beings is due to nutrition rather than arguing for this fact, quite trivial for a scholar of Aristotelian psychology.[109] As Avicenna clearly states in his psychology (*De anima* II, 1), once the nutritive soul in a living being is destroyed, so is its capacity to survive.[110]

Elsewhere, viz. the *Canon*, Avicenna did feel the need to address the question about the primary association of life to vital spirits and the role of natural spirits and nutrition. While in a text of Aristotelian psychology any discussion on the direct causal association between life may be taken for granted, the medical context proves remarkably different here. Given the close link between life, vital spirits and vital faculty, the connection between life and nutrition needed to be somehow put under clearer investigation. Despite these differences in terms of method and content (the vital faculty is never discussed in Avicenna's psychology), Alpina has shown that Avicenna's conclusions on what is a necessary and sufficient condition for life are quite consistent with each other. In the first case, in the *Canon* what is primarily associated with life is the vital faculty, which Avicenna pleads to relegate to natural philosophy. As a matter of fact, though, he never dealt with it there. On the other hand, the only necessary and sufficient condition which stands for the preservation of life always equates—in both sciences—with nutrition.[111] Before jumping to some conclusions about

108 Avicenna, Canon I fen 1 doctr. 6 c. 4, ed. Venezia 1507, fol. 24[va] [ed. New Delhi 1981–1996, vol. 1, 127.7–11] "Et si virtus nutritiua inquantum est nutritiua esset ipsa virtus que preparat ad sentiendum et mouendum: vegetabilia preparata forent ad sensum et motum recipiendum. Restat ergo sit vt preparans alia res que propriam sequatur complexionem que vocatur virtus vitalis. Et ipsa quidem est prima virtus que in spiritu prouenit cum spiritus ex subtilitate humorum procedit: deinde spiritus secundum philosophum Aristotelem per eam recipit principium primum et animam primam ex qua alie profluunt virtutes."

109 Alpina, "Is Nutrition a Sufficient Condition for Life?," 234: "[...] nothing can be inferred about what the relation of nutrition to life is, if any, according to Avicenna. In the context of *Nafs*, however, this might be not surprising: for, in line with Aristotle's view, Avicenna considers the soul as the principle of the activity of life in organic bodies, nutrition included, and perhaps it would have been trivial to specify its connection with life."

110 Avicenna, Liber de anima II c. 1, ed. Van Riet 1972, 105.25–29 [ed. Rahman 1959, 52.13–53.1].

111 Alpina, "Is Nutrition a Sufficient Condition for Life?," 38–39.

the relation between spirit, soul and life, some considerations on the role of the heart, the pulse and breathing seem in order here.

4.5 *Pulse and Respiration: a Galenic Heritage*

Avicenna's anatomy of the heart and respiration follows Galen and the Galenic medical tradition by dividing the heart into two chambers maintaining the existence of the cardiac septum as a sort of middle ventricle.[112] Like his predecessors, Avicenna maintained the usual anatomy and physiology of the breathing process: spirit comes through the arterial vein and reaches the heart's left ventricle. Vital spirit travels then from the heart into the arteries, causing the pulse, originated by the sympathetic vibration of the arterial tunics, a position extremely close to Galen's.[113]

Avicenna's anatomy of the heart assumes, like Galen did, that part of the blood and natural spirit produced in the liver from vapors (by digested food) filters from the right ventricle of the heart through the cardiac septum into the left ventricle of the heart. Part of the natural spirit passing into the heart or the lungs via the veins would then become materially part of the vital spirit spread from the heart into the body via the arteries. Following in the footsteps of the Galenic medical tradition, Avicenna maintained the role of the veins as the carriers of blood, produced by the concoction of food and of natural spirit, and as organs of the nutritive process which derives its nourishing power from the heart.[114]

Avicenna tackled the cooling function of the lungs only in breathing animals and humans, with the sole exception of fish. Avicenna's account does not cover, for instance, the case of more imperfect animals such as bloodless, non-breathing animals.[115] As Avicenna argues in the *Canon*, the utility of respiration is to nourish the spirit with matter and temperate its complexion.[116]

---

112 Avicenna, Canon III fen 11 tr. 1 c. 1, ed. Venezia 1507, fol. 257vb [ed. New Delhi 1981–1996, vol. 3/1, 399.20–21]: "Et in ipso (*sc.* corde) sunt tres ventres, scilicet duo ventres magni: et venter quasi medius quem Galienus nominavit foveam (Ar.: *dihlīz*) aut meatum non ventrem ut sit ei receptaculum nutrimenti [...]."

113 On the anatomy of arteries in the *Canon*, see: Avicenna, Canon I fen. 1 doctr. 5 summa 4 c. 2, ed. Venezia 1507, fols. 20^{va-vb} [ed. New Delhi 1981–1996, vol. 1, 112–113]; on the veinous artery (*arteria venalis/šaryān warīdī*) and on the other arteries, Canon, I fen 1 doctr. 5 c. 3–4–5, ed. Venezia 1507, fols. 20vb–21va [ed. New Delhi 1981–1996, vol. 1, 114–116].

114 On the anatomy of veins, see: Avicenna, Canon I fen 1 doctr. 5 summa 5, c. 1–5, ed. Venezia 1507, fols. 21va–23ra [ed. New Delhi 1981–1996, vol. 1, 116–121].

115 On the anatomy of lungs, see: Avicenna, Abbreviatio de animalibus XIII c. 2, ed. Venezia 1508, fols. 51^{ra-rb} [ed. Cairo 1970, 275–282].

116 Avicenna, Canon III fen 10 tr. 1 c. 1, ed. Venezia 1507, fol. 239va [ed. New Delhi 1981–1996, vol. 3/1, 317.7–8]: "Et iuvamentum huius aeris preparati est ut temperet sua euentatione

This position echoes the classic Galenic idea that outer air enters the composition of *pneuma* both to cool down the innate heat in the heart and nourish it. Given its strong focus on breathing animals, Avicenna's account of respiration leaves out the cooling functions that occur in bloodless animals, for instance. Only by confronting Avicenna and Aristotle Medieval readers would have been able to gain a more general perspective on the physiological functions (pulse, cooling, spread of vital heat, thermic balance) primarily associated with life in both humans and animals and ascribed in medical literature to the problematic 'vital faculty'.

On the other hand, the production of natural spirit, its refinement through the cardiac septum and pulmonary vessels, and its material entrance in the composition of vital spirits stresses a functional unity between the operations of the living being, which cannot be thought of separately. While respiration is crucial to keeping a living being alive and regulating its temperature, spirit produced by means of digestion is as important a component of the life-giving spirit as is the spirit drawn in from the lungs: without both of these, blooded animals cannot subsist. In Avicennian psychophysiology, the living beings in which we get a fuller picture of the relationship between soul, soul powers, spirit and the most basic conditions for the existence and subsistence of life are breathing animals. To understand how this is the case with bloodless animals, Albert would have to turn to Aristotle.

All in all, the main traits of originality in Avicenna's spirit theory reside in the integration of the post-Galenic scheme of three spirits into the framework of Aristotelian zoology, which he further grafted on a psychology strongly dominated by a Neoplatonic and emanative structure clear in works such as *On cardiac remedies*. From the analysis of Avicenna's psychological and medical writings, we can single out at least three basic physiological conditions for life: the spread of vital heat, the cooling down of the body, and the nutritive activities. The role of vital spirits is to provide a thermic balance between the supply of a suitable amount of vital heat to all bodily parts through arteries and a quantity of cooling material (air) able to stave off the self-destruction of the living being. How this cooling down takes place in non-breathing animals, though, it is not quite clear from Avicenna's zoological account. Moreover, breathing involves, according to Avicenna, not only cooling and the spreading of heat but also the retention of strength (as in Aristotle), directly correlated to the presence of voice in living beings, and the wasting away of harmful vapors. These are concepts which are typical of a medical tradition (the Galenic), which Avi-

caliditatem cordis: et ut restauret spiritum cum substantia que est magis vincens in complexione eius [...]."

cenna tries—sometimes embryonically, as in the case of respiration—to fit into a broader, zoological context consistent with Aristotle's science of nature.

The case of natural spirits is perhaps less complicated, for the existence of spirits arising from digestive processes is common to all animals and humans and, in general, allows the subsistence of life in living beings through the process of nutrition, which Avicenna divides into four stages: digestion in the mouth, in the stomach, in the liver and in the veins. In the end, the key role of natural spirits is to carry that vital heat so crucial for nutrition and the subsistence of life into living beings. While the connection between natural spirits is not spelled out at length in Avicennian works, they nonetheless represent a cornerstone of Avicenna's psychophysiology of human and animal nutrition.

In Avicenna's psychophysiology, the concept of spirit appears to be, on the one hand, a material substance, receptacle of the soul's embodied activities (except for intellect) endowed with a nature explainable according to the Galenic theory of humors and complexions. Unlike the previous medical tradition, though, Avicenna inserts spirit into a complex system of natural teleology based on a Neoplatonic and emanative structure of the world with God at the top. For this reason, Avicenna's psychophysiology emerges as an (accomplished) attempt at connecting the material and anatomico-physiological explanations of life-related phenomena with a more general outlook on the world of living beings as ruled by divine providence. This set of philosophical doctrines will prove to be of just as vital relevance to Albert.

5 *Actus animae secundum se*: Life, Soul, and Spirit in Albert the Great

As highlighted in the first chapter, Albert's early theological works represent a true goldmine for those interested in the study of Albertinian psychophysiology. By now, the studies of Henryk Anzulewicz, Thérèse Bonin and, earlier, Louis-Bertrand Geiger[117] have successfully portrayed the psychological and metaphysical framework of Albert's mature concept of life and its relation to his early theological works (especially *De homine*, *De quattuor coaequaevis*, *Commentary on the Sentences*).[118] On the other hand, Bono's investigation into

117 Henryk Anzulewicz, "Zur Theorie des menschlichen Lebens nach Albertus Magnus. Theologische Grundlegung und ihre bioethischen Implikationen," *Studia Mediewistyczne* 33 (1998), 35–49; Louis-Bertrand Geiger, "La vie acte essentiel de l'âme—l'*esse* acte de l'essence d'après Albert le Grand," in *Études d'histoire littéraire et doctrinale*, Montréal-Paris 1962, 49–116; Bonin, "The Emanative Psychology," 45–57.

118 Other relevant contributions on Albert's psychology, especially the nature of the human

the Medieval language of life has rightfully focused on the role of spirit as one of the cornerstones of Albert's psychophysiology.[119] To get a fuller picture of the role of spirit as the "proper vehicle of life" (*vehiculum vitae proprium est spiritus*), here I will try to integrate an analysis of the role of spirit in physiological life-related activities with Albert's emanative psychology.

5.1 Anima, vivere, vita: *Albert's Early Emanative Psychology of Life*

In his early theological works, Albert recurrently stated that life had to be understood, according to Aristotle, as "being among living beings and perfection" (*vivere viventibus est esse*).[120] According to Albert, life is to be understood as a "continuous act of the soul in the body" (*actus animae continuus in corpus*), following the formulation of the *De motu cordis*.[121] This means that life is an act of the soul which precedes any other activity of the soul, even those of the nutritive faculties. From this point of view, Albert follows closely, even more than the influential *De causis* and its tripartite scheme (*esse-vivere-intelligere*),[122] Alfred of Shareshell's *De motu cordis*. According to this text, life is the metaphysical and the physiological precondition for psychic activities to take place (*vitam quasi praeviam ceterae sequuntur, sine qua ceterarum nullam inesse possibile est*).[123] The priority of being over life, and in turn, over life-related operations, is already introduced in his early works beginning with the *De sacramentis*, where the priority of life over sensation

soul, are: Edmund Runggaldier, *Die Menschliche Seele bei Albertus Magnus: Ein nicht-reduktionistischer Beitrag zum Leib-Seele-Problem* (Lectio Albertina 11), Münster 2010, 7–54; Theodor Köhler, De quolibet modo hominis: *Alberts der Großen philosophischer Blick auf den Menschen* (Lectio Albertina 10), Münster 2009, 1–37; Georg Wieland, *Zwischen Natur und Vernunft: Alberts des Großen Begriff von Menschen* (Lectio Albertina 2), Münster 1999, 3–29; Loris Sturlese, "Il razionalismo filosofico e scientifico di Alberto il Grande," in *Storia della filosofia tedesca nel Medioevo: Il secolo XIII* (Accademia toscana di scienze e lettere La Colombaria 149), Firenze 1990, 69–125.

119 Bono, "Medical Spirits," 119–128.

120 Aristoteles, De anima II 4 (415b12); Aristoteles Latinus, De anima [A.L. XII.1], ed. Decorte-Brams (online): "Causa enim ipsius esse omnibus substantia est; vivere autem viventibus esse est."

121 Alfredus Anglicus, De motu cordis c. 8 n. 1 (ed. Baumker 1923, 30.19); Anzulewicz, "Zur Theorie des menschlichen Lebens," 38.

122 Katja Krause–Henryk Anzulewicz, "From Content to Method: the *Liber de causis* in Albert the Great," in *Reading Proclus and the Book of Causes. Volume 1. Western Scholarly Networks and Debates*, ed. Dragos Calma, Leiden 2019, 180–208.

123 Alfredus Anglicus, De motu cordis c. 1 n. 2 (ed. Baeumker 1923, 6.7–12): "Vivit igitur et vegetatur animal, sentit et movetur, discernit et intelligit, et quae ex his oriuntur non facile numerabilia habet efficere. Et harum potentiarum haec prima distantia est, quod vitam quasi praeviam ceterae sequuntur, sine qua ceterarum nullam inesse possibile est."

and its posteriority to being (*esse*) is compared to the priority of baptism over any other sacrament.[124] The same scheme is adopted in the *De resurrectione* (*omnes sequentes supponunt vivere, et ipsum non supponit aliquem sequentium*),[125] and the *De quattuor coaequaevis*.[126] The latter proves to be a particularly enlightening text, as it clearly argues that life needs to be considered as a second cause if compared to being (*tamquam causae secundae*), a distinction inspired again by the "Philosophus" of the *Liber de causis*. For this reason, the psychological structure within which life is situated seems to consist in 1) being, 2) life and 3) activities of the soul, a position which results from a fusion of Neoplatonic psychological sources such as the *De causis* and, first and foremost, the *De motu cordis*. An interesting parallel can be found in the *Summa Halensis*, which shares influence from the *De motu cordis* on this point.[127]

Similar observations on the primary connection between life and vital spirits can be found, at that time, in Peter of Spain's *Tractatus de longitudine*, a paraphrastic rearrangement of Aristotle's *De longitudine et brevitate*. Like Albert,

---

124 Albertus Magnus, De sacramentis tr. 1 q. 5 (Ed. Colon XXVI, 12.42–65).

125 Albertus Magnus, De resurrectione q. 2 a. 3 (Ed. Colon. XXVI, 341.83–342.17); Krause-Anzulewicz, "From Content to Method," 185–186. Krause and Anzulewicz aptly speak of "formal decontextualization" referring to Albert's use of the *De causis* in this passage.

126 Albertus Magnus, De quattuor coaequaevis tr. 2 q. 4 a. 1, ed. Auguste Borgnet, Paris 1895 (Ed. Paris. XXXIV, 360a): "Ad hoc quod obicitur, quod vivere viventibus est esse, dicendum, quod verum est per se loquendo: quia vitae actus est vivere: vita autem actus est spiritus: vivere autem est actus vitae tamquam causae secundae, ut dicit Philosophus, qui sibi praeponit esse quod est actus entis secundum causam primam quae est essentia. Unde cum dicitur, vivere viventibus est esse, intelligendum est de esse specificato per causam secundam, id est, per formam contrahentem specificatam quae vita est, et non esse quod est actus causae primae, et secundum quod est prima, id est, formae generalissimae quae est essentia. Et hoc patet ex principio commenti super librum *de Causis*, ubi expresse quasi per eadem verba istud exponitur."

127 Summa fratris Alexandri (Summa Halensis) II inq. 4 tr. 3 q. 1 tit. 2 a. 3 p. 1 (ed. Quaracchi II, 650): "Unde sicut actus videndi exit ab anima mediante virtute sive potentia visiva, actus intelligendi mediante virtute sive potentia intellectiva, similiter, ut videtur, et actus vivendi mediant aliqua potentia sive virtute quae sit propria huius actus; sed nulla potest esse haec nisi vita, quae est praevia omnium actuum potentiarum et a qua omnes habent quod possunt exire in actus suos, quae, secundum quod habetur libro *De motu cordis*, est potentia praevia aliarum; dicit enim idem: 'Vivit et vegetatur animal, sentit et movetur, discernit et intelligit, et quae ex eis oriuntur, non facile numerabilia, efficere habet, et harum potentiarum haec prima distantia est quod vitam quasi praeviam ceterae sequuntur, sine qua ceterarum nullam inesse possible est.' Cum ergo sit praevia aliarum potentiarum et una illarum, ut patet ex verbo praedicto, relinquitur quod ab ipsa est actus vivendi." This conception of life as a power (*potentia*) that is prior to any power of the soul is explicitly endorsed in the solution (651).

Peter is strongly influenced by Avicennian psychology, included Avicenna's emanative structure of soul powers. In the *Tractatus*, Peter considers life to flow from the soul (*anima qua a vita emanat*), to be the primary foundation of other life-related faculties such as the nutritive and to be primarily linked to the action of vital spirits.[128] Given the closeness of Peter's extensive (almost 6 pages) psychophysiology of life, the link between an emanative psychology with medical physiology would well be worth a comparative study between the authors to understand possible mutual influence or common sources. Dag Hasse has shown that, among 13th century authors, Albert seems to place a particular emphasis on the doctrine of the emanation of soul powers from the soul which he considers, following Boethius, a whole composed of soul essence and soul powers (*totum potestativum*).[129] This emanationist aspect of Albert's psychology is much less pervasive (if not absent) in the psychology of Albert's pupil Thomas Aquinas.[130]

According to Albert, being equates with the coming-to-be of the form which, according to the famous Avicennian-Boethian adage, gives being (*forma dat esse*) to living beings. As with Avicenna, the use of the concept of 'form' to define the human soul proves philosophically unsatisfying for Albert, who wants to maintain the immortality of the human rational soul. For this reason, he prefers to qualify, as Dag Hasse has shown, the soul as 'perfection' rather than 'form' (*melius dicitur perfectio quam forma*) to stress its separability from the body and its immortality.[131] On the other hand, Albert hesitates to follow Avicenna in defining human soul as a self-subsisting substance accidentally united to human body. This is a position which Albert cannot endorse, since it would undermine the psychosomatic integrity of the single human being, especially in view of Judgment Day.[132] To grant psychosomatic unity to the human being, Krause has shown how Albert proposes an argument for the substantial union between human body and soul based on a principle of numerical

128 Peter of Spain, Tractatus de longitudine et brevitate vite, ed. Manuel Alonso, Pedro Hispano: Obras filosóficas, III (Instituto de Filosofía 'Luis Vives.' Serie A Núm. 4), Madrid 1952, I c. 1, 415–422. On spirits in Peter, see the quaestiones in the Madrid manuscript attributed to him: Peter of Spain, Questiones super libros De animalibus Aristotelis XVIII, qq. 1–4, ed. Francisca Navarro Sanchez, Farnham-Burlington 2015, 397–400.

129 Hasse, "The Early Albertus Magnus," 243–244.

130 Bonin, "The Emanative Psychology," 53.

131 Albertus Magnus, De homine (Ed. Colon. XXVII/2, 32.45).

132 Albertus Magnus, De homine (Ed. Colon. XXVII/2, 49.11–17). On Albert's criticism of Avicenna, see: Massimiliano Lenzi, "Alberto e Tommaso sullo statuto dell'anima umana," *Archives d'histoire doctrinale et littéraire du Moyen Âge* 74 (2007/1) 32–33.

proportionality between cause and effect and rooted in what she calls "Albert's absolute psychological creationism" which will be later and briefly discussed in chapter 3.

According to Albert, Resurrection is inserted into the Neoplatonic scheme of the *exitus-perfectio-reductio*, which extends the need of the return to the Creator (*reductio*) both to body and soul. This bodily disposition towards Resurrection is given to the body from the God-created (*a Deo*) nature of the rational soul, whose nature is sharply contrasted with the origin of animal and plant soul from natural agents such as the semen (*a natura*). For this reason, animal and plants will not be able to resurrect and return to the Creator, as they are the mere product of natural and non-divine agents. The human soul seems therefore necessarily meant to return to its creator, but, unlike Avicenna, in all its psychosomatic unity (body and soul).[133] Consistent with his view on the need for the human body to be rejoined to its soul on Judgment Day, Albert stresses that after the body's death the annihilation of the sensitive and nutritive faculties concerns their operations but not their essence.[134]

Albert's theory of soul in the early works, as in his later works, wavers between a consideration of the soul as a substance in its own right (*substantia quaedam per se subsistens*) and its essential understanding as an act or form of the body in the case of plants or animals.[135] Dag Hasse has shown that, according to Albert, the powers of the soul, are ontologically distinguished from it, flow, as for Avicenna, from the essence of the soul itself, building up with it a *totum potestativum*, a concept borrowed from Boethius' *De divisione*.[136] This position is in keeping with Albert's mature psychology and sets Albert notably apart from the *Summa Halensis*, the text which, among his contemporaries, came closest to espousing a doctrine of the concept of life as prior to any activity of the soul. As remarked by Magdalena Bieniak, the psychological doctrine of the *Summa* points toward a plurality of forms rather than a unity, as Albert also intended.[137] Rather, Albert's psychological structure of the living being seems to be closer to Peter's.

---

133 Krause, "Albert the Great on Animal and Human Origin," 212–213.

134 Albertus Magnus, De homine (Ed. Colon. XXVII/2, 572.27–30): "Ad aliud dicendum quod anima sensibilis et vegetabilis corrumpuntur in suis operibus, eo quod sunt potentiae affixae organis, et non corrumpuntur per se secundum quod sunt in homine."

135 On Albert's anthropology in his early works, see: Henryk Anzulewicz, "Anthropology: The Concept of Man in Albert the Great," in *A Companion to Albert the Great: Theology, Philosophy and the Sciences*, ed. Irven M. Resnick (Brill's Companions to Christian Tradition 38), Leiden 2013, 335–340 and the bibliography therein.

136 Hasse, "The Early Albertus Magnus," 237–238.

137 Magdalena Bieniak, "The Soul-Body Union in the *Summa Halensis*," in *The Legacy of Early*

In the end, Albert's psychology of life rests to this point on the threefold distinction between being, life and psychic activities and on a quidditative distinction between the unperishable human soul and corruptible animal and plant souls. While it is clear that being, which equates to form (or, more loosely, soul), *is* ontologically different from life, it is still not clear *how* being (soul) connects to life. As highlighted by the study of Louis-Bertrand Geiger, in his early theological works Albert adopts a model he would follow and develop in his Aristotelian paraphrases. In the *De quattuor coaequaevis*, Albert argues that life equates with the Aristotelian expression *viventibus est esse* according to a conception of being that does not equate with the more general form (*forma generalissima*), which represents the soul. Rather, "being (*esse*)" must be understood as a specific form (*per formam contrahentem specificatam quae vita est*) that derives from the soul as its causal principle, a 'quasi-effect' of the soul which becomes such once it is received in the body: hence the distinction between soul, life (*vita*) and the act of living (*vivere*). The three-layered distinction between soul essence/life (*vita*)/act of living (*vivere*) is rooted upon a Neoplatonic and emanative scheme which appears clearly in the *Commentary on the Sentences*: the form of life flows from the essence of the soul and, by adhering formally to the living being (*formaliter adhaeret vivo*), causes the beginning of life in it.[138]

This distinction between the 'life of' a certain body and 'life in itself' as independent of the body is iterated in another passage from the *Commentary on the Sentences*. In this context, Albert distinguishes the life of the soul and of the Angel (*angelus*) from that received in the hylomorphic compound as *actus animae continuus in corpus*, a definition given by Alfred of Shareshell in his *De motu cordis*. Until now, Louis Geiger has proposed perhaps the most convincing reading of Albert's argument: life (*vita*) is an act by which the soul (*spiritus*) is in something (*quod est*) that is independent of the body, and this something (*quod est*), independent of any existence in a hylomorphic compound, is the form of life itself. In other words, the *actus spiritus in quo est* represents the life of the angel and of the human soul, existing independently of a natural body.[139] By

Franciscan Thought, ed. Lydia Schumacher (Veroffentlichungen des Grabmann-Institutes zur Erforschung der mittelalterlichen Theologie und Philosophie 67), Berlin 2021, 37: "The author of the *prima secundae* of the *Summa Halensis*—perhaps Alexander of Hales himself—was among the partisans of the plurality of forms."

138 Albertus Magnus, Super II Sententiarum dist. 26 art. 5, ed. Auguste Borgnet, Paris 1894 (Ed. Paris. XXVII, 451b).

139 Albertus Magnus, Super IV Sententiarum (dist. 23–50) dist. 46 a. 5, ed. Auguste Borgnet, Paris 1894 (Ed. Paris. XXX, 634a–b): "Dicitur vita *actus quo est spiritus in quod est* et hoc modo dicitur anima vivere et Angelus vivere. Alio modo dicitur actus animae continuus

contrast, the definition of life as *actus animae continuus in corpus* applies only to hylomorphic compounds. Therefore: the act of living which, according to the scheme of the *Liber de causis* and the *De motu cordis*, precedes any operation of the soul's faculties and lasts differently according to the different quiddities (mortal, immortal) of souls is a compound of three parts: the soul's essence, life as an emanation from the soul and independent from any physical body, and life received in a body. This strongly Neoplatonic doctrine of the life of the soul constitutes the backbone of the mature Albert's understanding of life, which is more systematically connected with the concept of spirit, a relevant concept already in Albert's early theological works.

In Albert's early works and especially in the *De homine*, the classification of spirits reflects the classical threefold classification into vital, natural and animal spirits which he could have inherited via the *Pantegni*,[140] the *Isagoge* of Johannitius or Avicenna. In the *quaestio* where Albert discusses the body-soul union (*De unione animae et corporis*), he upholds the view that the soul does not need any medium to act in the body, with which it is joined immediately as form to matter. Rather, the soul needs an operational medium, which is identified by Albert in the spirit. In this classification, strongly influenced by his medico-philosophical sources, Albert argues that vital spirit produces life and pulse in the body *qua* instrument of the soul (*vitalis vero est ille, per quem anima a corde vitam et pulsum operatur in toto corpore*).[141] A similar position is found in the *Commentary on the Sentences*, which emphasizes the metaphysical background of the diffusion of life into the body through vital spirits, qualifying this process as *influxus*.[142] In another passage from the *De homine*, Albert clearly states that the first activity of the spirit is to produce life in a living being, and only in a second moment enables the carrying out of the activity of nutrition (*opus spiritus non erit primum nisi vita, et secundum alimento uti et augmentare*

in corpus: et hoc modo dicitur animal vivere: et hoc modo omnis vita creata exemplariter et efficienter est a vita prima." The italicized passage emends the text in Borgnet's edition (*actus animae quo est spiritus in quo est*) according to the suggestion of Geiger: Louis Bertrand Geiger, "La vie, acte essentiel de l'âme," 63 n. 57: "Nous avons la distinction entre la vie de l'ange et de l'âme spirituelle, d'une part, composés uniquement de *quo est* et de *quod est*, et celle des corps vivants, d'autre part, composés de matière et de forme."

140 Though they are not quoted explicitly by Albert, their knowledge, especially concerning the chapter on spirits, is witnessed already in the *De homine*: Albertus Magnus, De homine (Ed. Colon. XXVII/2, Appendix III, Auctores a nobis allegati, 645).

141 Albertus Magnus, De homine (Ed. Colon. XXVII/2, 571.56–572.7).

142 Albertus Magnus, Super II Sententiarum, dist. 26, art. 5, ed. Auguste Borgnet, Paris 1894 (Ed. Paris. XXVII, 452a): "[…] spiritus vitalis vehiculum est vitae, quae continue influitur corpori ab animalis speciali organo, sicut cordis." On the characterization of life as an *influxus* from the heart: Albertus Magnus, De homine (Ed. Colon. XXVII/2, 52.38–49).

et generare et sic de aliis).[143] All in all, Albert's reading of the concept of life as an act of the soul which is prior to nutrition and corresponds, in Galenic terms, to the vital faculty can also be found, with the due psychological differences (plurality of forms vs unity of the substantial form, life as a potency vs life as an *actus*) in the second part of the *Summa Halensis*.[144] As shown previously, this is a doctrine found in Peter of Spain's *Tractatus de longitudine* too, with the notable addition of close doctrinal similarities between Albert and Peter's *Tractatus* (and not the whole of Peter's works).[145]

According to Albert, Alfred, the *Summa*, and Peter, life represents from a physiological point of view an even more basic physiological activity than nutrition, carried out by the vital spirits issuing from the heart. While this bit of the psychological explanation of the priority of life over nutrition and all other soul activities may be clear, a clarification of the relationship between this concept of life and the definition of life as "act of the soul according to itself" (*actus animae secundum se*) and not according to a single potency requires a clarification.[146]

While life (also, pulsation) may precede ontologically, the activities of the soul and life and pulse may the activity of nutrition, it is nutrition alone which grants subsistence to the living being.[147] The logical possibility of the subsistence of life without nutrition is discussed briefly by Albert in the part of the *De homine* dedicated to sleep and its relation to nutrition. Following Avicennian and Aristotelian zoological examples, Albert questions the existence of life without nutrition by putting forward the examples of snakes that survive for a long time without eating, or hibernating animals, such as the bear, which hibernate for a long time without eating and show apparently no sign of nutritive activity.[148] Albert solves the issue by recalling the Aristotelian *De insomniis*, according to which the activity of the nutritive faculty is clearly linked to sleep

143 Albertus Magnus, De homine (Ed. Colon. XXVII/2, 155.47–49; 155.55–58): "Opus spiritus non erit primum nisi vita, et secundum alimento uti et augmentare et generare, et sic de aliis [...] Spiritus, qui est in corde, est instrumentum et medium, per quod anima diffundit vitam in corpore, et vita non est actus ab ipso nisi sicut ab instrumento [...]."

144 Summa fratris Alexandri (Summa Halensis) pars 2 inq. 4 tr. 3 quaest. 1 tit. 2 a. 3 part. 1 (ed. Quaracchi II, 651–652).

145 For a more general outlook on Peter's psychology, especially on the Avicennian influence, see: Hasse, *Avicenna's* De anima, 55–60.

146 Albertus Magnus, De homine (Ed. Colon. XXVII/2, 48.51–52; 53.21–24; 54.36; 144.52; 552.58).

147 Albertus Magnus, De homine (Ed. Colon. XXVII/2, 54.35–37; 54.45–68).

148 Albertus Magnus, De homine (Ed. Colon. XXVII/2, 106.5–16). See: Aristotle, Historia animalium, VIII, 4, 594a21; Aristotle, Historia Animalium, III, 14–15 (599a20–33); Avicenna, Abbreviatio de animalibus VIII c. 2, ed. Venezia 1508, 37ra.

as its efficient cause (digestive vapors cause sleeping) and its temporal end, viz. sleep ends when digestion is completed. For this reason, the fact that hibernating animals sleep for a long period does not imply that their nutritive faculty is not active, let alone animals, such as snakes, which are awake.[149]

Therefore, if the activities of the nutritive soul are physiologically based on the natural spirits,[150] it follows that in any animal the necessary and defining condition of life with respect to being (*esse*) is the action of the nutritive faculty and not only of those parts of it connected to vital spirits. At a physiological level, vital spirit equates with life or *esse* according to the sense outlined above and based on Albert's reading of the *De motu cordis*, while natural spirits constitute the instrument of the soul's activities, posterior to life itself. The correspondence between physiological and metaphysical/psychological is clearly highlighted in the *Commentary on the Sentences*, where it is clarified that the soul gives the being to the body, a being (*esse*) that becomes actual when the body is living through vital spirits, which provide it with the suitable amount of vital heat and render it capable of subsisting thanks to the activities of the soul, nutrition in the first place.[151]

While the role of vital spirits and natural spirits in Albert's psychology and definition of life seem to be quite clear from his early works on, there are at

---

149 Albertus Magnus, De homine (Ed. Colon. XXVII/2, 106.31–37): "Solutio: dicimus quod nutritiva semper agit in vigilantibus, et dormientibus etiam magis, sicut dicit Philosophus in primo *de somno et vigilia* sic: 'Opus suum facit nutritiva particula in dormiendo magis ac vigilando. Nutriuntur enim tunc magis et augentur tamquam nihil egentia ad hoc sensus'." See: Aristotle, De somno 1 (454b32–455a4).

150 Albertus Magnus, De homine (Ed. Colon. XXVII/2, 96.71–97.2): "Anima enim est influens impressiones in corpus, ut iam habitum est, et illae impressiones sunt receptae a corpore et dicuntur virtutes affixae membris, quae non sunt substantialiter partes animae, sed fundantur in spiritibus naturalibus et calore et complexione membrorum, et ideo variantur et debilitantur et iuvantur secundum variationem membrorum."

151 Albertus Magnus, Super I Sententiarum (dist. 1–25) dist. 8 art. 26, ed. Auguste Borgnet, Paris 1893 (Ed. Paris. XXV, 260a): "Ad aliud dicendum, quod spiritus est instrumentum animae quoad operationes primas ipsius, quae sunt vitalis, naturalis, et animalis: per spiritum enim vitalem influit vitam, et pulsum: per naturalem autem opera naturalia, ut nutrire, augere, et generare: per animalem autem animalia, ut sunt sensus, et motus: et ista sunt eius, secundum quod est motor corporis: sed secundum hoc quod dat esse, non habet medium. Et si obicitur, quod secundum Philosophum vivere viventibus est esse: ergo vivere cujus instrumentum est spiritus vitalis, est esse: et ita utitur eo ut dat esse. Dicendum, quod hoc impossibile est: quia non utitur instrumento nisi quod est: ergo anima utens instrumento est jam in corpore ante instrumentum per rationem formae: et ideo, cum dicitur, quod vivere viventibus est esse, intelligitur de actu vitae quae est anima, cujus effectus dat esse animale, et hoc est esse vitae et vivere: sed non intelligitur de vivere quod est per spiritum et pulsum."

least a couple of issues which need be settled: in the first place, a clearer view of the role of spirit in animals, especially bloodless, lungless animals, which is at times unclear in Avicenna too. In the second place, its relation to the reception of life in the human body, qualified as a *fluxus* or *influxus* from the soul, a phenomenon whose connection to spirit remains opaque. In Albert's Aristotelian paraphrase, these issues find a clearer and explicit solution.

5.2 "Necesse est quod in omni vivo sit generatio spiritus": *Albert's Mature Account of Spirit and Life*

Albert's Aristotelian paraphrases mark a turning point in his intellectual development, for he took on the burdensome task of paraphrasing the whole of the Aristotelian *corpus*, a task which earned him the nickname of *Doctor Universalis*. As will become evident in what follows, Albert's mature psychophysiology moves significantly away from his early production only with respect to the physiological role of spirit in the world of living beings and not with respect to his general understanding of the psychological structure of living beings. Therefore, it is no wonder that Albert's approach to the concept of life in the *De anima* and in the *Parva Naturalia* owes much to his earlier theological works. In his mature production, Albert maintains central doctrinal cornerstones of his early psychology:

1. A quidditative distinction between, on the one hand, the immortal human soul and, on the other hand, the mortal souls of plants and animals;
2. An ontological distinction between soul's essence and soul's powers, understood in their unity as a *totum potestativum* of powers which flow from the essence of the soul;[152]
3. The function of the soul as a giver of being (*forma dat esse*) to the living being;[153]
4. The composition of life and soul as *quo est* (*esse*) and *id quod est*;[154]
5. The distinction, inherited from the *De causis* and *De motu cordis*, between being, life and activities of the soul (hence, living) according to this order of ontological priority;

The Neoplatonic interpretation of life as "act of the soul" which entails an emanative process based on a logical distinction between three different levels of life itself (in the soul, *in fluxu*, received in the body) is repeated in his *De anima* and in the *De morte et vita*, a paraphrase of Aristotle's *De longitudine et brevi-*

152 Bonin, "The Emanative Psychology," 48–49.

153 Albertus Magnus, De anima II tr. 2 c. 1 (Ed. Colon. VII/1, 84.67–80).

154 Thérèse Bonin, *Creation as Emanation: the Origin of Diversity in Albert the Great's* On the causes and the Procession of the Universe, Notre Dame 2001, 24.

tate.[155] The former (*De anima*) presents the same hierarchical and causal explanation of life according to the triadic order of: 1) soul, 2) life *qua* act originated from the soul alone and presupposed in any psychic activity (*supponitur ut fundamentum in omni opera sua*),[156] and 3) life *qua* received in the body (*prout est in corpore*). The latter (*De morte et vita*) offers some additional and valuable insights into Albert's theory of life in living being. In keeping with his early theology, most notably the *Commentary on the Sentences*, Albert singles out an imperishable (*nec huic vitae directe opponitur mors*) and a perishable (*haec est cui mors opponitur*) kind of life, which recall the life of angels and humans in his early work. While the first kind of (human) life flows from the soul (*in fluxu ab anima ipsa*) and exists independently of its bodily reception, the second is always (*universaliter*) subject to death *qua* life received in the body.[157] Life seems to be implicitly understood (*implicite intelligitur*) as inhering in the soul *qua* cause and principle of life, but is nonetheless something more specific than the mere soul: rather, it is a form flowing from the soul itself.[158] The distinction between a perishable and imperishable life, according to Albert, must not to be understood as an ontological distinction (*secundum intellectum et non in re*), though, since life is one (*una*) just like light, daylight and light (*lux*, *lumen* et *lucere*) are of the one and same nature.[159] The introduction of this twofold concept of life has a purpose which can be detected in Albert's own words: namely,

---

155 This logical distinction is presented in the opening digression of the *De morte et vita*, a paraphrase of Aristotle's *De longitudine et brevitate*, and in the *De anima*: Albertus Magnus, De anima II tr. 1 c. 6 (Ed. Colon. VII/1, 73.74–74.11). On the interpretation of these passages, see: Geiger, "La vie, acte essentiel," 58–62, 93–97.

156 This wording is remarkably close to Albertus Magnus, De resurrectione q. 2 art. 3 (Ed. Colon. XXVI, 342.15–17): "[…] omnes sequentes supponunt vivere, et ipsum non supponit aliquem sequentium."

157 Albertus Magnus, De morte et vita tr. 1 c. 3 ed. Auguste Borgnet, Paris 1890 (Ed. Paris. IX, 347b, 348a–b): "[…] sed vita est ipsa anima diffusa in eo quod vivit et in toto et in partibus. […] Vita enim secundum prius habita, licet non sit circa quaedam separata, tamen implicite intelligitur in causa vitae: sed hoc est in anima: et huic vitae non opponitur mors. Et vita quae est in fluxu ab anima ipsa: et nec huic vitae directe opponitur mors: eo quod illa et magis est animae quam corporis, sicut fluxus luminis plus est solis quam aeris recipientis. Et est tertia vita participata a corpore, adhaerens ei sicut forma formato: et haec est cui mors opponitur: hanc enim privat, et non aliam universaliter. Si enim quaedam anima perit cum corpore, hoc tamen universaliter non convenit omni animae."

158 Geiger, "La vie act essentiel," 62: "La vie dont vit le corps n'est donc, formellement parlant, ni l'âme elle-même, ni la vie dont l'âme vit formellement, mais justement un act qui émane de l'âme et qui, reçue dans le corps, y produit, avec l'*esse* du vivant, les ensémble des opérations spécifiques des êtres animés."

159 Albertus Magnus, De morte et vita tr. 1 c. 3, ed. Auguste Borgnet, Paris 1890 (Ed. Paris. IX, 348b).

to justify why there are some souls (human souls) which survive the body's death. These souls perish, unlike plants and animals, only in their operations (*secundum actus*) but not in their substance (*secundum substantiam*).[160] From a doctrinal point of view, the opening digression of the *De morte et vita* further develops the distinction between the three different aspects of life, clarifying their purely logical and not ontological meaning. Unfortunately, Geiger ended his excellent discussion on the conception of life in Albert without taking into account a further and final passage from Albert's *De spiritu et respiratione*, a key to the role of spirit in Albert's psychophysiology of life:

> For this reason, some say that life in living beings is of three kinds; one is indeed that which is the cause of life, and this is the life of the soul. The second is that which is the flow of this life to the living being, and this is the life of the soul's instrument, which is the spirit. The third is life formally received by the body and the parts of the animated body; and this is what is called living flesh, an ear or a living eye.[161]

In this passage, Albert clearly admits the existence of a fourth logical part of the definition of life, viz. the life proper of the spirit *qua* instrument of the soul (*vita instrumenti animae quod est spiritus*), which represents the key connecting point between life as metaphysical instance independent of the body (*in fluxu ab anima ipsa* or *actus spiritus in quo est*) and the reception of life in the body. In other words, spirit ensures that life is correctly received in the body and allows for the successful exercise of the different activities of the soul, just like a hammer correctly used by a craftsman shapes the artifact according to the formal and morphological configuration he intends.[162] The crucial role of

---

160 Albertus Magnus, De morte et vita tr. 1 c. 3, ed. Auguste Borgnet, Paris 1890 (Ed. Paris. IX, 348b): "Licet autem sic tripliciter variemus vitam, in hoc non est nisi secundum intellectum, et non secundum rem: quia secundum rem pro certo non est nisi una et eadem: et sic in quibusdam est corruptibilis secundum substantiam et actus suos quos habet in corpus. In quibusdam autem secundum actus tantum, et non secundum substantiam, sicut in hominibus. In quibusdam autem nec secundum actus nec secundum substantiam, ut in coelestibus [...]"

161 Albertus Magnus, De spiritu et respiratione, I tr. 1 c. 8, ed. Auguste Borgnet, Paris 1894 (Ed. Paris. IX, 226a): "Et ideo quidam dicunt triplicem esse vitam in his quae vivunt: unam quidem quae est causa vitae: et haec est vita animae. Alteram autem quae est fluxus huius vitae ad vivum: et haec est vita instrumenti animae quod est spiritus. Et tertiam quae est vita formaliter recepta a corpore et partibus animati corporis corporis: et haec est qua dicitur caro viva, et auris, et oculus vivus." Translation is mine.

162 Albertus Magnus, De spiritu et respiratione, I tr. 1 c. 3, ed. Auguste Borgnet, Paris 1894 (Ed. Paris. IX, 217a–b): "Est autem motus spiritus ab anima vivificante et regente corpus,

spirit as an instrument of the soul *qua* carrier of the form of life will be more evident, as we shall see in the next chapter in generation. In that context, spirit is said to receive the form of life at the moment of animation and to spread this form out in the body, producing the act of living (*vivere*) alongside all its related psychic activities.[163] But this is for us to discuss later, in the broader and much more complex context of embryology.

For present purposes, it is crucial to bear in mind that, if the form of life is that which contains in itself the forms of all the activities proper of the embodied powers of the soul (hence excluding intellect),[164] it explains why spirit can be defined as the "proper vehicle of life" (*vehiculum vitae proprium*) both from a physiological but also from a psychological and metaphysical point of view. On the one hand, because of its physical subtlety, spirit is the most appropriate instrument for the soul to spread vital heat and the form of life from the heart into the whole body.[165] In the embryo, the diffusion of heat carried by spirit

---

et movens per easdem virtutes corpus per omnia sicut faber et malleus movent et influunt materiae eamdem formam: faber quidem ut artifex, malleus vero ut instrumentum."; Albertus Magnus, De spiritu et respiratione, I tr. 1 c. 4, ed. Auguste Borgnet, Paris 1894 (Ed. Paris. IX, 220b): "[...] sicut enim unum instrumentum quod est malleus, habet faber ferrarius, per quod omnes inducat in ferrum formas artis suae: ita hoc unum instrumentum junctum est animae, per quod in toto corpore inducit omnes universaliter formas et operationes vitae: et sicut in arte fabrili est unus artifex et una ars et unum instrumentum, sed multae formae artis et multi motus instrumenti, et per consequens multa artificiata: ita etiam in physicis una est anima et unum instrumentum et multae perfectiones organorum constitutae ex illis, ut visus in oculo, digestio in hepate et stomacho, et sic de aliis naturalibus et animalibus organis et perfectionibus eorum. Haec igitur est causa, quod proprietates aliquas terrae non habet spiritus, aut minimum et insensibile habet."

163 Albertus Magnus, De spiritu et respiratione I tr. 1 c. 10, ed. Auguste Borgnet, Paris 1894 (Ed. Paris. IX, 229b): "[...] una forma qua ab anima informatur spiritus quae *forma vitae* vocatur, et est in spiritu, non tamquam in subjecto vivente formaliter, sed est in eo sicut ars, et artis forma est instrumentum per quod inducitur in materiam."

164 Albertus Magnus, De spiritu et respiratione I tr. 1 c. 10, ed. Auguste Borgnet, Paris 1894 (Ed. Paris. IX, 229b): "Sic igitur illa forma vitae potentia distinguibilis est et contrahi potest ad omnem virtutem quae operatur aliquem actum vitae, sicut est nutritiva, augmentativa, et sensitiva, et secundum locum motiva." The form of life, potentially capable of performing all the activities of the soul powers, is first understood as "life" more generically located in the heart, see: Albertus Magnus, De causis et processu universitatis II tr. 2 c. 25 (Ed. Colon. XVII/2, 118.66–70): "Sicut radii distinctionem accipiunt a sole procedentes, quamvis in ipso globo solis sint lux indistincta. Et sicut virtutes animae a corde procedentes distinguuntur et incidentes membris esse distinctum conferunt, quamvis in ipso corde sint vita indistincta."

165 Albertus Magnus, De spiritu et respiratione I tr. 1 c. 4, ed. Auguste Borgnet, Paris 1894 (Ed. Paris. IX, 219b): "Est autem hic spiritus tenuis et clarus quantum ad materiam: et est quidem tenuis propter penetrabilitatem, ut in omnes partes corporis per subtiles potentias arteriarum possit penetrare."

from the moment of animation onwards is crucial for the correct morphological formation of the embryo; in a formed living being, a correct transmission of vital heat makes for a correct balance of body temperature and allows the subsistence of life through nutrition, which in turn requires suitable amounts of vital heat.[166] On the other hand, the connection between a metaphysical emanative framework and a physiological degree of preparedness for matter to receive a specific form may reasonably speak in favor of an influence exerted by Avicenna's *De medicinis cordialibus*, known by Albert at the time of composition of his *De spiritu et respiratione*.[167] While Albert's position on the nature of spirit in the psychophysiology of life cannot be reduced to Avicenna's and indeed results from a synthesis of a plurality of sources (*De causis, De motu cordis, De differentia, De natura hominis*), the general coordinates of Albert's and Avicenna's spirit theories seem to agree on the more general psychological and metaphysical framework within which the life-giving action of spirit(s) is situated. Moreover, both authors emphasize how spirit can become a life-giving substance once that matter receives an appropriate degree of 'appropriate preparation' (according to Avicenna) or, in Albert's case, *inchoatio formae*. These notions, though, will be handled more in depth in the embryological context.

A second and key difference between Albert and Avicenna emerges again in the *De spiritu*, the work inspired by Costa Ben Luca's *De differentia spiritus et animae*. Having explained the object of the science of spirit and respiration (I, 1), and having exposed (I, 2) and critically reviewed (I, 3) the past conceptions of spirit, he moves on to analyze the true nature of spirit (*quid sit spiritus animalis*

166 As we will see in the next chapter, Albert seems not to be so sure about following Aristotle and Avicenna on the timing of embryonic animation (40 days): as a matter of fact, there are no passages throughout Albert's works which suggest he adhered to Aristotle's and Avicenna's standpoint.

167 Albertus Magnus, De motibus animalium I tr. 2 c. 4 ed. Auguste Borgnet, Paris 1890 (Ed. Paris. IX, 275a–b): "[…] hunc frequens Avicenna, in epistola *de Dispositionibus cordium et speciebus eorum* dicit, quod omnis delicia sive delectatio est apprehensio perfectae acquisitionis propriae virtutis sicut sentire dulce et odorem bonum. Delectatio est sensui percipere vindictam virtuti irascibili, et percipere speciem virtuti cogitativae et aestimativae, et theoremata virtuti intellectivae." Avicenna, De medicinis cordialibus, ed. Van Riet 1968, 192.67–70 [ed. Bābā, 277.11–13]: "Gaudium autem est delicia. Omnis autem delicia est apprehensio perfectae acquisitionis propriae virtuti apprehendenti, sicut sentire dulce et odorem bonum sensui, percipere vindictam virtuti irascibili, et percipere spem virtuti cogitativae et aestimativae." Note that the *De motibus animalium* is the work which follows immediately the *De spiritu* in Albert's division of sciences in the *Parva Naturalia* and is thus belied to have been composed in the same period (1256–1257) as the *De spiritu*. However, Albert's use of Avicenna's *De medicinis cordialibus* deserves a detailed study of its own.

secundum veritatem). The adjective *animalis* here is not to be strictly referred to the *spiritus animalis* of the medical tradition; rather, it has to be understood as an adjective derived from the substantive *anima*. In other words, Albert wants to explain how the spirit relates to the soul in what he calls *viva*. But what are the living beings endowed with spirit? Instinctively, one may think of humans and animals by adhering to the Aristotelian-Avicennian doctrine. Albert's doctrine, based on a demonstrative argument (*ratio*) and a sign (*signum*), takes an innovative direction:

> [The fact] that such spirit exists in every animate being is shown by a demonstration and a sign. The demonstrative argument is that, since the cause of any life (*omnis causa vitae*) is heat and moisture and moisture is always, together with heat, subject to vaporization and conversion into spirit (*cum calido est vaporativum et convertibile in spiritum*), then it is necessary that spirit be generated in any living being. By contrast, the sign that we see in every living being is that, when coldness presses it, at that moment its extremities shrink (*tunc contrahuntur extrema exteriora*). This is like when, during a time of cold, flowers shrink at night and animal bodies become smaller and paler. The cause of this cannot be anything but the backflow and compression of the spirit into the inner parts (*causa non potest esse nisi conversio spiritus et compressio ad interiora*). On the other hand, when the sun is hot, the outer parts begin to swell up and become filled with food, the cause of which cannot be anything but the spirit's outflow into the outer members. The spirit carries the nutriment with itself and, for this reason, they (sc. living beings) dilate and begin to swell. On the other hand, it can be argued that this [happens] at a single time when it is hot and moist, [and] that in this manner the spirit of plants and animals is drawn out, that almost everything which by that time swells up is made as if it were becoming pubescent. In plants swelling buds, and in animals filled seminal vessels, burst into fetuses and fertility.[168]

---

168 Albertus Magnus, De spiritu et respiratione I tr. 1 c. 7, ed. Auguste Borgnet, Paris 1894 (Ed. Paris. IX, 224b–225a): "Quod autem talis spiritus in omnibus sit animatis, ostenditur et ratione et signo. Ratio quidem est, quia cum omnis vitae causa sit calidum et humidum, humidum autem semper cum calido est vaporativum et convertibile in spiritum, necesse est quod in omni vivo sit generatio spiritus. Signum autem est, quod in omni vivo videmus hoc quod quando frigus comprimit eum extra, tunc contrahuntur extrema eius exteriora, sicut tempore frigoris contrahuntur flores in nocte, et corpora animalium minora et pallidiora inveniuntur: cujus causa non potest esse nisi conversio spiritus et compressio ad interiora. Tempore autem calidi solis exteriora tumescunt et replentur nutrimento:

The conclusion of the necessary existence of spirit in all living beings is conducted in accordance with the following line of reasoning: life is always caused by heat and moisture, since whenever heat acts over moisture during the digestive process it is always converted into vapor and spirit. Since plants possess nutritive soul and digestion too, then spirit must exist in plants as well. This is confirmed, in Albert's view, by the fact that, whenever a living being is surrounded by cold, it shrinks, like flowers do at night, because of the spirit's backflow. In another passage of the *De spiritu*, Albert complements his analysis by saying that, by contrast, plants blossom and swell up when spirit rushes to the outer limbs of the living being supplying it with food, a fact that can be well observed especially during the seasons and, more generally, in daylight.[169]

From a purely lexical point of view, Albert's use of the term *viva* to designate living, ensouled beings (plants, animals and humans) is consistent with his botany (*De vegetabilibus*).[170] From a doctrinal point of view, it is difficult to understand whether Albert developed his idea from a precise source or, as often, construed his own theory by assembling a wide range of sources and synthesizing them into a doctrinal whole. In terms of potential sources, endorsements of the existence of spirit in plants can be found in Alfred of Shareshell's *De motu cordis*,[171] and David of Dinant's *Quaternuli*.

cujus causa non potest esse nisi evacuatio spirituum ad exteriora, qui secum ad exteriora trahunt nutrimentum, et ideo aperiuntur et intumescunt. Declaratur autem hoc uno tempore quod est calidum et humidum, quod ita spiritus plantarum et animalium evocat, quod omnia fere tunc tumescentia quasi pubescentia efficiuntur, et turgentibus gemmis in plantis, et animalia impletis vasis seminariis erumpunt in foetus et foecunditatem." Translation is mine. For a similar, sleep-like phenomenon in plants: Albertus Magnus, Ethica I tr. 9 c. 4 ed. Auguste Borgnet, Paris 1891 (Ed. Paris. VII, 144a).

169 Albertus Magnus, De spiritu et respiratione I tr. 1 c. 9, ed. Auguste Borgnet, Paris 1894 (Ed. Paris. IX, 227b–228a). On similar passages in the *De vegetabilibus* on plants' sleep, see: Amalia Cerrito, "Disclosing the Hidden Life of Plants. Theories of the Vegetative Soul in Albert the Great's *De vegetabilibus et plantis*," in *Vegetative Powers. The Roots of Life in Ancient, Medieval and Early Modern Natural Philosophy*, ed. Fabrizio Baldassarri/Andreas Blank (International Archives of the History of Ideas Archives internationales d'histoire des idées 234), Cham 2021, 110–112.

170 Albertus Magnus, De vegetabililibus libri VII, ed. Ernst Meyer and Carl Jessen (Berlin, G. Reimer: 1867; repr. Frankfurt am Main: Minerva 1982), II tr. 1 c. 1 nn. 4–5, 104–105: "Plantae tamen nomen ab actu artis imponitur qui est plantare; et ideo hoc nomen non convenit plantae secundum suum genus, sed secundum accidens. Et si volumus nomen fingere secundum genus, dicemus, quod illud vocamus plantam proprie, quod vocatur vivum vel animatum occultum."

171 Alfredus Anglicus, De motu cordis c. 14 n. 8 (ed. Baumker 1923, 74.18–24): "In simpliciter autem vegetabilibus duae tantum celebrantur digestiones, una scilicet in radicibus, altera in corpore et ramis quae epatis vicem tenent. His igitur idem est vivere et vegetari, a

Whereas the former has a position of its own, the second reports the Hippocratic *De natura pueri*[172] despite this fact, the *ratio* and *signum* put forward by Albert to argue the necessary presence of spirit in living beings prove to be completely independent of both sources. Moreover, while Albert's knowledge of Alfred's *De motu cordis* is certain, Albert's dependence on the *Quaternuli* for his doctrine of spirit in plants represents more of a fascinating suggestion rather than a documentable fact.[173] A third possible source for inspiration does not rest on a specific source but is rather represented by the concept of natural spirit itself and, in particular, its existence in lungless (also, bloodless) animals, which he can infer—for this point is not so explicit—from Avicennian spirit theory. A fourth and last possible, source, though, and perhaps the closest to Albert's *ratio*, is a passage from the pseudo-Aristotelian *De plantis* which insists

---

vitae scilicet spiritu qui radices inhabitat et semen—in semine tamen nondum terrae commisso vis vegetationis sopita alimenti privatione quiescit—; idem enim, sed aliter se habens, vivificat et vegetat."

172 David de Dinanto, Quaternulorum Fragmenta, ed. Elena Casadei, *I testi di David di Dinant: filosofia della natura e metafisica a confronto con il pensiero antico. Introduzione ed edizione dei testi di Elena Casadei* (Testi, Studi, Strumenti, 20) Spoleto 2008, 209: "Dicit autem Ypocras materiam herbe teneram esse et levem et a modico spiritu facile elevari supra terram; et cum iam non possit a modico spiritu elevari, redit spiritus intra terram nutrientem eam attrahitque nutrimentum, ex quo radix formatur. In arboribus vero dura est materia nec valet elevari a terra nisi a multo spiritu. Constat autem spiritus ex aere et terra ideoque spissior est aere; et est caloris susceptivuus propter terram sibi admixtam, sicut est videri in flamma, quae est nichil aliud quam fumus terreus accensus. Non enim accenditur fumus, si sit aqueus. Spiritus igitur attrahit a terra nutrimentum, quo formatur radix, et nutritur ipse spiritus. Cum postmodum multus fuerit spiritus, elevat arborem a terra. Scire autem oportet, quod omnis planta venulas habet in longum sitas, per quas transit spiritus trahens secum intermedium nutrimentum plantae, quod terreum est nec potest sursus ⟨elevari⟩ ad nutriendum plantas. Est enim spiritus sive ignis vehiculum terre et aquae, ut est videri in aqua bulliente ad ignem." As proven in many scholarly contributions, Albert used David's *Quaternuli* already from the *De homine*: there, among the quotes from David, Albert reports his views on sight contained in the part of fragment G (*Liber de visu*) in which this passage is also contained. Since the *liber de visu* belongs to the same fragment G in which the above-mentioned quote is found, Albert's acquaintance with the passage must not be excluded *a priori*. On the other hand, given the non-unitary nature of fragment G, this knowledge is difficult if not impossible to prove. On Albert's knowledge of David, see: Henryk Anzulewicz, "Person und Werk des David von Dinant im literarischen Zeugnis Alberts des Grossen," *Mediaevalia Philosophica Polonorum* 34 (2001), 15–58; Elena Casadei, Introduzione, in: I testi di David di Dinant, 4–19.

173 Further, David attributes to *pneuma* a nature composed of watery and earthy nature, which Albert flatly rejects: Albertus Magnus, De spiritu et respiratione, I tr. 1 c. 4, ed. Auguste Borgnet, Paris 1894 (Ed. Paris. IX, 220a): "[...] sed parum vel nihil habet de proprietatibus terrae."

on the smoky nature of the exhalations produced by the action of vital heat over moisture.[174]

In a recent contribution on spirit and plant formation, Marilena Panarelli has convincingly shown how Albert's concept of spirit in plants blends together insights and intuitions derived from different traditions, viz. Aristotle and the Aristotelian (*De plantis*) tradition and, we may add here, the medical tradition (notion of natural spirits) and other possible philosophical sources (*De motu cordis*, David of Dinant).[175] To do so, Albert roots his ideas on spirit on the fact that digestion is the product of heat over moisture, thus making the existence of spirit necessary in all living beings. On the other hand, though he is aware that spirit is more necessary in breathing animals than in plants. Commenting upon the passage from the *De plantis* which denies the existence of spirit in plants, he writes:

> The fact that a plant does not possess such spirit as is attracted by inspiration and respiration, as Anaxagoras stated, can be proven by the fact that there are many animals that do not have inspiration and respiration. However, this spirit belongs more to animals than to plants, because animals are warmer and partake of higher elements than plants, which are earthy and cold. This is why animals need animal spirits more than plants.[176]

Having explained why spirits exist in all living beings, Albert goes on to explain that, to cause life in a living being, spirit needs to possess some specific properties: it is fiery, airy and watery but not earthy, as held, for instance, by David of Dinant. According to Albert, spirit, which is a mixed substance, is fiery given its

174 Nicolaus Damascenus, De plantis II n. 148 (ed. Drossaart Lulofs/Poortman 1989, 540): "Ut in balneo calor humorem attrahit et in vaporem convertit qui elevatur, et quando superfluus fuerit vertetur in guttas, similiter quoque in animali et planta superfluitates ascendant ab inferioribus ad superiora et descendunt a superioribus ad inferiora in actionibus suis."

175 More on this in: Marilena Panarelli, "*Spiritus* and plant formation. Albert the Great's Restatement of the Pseudo-Aristotelian *De plantis*," *Rencontres de Philosophie Médiévale* (forthcoming).

176 Albertus Magnus, De vegetabilibus I tr. 1 c. 6 n. 34 (ed. Meyer-Jessen, 21): "Quod autem planta non habeat spiritum talem, qui attrahitur per inspirationem et respirationem, sicut dixit Anaxagoras, ex hoc probare possumus, quod multa animalia invenimus, quae inspirationem et respirationem non habent; cum tamen talis spiritus magis animalibus conveniat quibuscunque quam plantis, eo quod animalia secundum se sunt magis calida et participantia superiora elementa quam plantae, quae sunt terrestres et frigidae. Propter quod animalia indigent spiritibus animalibus plus quam plantae". Translation by Panarelli in Panarelli, "*Spiritus* and plant formation."

ability to penetrate all the bodily members thanks to the impulse received in the blood vessels. Spirit is watery since it is able to receive and retain (as moisture does) the impression of forms of objects, it is airy because of its thinness and because heat and moisture, principles of life, can be found in air. If wateriness is the characteristic which enables spirits to retain forms, their tenuity, their nature as *corpus subtile* enables them to receive the form of sensation but also, most importantly, the form of life impressed from the soul on the cardiac spirit, crucial in the process of generation.[177]

According to Albert, vital, natural and animal spirit are related to each other—in this order—in the same manner that vegetative, sensitive and rational soul are. This metaphor, clearly borrowed from Aristotle's *De anima* and repeated in Albert's own paraphrase, emphasizes the strict mutual dependence among the three spirits qualified as formal (*formaliter*): without vital spirit, there exists no natural spirits, and without those two there exists no animal spirit.[178] Still, this division is not clearly detectable in all living beings: for instance, in plants vital and natural spirits cannot be distinguished (*in planta non separari vivificum a naturali*). In animals and humans, this distinction is detectable. According to Albert, the ranking of these three spirits must be understood as formal (*omne posterius in ordine formali essendi tamquam fundamentum supponit et relinquit suum antecedens*).[179] As we shall see in the next chapter, the first spirit to come about is the vital spirit located in the heart of the embryo, whether of the animal, of the chick or in the roots of the plant, and is functionally analogous to a human and animal heart.

5.3 *Albert on Spirit in Plants: Between Tradition and Universal Physiology*

Albert's doctrine on spirit in plants represents an unprecedented *unicum* in the history of philosophy because of, among other things, the organic framework within which this theory is developed. While, as highlighted, Alfred of Shareshell and David of Dinant presented two single passages hinting at the

177 Albertus Magnus, De spiritu et respiratione I tr. 1 c. 4, ed. Auguste Borgnet, Paris 1894 (Ed. Paris. IX, 219b–220b).

178 Albertus Magnus, De spiritu et respiratione I tr. 2 c. 5, ed. Auguste Borgnet, Paris 1894 (Ed. Paris. IX, 237b): "Videtur autem talis esse determinatio horum spirituum, qualis est et animae. *De anima* enim diximus, quod sua determinatio est determinatio figurae: quoniam sicut in figuris trigonum est in tetragono, ita est vegetativum in sensitivo, et sensitivum et vegetativum in intellectivo. Omni eodem modo videtur vivificus esse spiritus in naturali, et naturalis et vivificus in animali in omni eo quod habet has spirituum differentias."

179 Albertus Magnus, De spiritu et respiratione I tr. 2 c. 5, ed. Auguste Borgnet, Paris 1894 (Ed. Paris. IX, 237b–238b).

existence of spirit in plants, Albert develops an entire psychophysiology of plant life closely connected to spirit theory. According to Albert, plants possess an analogue to the heart, situated in the plant's roots,[180] and vein-like channels which exist for the sake of carrying the nutritive substances in the plant. These channels have a twofold origin, at least in the more complex plants. Some may originate either from the roots, like veins originate from the heart, or from the marrow.[181] Unlike veins and arteries in humans, marrow and vein-like channels seem to be directly connected, since the former often spring out of the latter, sometimes in a star-like fashion (*radiosa sicut stella*).[182] These channels undertake the key function of carrying the life-giving essence (*humor*), considered by Albert as an analogous to blood,[183] and the spirit which is transported within the sap itself. These parts (heart, vein-like channels, the marrow), alongside other key parts such as the bark and the knots, are defined as the "essential and integral parts" (*partes integrales essentiales*) since they constitute the basic physiological support for life's subsistence in plants. Unlike blooded and breathing animals, though, plants do not really possess a clear distinction between veins and arteries, since they do not have pulse-related spirit (*spiritus pulsatilis*), pulsation or pulsing vessels (*venae pulsatiles*). This medical termi-

180 Albertus Magnus, De spiritu et respiratione I tr. 1 c. 6 ed. Auguste Borgnet, Paris 1894 (Ed. Paris. IX, 223b): "In plantis autem et animalibus cor non habentibus est membrum quod est loco cordis. Cum enim sic levatur ad principium vitae, tunc effumat in ipsum principium vitae, et ibi formatur ad hoc quod vehat vitae virtutes et conjungatur diminutio spiritui et restauret ipsum." Albertus Magnus, De vegetabilibus V tr. 1 c. 2 n. 18 (ed. Meyer-Jessen, 296–297): "[…] sed a radice, quae est loco cordis in plantis, non videtur illi aliquid simile aliqua plantarum, nisi medulla aut concavitas, quae nunc dicta est; et ideo semen ut frequenter in plantis sequitur naturam radicis, quia per vicarium radicis, usque ad summum plantae deductum proprietates radicis accipit, per vicarium ejus et naturam." De vegetabilibus V tr. 1 c. 3 nn. 26–27 (ed. Meyer-Jessen, 300); Amalia Cerrito, "Disclosing the Hidden Life of Plants," 114–115.

181 Albertus Magnus, De vegetabilibus I tr. 2 c. 1 n. 114 (ed. Meyer-Jessen, 58): "Vena autem in nulla plantarum est vere, et secundum perfectam venae rationem. Sed sunt viae rectae, in quibus decurrit nutrimentum […] Et non invenitur planta, quae his careat, nisi illa, quae fere in toto est cortex; et tamen etiam in illa aliquid simile viis nutrimenti invenitur, sed minus quam in aliis. Inveniuntur autem istae venae a duobus principiis in plantis oriri. In his, quae nodosae non sunt, et sunt rarae substantiae, extenduntur viae illae a radicibus in altum, per stipitem et per ramos et per folia divisae et sparsae. Aliquando autem derivantur ex medulla, et protenduntur ad superficiem plantae; et hoc cognoscitur, quia, quando inciduntur plantae, apparent in talibus lineae albae per modum stellae a medio medullae ad corticem plantae protensae; sicut est in vite et aliis nodosis, praecipue in nodo, et juxta nodum."

182 Albertus Magnus, De nutrimento et nutrito tr. 1 c. 2 (Ed. Colon. VII/2a, 4.28–29).

183 Albertus Magnus, De vegetabilibus II tr. 1 c. 3 (ed. Meyer-Jessen, 114–116); Wöllmer, "Albert the Great and His Botany," 231–232.

nology, clearly inherited from the medical tradition of the Islamic world where the concept is rendered "*urūq al-ḍawārib*" (pulsating blood vessels),[184] underscores on the one hand the possible similarity in organic structure between plants and animals, while clarifying at the same time the differences in the physiological structure among living beings and the physiological distinction between the different kinds of spirits. In spite of this, Albert assigns a primarily life-giving function to the sap and the spirit which pulsate within the marrow.[185] Albert's unifying account of botanical psychophysiology sometimes bristles with tensions. In his account of digestion of the *De vegetabilibus* he states, following the pseudo-Aristotelian *De plantis*, that digestion takes place in the earth (*in terra sub radice*). This remark is consistent with Aristotle, who considered earth in plants as having a function analogous to the stomach in animals and humans. The second step of digestion takes place, according to Albert, in the plant's marrow (*in medulla plantae*).[186] This account of digestion clashes with the account provided in the *De nutrimento et nutrito*, where Albert distinguishes four different stages of digestion in plants which take place in earth (*in terra*), in the roots (*in radice*), in their trunk (*in stipite*) and in the specific organs, closely following the Avicennian scheme of four digestions.[187] While these accounts may seem at odds with each other, they may be reconciled, as convincingly argued by Panarelli, based on a difference in terms of explanatory purpose: Albert's botanical account of digestion explains more in general how digestion occurs in plants, while the *De nutrimento* explains in greater detail what are the necessary steps to produce the generation of each plant part.[188]

Through the analysis of physiological differences, one can still see how Albert's physiology of plant life represents an attempt at building the most unified account of activities of the nutritive soul and its organs (heart, vein-like

184 Albertus Magnus, De vegetabilibus II tr. 1 c. 3 n. 50 (ed. Meyer-Jessen, 122): "Sed vias pulsatiles plantae non habent, quia spiritum pulsatilem non habent." On the Arabic terminology: Hans Hinrich Biesterfeldt, "Zur medizinischen Terminologie des arabisch-islamischen Mittelalters," in *La formation du vocabulaire scientifique et intellectuel dans le monde arabe. Sous la direction de Danielle Jacquart* (Études sur le vocabulaire intellectuel du Moyen Âge/ CIVICIMA 7), Turnhout 1994, 66–90.

185 Wöllmer, "Albert the Great and His Botany," 233; Albertus Magnus, De vegetabilibus II tr. 1 c. 3 n. 51 (ed. Meyer-Jessen, 122).

186 Marilena Panarelli, "How Do Plants Live and Grow? Radical Moisture and Digestion in Albert the Great's *De animalibus*," *Quaestio* 20 (2020), 355.

187 Albertus Magnus, De nutrimento et nutrito tr. 1 c. 3 (Ed. Colon. VII/2a, 6.51–7.4); Panarelli, "How Do Plants Live and Grow?," 355–356.

188 Panarelli, "How Do Plants Live and Grow?," 356: "it does not seem contradictory that Albert, on the one hand, talks about only two steps of digestion in plants and, on the other, speaks about digestion in general in order to explain the generation of each plant part."

channels, stomach) in all living beings as possible, in spite of some tensions within his intellectual production. The introduction of the existence of spirit in plants represents a true difference between Albert, the Galenic medical tradition and even Aristotle. In its doctrinal genesis, spirit was primarily and etymologically associated with breathing and then, in the case of animal spirit, with sensation, cognition and motion. At the broadest level, in the Galenic medical tradition spirit was considered the byproduct of the digestive process and the instrument of the nutritive faculties in humans and animals, but certainly not in plants. From being the vehicle of life and life-related activities in humans and animals, spirit turns into the vehicle of life in all living beings.[189]

5.4 Vita abstrahitur a vegetatione*: Revisiting a Biological and Metaphysical Problem*

As seen above, Albert's mature concept of life from a metaphysical and psychological point of view appears to be quite consistent with his early works (though more explicit on the metaphysical role of spirit), while his physiology undergoes the above-mentioned change of the introduction of spirit into plants. This notable shift in Albert's understanding of spirit pushes toward a more unifying combination between his emanative psychology and his physiology of soul activities through vital heat and spirit. In his *De spiritu et respiratione*, Albert states again the existence of activities in the animal body that are independent of the corruption of the nutritive organs, and hence prior to nutrition. These basic activities would grant, even though only briefly (*licet non diu in vita salventur*),[190] the survival of the living being, a fact that leads Albert to argue that life is independent of nutrition and nutrition-related activities (*vita abstrahitur a vegetatione*) not only at a psychological/metaphysical level but also at a physiological level.[191] For this reason, the definition of life as an act of the soul

189 Albertus Magnus, De spiritu et respiratione I tr. 1 c. 1, ed. Auguste Borgnet, Paris 1894 (Ed. Paris. IX, 214a): "[…] patet vitam esse fundamentum et vegetationis et sensus et intellectus et motus. Vehiculum autem vitae proprium est spiritus."

190 Albertus Magnus, De spiritu et respiratione I tr. 2 c. 5, ed. Auguste Borgnet, Paris 1894 (Ed. Paris. IX, 238b): "Videtur enim forte ex hoc alicui, quod etiam quidam spiritus et quaedam potentiae debent operari vivifice et non naturaliter hoc autem stari non potest: quia omne animatum terrestre, sive membrum, sive totum corpus necessario nutrititur si debet in esse salvari."

191 Albertus Magnus, De spiritu et respiratione I tr. 1 c. 1, ed. Auguste Borgnet, Paris 1894 (Ed. Paris. IX, 213b–214a): "Omnis enim operatio animae est operatio vitae: sed tamen vita abstrahit a vegetatione, et sensu, et intellectu, et est ante unumquodque istorum secundum naturam. Nos enim videmus adhuc vivere animalia quando corruptum est membrum per quod fit nutrimentum: et quando non nutriuntur, adhuc vivunt, licet non diu in vita salventur." See also: Albertus Magnus, Quaestiones super De animalibus III q. 4 (Ed.

prior to any activity of the soul's faculties can apply to heavenly souls—which have no nutritive soul—as well.[192]

The difference between the two cases, though, is substantial: without nutrition, animals do not subsist for long, while celestial bodies do. Albert's reference to an empirical knowledge (*nos videmus*) of the physiological activities separable from nutrition rings a bell with a passage from the *De motu cordis*, once again making his readers very careful about the remarks connected to his own 'experience' as an epistemic tool to achieve the truth.[193] While dressed up as an argument from Albert's own experience, there is reason to believe that, in fact, these considerations on the subsistence of life are once again based on the impressive bookish knowledge Albert was so keen to leverage. Moreover, this passage raises another question: what kind of animals are meant here? Bloodless or blooded? How old are they? Does this failure have anything to do with aging or a violent event?

Apart from the manifold questions which a passage like this may spark, it is unquestionable that from a logical point of view Albert is trying in this passage to separate a more basic form of life from nutrition. The more basic life-supporting activities appear to be carried out by the action of vital spirits,

---

Colon. XII, 125.39–45): "[…] in animali, cum sit substantia una plures habens partes diversas virtutes habentes, necesse est unam esse virtutem, quae sit principium aliarum et a qua aliae recipiant influentiam. Ista autem virtus vitalis est, quia sine virtute vitali non operatur nutritiva in animali."

192 Albertus Magnus, De spiritu et respiratione I tr. 1 c. 1, ed. Auguste Borgnet, Paris 1894 (Ed. Paris. IX, 214b): "Vita est enim actus ab essentia animae procedens, et nutrire est actus cujusdam particulae ipsius in omni genere mortalium animatorum. Si enim ad immortalia respiciamus, invenimus vitam sine vegetatione et sensu cum solo intellectu et motivo secundum locum: propter quod etiam in *Coelo et mundo* diximus superiora vivere nobiliori vita quam inferiora: est enima talis vita perfecta et perpetua, nullo indigens omnino eorum quae sunt extra se. Patet igitur vitam, cujus vehiculum primum spiritus est, esse ante omnem aliam animae operationem." For the priority of life over nutrition in celestial souls as in living beings (*omnia viventia*), see: Albertus Magnus, De causis et processu universitatis a prima causa I tr. 2 c. 2 (Ed. Colon. XVII/2 27.1–4): "Adhuc, vitae principium non est ante quod invenitur vita. Vita autem in intelligentiis invenitur ante sentire et vegetari. Sentire igitur et vegetari principium vitae non est, quod in omnia viventia primum influat vitam."

193 Albert has a history of rendering his "experience" in terms of bookish passages: the major candidate for this passage seems to be: Alfredus Anglicus, De motu cordis c. 1 n. 4 (ed. Baumker 1923, 7.14–8.5): "Vitae item actus necessario continuus est, ceterorum vero minime. Intercipiuntur enim a natura aut aegritudine sensus, motus, intellectus et similia, ut somno, apoplexia, his deinde solutis in statum redeuntes. **Videtur etiam et vegetabilitatem aegritudo interrupere, senectus corrumpere; et ob hoc quantitatem et partes corporis animalis taxavit natura. Non enim vivere et vegetari eadem illi potentia est.**" Emphasis is mine.

as he confirms in his chapter on the arteries in the *De animalibus*,[194] and are independent from the activity of nutrition itself. Clearly, the class of animals where the distinction between pre-nutritive functions (life, pulse, breathing) and nutrition-related functions can be better observed is comprised of blooded animals and humans, both endowed with lungs. This distinction proves to be much more difficult in bloodless (lungless) animals and plants: to analyze the psychophysiology of life in animals and humans, it is therefore easier to start with the latter.

5.5 *Albert on Heart and Lungs: Harmonizing Aristotle and Avicenna*

Among animals, humans certainly possess the most complex system of life-related organs and activities and therefore represent the key to understanding the nature of the psychophysiology in less complex animals such as the bloodless ones, as Albert himself will argue. The *De animalibus*, which devotes a lot of space and attention to the world of living beings, is quite detailed in its account of the physiological structure of life support. The descriptions, as noted by Nancy Siraisi, are quite often based on the chapters from Avicenna's *Canon*.[195] This dependence on Avicenna, blended with attempts at harmonization with Aristotle's own views, can clearly be observed in Albert's doctrine of respiration and cardiac functions in relation to life. In his zoology, Albert explains how the pulmonary vein (*arteria venalis*) is the vessel through which air passes from the lungs into the left heart ventricle and then becomes vital spirit. Having received the form of life residing in the heart,[196] the vital, now life-giving spirit is pumped through the arteries into the whole body to make it alive.[197] It must not be forgotten, though, that vital spirit is nourished by the natural spirit, which passes through the cardiac septum, or through small pulmonary vessels, from the pulmonary arteries (*vena arterialis*), apparently an artery but with the same function as a vein, viz. carrying food, blood and natural spirit (hence the name *vena arterialis*).[198] Vital spirit in humans seems therefore to be the vehicle of the most basic life functions, although it needs the natural spirit to carry

194 Albertus Magnus, De animalibus I tr. 2 c. 20 n. 381 (ed. Stadler, I, 135.30–33): 'Quia autem arteriae sunt viae spiritus, et spiritus est vehiculum vitae et virtutis, ideo etiam arteriae viae sunt vitae et virtutum, quae vitae habent operationem et potentiam […]"

195 Siraisi, "The Medical Learning," 400, 408.

196 On the reception and localization of the *forma vitae* in the heart: Albertus Magnus, De animalibus, I tr. 1 c. 5 n. 74 (ed. Stadler, I, 28.10–16).

197 On the anatomy of the veinous artery (*vena arterialis*): Albertus Magnus, De animalibus, I tr. 2 c. 20 n. 404 (ed. Stadler, I, 144).

198 Albertus Magnus, De animalibus XIII tr. 1 c. 1 n. 4, (ed. Stadler, I, 894.34–40).

on his activity properly. This is a conclusion that is both in keeping with Albert's early theological works, his *De spiritu*, and his Arabic sources such as Avicenna. Much as a clear Avicennian influence may be detected, Albert's anatomy of the heart also shows a clear attempt at harmonizing between the medical and the Aristotelian anatomy of the heart.

In his *De animalibus* and in the *De spiritu et respiratione*, Albert adopts a three-chambered view based on Aristotle, which fits in quite well with the Avicennian division of the heart into two ventricles and the septum, which becomes the third and middle ventricle. This harmonization is self-evident from a lexical point of view, for Albert defines the middle chamber of the heart with the exact same term (*thalamus*) deployed in the *vetus* of Aristotle's *De somno*, while Avicenna speaks of *quasi ventriculus* to define the septum.[199] In particular, the middle ventricle of the heart takes on a clear function, according to Albert, namely the impartation of the power of nourishing to the nutrimental substance, drawn in from the stomach and the liver through the *vena ventrem habens* (vena cava inferior). In this manner, the natural spirit receives the power of nourishing (*accipit virtutem nutriendi*) the body.[200]

199 Aristotle, De somno 3 (458a16–19); Aristoteles Latinus, De somno c. 3 (ed. Drossaart-Lulofs 1943, 10. 1–4): "Horum quidem qui in corde utriusque thalami communis qui est medius; quorum autem uterque haurit ex utraque vena, ex ea quidem quae vocatur magna et quae adorti: in medio vero fit discretio." See: Albertus Magnus, De spiritu et respiratione, I tr. 1 c. 9, ed. Auguste Borgnet, Paris 1894 (Ed. Paris. IX, 226b): "Per venam enim quae est in imo hepatis sive convexo ipsius quae ad medium thalami cordis dirigitur, substantia spiritus bulliens spiritum elevat versus cor, vel versus id quod est loco cordis in his quae cor non habent et hepar: et forte etiam venis carent. Statim autem ut ibi est cibus et spiritus in omnibus cor habentibus, discursio fit ad extremos thalamos, ita quod dextro plus attribuitur sanguinis et grossior et minus de spiritu, sinistro autem plus attribuitur spiritus et minus sanguinis, qui tamen est sanguis clarus et subtilis per spiritualium praecipue nutrimentum. Claritatem autem habet et subtilitatem propter permixtum sibi multum spiritum. Subtilitatem autem habet propter ictus spiritus, et eam quae facta est in ipso spiritu ejusdem sanguinis sublimationem. A sinistro autem cordis thalamo arteriae producuntur per totum corpus, in quibus continue spiritus pulsat vitam a fonte vitae qui est cor: trahitur autem ad sinistrum spiritus in habentibus cor et sanguinem; quoniam cum hepar sit in dextro, repletur fumo malo spiritus, si ad dextrum ventrem sive thalamum cordis distribueretur: hoc autem maximum in vitae operationibus generaret impedimentum."

200 Albertus Magnus, De animalibus, III tr. 2 c. 4 n. 125 (ed. Stadler, I, 332.6–8): "Alia autem pars kyli de substantia est quattuor humorum dictorum, et haec a medio cordis thalamo virtutem accipit nutriendi, sicut in praehabitis ostendimus." Albertus Magnus, De animalibus III tr. 1 c. 4 n. 35 (ed. Stadler, I, 292.32–36): "Epar igitur hoc quod sugit per meseraicas defixas in sima parte sui, transmittit ad cor, ut ibi compleatur sub medio thalamo, ut dicit Aristoteles in *Somno et Vigilia* et in libro *de Nutrimento*, et similiter facit cerebrum."

The vena cava inferior enters the middle ventricle of the heart,[201] where the smokiest and most harmful vapors are separated from the most subtle and pure part of the spirit, passed onto the left ventricle of the heart where, together with the air drawn into the lungs through the arterial vein (*vena arterialis*), it forms the vital spirit.[202] On the other hand, the spirit, which has received its nourishing function in the middle heart chamber, is then sent out to nourish the lungs through the veinous artery (*arteria venalis*) and to the other bodily members such as the brain or other bodily areas.[203] The outcome of Albert's inquiry is a simple one: if there is some basic function prior to nutrition defined as life, it is inevitably connected with vital spirits,[204] but life *qua* subsistence of the living being depends necessarily on both natural and vital spirits.[205] This is because the material composition of vital spirit comes partially from the natural spirit. If digestion is impaired, and therefore the quantity of heat and spirit provided by the digestive process is below the minimum threshold, this triggers failures in the bodily organs, eventually causing death. According to Albert, the restoration (*restaurum*) of the material composition of spirits depends more on the vapors produced by digestion rather than those produced by respiration.[206] The teleology of respiration is addressed by Albert in his comment on

201 Albertus Magnus, De animalibus, I tr. 3 c. 4 n. 592 (ed. Stadler, I, 211.18–23). Note that in Avicenna the *vena ventrem habens* is joined to the heart in its right ventricle and not to the septum or the middle ventricle as in Aristotle: this is another piece of evidence that Albert is trying to harmonize Avicenna's and Aristotle's accounts of the heart's anathomy.

202 Albertus Magnus, De animalibus, I tr. 3 c. 4 n. 593 (ed. Stadler, I, 211.32–212.2).

203 This process is described in detail in Albertus Magnus, De animalibus I tr. 2 c. 21 (ed. Stadler, I, 143–147).

204 Albertus Magnus, De spiritu et respiratione I tr. 2 c. 3, ed. Auguste Borgnet, Paris 1894 (Ed. Paris. IX, 235b): "Cum autem vitalis spiritus dicitur, intelligitur spiritus indinstincte et communiter operans vitam, quae quidem vita viventibus est esse et perfectio."

205 Albertus Magnus, De spiritu et respiratione, I tr. 2 c. 2 ed. Auguste Borgnet, Paris 1894 (Ed. Paris. IX, 234a–b): "Dubitabit autem forte aliquis, quare nominetur spiritus naturalis qui per nutrimentorum spirat regionem, et nutrimento et digestioni deservit. Adhuc autem quare spiritus ille non discurrit in arteriis sicut spiritus vitalis, et ad quid sit necessarius spiritus ille. Si enim naturalem vocant, quia naturae sit instrumentum, non videbitur esse verum: quia non naturae, sed animae nutritivae instrumentum est spiritus ille: cujus signum expressum est, quod non tantum unum operatur, sed multa, nutrimentum vertendo, et augmentando, et huiusmodi multa quae sunt digerere, attrahere, expellere, retinere, vivere, de quibus alibi dictum est."

206 Albertus Magnus, De spiritu et respiratione I tr. 1 c. 4, ed. Auguste Borgnet, Paris 1894 (Ed. Paris. IX, 220a): "[…] propter hoc etiam quoddam restaurum sumit ex aere per anhelitum, sed specialiter per evaporationem nutrimentalis humidi restauratur, sicut in *Somno et Vigilia* dictum est."

Aristotle's chapter on lungs (*DPA* III, 6). In this passage, he agrees with much of what Aristotle had said, although adding some information from his medical doctrines.

Until now, we have observed that *breathing* is considered by Albert, together with pulse, to be the most basic life-related activity. In *De animalibus* XIII, 6, Albert investigates *why* animals and humans have pulsation.[207] Note that this is an addition to Aristotle's text, which nowhere mentions pulse in this context. The only remarks on pulsation are found in the lost *De respiratione*, unavailable to Albert. Rather, the addition on the correlation between pulse and breathing further confirms that Albert comes after the integration of medical teachings with Aristotelian psychophysiology, precisely as in the case of the heart's anatomy. According to Albert, respiration is for the sake of cooling down those land animals endowed with a great amount of blood and vital heat, and other animals such as dolphins and whales.[208] Fish, as in Aristotle, cool down through the cooling action of water through gills.[209]

In bloodless animals, the action of cooling down exerted by the natural spirit produced in digestion is sufficient, if combined with the cooling provided by external air, to temper the heat of the spirit placed in the heart.[210] The lack of respiration and, as a consequence, of a spirit charged with a function analogous to arteries, viz. the spreading of (vital) spirit derived from respiration, leads to a lack in certain functional parts in these living beings. Arteries, specifically designed to spread the kind of spirit derived from breathed in-air, blood and heat, are absent in bloodless animals and plants for this very reason.[211] Rather, these living beings present only channels analogous to veins, meant to transport nutrition into the whole body and helped by natural spirits. To be clear,

207 Albertus Magnus, De animalibus XIII tr. 1 c. 6 n. 46 (ed. Stadler, II, 913.15–19): "Pulsus autem cordis praecipue in homine accidit, eo quod passiones timoris et spei et gaudii et tristitiae magis movent cor hominis quam aliorum animalium: et ideo licet pulsus sit in omnibus animalibus, tamen differentiae pulsus non ita manifestantur in aliis sicut in homine."

208 Albertus Magnus, De animalibus XIII tr. 1 c. 6 n. 47, (ed. Stadler, II, 913.39–914.2).

209 Albertus Magnus, Quaestiones super de animalibus I q. 4 (Ed. Colon. XII, 82.17–21): "Animalia vero, quae non habent proprie cor nec pulmonem, sed aliquid materiale proportionale cordi sicut quidam pisces, attrahunt aquam propter temperamentum illius sicut habentia pulmonem aerem."

210 Albertus Magnus, De animalibus XIII tr. 1 c. 6 n. 45 (ed. Stadler, II, 912.34–38): "Animal autem carens sanguine non habet pulmonem, eo quod sufficit ei hoc quod evenit sibi de spiritus naturalis qui in ipso pulsat infrigidatione. Animal autem spirans infrigidatur per spiritum aëreum qui ingreditur in ipsum: propter quod etiam haec dicimus, pulmonem habere."

211 Albertus Magnus, De spiritu et respiratione I tr. 1 c. 7, ed. Auguste Borgnet, Paris 1894 (Ed. Paris. IX, 224a).

the vessels analogous to veins do transport vital heat in bloodless animals and plants, thus conflating into a single part (the vein-like channels or bundles) what in breathing animals is the function absolved both by veins and arteries, viz. the nourishing of (veins) and the spreading of a suitable amount of vital heat through the body (arteries) to enable the soul's activities, first and foremost the process of nutrition.[212] The main difference between animal species is given by the need for hotter animals to develop a specific functionality, viz. breathing, meant to cool down their greater amount of vital heat. In bloodless animals and plants, the cooling function is not a function of the living being produced by specific organs such as lungs or gills but, rather, a product of exogenous factors such as the surrounding climate. For this reason, these living beings do not have a need for the existence of arteries.[213]

According to Albert, lungs are the organ that represent the instrument able to bring about pulse by introducing spirit, pumped into the body, and respiration (*pulmo est flabellum et organice deserviens pulsui cordis et spirationi*). Respiration is meant as a cooling device for the heat of the vital faculty (*ad refrigerium virtutis vitalis qui est in corde*),[214] one which comes mainly from the heart and the liver. This vital faculty is that which enables all the other psychic activities to take place.[215] This position represents an attempt to fuse the Aris-

212 Albertus Magnus, De spiritu et respiratione I tr. 1 c. 9, ed. Auguste Borgnet, Paris 1894 (Ed. Paris. IX, 227a–b): "Cum enim actus vitae communiter inventus sit in omnibus animatis, et sit vita proportionaliter in eis perfectior et minus perfecta, oportet quod et principia et instrumenta vitae omnibus proportionaliter insint: et ideo licet non sit in plantis talis spiritus et alimenti distributio, sicut in habentibus cor et sanguinem, tamen necesse est quod calor naturalis in eo quod est fons et principium caloris, sicut radix est in plantis, aliquid de humido resolvat in spiritum in quo virtus vitae quae est in radice devecta et succum vegetet et praeparet, et totam plantam pulsando moveat ad vitae vigorem."

213 Albertus Magnus, Quaestiones super de animalibus III q. 3 (Ed. Colon. XII, 126.19–28): "Dicendum, quod arteriae sunt necessariae, quia ad conservationem animalis duo requiruntur, scilicet restauratio deperditi et continuatio vel conservatio acquisiti. Sed propter restaurationem deperditi ponuntur venae, quae sunt quietae et non pulsatiles, ut per eas sanguis sive nutrimentum ad singula membra deferatur. Sed propter conservationem acquisiti ponuntur arteriae, quae sunt venae pulsatiles, per quas virtus vitalis et calor naturalis et spiritus, sine quibus non fit operatio, a corde derivantur ad singula membra."

214 Albertus Magnus, De animalibus, XIII tr. 1 c. 6 n. 45 (ed. Stadler, II, 915.5–8).

215 Albertus Magnus, De animalibus, XX tr. 2 c. 3 n. 76 (ed. Stadler, II, 1313.14–21): "Ut autem expressius ista pateant, revocentur ad memoriam antedicta de virtute cordis, quoniam cordis virtus quae prima est in animali, dat omnibus aliis virtutes tam vegetandi quam sentiendi, et ad cor omnis alia virtus tam naturalis quam animalis habet referri. Similiter motum cordis qui secundum diastolen et systolen est, participat omne quod est in corpore : quia sine illo nec vitam haberet neque spiritum vitalem. Adhuc cor primam speciem confert nutrimento qua movetur ad formam cuiuslibet membri nutrimentum accipientis."

totelian teleology of respiration (cooling down vital heat), medical views on it (cooling down the heat from the vital faculty/vital spirits) and the medical correlation between the mechanism of pulse (systolic/diastolic motions of the heart) and respiration. Respiration, in the end, is necessary because of the high amount of vital heat in breathing animals,[216] and it appears to be correlated to pulsation. In the second book of the *De spiritu et respiratione*, the correlation is less specific than in medical texts, which explicate the relationship between inspiration-systole and expiration-diastole. Anyway, Albert argues that pulse has a life-giving purpose insofar as it pushes the spirit generated in the heart into the whole body (*spiritum suum qui generatur in ipso, propellit ad membra cum vita quam advenit*).[217] In this sense we can see how the lungs act as servant to the mechanism of pulse, an expression remained quite vague in *De animalibus* XIII, 6. The air drawn in the lungs via inspiration is key to produce pulsation and spread life into the body.

Bloodless animals and plants, by contrast, do not need breathing, for surrounding air already mitigates the vital heat intrinsic to living beings. While there are no explicit links between the cooling down of vital heat, outer air and spirit in plants, as in the case of bloodless animals, there is still some evidence which suggests so. In some passages of the *De natura loci*, though, Albert argues that plants which are transplanted from one climate to another may well perish because of changed biological conditions, among which surrounding climate looms large.[218] The biological rationale behind Albert's line of reasoning is quite clear: the effectiveness of vital functions in plants depends on both endogenous factors (vital heat intrinsic to the plant) and exogenous factors (climate, which mitigates the plant's heat). If one of these two factors drastically changes, the delicate thermic balance so crucial to the plant's survival is jeopardized, if not irreversibly compromised. Among the activities of the vital faculty feature, together with respiration, pulse. In his *Quaestiones super de animal-*

216 Albertus Magnus, De animalibus, XIII tr. 1 c. 6 n. 52 (ed. Stadler, II, 915.24–31): "Oportet enim interiora animalis esse sicut fornacem, in qua fit calor naturalis: et oportet quod sit ille calor conservatus in hiis duobus membris (*sc.* heart and liver): eo quod calor ille regit totum corpus, sicut princeps regit regionem quae est regnum ipsius. Epar autem est ad coquendum et digerendum. Omne igitur animal habens sanguinem, hiis duobus indiget membris: propter quod omnia huiusmodi animalia cor habent et epar. Animal autem spirans tertio indiget membro quod est pulmo."

217 Albertus Magnus, De spiritu et respiratione II tr. 2 c. 3, ed. Auguste Borgnet, Paris 1894 (Ed. Paris. IX, 249a).

218 Albertus Magnus, De natura loci tr. 1 c. 1 (Ed. Colon. V/2, 2.4–16); tr. 1, c. 3 (Ed. Colon. V/2, 4.15–23); tr. 2 c. 2 (Ed. Colon. V/2, 24.53–69).

ibus, Albert admits that there are two kinds of pulse: one proper to animals with lungs and one proper to animals without them but which still have a heart.[219]

According to Albert, the principle of the lungs' motion is represented by the heart, which dilates when the lungs draw in air. The cardiocentric nature of the motion of lungs reinstates once again the primacy of the heart over other bodily members. Conceptually, the heart proves to be the receptacle of the form of life which, once received, provides the vital spirit with life-giving function and natural spirit with its nourishing function. It represents the point of junction in the human body between Albert's complex emanative psychology and detailed physiology. These considerations on the role of the heart, vital spirits and natural spirits can well be used to achieve a better understanding of the more difficult psychophysiology of life in bloodless animals.

5.6 *Bloodless Animals and Connate Spirit* (spiritus complantatus)

According to Albert, it is impossible for natural philosophers to distinguish the kind of spirit performing vital and nutritive functions in bloodless animals.[220] In the second book of his *De spiritu et respiratione*, Albert devotes an entire *digressio* to deal *ex professo* with what Aristotle had called "connate spirit" (*spiritus complantatus*), as in the Graeco-Latin translation of Aristotle's *De somno*.[221] In this section, Albert draws a comparison which should help us better understand similarities and differences between spirit produced by respiration and connate spirit.

According to Albert, connate spirit is found in animals with very little or no blood, which have less vital heat than other animals and are therefore cooler near the heart (*oportet quod habeant frigiditatem circa locum vitae et digestionis*). Since they have little heat, their moisture is phlegmatic and thick and, for this reason, respiration is simply not suitable (*non confertur*) for them, for it would cool down their vital heat excessively and slowly freeze them to death. This is the case, on earth, for snakes and reptiles and, in water, for the beaver

219 Albertus Magnus, Quaestiones super de animalibus XIII q. 12 (Ed. Colon. XII, 244.61–72).

220 Albertus Magnus, De spiritu et respiratione I tr. 2 c. 6, ed. Auguste Borgnet, Paris 1894 (Ed. Paris. IX, 239b).

221 Aristotle, De somno 3 (456a30–b6); Aristoteles Latinus, De somno c. 2 (ed. Drossaart Lulofs 1943, 6.8–10): "Non habentibus vero sanguinem et intersectis et non suscipientibus spiritum in eo quod est proportionaliter *conplantatus spiritus* dum exsufflat et collocatus videtur." The translation *conplantatus* for the Greek σύμφυτον appears only in this Graeco-Latin translation. In the rest of the zoology, the term σύμφυτον is translated as *naturalis* as a result of the Arabic "natural spirit" (*rūḥ ṭabāʿyī*): see: Aristoteles Latinus, De generatione animalium (Aristotle Semitico-Latinus 5/3), ed. van Oppenraij 1992, Index Arabo-Latinus, 337.

(*castor*), the sea dog (*lupus marinus*) and the seal (*canis marinus*).[222] Albert distinguishes these animals from other species of breathing marine animals such as dolphins, whales and others that would die if they tried not to breathe and just live underwater like other sea-dwellers.[223] At this point, one may wonder about the difference between the spirit supplied by respiration and connate spirit. To explain similarities and differences between the two concepts of spirit, Albert argues that such difference is fivefold:

1. Connate (*complantatus*) *pneuma* arises by pulsating from the inside of the individual, while breathing-related spirit is a product of respiration, hence of drawn-in air;
2. The motion of connate spirit begins in the heart, whereas non-connate spirit is drawn in from the motion of lungs;
3. While attraction and expulsion of connate spirit are balanced, respiration, needed by those animals also for their voice, expels more air than what it actually draws in;
4. The goal of respiration is to draw spirit into the lungs and make it suitable for being received in the rest of the living being, especially in the heart. In bloodless animals, connate spirit already can spread into the body without the need for previous cooling down, since their amount of heat is smaller. The same does not hold true for breathing animals.
5. Spirit from outer air purges lungs from corrupted humors, whereas connate *pneuma* simply pulsates and by doing so gives life to the body itself.[224]

---

222 Albertus Magnus, De spiritu et respiratione II tr. 1 c. 1, ed. Auguste Borgnet, Paris 1894 (Ed. Paris. IX, 242a): "Relicto igitur quod cum frigida sint illa, ut dictum est, et non conferret eis respiratio et inspiratio. Ea autem quae semper sunt aquatica, sunt talia: et ideo omnia aut pauci, aut nullius sunt sanguinis: et est sanguinis eorum frigidus et aquosus. Ea autem quae diu quidem sub aqua, non tamen continue sub aqua esse possunt, sive sint volatilia, sive gressibilia, ut lucus, castor, et lupus marinus, et canis marinus, sive etiam reptilia, ut quidam serpentes, proportionaliter procul dubio etiam oportet quod habeant frigiditatem circa locum vitae et digestionis."

223 Albertus Magnus, De spiritu et respiratione II tr. 1 c. 1, ed. Auguste Borgnet, Paris 1894 (Ed. Paris. IX, 242b): "Animalia autem continue respirantia quae in aquis cito submerguntur, calida sunt valde circa loca digestionis et vitae, et sincero et frigido aere indigent ad refrigerium pectoris, quem cum non habent, moriuntur corde incendente ex calore suo naturali, et ex eo qui ex motu suo nascitur humidum quod est in ipso, et spiritum consumente utrumque, et ipsum exstinguitur, cum sui caloris destituatur subiecto et vehiculo."

224 Albertus Magnus, De spiritu et respiratione II tr. 2 c. 1, ed. Auguste Borgnet, Paris 1894 (Ed. Paris. IX, 246b).

In addition to these five differences, there are common features between connate spirit and spirit in the proper sense of the word. According to Albert, spirit derives materially from nutrimental and radical moisture, exactly as in the case of breathing animals.[225] Like other animals, bloodless animals or animals with little blood present the same psychological scheme (being-life-soul activities) of all living beings, and analogous physiological structure, for they possess an analogue to the heart and to the blood, and analogous parts to the veins.[226] Because of their different physiological makeup (*organizatio*), their life-related activities prove to be much more difficult to detect than in bigger and hotter, breathing animals, and much more similar, by contrast, to plants.

Conclusions

Spirit represents one of the cornerstones of the psychophysiology of 13th century authors already from the theological treatises on the human soul. Together with the primary qualities, spirit was employed to explain the concrete instantiation of the different psychic powers in the human body and the body-soul union. Its role as life-giving substance, instrument of the soul and vehicle of its powers, was developed by Albert within an Avicennian psychology of faculty theory based on the Neoplatonic notion of *fluxus*. Albert's psychology of the soul's faculties differs notably from the traditional doctrine of the intermediaries and the trend toward a pluralism of substances in living beings developed by Philip the Chancellor at Paris, where Albert himself studied theology and wrote his *De homine*. This strong influence of the Neoplatonic metaphysics of *fluxus* in a psychological context, mainly aimed at stressing the unity of the soul, will remain a consistent pattern throughout Albert's intellectual production and will distance himself from many of his contemporaries in Paris at the time. At that time, the only author who could match Albert's deep interest into psychophysiology and his strong attraction to the Avicennian emanationist model of psychology was Peter of Spain.

---

225 Albertus Magnus, De spiritu et respiratione II tr. 1 c. 2, ed. Auguste Borgnet, Paris 1894 (Ed. Paris. IX, 243a–b): "Videtur tamen dicendum, quod talia spiritum habent ex natura sibi innatum et complasmatum, quem complantatum vocat Aristoteles. Spiritus autem ex natura complantatus est, qui ex humido cibali et radicali nascitur, et iuxta diaphragma in regione vacua retinetur."

226 Albert also deals *ex professo* with bloodless, non-breathing animals in Albertus Magnus, De animalibus, IV (ed. Stadler, I, 357–406).

From a psychological point of view, Albert's account of existence in living beings, based on a combination between *De causis* and *De motu cordis*, maintains that life is an act of the soul inasmuch as it represents a quasi-effect of the soul flowing from it (*in fluxu*), received in the body once this body possesses a suitable disposition. By distinguishing at a merely logical level the concepts of being/life/activity among the soul's powers, Albert maintains this psychological structure of the living being consistently throughout his intellectual production. Strongly permeated by Arabic Neoplatonism and medical sources, Albert establishes the metaphysical and physiological conditions for the existence of life in living beings which can be separated and is prior to the exercise of any activity of the soul. This reading, strongly based on Alfred's *De motu cordis*, is also held by the author of the second part of the *Summa Halensis*, which, on the other hand, displays key doctrinal differences (plurality of forms vs unity of the substantial form) compared to Albert's psychology. According to Albert, just like being (form/soul) is prior to life and life is prior to nutrition and sensation, so the activities of the vital spirits (pulse, respiration) are prior to nutrition and represent the concrete physiological instantiation of life as an *actus* of the soul according to itself and not its powers.

This primary association of the vital spirits to life, pulse and respiration can be observed already in Albert's *De homine* and *Commentary on the Sentences* in a way that seems clearly inspired by his medico-philosophical sources such as Costa ben Luca, Nemesius, Alfred of Shareshell and Avicenna. All in all, spirit in the *De homine* and the *Commentary on the Sentences* represents a more anthropocentric notion aimed at the concrete and physiological explanation of the soul's activities in the body. Still, Albert's early works provide a valid physiological and metaphysical framework for understanding his mature spirit theory for (at least) a couple of reasons. In the first place, the general architecture of the psychology of living beings, based on an Avicennian-Boethian interpretation of Aristotle's faculty theory according to an ontological distinction between soul's essence and soul's faculties, proves to stand the test of time and be in keeping with Albert's Aristotelian paraphrases. Second, this general architecture of faculty theory is crucial to understanding the relationship between, on the one hand, Albert's theory of the soul, which distinguishes between the human immortal soul and the mortal souls of plants and animals, and the different faculties of the soul, and on the other hand the role of spirit as instrument of the soul or vehicle of the soul's faculties in the broader context of the psychophysiology of the living being.

While it is indeed oriented towards a human perspective, for it is mainly concerned with the human soul, Albert's early theology already outlines the complex metaphysic of life, divided into soul, life '*in fluxu ab anima ipsa*' and life received in the body itself, first developed in the *De quattuor coaequaevis*, *Commentary on the Sentences* and later replicated in his *De anima*, *De morte et vita*, *De causis et processu universitatis*, viz. Albert's paraphrase of the *De causis*, and *De spiritu et respiratione*. This latter is a crucial work for Albert's spirit theory, as it contains a combination of psychological emanationism with a detailed account of the elemental properties of spirit and medical doctrines strongly influenced, as we have seen, by Avicenna's *De medicinibus cordialibus*, a hitherto neglected source.

The *De spiritu et respiratione* is the main work where Albert explains *ex professo* the role of spirit in the complex psychophysiology of life, hinged on the idea that spirit represents the physical receptacle of the emanated form of life. Once that spirit possesses a suitable material configuration or preparation, it is ready to welcome the form of life and spread into the whole body, thus acting as the first bodily junction between the abstract form of life and the physical body. The resulting picture is that, according to Albert, life is logically distinguished into four parts: life in the soul, life flowing from the soul, life received in the spirit, life received in the body. Because of its function as instrument of the soul or vehicle of life, spirit spreads enough vital heat in the body to enable all life-related operations. Because of the cooling down of the vital heat through respiration or through external cooling (as in bloodless animals and plants), vital spirit both avoids the self-destruction of the living being and provides the right balance of heat necessary for nutrition to take place. To do so, vital spirit causes certain phenomena already, viz. pulse and—where possible—respiration, which are even prior to nutrition, as Albert argues recalling the 'experience' of living animals with corrupted nutritive organs. The conclusion of this apparently empirical (but, most likely, bookish) argument is that the subsistence of life in living beings is rooted on the combined action of vital and natural spirits, the minimal conditions for life's preservation.

The main doctrinal point of the *De spiritu et respiratione*, though, is represented by Albert's introduction of the concept of spirit in plants. Perhaps inspired by passages scattered in previous authors (the *De plantis*, less likely the *De motu cordis* and David of Dinant's *Quaternuli*) and by the traditionally medical concept of natural spirits (byproduct of the action of heat over moisture, independent of any breathing-related activity), Albert seamlessly weaves together the physiology of life in plants, animals and humans relying on the concept of spirit as "vehicle of life". In his attempt to build a unified pattern of psychophysiology, Albert construed a close comparison between plant and animal physiological structure, emphasizing the similarities between the same life-related processes, as in the case of nutrition, even though this account leads to some conceptual tensions between his *Parva Naturalia* (*De nutrimento et nutrito*) and botany (*De vegetabilibus*) on the account of digestion.

His complex 'universal physiology'[227] tries to construe a harmony between medical and philosophical doctrines which seems to be evident in his handling of lungs, teleology of pulse and respiration and, most relevantly, cardiac anatomy. Blending Aristotle with Avicenna, Albert creates a homogeneous model of *scala naturae* in which all living beings possess a univocal instrument (spirit) or vehicle of the soul's faculties *qua* carrier of vital heat. Spirit is considered an instrument only when it is connected to an active and formal principle (the soul), a causal principle of life, while it can only be regarded as a vehicle of life itself. This is because life, according to Albert, flows from the soul as a quasi-effect of it and, therefore, is not placed on the same level of active causality with the soul in Albert's Neoplatonic and emanative psychology. Since life is not capable of actively using the spirit itself as its instrument, it is carried by it into the whole body, as something is passively carried by a vehicle with a clear starting point in the body, the heart or the analogous to it. Those faculties are different from the soul in the same manner that life is: they are the product of the soul's emanative and life-giving capacity and not primary causal agents of their processes (nutrition, sensation, motion). Intellectual activities are, by contrast, essentially independent of the body but can be still affected by impairments in the activities of the sensitive soul.

The role of spirit as the proper vehicle of life further strengthens, on the one hand, its connection to the causal chain of emanation starting from the soul and, on the other hand, its dependence on correct bodily dispositions for life to come about. In Avicenna's language, this idea would come close to "appropriate preparation" and, in Albert's language, to *inchoatio formae*, both crucial yet very different concepts for explaining the coming-to-be of the individual. The material composition of spirit depends, in keeping with Aristotle, on the amount of blood and vital heat, much greater in animals endowed with lungs than in non-breathing or bloodless animals or plants. Even though these different models of living beings may not share the *same* physiological makeup and functions, they do possess *analogous* sets of material substances (vital heat, spirit, blood) and organs (heart, veins, arteries). These organic structure of living beings are purportedly designed to enable the reception of the form of life received in the body, more specifically in the heart, at the moment of birth, and to enable subsequent psychic activities to take place.

227 Cadden, Albertus Magnus' Universal Physiology, 231–248.

CHAPTER 3

Spirit, Generation, and Natural Teleology—Part 1: Human Generation

The entry of the Aristotelian philosophy of nature into the Late Middle Ages (12th–13th century) contributed to a remarkable shift in the consideration of nature and natural phenomena. Over the previous centuries in the Early Middle Ages, nature had been mainly viewed as a sign of God's providential creating action and not as an object of a per se legitimate rational and scientific inquiry with its own laws and principles. According to a famous Medieval analogy, any investigation into natural phenomena would have been analogous to the study of the traces left by a bear on the snow. These traces acquired a value insofar as they allowed to uncover the providential action of God, which becomes visible by transcending the surface of natural phenomena. Aristotle's philosophy of nature gradually changed this paradigm by promulgating the famous motto "nature does nothing in vain" (*De anima*, III, 12 430a31). The spread of Aristotle's philosophy of nature and its study of the laws that rule it (i.e. "natural law") legitimated a rational inquiry into the natural word independent of any allegorical or symbolic value derived from divinity. In the Latin Middle Ages, the motto *natura nihil facit frustra*[1]—often found with different and nuanced wordings—was invoked to explain natural teleology, viz. how nature had to be conceived as a set of ordered causes explainable in philosophical terms and not merely as a sign of God's often imperscrutable providence.

This orientation is well exemplified by the famous adage *opus naturae est opus intelligentiae*, coined by Albert the Great and spread by his pupil Thomas Aquinas. This adage, though, goes well beyond Aristotelian natural teleology by emphasizing that the teleological processes at play in nature are the work of a superior intelligence, viz. God. This axiom makes its early appearance in Albert's *De bono* and the second book from his *Commentary on the Sentences*. There it is,[2] tellingly, attributed to the father of natural teleology, viz.

1 Auctoritates Aristotelis, n. 168 (ed. Hamesse 1974, 188.83–84).

2 Ludwig Hödl, "*Opus naturae est opus intelligentiae*: Ein neuplatonisches Axiom im aristotelischen Verstandnis des Albertus Magnus," in *Averroismus im Mittelalter und Renaissance*, ed. Friedrich Niewöhner/Loris Sturlese, Zurich 1994, 133.

 DOI:10.1163/9789004706743_005

Aristotle.[3] Later on, Albert would ascribe the paternity of this axiom to Averroes.[4]

Among the marvels of the natural world, the process of generation counted certainly among the most hotly debated, for many reasons. In their comments to Aristotle and in their treatises on the human soul, Medieval authors had to deal with three main sets of problems. The first concerned mainly psychology and revolved around the causal explanation of the origin of the soul and the temporal determination of the beginning of life, viz. the days which intercur between successful fecundation and the animation (or ensoulment) of the embryo. The latter issue was a question with obvious and delicate ethical and legal consequences, especially in discussions on abortion. On the other hand, the second set of intricate questions concerned the physiology of fecundity and embryology, viz. the determination of the causal relationship between the different agents at play in the process such as semen, menstrual blood, spirits and humors. The third and last problem, more specific to zoological commentaries, was represented by the Galenic-Aristotelian quarrel over the origin of veins from the heart (Aristotle) or the liver (Galen). The arguments inherited from these quarrels and transmitted by Avicenna and Averroes built much of the background of Medieval discussions on the psychophysiology of generation.

Like his contemporaries, Albert sought to outline the origin of the soul and the material explanation of the process of conception and, more generally, of embryology in terms of laws of natural causation and material substances, among which spirit. Moreover, he addressed the theological and juridical concerns revolving around the theme of induced abortion, prompted by the famous passage in Exodus (21, 22–25).[5] In his approach to embryology and, more generally, of the generation of ensouled beings, Albert tried to be staunchly faithful to the key tenets of an Aristotelian philosophy of nature (hylomorphism) and the psychological structure of living beings (cardiocentrism). To succeed in his task, he strongly relied on his key Arabic sources, Avicenna and Averroes. As we shall see, these attempts produce appreciable results, even if sometimes with relevant tensions, as in the case of the interac-

3 James A. Weisheipl, "The Axiom 'Opus naturae est opus intelligentiae' and its Origins," in *Albertus Magnus—Doctor Universalis 1280/1980*, ed. Gerbert Meyer OP/Albert Zimmermann, Mainz 1980, 451.

4 Hödl, "*Opus naturae est opus intelligentiae*," 133.

5 Exodus, 22, 21–24: "If men quarrel, and one strike a woman with child, and she miscarry indeed, but live herself: he shall be answerable for so much damage as the woman's husband shall require, and as arbiters shall award. But if her death ensues thereupon, he shall render life for life. Eye for eye, tooth for tooth, hand for hand, foot for foot [...]."

tion between the fluids at play (masculine semen, feminine semen, menstrual blood, radical and seminal moisture) in generation.

Albert wrote extensively on generation. For this reason, it is nearly impossible to synthesize his take on the causal link between spirit and generation in a single chapter without its being of mammoth proportions. For this reason, we will break down Albert's view on the role of spirits in generation into two sub-chapters (3, 4). In the first (3), we will highlight 1) the role of spirit in the generation of humans and of viviparous animals, of which humans constitute the most perfect example, and 2) the origin of life in relation to the sensitive issue of induced abortion. Additionally, we shall assess 3) the unintended outcomes of generation, viz. deformities, which Albert calls "errors of nature" (*peccata naturae*), and less frequent phenomena such as the birthing of twins. In chapter 4, we will consider the generation of less perfect animals such as egg-laying animals, plants and spontaneous generation. On the one hand, we will reconstruct how the notion of spirit connects Albert's psychophysiology of generation to his natural teleology, synthetized by his axiom *opus naturae est opus intelligentiae* and explicitly linked to spirit in the *De unitate intellectus*: "opus naturae est opus intelligentiae et spiritus qui est in semine hominis, est intelligentiae instrumentum, vel etiam causae primae."[6] The aim of this chapter will be to shed light on the meaning of this connection, comparing the zoocentric understanding of spirit in Albert's early works to his more extensive understanding typical of the Aristotelian paraphrases.

Enlarging on the excellent studies by Adam Takahashi on spirit in Albert's theory of human and animal generation,[7] the two subsections of this chapter will pinpoint how spirit represents the material vehicle of formative power in both his early and mature works. While Takahashi focuses only on the latter and studies human, animal, and mineral generation, we shall try complement his account by focusing on Albert's early theological works and comparing his study of generation in humans, animals, and plants, the latter being the only instance of living beings left out by Takahashi. We shall leave aside the case of minerals, which do not possess spirit. Throughout this chapter, we shall endeavor to show how the notion of formative power is key to understanding the manifold types and outcomes of generation. An accurate explanation of the concept of formative power is strictly connected to the interpretation of

---

6 Albertus Magnus, De unitate intellectus pars 2 (Ed. Colon. XVII/1, 16.89–17.1): "[…] opus naturae est opus intelligentiae et spiritus, qui est in semine hominis, est intelligentiae instrumentum, vel etiam causae primae."

7 Adam Takahashi, *Interpreting Aristotle's Cosmos: Albert the Great as a Reader of Averroes*, Nijmegen, 2017, 165–187; Adam Takahashi, "Nature, Formative Power and Intellect in the Natural Philosophy of Albert the Great," *Early Science and Medicine* 13 (2008) 465–468.

Aristotle's statement that "man begets man and the sun" (*Phys.*, II, 2, 194b13),[8] which underlies the doctrinal explanation of all generation-related phenomena. The link between formative power and what has been called the principle of 'synonymy' or 'formal replication'[9] represents the key to understanding the material (spirit, semen, menstrual blood, climatic influences etc.) and immaterial (God, soul) causes involved in the process of generation and which, in Albert's thought, stem from a Neoplatonized reading of Aristotle that owes much to Avicenna and Averroes.

1 "The *pneuma* and the Nature within It": Aristotle on Generation

"Nature does nothing in vain" (DA III.9, 432b20) represents perhaps the expression that best and most famously sums up Aristotle's teleology of nature. Unlike his predecessors, Aristotle decidedly denied that the only factors at play in generation-related—and, in fact, all natural—processes were represented by their material and efficient causes only. The formal causes, but most importantly, that for the sake of which (τὸ τοῦ ἕνεκα) these processes occurred, represented an essential feature of natural phenomena as much as efficient and natural causation.[10] The mingling of menstrual blood and masculine semen explains how and that the woman becomes pregnant, but certainly does not explain that for the sake of which this occurs, viz. the preservation of the species. To do that, it is necessary to refer to the key role of final causation in natural processes according to the concepts of end (τέλος) and chance (τύχη). This is because, in Aristotle's view, some animals appear to be generated through usual sexual intercourse and why others, especially marine animals, appear to be the product of asexual or spontaneous (ἀπὸ ταὐτομάτου) generation, although the latter will prove to be anything but an unproblematic concept.

In his explanation of generation, Aristotle takes male semen to be the formal and efficient cause, the female menstrual blood to be its material cause, and the preservation of the species and the plant-, animal-, or human-to-be its final cause. The primary cause in generation is indeed nature, which performs

8 Auctoritates Aristotelis, Physica, II, n. 65 (ed. Hamesse 1974, 145): "Homo generat hominem et sol."

9 The expression "formal replication" is borrowed from: James Lennox, "Teleology, Chance, and Aristotle's Theory of Spontaneous Generation," *Journal of the History of Philosophy* 20 (1982/3), 226.

10 David Charles, "Aristotle on Hypothetical Necessity and Irreducibility," *Pacific Philosophical Quarterly* 69 (1988) 1–53.

purposeful, end-directed processes which do not need previous deliberation.[11] Consistently with this view, we can observe how Aristotle takes that, despite not being produced by a voluntary act, the mixing of masculine semen and menstrual blood can reflect an end-directed process. This time, the final cause is the species' preservation. Aristotelian scholarship has stressed the many differences between the crafting of an artifact and biological reproduction, and these can neither be neither fully exhausted nor completely overlooked here.[12] On the one hand, natural processes of generation presuppose the intrinsic nature of the source of substantial change (masculine semen), while artificial processes are generated from extrinsic causes inducing substantial changes (the craftsman or carpenter). Matter in natural processes is already arranged for the process, while it needs to be properly put together to produce the artifact. Both processes have a common end, represented by the physical form and the teleological functions of the artifact or of the living being, which transcend the merely material features of the final product. By contrast, the carpenter does not materially figure in the composition of their artifact. In the interpretation given by Sophia Connell, which builds the backbone of this whole Aristotelian section on generation, the carpenter represents the principle that guides the production or motion of such artifact towards a specific end, viz. the new living being.[13]

According to Aristotle, generation usually occurs between two individuals of the same species; cases of hybrids like the mule are very likely to be sterile and thus uncapable of preserving the species of neither parent.[14] Aristotle argues in the *Physics* that this is the reason why the principle which guides generation is that the species is preserved via reproduction between individuals of the same species. In other words, "man begets man". Being identical in species,

---

11 Aristotle, Physics II 8 (199b26–31).

12 For a more detailed discussion of the art-nature analogy and its different interpretations, see at least: Connell, Aristotle on Female Animals, 123–127 and bibliography therein.

13 On the carpenter metaphor, see: Connell, *Aristotle on Female Animals*, 183. Another and less fortunate metaphor which Aristotle uses to explain the role of feminine matter in generation is that of automatic puppets. These puppets are created by an external source (viz. the craftsman), which can trigger their action from the outside, creating a chain of causes. Sophia Connell has shown how this comparison has the point of likening the female contribution that is being likened to a material substance (as the automatic puppet) poised to react in complex and specific ways to a stimulus: Aristotle, De generatione animalium II 1 (734b5–15). On the automatic puppets metaphor, see: Connell, *Aristotle on Female Animals*, 159–160, 200.

14 Jocelyn Groisard, "Hybridity and Sterility in Aristotle's Generation of Animals," in *Aristotle's Generation of Animals. A Critical Guide* (Cambridge Critical Guides), ed. Andrea Falcon/David Lefébvre, Cambridge 2018, 153–170.

or at least compatible (as in the case of mules), does not exhaust the set of necessary and sufficient conditions for generation to take place, though. Since the full Aristotelian formula reads "man begets man and the sun", it is necessary to understand whether Aristotle thinks that seminal heat is sufficient for a successful process of generation to occur. This is precisely where the notion of *pneuma* comes into our discussion.

Aristotle regards semen as the formal and efficient cause in generation because of its being the instrument of the father's soul.[15] Moreover, he considered menstrual blood as playing the role of material cause of generation, as he was unaware of the existence of ovaries in females. To explain how masculine semen and feminine menstrual blood interact, he used the famous metaphor of the process of cheese-making: masculine semen informs menstrual blood in the same manner in which rennet makes the substance of milk set. In both cases, this happens without the formal principle entering the material composition of the substrate of its shaping, formative action.[16] The relation between *pneuma* and the principles of generation, especially masculine semen, emerges from an equally famous and controversial passage from Aristotle's *De generatione animalium*:

> (1) Now, it is likely that every faculty of soul has to do with a body, different from, and more divine than, the so-called four elements. Souls differ from one another as being more noble and ignoble, so also the nature of this body differs. (2) For in all cases there is in semen that which makes it fertile, the so-called hot. This is not fire or any other such power, but that

15 Aristotle, De generatione animalium I 22 (730b8–32); Aristotle, De generatione animalium II 1 (734b20–735a29); II 4–5 (740b25–741a14). Despite the analogical strength of this metaphor, the irreducible difference between art and nature is that in nature the source of change in generation is an inherent, intrinsic cause, while in art the source of change is extrinsic; see: Aristotle, De generatione animalium II 1 (735a4). In Aristotle's as in the following accounts, the word sperm and semen are often used interchangeably, even though in our eyes this causes a tension between the four different aspects of semen: (1) semen as spermatic fluid, (2) semen as the singular gamete, (3) semen as the entire genetic heritage of each of the two parents and (4) semen as the single chromosome charged with sex differentiation.

16 Aristotle, De generatione animalium II 4 (739b21–26): "The action of the semen of the male in setting the female's secretion in the uterus is similar to that of rennet upon milk. Rennet is milk which contains vital heat, as semen does, and this integrates the homogeneous substance and makes it set. As the nature of milk and the menstrual fluid is one and the same, the action of the semen upon the substance of the menstrual fluid is the same as that of rennet upon milk." English Translation by: Connell, *Aristotle on Female Animals*, 185.

> which is enclosed in the semen and in the foamy pneuma, i.e. the nature in the pneuma, analogous to the element of the stars. (3) Because of this, fire generates no animal, and we do not see anything taking shape either in wet or dry fire. But the heat of the sun and that of animals [does generate], not only through semen, but also any chance residue different in its nature; this also has the principle of life. From these things it is clear that the principle of heat in animals is neither fire nor does it come from fire. (4) The body of the semen, in which the principle of soul travels, on the one hand being separate from body, i.e. for those animals that enclose something divine [within them] (the so-called intellect is such), and on the other hand being inseparable: this body of the semen dissolves and evaporates, having a moist and watery nature.[17]

In this context, *pneuma* appears as the material substrate of the vital heat necessary for generation and embryonic formation. Both fig-juice curdling milk and masculine semen need a tool that is suitable to form cheese and embryonic matter. In both cases, the agents use the instrument to shape the respective matter remaining extrinsic to it and triggering an intrinsic process of substantial change which lasts until the embryo-to-be reaches the degree of bodily organization aimed at in the process.[18] The role of spirit as a carrier of vital heat and, which is in turn a key to generation and the conservation of the species, can explain, at least according to Freudenthal's explanation, why *pneuma* took on the adjective of "divine body" (θεῖον σῶμα) elsewhere in the Aristotelian corpus. Since "divine" stands for something eternal (like the animal species generation is supposed to preserve), it makes sense to define *pneuma* as "divine".[19] It must be remembered, though, that *pneuma* is not an entity capable of producing life in itself, but only when directed by nature—as we shall see in the case of spontaneous generation—if not by the combination of nature and the soul of the father.[20]

According to Aristotle, life comes about at different moment in different species: 40 days through pregnancy in humans, 20 days in oviparous animals (chicken especially), while there are fewer clear indications of the beginning of animation in plants. In the Middle Ages, many readers of Aristotelian

---

17 Aristotle, De generatione animalium II 3 (736b30–737a17). Translation by: Connell, *Aristotle on Female Animals*, 183.

18 Aristotle, De generatione animalium I 21 (729b19); I 22 (730b11); II 3 (737a8–16); Connell, *Aristotle on Female Animals*, 301.

19 Freudenthal, *Aristotle's Theory of Material Substance*, 106–114.

20 Connell, *Aristotle on Female Animals*, 220.

psychology—among which Albert was one—understood the moment of the full formation of plants and animals as the moment at which the soul was educed from matter, yet not extrinsically created by an extrinsic principle, as in Plato. By contrast, intellect *qua* capacity independent of the body was considered as not having been educed from matter but, rather, as entering the embryo "from the outside" (*intellectus ex extrinseco*/νοῦς θύραθεν) following a creationist interpretation of this Aristotelian passage.[21] Although this was quite a common reading of Aristotelian embryology, it is not undisputable neither as respects the divine origin of intellect (which Aristotle would have not subscribed to) nor in terms of coming "from the outside," which has been challenged by Enrico Berti.[22]

A second issue in Aristotelian zoology is represented by the clause of the above-mentioned passage: "the natural principle in the pneuma, being analogous to the element of the stars". Now it may be reasonable to wonder, given that Aristotle takes *pneuma* to grant the survival of the species by carrying vital heat, whether this 'divine' nature of *pneuma* comes from a purely sublunar elemental characterization or whether there is more to it and this 'more' is somehow connected to the element of the stars, viz. *aither*. Given this passage, one may start to wonder whether seminal heat is composed of purely material, seminal heat or does it need to be complemented somehow by the function and nature of *aither*. When discussing this issue in book IV of the *Meteorologica*, Aristotle favors an explanation of the material composition of seminal heat which rests on the purely sublunar components of heat and moisture. There is no indication of any influence of heat produced by the motion of heavenly bodies. The heat of the surrounding environment can rather appear as a concomitant cause which offers a favorable environment to reproduction, but does not enter the material composition of seminal heat itself. In a like manner, Sophia Connell has stressed how there is no need to posit a strong causal dependence between *aither* and *pneuma* in Aristotelian zoology: the former does not provide the latter with its life-giving force, which is rather due to nature and soul.[23] In Aristotle's view, the sun constitutes the principle granting subsistence of life in the sublunar realm, posing favorable conditions but not entering directly on the elemental composition of seminal heat carried by the

21 Aristotle, De generatione animalium, II 3 (736b27–29).

22 Enrico Berti, "Natura e generazione degli animali in Aristotele," Kriterion. *Revista de filosofia* 51 (2010), 489–504. According to Berti, the intellectual faculty derives from *pneuma* contained in the sperm once it reaches the brain (De generatione animalium II 3, 737a7–16).

23 Connell, *Aristotle on Female Animals*, 190.

pneuma. By contrast, this will be a feature typical of the Arabic philosophical tradition which will be so influential in Albert's thought.

Pneuma is therefore the instrument through which nature and soul can craft embryonic matter by carrying the vital heat seated in the semen during generation. In itself, spirit is a cause that becomes such only because of the guiding role of nature and soul. This tenet will hold true for all subsequent authors (Avicenna, Averroes, Albert). The action of spirits and their elemental composition, especially in terms of vital heat, is nevertheless crucial, and it determines the concrete morphological structure of the embryo, viz. the determination of the morphological traits and sex of the embryo. The endowment of the child-to-be is determined by the interaction of the causal agents at play in generation (semen, *pneuma*, menstrual blood) which now need be examined.

1.1 *Embryonic Morphology and Sex Differentiation*

Aristotle considers masculine semen to be the formal agent, and menstrual blood the patient and material cause, within his hylomorphic understanding of generation. In his account of generation, Aristotle maintains that active agency exists only on behalf of the father through his semen. Feminine menstrual blood has, by contrast, no (active) say either in the determination the offspring's sex or in its resemblance to its parents if the vital heat of the father is able to shape embryonic matter. This is because Aristotle's account of generation does not treat the masculine and feminine contributions to birth equally, for he often defines females as "disabled"[24] or "like a natural mutilation".[25] These are strong words, which need to be understood against Aristotle's general outlook on generation. As he himself states, females are as necessary to preserving the species as males.[26] Rather, Jessica Gelber has shown that the claim that women are "disabled" means that they are less able to concoct their digestive residues because of their colder blend and this, in the end, produces a material substance (menstrual blood) that is more impure than masculine semen.[27]

This partial rectification of Aristotle's statements does not change his consideration of females as a deviation from the teleology of generation, which

24 Aristotle, De generatione animalium II 3 (737a28).

25 Aristotle, De generatione animalium IV 7 (775a15–16).

26 Jessica Gelber, "Females in Aristotle's Embryology," in *Aristotle's Generation of Animals: A Critical Guide*, ed. Andrea Falcon and David Lefébvre, Cambridge 2018, 172. Aristotle, *De generatione animalium*, II 1 (732a1–3). On the need for females in the perpetuation of the species: Aristotle, *De generatione animalium* II 1 (731b8–24).

27 Gelber, "Females in Aristotle's Embryology," 172–173.

presupposes the birth of masculine offspring as the best possible outcome. Femaleness is not, strictly speaking, a defect of nature but, in the words of Sophia Connell, "is a sort of second-ranking goal for an individual animal".[28] Moreover, when Aristotle examines the mechanism of sex differentiation and the resemblance of offspring to their parents according to the so-called principle of proportion (συμμετρία), he does so by emphasizing the active, formal and efficient role played by the masculine semen.[29] While generation is codependent on both principles, those principles are not treated on an equal footing, for the masculine principle is the capacity to act, while the feminine principle is the principle of being acted upon.[30]

Semen or sperm is, according to Aristotle, the active, formal and efficient principle of generation in that it possesses a higher degree of heat than menstrual blood and is directly correlated to sex determination, since men are hotter than women in their blend. According to him, the resemblance of the offspring to their mother (or its female relatives or ancestors) can be explained in one way: feminine matter resists the formative action of the masculine semen. The resistance of feminine matter to the formative action of masculine semen can therefore explain why we have daughters, sons resembling their mothers, or daughters resembling their mothers or female ancestors. This fact is explained, according to Aristotle, in terms of resistance offered by the matter provided by the mother. This depends also on the efficiency in the agency of the semen of the father, which possesses corresponding movements and potencies (κινήσεις, δυνάμεις) at different levels: *qua* animal, *qua* member of a certain species, *qua* male, *qua* individual male with certain anatomical features.[31]

During the formation of the embryo, these motions battle it out with the opposite motions in menstrual blood (defined by Aristotle as "potential")[32] and according to the degree of so-called "mastery" achieved over the feminine motions.[33] The results can vary, from sons resembling their mothers to daughters resembling their fathers, to children resembling both or none of their parents; sometimes these resemblances to parents concern only partic-

28 Connell, *Aristotle on Female Animals*, 286.

29 Aristotle, De generatione animalium IV 4 (767a14–35).

30 For a detailed account of the principle of συμμετρία, see: Connell, *Aristotle on Female Animals*, 270–276.

31 Aristotle, De generatione animalium IV 3 (767b23–29).

32 Gelber, "Females in Aristotle's Embryology," 184.

33 The concept of "mastery" of heat over moisture is a crucial concept in Aristotle's *Meteorologica* IV, which explains the successful operation of the active over the passive qualities: Aristotle, Meteorologica IV 1 (379a1).

ular anatomical features (nose, eyes etc.).[34] In case none of the motions (viz. somatic traits) inherited from parents prevail, one can observe a relapse to the features of one's ancestors (grandparents). Sophia Connell has remarked the problematic limitations of this theory since, according to Aristotle, the somatic features that can be inherited from a certain sex concern only the hereditary line of the same sex. If a child does not resemble his father, he can resemble only his paternal grandfather and not his paternal grandmother. By the same token, a daughter can resemble only her female maternal ancestors.[35]

Even though Aristotle refused to link sex differentiation directly to primary qualities such as heat and coldness, he nonetheless argued that this would be "somewhat reasonable," probably in view of the above-mentioned difference of blend between males and females.[36] For Aristotle, to argue that primary qualities are the cause of sex differentiation would be grasping the cause from too far away. However, this does not change the fact that primary qualities are directly correlated to sex determination in two respects: males are hotter than females in blend and the likelihood of generating daughters is directly proportional to either old age or eastern winds, which are colder.[37] From this perspective, females, or sons resembling their mothers, do not constitute anomalies of nature but only one possible result of the interaction between the fluids at play. On the other hand, though, their manifestation depends on the failure of the nature of the masculine semen to gain complete mastery over the feminine motions, so their birth appears to be a partial deviation from natural teleology. From a chronological point of view, the interaction between these opposite motions takes place during the 40 days necessary to full embryonic formation (20 in egg-laying animals such as chicks).

All in all, the instrumental and efficient role of *pneuma* emerges as one of the notions crucial to explain generation from a mechanical, efficient point

34 For Aristotle, the motions in the feminine menstrual blood are defined as "potential," in the sense that they are actualized on condition that the motions of masculine semen fail to gain mastery over them: for this interpretation: Gelber, "Females in Aristotle's Embryology," 183; Aristotle, De generatione animalium IV 3 (768ab28–31).

35 Aristotle, De generatione animalium IV 3 (768a10–21); Connell, *Aristotle on Female Animals*, 292–299.

36 Aristotle, De generatione animalium IV 1 (765a35–b6).

37 On daughters, Aristotle, De generatione animalium IV 2 (766b27–b37); De generatione animalium IV 3 (767b11–12). On eastern winds, Aristotle, Historia Animalium VI 19 (573b32–574a1); on the above-listed passages, Gelber, "Females in Aristotle's Embryology," 179–187.

of view *qua* carrier of seminal heat. It is indeed true that *pneuma* explains the different developmental stages of the embryo, but its motion acquires an end-directed nature because of the father's soul as its steering principle. In other words, *pneuma*, together with heat, masculine semen and feminine menstrual blood does provide a physical account of the process of generation, but the mechanical account of generation alone does not and cannot explain the specificity of the fluids without reference to the final cause of generation, viz. the future animal to be and its specific soul (plant, animal or human). That is to say that the complex interaction of fluids finds it explanation thanks to the directing action of nature and soul towards the final cause of generation, viz. the preservation of the animal species. In this manner, the mechanical motions of semen, menstrual blood and *pneuma* become aimed at the formation of an animate being of the same species and not of a random conceipt. From this point of view, *pneuma* emerges as the material substance that, operating as an instrument of the father's semen, guides the formation of the new embryo, although its role is far from being fully explicitated by Aristotle.

Like any crafted instrument, *pneuma* is confronted with the matter of its own artifaction. The material substrate of its action may be excessive or too small, difficult to manipulate and so on, and the final outcome may be unintended or less common than in most cases. This series of problematic aspects concerning the action of form upon matter emerge particularly in Aristotle's account of monstruous births and less common results of reproduction, such as the birthing of twins.

1.2 *Errors of Nature or Pure Chance? Twins and Deformities*

The first case of "errors" (ἁμαρτήματα) of nature to be considered concerns embryos with a redundancy or deficiency of parts. Aristotle ascribes those irregularities to problems in the "foetations" (κυήματα),[38] i.e. the matter attracted from the heat of the semen into the uterus. This matter can be either defective or overabundant and can thus cause redundancy or deficiency not only in terms of material composition but also of functionality of the organs.[39] A second case of irregularities of nature is constituted by monsters that result usually from the mating of animals belonging to different species. Except for the case of compatible species such as horses and donkeys, which generate mules or hinnies depending on the sex of the parents, fecundation between

38 Aristotle, De generatione animalium IV 4 (770a8).

39 Aristotle, De generatione animalium IV 4 (772b13–26).

incompatible species produces monsters.[40] A third and curious case of natural irregularity in generation is represented by the birthing of twins.

To begin with the first case, Aristotle views deformities as a partial failure of the mastering action of heat over moisture because of material realities such as (1) excess or defect of matter and (2) defects in the primary qualities of the parent's blend. This decrease is directly correlated to the decrease of heat over time due to aging. Lennox has shown that, if we consider Aristotle's account of chance and teleology in his *Physics*, deformities, monstrosities and twins can well be counted among the number of results of reproduction that have a material basis but are less usual because they reach their ultimate goal (generation) in a manner that is not usual.[41] Continuing on the art-nature analogy, we can think of twins or deformities as craft products that are rare (in the case of twins) or, in the case of deformities, partially failed results of an artificial process where the instruments of the soul (such as heat and *pneuma*) have failed to carve out of matter the expected artifact.

The remarkable aspect of Aristotle's doctrine of deformities and monstrosities is its grounding of such phenomena in the purely elemental, sublunar components of generation. Nowhere does the Aristotelian account of natural teleology invoke the benign or malignant influences of stars or planets as a means of explaining monstrosities and deformities. By contrast, this latter approach will be shared by the likes of Avicenna and Averroes, who both influenced Albert the Great.

To illustrate the irregularity of their occurrence in nature, Sophia Connel has highlighted a peculiar metaphor deployed by Aristotle to prove his point, viz. the metaphor of the "smoky grape".[42] The peculiarity of this kind of grape lies in the heterogeneous color of its grapes: sometimes it produces white grapes, but its production of black ones is not really seen as contrary to nature, for its occurrence is not an isolated one.[43] While this does not occur in most cases, it is indeed a natural outcome of a natural process, in this case the growth of grapes. Likewise, the birthing of twins is a less usual but still natural outcome of reproduction. Aristotle observes the case of twins not only in humans but also

40 Groisard, "Hybridity and Sterility," 153–170.

41 Lennox, "Teleology, Chance," 230: "In chapter 5 (sc. of the second books of the *Physics*) [...] he next insists that it is possible that some generations are both unusual and for the sake of something, namely, whichever could have been done by thought or by nature."

42 Aristotle, De generatione animalium IV 4 (772a36–772b).

43 Aristotle, De generatione animalium IV 4 (770b13–22); Sophia Connell, "Aristotle's Explanation of Monstruous Births and Deformities," in *Aristotle's Generation of Animals. A Critical Guide*, ed. Andrea Falcon and David Lefébvre (Cambridge Critical Guides), Cambridge 2018, 212.

in twin chicks in a single egg. The birthing of these twins is made possible by the presence of a double yolk (which has an analogous function to feminine menstrual blood) in the material composition of the egg.[44] A final example shows how twins can witness to conjugal unfaithfulness, as there may be twins that resemble both the mother's husband and her lover. This may be due to a precise biological phenomenon, viz. that of superfoetations, which occur when a woman is impregnated twice during the same cycle.[45] The example used by Aristotle is a famous one, viz. the impregnation of Alcmena by Zeus: the result is that she gave birth to twins: Heracles, who resembled his father Zeus, and Iphicles, son of Alcmena's husband Amphytrion.[46] In the end, Aristotle views twins as a rare and unexpected outcome of reproduction which are still to be included in the roster of products of natural teleology.

Up to now, we can observe how the role of *pneuma* in the Aristotelian theory of generation is much more explicit (although almost equally problematic) than has been observed in the previous chapter. Nonetheless, it is not exempt from hermeneutical difficulties. As a carrier of the semen's hot air, spirit produces the information of the embryo until it is fully formed. Like rennet in the case of cheese, the masculine semen (and *pneuma* in it) do not enter the material composition of the embryo-to-be. In Aristotle, *pneuma* helps the process of "man begets man and the sun" by supplying the whole amount of vital heat inherited from the father's semen, crucial to the hylomorphic mechanism at play in fertilization. In no way, though, can *pneuma* be regarded as the instrument of a higher intelligence than the human soul or as having been affected in its material composition by astral influences. The teleological function of *pneuma* in the Aristotelian doctrine of generation is simply to grant subsistence to a certain animal species by carrying the amount of seminal heat necessary to form embryonic matter. This formative action of the semen may meet with failures (deformities), or unintended or less usual (twinning) results, but it remains teleologically directed by nature and the father's soul. Moreover, resistance exerted by embryonic matter to the formative action of heat and *pneuma* explain the generation of feminine offspring and the resemblance of children to their mothers and/or maternal ancestors.

In Aristotle, while nature is generally governed by the teleological principle that "nature does nothing in vain", such nature is not permeated by the providentialism that inspired the Neoplatonic schemes so influential in Albert's sources, such as Avicenna. The only trace of relationship between Aristotle's

44 Aristotle, Historia Animalium VI 3 (562a24–562b2).
45 Aristotle, Historia Animalium VII 4 (582a12–17).
46 Aristotle, Historia Animalium VII 4 (584a13–15).

God and the world of sublunar beings is found in his assertion that God operates as the final cause of the world's motion, having nothing to do with the process of generation of any living beings. Through key authors such as Avicenna, deeply permeated by Neoplatonic readings of Aristotle's philosophy of nature, the Aristotelian teleology of natural processes was gradually framed into a metaphysical and cosmological structure of the cosmos highly palatable to any adherent of the Islamic religion or of the Abrahamic monotheistic religions in general. When it comes to generation, the Avicennian structure of the cosmos seeks to weld together two different branches of knowledge: natural philosophy and medicine. To do so, Avicenna proceeds according to specific epistemic standards, for otherwise the results of his investigation might clash with the key tenets of Aristotelian philosophy of nature so crucial for his philosophical endeavor.

2 First Teaching vs. Anatomical Practice: Avicenna on Generation

The first chapter showed how Avicenna's epistemological approach to medicine draws clear epistemic boundaries which medicine, when employed in works of natural philosophy, is not allowed to trespass. This is the case, for instance, with Avicenna's rejection of the Galenic theory of feminine "semen". Those considerations, though, concerned Avicenna's approach to medical learning primarily in view of its method rather than its doctrinal content, which will be scrutinized in greater detail here. As in the Aristotelian section, it would be good philosophical practice to preface the physiological explanation of generation and its relation to the concept of spirit with some remarks on the doctrine on the origin of the soul.

2.1 *Avicenna's "Giver of Forms" and the Origin of the Soul*

Unlike Aristotle, who had decidedly rejected any Platonic view on the origin of the soul as the product of causation by a separated principle, Avicenna considers the soul as being created by God through the mediation of the tenth and lowest celestial intelligence, called the "Active Intellect" or "Giver of forms" (*dator formarum/wāḥib al-ṣuwar*). While the former name (Active Intellect) refers to its gnoseological function, the latter refers to its ontological role as emanative principle of all sublunary forms. These forms range from simple elements to minerals and the more complex forms (souls) of plants, animals and humans.

According to Avicenna, souls are introduced, or better, flow into matter upon its acquisition of a suitable degree of preparedness (*praeparatio appropri-*

ata/*isti'dād ḫāṣṣ*). Preparedness is achieved through the combined and reciprocal interaction of the heat combined from heavenly influences and seminal heat, so-called "formative power" (*virtus informativa*/*quwwa muṣawwira*), and the primary and complexional qualities of the feminine semen.[47] Avicenna's position on the notion of "appropriate preparation" is not exempt from possible charges of "pluralism of forms", though. Andreas Lammer has recently shown how this risk is implicit in the Avicennian assumption that matter always possesses a "form of corporeity" (*forma corporeitatis/ṣūra ğismiyya*) common to all natural bodies which accounts for their numerical commonality as indivisible substances.[48] Yet, he has explained how Avicenna's claims—especially in the *Metaphysics* (*Ilāhiyyāt* IX, 5)—that the existence of the form of corporeality is not sufficient to account for life in living beings but must be viewed against the backdrop of the souls of plants, animals and humans. Avicenna's form of corporeality denotes 'form' in a loose sense, since its function is to signal that, from a logical—and not ontological—point of view, there are different layers which make up the structure of the forms of living beings.[49] In humans, this degree of preparation is reached when the formative power completes the formation of the embryo (40 days) or of the chick (20 days). The distinction between *forma corporeitatis* and 'form' in the sense of 'soul' would then seem more of a logical than an ontological entity. In this manner, Avicenna's position is much less prone to criticisms that the forms of sublunar things represented an accidental component of the hylomorphic compound given by matter and corporeal form. This was precisely the gist of Averroes' critique against Avicenna.[50]

According to the account laid out so far, the "Giver of Forms" must be regarded as the proximate efficient cause of all sublunar forms and, hence, souls. This is because souls belong to the class of beings defined as "possibly existent in itself and necessary existent through another", as opposed to the one

47 On Avicenna's notion of preparation and its possible comparison to Albert's theory of *inchoatio formae*, see: Olga L. Lizzini, "Flusso, preparazione appropriata e *incohatio formae*: brevi osservazioni su Avicenna e Alberto Magno," in *Medioevo e filosofia: per Alfonso Maierù*, ed. Massimiliano Lenzi/Cesare A. Musatti/Luisa Valente (I Libri di Viella 150), Roma 2013, 129–150.

48 Andreas Lammer, *The Elements of Avicenna's Physics. Greek Sources and Arabic Innovations* (Scientia Graeco-Arabica 20), Berlin 2018, 165–179.

49 Lammer, *The Elements*, 174: "Avicenna, against the more widespread opinion, actually denies the multiplicity-thesis, as each and every natural thing consists of only one chunk of matter and only one form, even though his form contains several layers of formal determination 'by way of generality and specificity'."

50 Lammer, *The Elements*, 169, especially n. 99.

and only "necessary existent" (*necesse esse*/ *wāǧib al-wuǧūd*), God. According to Avicenna, while the "Giver of Forms" represents the proximate and efficient cause of plant, animal and human souls, the ultimate and remote efficient cause of the origin of any sublunar soul is God, who acts through the mediation of the "Giver of Forms".[51] This example of mediated creation has a relevant antecedent in Arabic Neoplatonism, namely proposition 3 of the *Liber de causis*, where the human soul is said to be created by the One by the means of the Intelligence.[52] In Avicennian psychology, God is not only the efficient principle of the soul but is also that which represents the ultimate end of human existence. Avicenna's teleology of nature is therefore ruled by a much broader and Neoplatonizing interpretation of Aristotelian teleology of nature based on the Neoplatonic concept of the One (God) as both the final principle and the end. This concept is well exemplified by the key role that Avicenna attributed to the Neoplatonic concept of "return" (Gr. *epistrophé*/Ar. *maʿād*). This concept will prove crucial in the next chapter to understanding Avicenna's position on the teleology of non-teleological process such as death. Until then, though, our investigation will revolve around a teleological process, the origin of life, whose domain of investigation lies at the intersection between natural philosophy and medicine.

2.2 *The Vehicle of Formative Power: Spirit and Avicenna's Theory of Generation*

In the first chapter, we saw how in the *Canon* spirit (*spiritus*/*rūḥ*) belongs to the concepts of natural philosophy (such as elements, humors etc.), whose discussion belongs to natural philosophy alone and whose essence and existence must be only assumed, and not discussed, by physicians. Given Avicenna's strictness in epistemic matters, the fact alone that large chunks of anatomy and physiology from the *Canon* are transplanted into *De animalibus* (books IX, XII–XIV) is quite telling. Perhaps this is a sign that his *Canon* had already been conceived in a manner that could reconcile Galenic medical learning, the fundamental tenets of Aristotle's philosophy of nature and Avicenna's own psychological doctrines. Such a supposition will prove true when considering

51 See at least: Julius Janssens, "The Notions of Wāhib Al-Suwar (Giver of Forms) and Wa-hib Al-ʿAql (Bestower of intelligence) in Ibn Sīna," in *Intellect et imagination dans la Philosophie Médiévale, Actes du 11. Congrès international de philosophie médiévale de la Société internationale pour l'étude de la philosophie médiévale* (S.I.E.P.M.)*: Porto, du 26 au 31 août 2002*, ed. Maria Candido Pacheco/José Meirinhos, 3 vols, Turnhout 2006, vol. 1, 551–562.

52 Cristina D'Ancona, "La doctrine de la création '*mediante intelligentia*' dans le *Liber de Causis* et dans ses sources," *Revue des Sciences philosophiques et théologiques* 76 (1992/2), 209–223.

Avicenna's treatment of human generation, which involves spirit as a key material substance in the process.

This pivotal role of spirit becomes clearer when considering the notion of formative power applied to human and animal generation. This power represents the ability of the masculine semen to form the feminine matter. The shaping of embryonic matter occurs according to a faithfully hylomorphic scheme inspired by Aristotle, although Avicenna's understanding of the interaction between masculine semen and feminine menses rests on a different doctrine developed from Aristotle. In his zoology and meteorology, the Stagirite had upheld that the heat borne by *pneuma* within the semen was to be identified with seminal heat alone. At least, it seems so. By contrast, Avicenna adhered to a conception of formative power which results from a combination of cosmic heat (caused by heavenly influences) and seminal heat itself.[53] In this respect, Avicenna follows the tendency of "astrologization" in Aristotelian biology introduced by Alexander's *De providentia* and inherited in the philosophy of the Islamic world.[54] What in Aristotle was a purely causal correlate, viz. favorable environmental conditions, here becomes an active agent in the determination of the amount of heat in the semen and, hence, of the nature of the embryo-to-be. For Avicenna, as for Aristotle, spirit is something "better than any corporeal substance" (*spiritus ille est melior omni substantia corporali/wa-l rūḥ afḍal ǧawhar ǧismāniyy*)[55] and a "divine body" (*ǧism*

53 Avicenna, Abbreviatio de animalibus XVI c. 1, ed. Venezia 1508, fol. 61[ra] [ed. Cairo 1970, 403.13–15; 403.16–17]: "[…] videtur quod ille calor veniens a supercelestibus habet effectum in virtute quod non posset habere calor ignis; et illa virtus est que dat vita et est proportionale virtutibus supercelestium (*yušabbihu an yakūna harāratu tatbaʿuhā quwwa lā tatbaʿu al-ḥarāra al-nāriyya, wa-anna tilka al-quwwa quwwa muǧība wa-munāsaba bi-waǧhin-mā bi-l-aǧsām al-samāwiyya, ḥattā yakūna lahā an taqbala al-hayā*). […] in spermate est substantia prima potens recipere hanc virtutem et est spiritus deferens hunc calorem. (*wa fī-l-minan ǧawḥar huwa ʿawwal ǧawharin yaqbaluhā* (sc. *quwwa*) *wa-huwa al-rūḥ allaḏī huwa ʿawwal ḥāmil li-haḏā al-ḥārr*)." Avicenna, Abbreviatio de animalibus IX 4, ed. Venezia 1508, fol. 42[va] [ed. Cairo 1970, 166.7–8]: "Tunc non est remotus quod sperma sit locus primus spiritus (*lā yaḫlū ʿimma ʿan yakūna al-minan kulluhu makān ʿawwal lahu*) vel quod sit in eo loco propter spiritum in quo ipse adunatur." In the first passage (XVI, 1), the Latin abbreviates the Arabic, but the conceptual framework is intact.

54 Gad Freudenthal, "The Medieval Astrologization of Aristotle's Biology: Averroes on the Role of the Celestial Bodies in the Generation of Animate Beings," *Arabic Sciences and Philosophy* 12 (2002), 111–137; Gad Freudenthal., "The Astrologization of the Aristotelian Cosmos: Celestial Influences on the Sublunary World in Aristotle, Alexander of Aphrodisias, and Averroes," in *New Perspectives on Aristotle's* De caelo, ed. Alan C. Bowen/Cristopher Wilberding (Philosophia Antiqua 117), Leiden 2009, 239–281.

55 Avicenna, Abbreviatio de animalibus XVI c. 1, ed. Venezia 1508, fol. 61[va] [ed. Cairo 1970, 404.2].

ilāhiyyī).[56] For Avicenna, though, the concept of spirit as "divine" body appears to be loaded with a whole different meaning than Aristotle had originally intended.

Avicennian *rūḥ* possesses a unifying principle for generation that is not based on mere seminal heat, for it combines seminal heat, celestial influences and, unequivocally, God's will.[57] According to Avicenna, every act of generation that takes place in the sublunary world, every kind of form which comes to be, unescapably represents the product of his will and is guided by his providential teleology. This will, a problematic concept in Avicenna's philosophy of nature, is realized by means of intermediaries like the "Giver of forms", the ultimate efficient cause of all sublunary forms, among which are souls. Unlike Aristotle's account of generation, who correlates favorable material conditions produced by the sun's action without causally linking planets and stars to the elemental composition of seminal heat, Avicenna's reading of the concept of formative power is much stronger than Aristotle, and richer in terms of causal agents involved. While in Aristotle favorable environmental conditions are only correlated to the process of generation, in Avicenna the heavenly world represents a direct causal agent in the elemental composition of the instrument (spirit) so crucial in the information of the future embryo. If we consider the background of Avicenna's overall theory of spirit, strongly based on the Galenic concept of complexion, the nature of the *rūḥ* placed in the semen fits well within Avicenna's emphasis on the correlation between different actions and different complexions of spirit:[58] to be suitable for generation, it needs to possess a heat specific to the task. The heat—and hence the complexional balance—placed

56 The Latin has "virtue" (*virtus*) instead of "spirit" (*rūḥ*) in the passage referring to the divine nature of *pneuma*: Avicenna, Abbreviatio de animalibus XVI c. 1, ed. Venezia 1508, fol. 61rb: "[…] et ista virtus est corpus divinum". The Arabic reads: Avicenna, al-Ḥayawān XVI c. 1, ed. Cairo 1970, 404.1: "And this spirit is a divine body (*wa hadā al-rūh ǧismun-mā ilāhī*)." Translation is mine.

57 Avicenna, Abbreviatio de animalibus XVI c. 1, ed. Venezia 1508 fol. 61ra [ed. Cairo 1970, 402.7–10]: "Et nunc loquimur in spermate utrum in illo sit virtus animalis et quoniam sperma movetur ad generandum creaturam per causam extrinsecam, sed sua natura propria que est causata nutu Dei (*bal bi-ṭabīʿatihi al-musaḫḫira bi-idni-llahi*): et erit illum principium (*mabdāʾ*) virtutis nutritive et generatio membrorum (*takwīn al-aʿḍāʾ*) […]"

58 Avicenna, Liber de anima V c. 8, ed. Van Riet 1968, 175.58–176.63 [ed. Rahman 1959, 263.15–19]: "Habet (*sc.* spiritus) autem complexionem propriam (*mizāǧ maḫṣūṣ*), quae mutatur etiam prout necesse est in ea esse diversitatem propter quam fit vehens virtutes diversas (*quwà muḫtalifa*): complexio enim qua concupiscitur non est apta ut per eam sentiatur, nec complexio quae est apta visibili spiritui est apta spiritui moventi; si autem complexio una esset, virtutes que sunt in spiritu essent una, et actiones earum una."

in spirit, in order for generation to occur, needs to be a combination of both sublunar (seminal) and heavenly (cosmic) heat.

There is more to the role of spirit in generation than just the possession of a suitable complexion or appropriate preparedness, though. In Avicenna's understanding of generation, spirit is a material substance placed into the semen which carries formative power, ensures that the matter (viz. the body) of the offspring-to-be is suitably organized and receives its specific form from God via the Giver of Forms. Moreover, spirit can be viewed as the connecting point of a teleological and providential system of nature combining sublunar and heavenly causes. This cosmos is ruled by a hierarchy of intelligences with God as the remote efficient and ultimate final cause at their top. This worldview applies to all kinds of creatures: males and females, humans and animals, twins and monsters, and sexually and spontaneously generated animals.

2.3 *Masculine and Feminine: The Principles of Avicennian Human Embryology*

Avicenna's account of generation is developed first in the *Canon* (composed ca. 1015) and then copied in his later zoology (ca. 1028). Despite some physiological differences in the copying process, the gist of his doctrine is essentially consistent and was destined to influence generations of philosophers and physicians by its blending Galenic doctrines with the principles of Aristotelian natural philosophy. As the consensus in scholarship has shown, in his embryology Avicenna made a relevant concession to Galen by introducing feminine semen into the process of generation. Both masculine and feminine semen result, according to him, from a process of four successive digestions: in the stomach, in the liver, in the veins and in the specific organs (testicles in males and ovaries in females). The Persian philosopher and physician, though, made it clear that the mode of agency of what Galen called feminine semen simply could not be on a par with male semen.

Rather, the agency of this fluid in generation had to be qualified as purely passive, endowed with a potency capable only of providing embryonic matter. In this respect, Avicenna transferred to the Galenic feminine semen what Aristotle had thought was the causal role of menstrual blood.[59] Following Aristotle,

59 Avicenna, Abbreviatio de animalibus IX c. 3, ed. Venezia 1508, fol. 42ra [ed. Cairo 1970, 161.18–19]: "Materia (ar. *nuṭfa*, "semen") mulieris que dicitur sperma in se non habet virtutem generativam (*quwwa muwallida*), sed potestatem tantum ut in ea fit generatio." The second part of the sentence (*sed potestatem* etc.) is a rephrasing of the Arabic: "If it (sc. feminine semen) was a power, this would have no active capacity at all (*wa-in kānat quwwa, fa-lā fiʿl lāhu al-battatan*) [ed. Cairo 1970, 161.19]."

Avicenna maintained a comparison between the role of formal and efficient causes for masculine semen in the process of embryonic formation and the action of rennet on milk.[60] The same metaphor of rennet received a different interpretation in the *Canon*. Unlike Aristotle, Avicenna endorsed the view that semen could somehow enter the material composition of the embryo-to-be, just like rennet could—to a small extent—enter the substance of cheese.[61] Beyond any reasonable doubt, this assertion conflicts with Aristotle's interpretation and puts into clear and serious danger the Aristotelian hylomorphic pattern according to which a formal or efficient cause cannot be a material cause at the same time. On the one hand, it is very difficult to understand whether Avicenna was consciously and willfully trespassing the Aristotelian hylomorphic scheme at play in generation. Given the scattered remarks dedicated to radical moisture in the *Canon* and its almost complete absence from Avicenna's zoology, the reasons for this patent infringement of Aristotle's hylomorphic scheme are even more difficult to be clarified, and the brief space we have at our disposal seems ill-suited for the task. To be sure, however, this will represent a relevant issue for future research on Avicennian zoology and medicine.

Up to now, we have devoted our attention to the Avicennian interpretation of the Galenic concept of 'feminine semen' and the consequent aporia provoked by his relation to the material composition of the embryo-to-be. An attentive reader of Aristotelian zoology like Avicenna, though, could not just overlook the causal role of menstrual blood. Avicenna's solution was to consider menses as a less ripened or less concocted version of feminine semen. Note that this distinction echoes to some extent Aristotle's differentiation between a more and less concocted version of menses: the first is deployed in fecundation, the second for embryonic growth. In Avicenna's psychophysiology, the first corresponds to feminine semen, the second to menstrual blood proper. According to Avicenna, the main function of menses was to provide food and liquid for

60 Avicenna, Abbreviatio de animalibus XVI c. 1, ed. Venezia 1508, fols. 61^{rb-va} [ed. Cairo 1970, 404.5–8]: "Et erit iste ventus spiritus qui transit et facit membra cum virtutibus quae sunt in eis et complet ea sicut coagulum factum in lacte, et operatur in partibus lactis qui sunt in illis, et profundatur in partibus non exiens pars casei, et caseus fit per illud; et sic istud sperma non est pars membrorum, sed est principium spiritus transeuntis ad membra et operatur illa […]."

61 Avicenna, Canon I fen 1 doctr. 5 c. 1, ed. Venezia 1507, fol. 8ra [ed. New Delhi 1981–1996, vol. 1, 61.6–8, 10–11]: "Sed secundum sermonem eius, qui de sapientibus verificauit, de spermate mulieris generatur, sicut generatur caseus de coagulo, et de spermate mulieris generatur, sicut generatur caseus de lacte […] Et quemadmodum unumquodque duorum, coaguli videlicet, et lactis, est pars substantie casei, qui fit ex eis, ita unumquodque duorum spermatum est pars substantie embrionis."

the embryo and milk for the woman's breasts.[62] Avicenna was not content to explain that menstrual blood acted as the source of nourishment for the embryo, though. He also wanted to explain whether this reddish fluid entered the material composition of the embryo, and how it could be assimilated into the whitish nature of the embryo. To solve this problem, Avicenna ascribed to feminine semen the capability of whitening (*dealbatio*) menstrual blood.[63] The introduction of feminine semen, though, defined as "moisture", produced more problems than solutions since, as noted by later physicians, Avicenna used the term *humiditas* (*ruṭūba*) to refer to feminine semen in the proper sense and to the moisture emitted by the woman during intercourse, which by definition does not enter the embryo-to-be.[64]

The direct upshot of Avicenna's denial of the active role of feminine semen had a direct bearing on his biological explanation of the morphological traits of the embryo-to-be. According to Galen, this resemblance was due to the agency of the feminine semen whenever it had prevailed over the masculine. In his embryology, Avicenna appealed to the Aristotelian principle of *inoboedientia materiae*. This principle claims that the offspring resembles its mother, its (feminine) maternal ancestors or is of feminine sex, one, two or all of these cases happen if and only if the matter of feminine semen resists to the formative power directed by the father's soul.[65] Galen's doctrine, by contrast, which put

---

62 Avicenna, Abbreviatio de animalibus IX c. 3, ed. Venezia 1508, fol. 42rb [ed. Cairo 1970, 160.10–11; 161.5]; Nahyan Fancy, "Generation in Medieval Islamic Medicine," in *Reproduction. Antiquity to the Present Day*, ed. Nick Hopwood, Rebecca Flemming, and Lauren Kassell (Cambridge: Cambridge University Press, 2008), 133: "properly speaking, it (sc. the feminine semen) is not semen, but rather a type of ripened menstrual blood." On menstrual blood as food for the embryo, see: Avicenna, Abbreviatio de animalibus IX c. 5, ed. Venezia 1508, 43rb [ed. Cairo 1970, 174.1–2]. On the process of transformation of menstrual blood into milk and food for the embryo, see: Avicenna, Abbreviatio de animalibus XV c. 3, ed. Venezia 1508, fols. 60$^{va–ra}$ [ed. Cairo 1970, 396–400].

63 Avicenna, Abbreviatio de animalibus IX c. 3, ed. Venezia 1508, fol. 42rb [ed. Cairo 1970, 162.14–18]: "[…] postquam coagulatio creature est ex isto duplici spermate (*min hāḏayni al-minanayn*), necessarium est ut decurrens sanguis menstruus assimiletur isti coagulationi. Et propter hoc permutatur ab eis in similitudine spermatis, ut aptum sit transire in nutrimentum." On this passage and on the process of *dealbatio*, see: Gabriella Zuccolin, *I gemelli nel Medioevo. Questioni filosofiche, mediche e teologiche*, Como-Pavia 2019, 24.

64 Zuccolin, *I gemelli nel Medioevo*, 21– 23.

65 Avicenna, Abbreviatio de animalibus IX c. 2, ed. Venezia 1508, fol. 41vb [ed. Cairo 1970, 154.11–14]: "Si erit ex parte fortitudinis virtutis informativae erit assimilatio patri; et si materia fuerit dominans supram operatorem ex toto assimilabitur matri." Avicenna, Abbreviatio de animalibus XVIII c. un., ed. Venezia 1508, fol. 63ra [ed. Cairo 1970, 421.6–15]: "Et fortasse erit materia ex parte complexionis digestae non recipiens complexionem masculinam: que est in corde cum qua assimilatur creatura in principium sui motus: et erit

male and feminine fluids on the same level of possible causation in nature, had the further downside of patently infringing upon the Aristotelian rule that there cannot be more than one formal cause (masculine semen) in the process of mixing. For this reason, Avicenna resolutely rejected Galen's theory in favor of one that could be consistent with Aristotelian principles of natural philosophy, at least when it came down to the feminine contribution to generation. Like Aristotle, Avicenna posited age as a factor directly correlated to sex differentiation: the older the parents, the colder the complexion of masculine semen and its formative power, and thus the higher the rate of feminine offspring.[66] According to Avicenna, the only case of asexuality in nature is represented by eels.[67] In Avicenna's theory of generation and sex differentiation, spirit (*spiritus/rūḥ*) appears therefore to play the pivotal role *qua* carrier of the formative power and its vital heat so crucial to determining both the sex and the morphology of the embryo-to-be without entering it.

2.4 *Heart and Veins: The Development and Morphologic Structure of the Human Embryo*

In the *Canon* and in *De animalibus*, which copied this section (IX, 5) from the *Canon* (III, 21, 3), Avicenna provides a detailed account of embryonic development. To gain a full picture of Avicennian embryology, this account will need to be combined with Avicenna's discussion on the origin of veins in the embryo, which features his second famous clash against Galen, this time on the topic of the origin of veins. Before that, let us take a look at the Avicennian medical account presented in the *Canon* (and the *De animalibus*), which singles out five different stages of embryonic development:

1. The concept acquires a foamy- and milk-like form, with three large vesicles to provide space for the principal organs (day 1–6);
2. A blood-like droplet and red thread appears, preparing the vessel for the spirit, i.e. the place of the heart (day 9 [8–10]);
3. Coagulation, preparation of the heart's substance (day 15 [13–17]);

ex parte complexionis humide et mollis recipiens lineamenta et extensiones secundum quas assimilatur patri: et fortasse erit reversus: et fortasse inobediens in ambobus: et tunc assimilatur matri quando vincerit virtus passiva."

66 Avicenna, Abbreviatio de animalibus, XVIII c. un., ed. Venezia 1508, fol. 63ra [ed. Cairo, 422.2]: "Senex (*šayḫ*) a quo diminuitur calor facit filias [...]."

67 Avicenna, Abbreviatio de animalibus IV c. 3, ed. Venezia 1508, fol. 343333va [ed. Cairo 1970, 65.10]: "Et quedam sunt in suo genere que non habent mares et feminas in suo genere sicut encheli." *Enchelus* is the Latin rendition of the Arabic *inklīs*, a transliteration of the Greek ἔγχελυς ("eel").

4. Formation of heart and, from it, liver and brain; appearance of the fetus as flesh (day 27);
5. Embryo fully formed with limbs, hands and feet (day 36–40).[68]

After the embryonic formation and its full animation after 40 days, Avicenna takes pains to detail the causal relationship between the timelapse of embryonic formation and animation and the moment of childbirth. To do so, he deploys the method of duplication: if formation and animation is completed in 40 days, embryonic motion comes about in another 80 days (for a total of 120) and the hypothetical birthdate is thus set around nine months (40 + 80 + 160 = 280). The final number is calculated by doubling the number of days necessary for the children to start moving. By contrast, if the embryo is formed prematurely (30 days), childbirth occurs in seven months (30 + 60 + 120 = 210); or, if the embryo is formed in 45 days, its birthdate is situated sometime around 10 months (45 + 90 + 180 = 315).[69] Among embryos, males are formed earlier (30 days) than females (40 days) because of their hotter complexion.[70] During the process of embryonic development, the embryo is enveloped by three membranes: the first and outer one is the surrounding membrane (*secundina* / *mašīma*), most likely corresponding to the chorion. This membrane contains veins and arteries from the uterine wall to the embryo, supplying the nourishment provided by menstrual blood. A second membrane, called *ascha* in the *De animalibus* (ar. *balāīn*) and *abgas* in the *Canon* (ar. *falās*), viz. the amnion, serves as a receptacle for sweat; a third and anonymous final membrane is designed to retain the moisture necessary for embryonic growth.[71]

From the moment of conception, the embryo possesses in potency all the different kinds of spirit, although not yet distinguished from each other.[72] As we

68 This scheme in: Demaitre-Travill, "Human Embryology," 426.

69 Avicenna, Canon III fen 21 c. 3, ed. Venezia 1507, fol. 361ra [ed. New Delhi 1981–1996, vol. 3/2, 829.5–11; 829.23–30]; Avicenna, Abbreviatio de animalibus, IX c. 5, ed. Venezia 1508, fol. 43rb [ed. Cairo 1970, 172.11–173.5, 173.10–19].

70 Avicenna, Canon III fen 21 c. 2, ed. Venezia 1507, fol. 361va [ed. New Delhi 1981–1996, vol. 3/2, 830.3–4]: "Et dixerunt: non invenitur in abortu masculus completus ante triginta dies et femina que compleatur ante quadraginta dies." On the completion of the masculine embryo in 30 days: Avicenna, Abbreviatio de animalibus, IX c. 5, ed. Venezia 1508, fol. 43rb [ed. Cairo 1970, 173.8–9]: "[…] et masculus est magis festinus ad hoc quam femina: et videtur minimum tempus in formatione masculi 30 dies […]."

71 Avicenna, Canon III fen 21 c. 2, ed. Venezia 1507, fol. 361ra [ed. New Delhi 1981–1996, vol. 3/2, 830.10–14]; Avicenna, Abbreviatio de animalibus, IX c. 5, ed. Venezia 1508, fol. 43rb [ed. Cairo 1970, 174.2–6].

72 Avicenna, Abbreviatio de animalibus, IX c. 4, ed. Venezia 1508, fol. 42rb [ed. Cairo, 168.18]: "Et spiritus naturalis et vitalis et animalis non sunt segregati in spermate […]" (*wa lam*

observed in the second chapter, human life in the *Canon* is primarily associated with the action of vital faculty and vital spirit: consistent with this position, Avicenna classifies the spirit involved in the process of embryonic formation, starting from the heart, as "vital" (*vitalis/ ḥayawāniyya*).[73] This claim is not uncontroversial, though, since it clashes with Galen's position on the origin of veins from the liver. This problem is discussed at length in his *Ḥayawān* (III, 1) and has recently been investigated by Tommaso Alpina, whose article will represent the backbone of our exposition of Avicenna's take on this controversial matter.[74]

To gain a full glimpse of Avicenna's approach to the topic, it is necessary to refer back to Avicenna's epistemological underpinnings for this discussion, presented in the last chapter of his psychology (*De anima*, V, 8). In that passage, Avicenna holds that cardiac spirit is the instrument through which the heart fashions the body's members.[75] This statement proves once again Avicenna's staunch adherence to an Aristotelian embryological cardiocentrism. Alpina has convincingly argued that the cross-reference to zoology is a clue that, according to Avicenna, this passage provides the doctrinal solution to the Galenic-Aristotelian quarrel over the origin of veins in *De animalibus*, III, 1.[76]

---

yakun al-rūḥ al-nafsāniyy wa-l-ṭabīʿīyy wa-l-ḥayawānīyy muftariqan fī l-minan); Avicenna, Canon III fen 21 c. 2, ed. Venezia 1507, fol. 360[vb] [ed. New Delhi 1981–1996, vol. 3/2, 827.19–20]: "Et veritas quidem dispositionis ipsius spumositatis est ex hoc quod virtus informativa (*quwwa muṣawwira*) movet illud quod est in spermate ex spiritu animali naturali et vitali ad muneram uniuscuisque eorum in qua fingitur […]."

73 Avicenna, Canon III fen 21 c. 2, ed. Venezia 1507, fol. 361[ra] [ed. New Delhi 1981–1996, vol. 3/2, 828.13–15]: "[…] necesse est ut creetur membrum a quo procedit calor naturalis et spiritus vitalis (*rūḥ ḥayawāniyya*) ante creationem nutrimentis ipsius."

74 Tommaso Alpina, "Exercising Impartiality to Favor Aristotle: Avicenna and the accomplished anatomists' (*aṣḥāb al-tašrīḥ al-muḥaṣṣilūna*) in *Ḥayawān*, III, 1," *Arabic Sciences and Philosophy* 32 (2022/2), 137–178.

75 Avicenna, Liber de anima V c. 8, ed. Van Riet 1968, 176.64–75 [ed. Rahman 1959, 263.20–264.8]: "Unde si anima una est, oportet esse membrum unum propter quod principaliter pendeat ex corpore et ex quo regat corpus et augmentet, et hoc fiat mediante hoc spiritu (*bi-tawassuṭ haḏā-l-rūḥ*); et ut prima actio quam facit anima sit ipsum membrum quo mediante diffunduntur eius virtutes in alia membra; et ut ipsum membrum sit generatum ante omnia membra, et sit primus locus ubi generatur spiritus: hoc autem est cor, sicut nobis ostendit qui certificavit hoc chirurgia vera (*wa-haḏā huwa al-qalb, yadullu ʿalā ḏalika mā haqqaqahu al-tašrīḥ al-mutqan*); postea autem hoc clarius faciemus in libro qui est *De Animalibus*. Oportet ergo ut anima principaliter pendeat ex corde; impossibile est autem prius eam pendere a corde et postea a cerebro: cum enim pendet a primo membro fit corpus animal; in secundo autem non operatur nisi mediante primo."

76 Avicenna, Liber de anima V c. 8, ed. Van Riet 1968, 181.46–54 [ed. Rahman 1959, 266.19–267.6]: "Ergo cor est principium primum (*mabdāʾ ʿawwal*) et ab ipso emanant virtutes

The reason we find a cross-reference between the *De anima* and the origin of veins in the *De animalibus* is that Avicenna regards the consequences of Galen's position to have deeper and much more troublesome consequences than just a change in the physiology of embryonic development. The Galenic doctrine, according to which the liver would be able to form veins (and likewise the brain and the nerves), would have put those two bodily members on equal footing with the heart, thus jeopardizing the unity of the substantial form. This position was totally unacceptable to Avicenna. To question the philosophical validity of and then ultimately reject Galen's position, Avicenna proceeded in two steps. First, he undermined the Galenic association between the thickness of an organ and its proximity to its source. To assert that the vena cava linking heart and liver actually originated from the liver, Galen had used the branch analogy: as the branch of a tree is thicker when closer to its trunk, such is the case of the vena cava inferior to the liver. Avicenna rejected Galen's argument on empirical grounds, arguing that the optical nerve becomes thicker when closer to the eye and that the consensus among physicians (Galen himself thought so) is that this nerve originates in the brain and not in the crystalline humor.[77] According to Avicenna, Galen's arguments lack demonstrative cogency, and the ability to overturn these arguments is a sign of demonstrative weakness on Galen's part. In his refutation of Galen's position, Avicenna then went on to attack the Galenic tenet that "the principle of the organ is where the principle of the faculty is" (*ubi est principium virtutis ibi est principium instrumenti*), asserting that the characteristics of the organ do not depend on proximity to their source,

---

ad cerebrum (*tafīḍu minhu ilà-l-dimāġ al-quwà*), quarum quaedam suas actiones perficiunt in cerebro et in partibus eius, sicut imaginatio et formalis et ceterae, quaedam vero emanant a cerebro ad alia membra quae sunt extra illud, sicut ad pupillam et ad musculos moventes. Et a corde emanat ad epar virtus nutriendi, deinde emanat ab epate ad totum corpus mediantibus venis, et nutrit etiam cor: virtutis enim principium est a corde, sed materiae principium est ab epate (*wa tafīḍu min al-qalb ilà-l-kabd al-quwwa al-taġdiyya, ṯumma tafīḍu min al-kabd bi-tawassuṭ al-ʿurūq fī gamīʿ-l-badan wa taġdū al-qalb aiḍan: fa-takūnu al-quwwa mabdāʾuhā min al-qalb, wa-l-māddatu mabdāʾuha min al-kabd*)." Note that the language is strictly connected to the key Avicennian notion of *fluxus* (*fayḍ*): the heart is taken as the principle of existence of members through the emanation of the power of the soul, just like the First Cause is the principle of the existence of everything through its causal action of emanation. More on this in Alpina, "Exercising Impartiality," 149–151.

77 Avicenna, Abbreviatio de animalibus III c. 1, ed. Venezia 1508, fol. 31[rb] [ed. Cairo 1970, 41.3–6]: "[…] neque incoveniens est dici quod arterias nasci a corde neque grossiores significant quod pars grossior sit origo: quoniam nervi optici addunt in grossicie quando coniunguntur cum cristallino humore et sunt subtiliores apud principium eorum […]". Alpina, "Exercising Impartiality," 165.

but rather on the fashioning activity of the formative power that makes the organ suitable to perform its own function. In other words, it is not proximity to their liver that determines the activity carried out by veins, but rather the virtue given to the veins by the heart, through the action of their source, i.e. formative power.[78]

In this manner, Avicenna demonstrated (*probare*/*barhana*) that the heart *qua* principle of the formative power represented the true principle of veins.[79] The demonstration of the primacy of the heart expounded in *De animalibus* III, 1, though, assumes rather than demonstrates that the heart is the principle (*principium*/*mabdāʾ*) of all the body's members. This idea is presented in *De anima* V, 8 and is consistent with his account of the embryological development of every body part from the heart as the seat of the soul, and through the acton of spirit *qua* soul's instrument in the embryo's formation.

Avicenna's psychology (*De anima*, V, 8) and zoology (*De animalibus*, III, 1) both view medical experience as valid when it fits the rules of natural philosophy and is therefore "true" (*vera*/*mutqan*) when certifying cardiocentrism (*De anima*, V, 8); likewise, "the accomplished practitioners of anatomy" (*aṣḥāb al-*

78 Alpina, "Exercising Impartiality," 171.

79 This Galenic idea (most likely from the *De placitiis Hippocratis et Platonis*), as used against Galen, appears in the Arabic: Avicenna, Ḥayawān, III, 1, 43.18–44.8: "Since all of this is possible, nothing of what the excellent physician (*fāḍil al-aṭibbāʾ*, *sc.* Galen) says is necessary, even though he would bet [on this] and put the money for the soothsayer of the temple as a payment for the one who establishes for him (*sc.* Galen) that nerves are from the heart. It is possible that someone comes to him, who establishes that (*sc.* that nerves are from the heart) against him in a dialectical manner (*min ṭarīq ǧadalī*), admitting that the origin of nerves is from the heart. Galen has already conceded that the principle of the organ is where the principle of the faculty [of that organ] is. Then, when this premise is conceded by him, there can be demonstrated (*barhana*) against him (*sc.* Galen) that the soul in the human being is essentially one, from which the other faculties flow, and that the first connection of that single essence [with the body] is where the first organ for life (*sc.* the heart) is. Then, he (*sc.* Galen's opponent) will close the gap to impose on him that blood vessels and nerves are from the heart, and Galen will unquestionably lose his money." Translation and explanation of this passage by Alpina in: Alpina, "Exercising Impartiality," 168. The Latin version, though, does not make it evident that Avicenna is arguing against Galen's position by relying on Galen himself: Avicenna, Abbreviatio de animalibus, III c. 1, ed. Venezia 1508, fol. 31[va]: "Non oportet verum dixisse Galenus in pignore suo quando posuit pignus contra custodem altaris quod daret multum aurum qui monstraret quod nervi nascerentur a corde. Et consequenter poterimus obviare probando hec et sumendo hic pro constanti quod ubi principium est virtutis, ibi est principium instrumenti. Et tunc quia hominis anima est una virtus a qua omnes procedunt virtutes corporis accidit quod in illo membro in quo conculcatur anima quod illud membrum sit primum membrum in vita, et tunc de facili probabitur quod cor est principium venarum et nervorum."

tašrīḥ al-muḥaṣṣilūna) who lack any partisan spirit (*taʿaṣṣub*) are considered role models inasmuch as they value medical experience on the condition that it does not trespass the epistemological criteria of Aristotelian natural science (*De animalibus*, III, 1).[80] This overall approach to human embryology, based on the notions of spirit as the vehicle of formative power, the causal roles of masculine and feminine semen and menses and ultimately rooted in Avicennian cardiocentrism and emanationist theory of soul faculties, bears many traits in common, as we shall see, with his account of oviparous generation. For the time being, though, we need to focus on his handling of human generation, especially in connection with irregularities of nature such as monsters and less frequent outcomes such as twins.

2.5 *Twins, Deformities, and Monstrosities in Avicennian Embryology*

Avicenna's study of natural anomalies and non-teleological outcomes of generative processes can be best understood focusing on the notions of formative power and embryonic matter as their causal principle. In particular, the notion of formative power is key also to shedding light upon the causal role of spirit in these processes. If the main teleological guide to the events in the world of sublunar living beings is represented by the procession or flow (*fluxus*/*fayḍ*) from God to the sublunar living beings, Avicenna needs to detail its relation to what appear as non-teleological or unusual outcomes of generation.

In the case of birthing of twins, Avicenna correlates their rare occurrence to the relationship between formative power and three possible factors: (1) "windiness" (*ventositas/rīḥ*), i.e. the impulse of spermatic emission, (2) the movements of the uterus during intercourse or (3) the so-called "superfetations". The first factor, viz. windiness, is considered a feature of masculine semen directly linked to spirit given its wind-like nature (*assimilatur vento/šabīh al-rīḥ*).[81] The strength of formative power is directly proportional to the capacity of the masculine semen to attract embryonic matter in the uterus: if this matter is correctly retained and its amount is sufficient, birthing of twins

80 Avicenna, Ḥayawān III c. 1 (ed. Cairo 1970, 44.11–14): "Then, the accomplished practitioners of anatomy confirm (*yašhadu aṣḥāb al-tašrīḥ al-muḥaṣṣilūna*) that the heart is the first generated organ. As for the other organs, the formative faculty belonging to the generative faculty needs, in perfecting its fashioning activity, the mediation of the faculty that is in the heart. Then, it penetrates from it into those [other] organs." This translation by Alpina in: Alpina, "Exercising Impartiality," 171.

81 Avicenna, Abbreviatio de animalibus IX c. 4, ed. Venezia 1508, fol. 42vb [ed. Cairo 1970, 167.1–2].

occurs.[82] The second factor considered by Avicenna in birthing of twins are the movements of the uterus (*matrix/raḥm*) during intercourse: when pleasure is felt during it, the uterus begins a motion capable of dividing the sperm. If this division is regular, it produces twins; when this division is imperfect, it produces monstrosities such as conjoined twins or chicks with two heads and a conjoined body.[83] The last cause for birthing of twins, i.e. superfetation, consists in the action of new sperm on the already fertilized embryonic matter: by dividing it into two different embryos, it can cause unpleasant consequences for unfaithful wives (as it was the case with Alcmena in Aristotelian zoology), since the embryos can resemble two different biological fathers, the husband and the lover (*amasius/ʿašīq*).[84]

Although in his study of birthing of twins Avicenna does not repeat the Aristotelian metaphor of the smoky grape, he still seems very interested in examining a broad roster of possible cases, such as conjoined twins in humans and in chicks, as in the case of twin chicks mentioned by Aristotle in *Historia Animalium*.[85] Concerning monstrosities and teratology, Avicenna touches briefly on the subject, attributing it to malignant celestial influences. In his view, monsters are caused by malignous astral influences which, by acting on the composition of formative power and spirit, cause the birth of children with ram-like heads or lambs with bull-like ones.[86] Avicenna, even though quite

82 Avicenna, Canon III fen 21 c. 2, ed. Venezia 1507, fol. 362[ra] [ed. New Delhi 1981–1996, vol. 3/2, 832.10–11]: "Et causa quidem in geminis est multitudo spermatis (*wa- l-sabab fī taw'ām kaṯratu al-minan*): ita ut non spargatur ad duos ventres matricis sparsione replente totum secundum singularitatem." Avicenna, Abbreviatio de animalibus, IX 5, ed. Venezia 1508, fol. 43[va] [ed. Cairo 1970, 187.12–13].

83 Avicenna, Abbreviatio de animalibus XVIII c. un., ed. Venezia 1508, fol. 63[rb] [ed. Cairo 1970, 423.12–13; 424.2–3]: "Et videtur quod causa in hoc est quod motus innaturalis (*ḥaraka ridā'iyya*) qui accidit spermati [...] et ita apparuit unus corpus cum duobus capitibus secundum istam causam [...]"

84 Avicenna, Abbreviatio de animalibus IX c. 5, ed. Venezia 1508, fol. 44[ra] [IX, 6, ed. Cairo 1970, 181.6]: "contingit mulieri quae habebat duo filios et unus assimilabatur marito et alius amasio (*aḥaduhumā yušabbihu al-zawǧ wa-l-aḫar al-ʿašīq*)." The Latin conflates the Arabic chapters IX, 5 and IX, 6 into a single chapter, viz. IX, 5.

85 Avicenna, Abbreviatio de animalibus XVIII c. un., ed. Venezia 1508, fol. 63[rb] [ed. Cairo 1970, 424.18–425.1]: "Sed dicendum est melius quod generatio duorum in una cellula est res monstruosa que dividit sperma: sic dividitur sperma in matrice mulieris humane in multos filios parvos."

86 Avicenna, Abbreviatio de animalibus XVIII c. un., ed. Venezia 1508, fols. 63[ra–rb] [ed. Cairo 1970, 423.5–8]: "[...] et si non recipiet humanitatem (*insāniyya*) recipit animalitatem (*ḥayawāniyya*) sicut in rebus monstruosis: ut quando filius hominis habet caput arietis et agnus habens caput thauris, quoniam virtus in ipso induxit in eo forma secundum figuras celestes que accidunt et vincunt."

skeptical about detecting astral influences, sees these influences as possible causal agents which interfere in generation and interfere with the mechanism of formal replication, thus producing monsters.

The occurrence of twins, anomalies and monstrosities shows how the teleology of nature, even though operating through proper instruments such as heat and spirit, is unavoidably subject to the constraints of material contingency. In these cases, the end-directed nature of twins, anomalies and monstrosities is subject to the typically Aristotelian concept of hypothetical or suppositional necessity, but in a way that is different from Aristotle's. This is because in the case of deformities, the physical elements of defect can depend not only on the primary qualities of semen or menstrual blood (correlated to age and a few other factors such as climate) but also on the malign action of heavenly influences, an idea completely absent from Aristotle's biology. This attitude toward the "astrologization" of the Aristotelian biology of generation is shared by another crucial Arabic source for Albert's discourse on generation, viz. Averroes.

All in all, spirit emerges as a crucial instrumental cause in Avicenna's doctrine of generation inasmuch as it is at the service of the semen to form correctly an embryo of the same species, just like instruments are vital for the artisan to bring out the desired form from matter and yet are separated from it. This separation between formal (semen) and material (feminine semen, menstrual blood) fluids faithfully adheres to Aristotle's hylomorphism in his zoology. By contrast, Avicenna's *Canon* presents an aporia regarding the entry of part of the masculine, viz. the so-called radical moisture, into the embryo-to-be. The outcome is a breaking of the usual Aristotelian hylomorphic scheme of generation, which causes a friction between the accounts of Avicenna's zoology and medicine respectively.

In Avicenna's doctrine of generation, spirit provides a physical story of generation which must be complemented by the steering action of immaterial causes such as nature and soul and, thus, cannot be traced to material agents alone. Unlike Aristotle, though, Avicennian spirit is the receptacle of a heat produced by combined causality (seminal and cosmic) reflected in the composition of the *virtus formativa*. Spirit *qua* carrier of the formative power's heat can certainly explain how the interplay between masculine semen, feminine semen and menstrual blood leads to the formation of an embryo with specific morphological features, sexual differentiation and resemblance to its masculine or feminine ancestors. In his embryology, Avicenna favors a staunchly Aristotelian cardiocentric approach to the question of embryonic development for philosophical reasons. According to him, defending cardiocentrism is strictly connected to defending the unity of the substantial form, read through the lense of the Neoplatonic concept of "flow" (*fluxus*/*fayḍ*) against the pluralistic implications of Galen's ideas.

On the other hand, spirit alone is unable to explain why the interaction of these material fluids produces a specific body with a determinate set of organs. Rather, spirit can explain that a determinate set of organs are formed in an embryo if and only if it is considered an instrument to the semen of the father granting the mechanism of formal replication. Avicenna's own ideas on the end-directedness of non-voluntary motion are stressed in his *Physics* and chime with Aristotle's approach: teleological processes do not always presuppose conscious willingness to be end-directed, as in the case of artistic performance or writing. In these cases, (over)thinking is not only pointless but, worse, an obstacle to the good performance of the artist.[87] The motion of spirit can therefore be considered similar to an artificial process: it is non-voluntary but is end-directed and has many possible outcomes (males, females, twins, deformities, etc.) due to sublunar or celestial (astral influences) factors. In addition to the astrologization of the theory of generation, a further element of difference between Avicenna and Aristotle consists in the origin of the soul. While the former considers soul to be actualized or educed from matter upon the achievement of a suitable organization of the body, Avicenna considers all sublunar forms (among which is soul) to be created via emanation by God through the Active Intellect or "Giver of forms". This view was heavily criticized by Averroes in his *Long Commentary on Metaphysics*, which represents a second relevant source for Albert's own theory of formative power.

3 Averroes on Heart, Spirit, and Generation: An Undervalued Albertinian Source

The contribution of Avicenna to Albert's views on generation has been widely acknowledged by scholarship. By contrast, much lesser attention has been devoted to the role of Averroes' philosophy of nature in Albert's thought on this subject. Much of the merit of drawing attention to the influence of Averroes in Albert's biological works is due to the doctoral dissertation of Adam Taka-

87 Avicenna, Sufficientia I c. 14, ed. Van Riet 1992, fol. 22rb [ed Cairo 1983, 72]. Translation and doctrinal analysis of this passage in Kara Richardson, "Two Arguments for Natural Teleology from Avicenna's *Šifāʾ*," *History of Philosophy Quarterly* 32 (2015/2), 125–126. On efficient and final causality in Avicenna, see also: Robert Wisnovsky, "Final and Efficient Causality in Avicenna's Cosmology and Theology," *Quaestio* 2 (2002), 97–123.

hashi.[88] In particular, he has insisted on the role of Averroes in the formation of Albert's concept of formative power. In this short section, we shall see how his theory of formative power connects to his understanding of spirit, a question dealt in detail by the contribution of Cristina Cerami we shall rely on.[89] Averroes' concept of spirit is slightly different from Avicenna's, since it postulates the existence of two spirits (vital, animal) only, like Costa ben Luca, Nemesius and other authors.[90] Moreover, Carmela Baffioni has noted that Averroes adheres to Aristotle's own views on generation in terms of the fluids involved. Unlike Avicenna, Averroes admits as male semen and female menses as the only fluids at play in generation, thus excluding the Avicennian reading of Galen's concept of 'feminine semen'.[91] Throughout his works, Averroes presents two main doctrinal points that are key for Albert's theories of generation: (1) the concept of formative power (*Long Commentary on Metaphysics*) and (2) the discussion on the origin of the veins provided in the chapter on the heart (*De corde*) from Averroes' commentary on the *De partibus animalium*. By contrast, Albert did not have access to Averroes' *Colliget* nor did he adopt the Aristotelian scheme of fluids in generation, favoring the solutions of Avicennian zoology and medicine.

The study of Dag Hasse on the Greek, Arabic, and Latin theories of spontaneous generation have shown that in the *Long Commentary on Metaphysics* Averroes displays a position on formative power that, under a certain respect, seems quite close to Avicenna's, for Avicenna considers formative power as the compound of both cosmic and seminal heat. In his Middle commentary on Aristotle's *Meteorology*, Averroes displays an opposite position, in which heat necessary for generation is only natural heat and does not contain a heavenly component.[92] In the *Metaphysics*, Averroes discusses spontaneous generation

88 Takahashi, *Interpreting Aristotle's Cosmos*, 187.

89 This topic emerges especially in Averroes' discussion of spontaneous generation: Cristina Cerami, *Génération et substance: Aristote et Averroès entre physique et métaphysique* (Scientia Graeco-Arabica 18), Berlin 2015, 622–665.

90 Averroes, Colliget, ed. Venezia 1552, fol. 9B [ed. Beirut 1999, 147.11–12]: "Spiritus duo sunt (*ruḥāni*), scilicet spiritus apparens in corde (*mušāhid fī al qalb*) et spiritus apparens in cerebro (*al-rūḥ al-mušāhid fī l-ra's*), sed spiritus in hepate non apparet per viam sensus (*wa-ammā al-kabd fa-laysa yaẓhuru fī-hā bi-l-ḥiss rūḥ*)."

91 Carmela Baffioni, "Further Notes on Averroes' Embryology and the Question of the 'Female Sperm'", in *Averroes and the Aristotelian Heritage*, ed. Carmela Baffioni, Napoli 2004, 159–172.

92 Dag N. Hasse, "Spontaneous Generation and the Ontology of Forms in Greek, Arabic, and Medieval Latin Sources," in *Classical Arabic Philosophy: Sources and Reception*, ed. Peter Adamson (Warburg Institute Colloquia 11), London-Turin 2007, 160. A different position, where cosmic heat is not involved in generation in Averroes' Middle Commentary on

as a counterargument to Aristotle's objections on the existence of separated forms. In this section, he attacks Themistius and other authors such as Avicenna for favoring a Platonic position which endorses an origin of the soul that is completely extrinsic to matter. This is the case, for instance, of Avicenna's theory of the mediated origin of the soul, created by God through the emanation of the "Giver of forms". Averroes, by contrast, follows the Aristotelian doctrine of eduction of soul from matter. In his *Long Commentary on the Metaphysics*, spontaneous generation occurs when the heat of the sun, mixed with that of the stars, generates heat in water or earth.[93] The heat that is concentrated and enclosed within the spirit then triggers the natural process that causes the coming-to-be (or actualization) of the substantial form. In this process, Averroes sees no need to postulate an external agent such as Avicenna's "Giver of Forms".

By the same token, all sublunar souls are educed from matter and do not originate from an extrinsic principle. Averroes' doctrine of the ontology of forms in spontaneous generation constituted a (successful) attempt to return to an Aristotelian stance, arguing that forms are never produced from outside agents but are rather actualized. The agents responsible for the actualization of such forms are man (also, animals), the sun, the stars and the heavenly bodies in sexual reproduction and, on the other hand, sun and the stars/heavenly bodies in spontaneous generation. Like Aristotle, Averroes maintained a firmly cardiocentric approach, in keeping also with the doctrines of his philosophical archenemy, Avicenna. Averroes' views on the cardiocentric nature of human embryology appear clearly in his chapter on the heart (*De corde*). This small piece of Averroean zoology will enjoy a separate circulation in the Latin Middle Ages and will represent a relevant doctrinal piece of Albert's discussion of

---

the Meteorology: Averroes, In Aristotelis Meteorologicorum libros IV commentarius (Aristotelis Opera cum Averrois Commentariis, vol. 5), ed. Venezia 1562, fol. 461r M: "Iam igitur declaratum est ex hoc quod calor est duplex, naturalis et extraneus; et quod generatio fit per calorem naturalem, corruptio autem per extraneum."

93 Averroes, In Aristotelis Metaphysicorum libros XIIII commentarius (Aristotelis Opera cum Averrois Commentariis, vol. 8), XII comm. 18, ed. Venezia 1562, fol. 299 I–K [ed. Bouyges 1973, 1501.17–1502.7]: "Unde Aristoteles dixit quod homo generatur ex homine et sole. Et factus est ille calor in terra et aqua ex calore solis admixtum cum calore aliarum stellarum. Et ideo sol et aliae stellae sunt principium vivi cuiuslibet in natura. Calor igitur solis et stellarum divisus in aqua et terra generat animalia per putrefactionem nata. Et universaliter omnia quae fiunt ex non semine. Non quod illic sit anima in actu inventa in orbe declivi et a sole sicut narrat Themistius. Et omnia ista sunt declarata in libris *De animalibus*." Other relevant passages from Averroes' zoology are discussed in: Freudenthal, "The Medieval Astrologization," 121–123. On generation and formative power in Averroes, see also: Cerami, *Génération et substance*, 500–534.

the Aristotelian-Galenic quarrel over the origin of the veins that has hitherto received very little scholarly attention.[94]

In the *De corde*, Averroes sets out to defend the Aristotelian view that the heart is the principle of digestion, of veins and of the formation of spirits. This is because of Averroes' staunch belief that the heart plays a central role in the diffusion of heat and production of blood (hematopoiesis).[95] To kick off his refutation of the Galenic theory, Averroes first raises concerns about the consequences of Galen's assumption. Putting heart on equal footing with brain, because of the latter's purported role as chief of perceptive operations, and liver, chief organs of the nutritive operations, would have denied the heart its nature as principal member (*membrum principale*) capable of exerting its power over the entire body. In this manner, the sphere of action of the heart would have been limited to the realm of pulse and, at broadest, to the kind of function associated with vital spirit. Dag Hasse has shown that, for a staunch Aristotelian like Averroes, this risk had to be avoided and the potential harmfulness of Galen's position had to be defused appropriately. According to the Corduban thinker, the natural upshot of considering the liver a member capable of forming human members would have put it, like the brain, on equal causal footing with the heart, thus questioning the unity of the substantial form and opening the door to a pluralist position. Averroes, who was committed to an interpretation of Aristotelian psychology based on the unity of the substantial form, insisted on the plurality of faculties though maintaining the soul's essential unity.[96] For this reason, his refutation of Galen's doctrine cannot be understood as separate from his psychology. This line of reasoning, which already applies to his philosophical archrival Avicenna, will be crucial for Albert the Great.

---

94 Richard de Vaux, "La première entrée d'Averroës chez les Latins," *Revue des Sciences philosophiques et théologiques* 22 (1933) 193–245.

95 Averroes, Capitulum de corde (Paris, Bibliothèque National de France, 16110, fol. 99v): "In corde autem est secundum Aristotelem nutrimenti cibalis et spiritus et venarum principium quatenus in uno eodem calor esset agens et patiens sanguinis et venae illius actionis et passionis essent fundamentum et cor omnium istarum esset origo." Albertus Magnus, De animalibus III tr. 1 c. 5 n. 43 (ed. Stadler, I, 296.1–5).

96 Averroes, Commentarium magnum in Aristotelis De anima libros, I, comm. 7, ed. Crawford 1953, 10.13–16: "Plato enim dicebat quod virtus intelligibilis est in cerebro, et concupiscibilis in corde, et naturalis, scilicet nutritiva, in epate. Aristoteles autem opinatur eas esse unam subiecto et plures secundum virtutes." Averroes' criticism against Plato and Galen is examined more thoroughly in: Dag N. Hasse, "*Plato Arabico-Latinus*: Philosophy—Wisdom Literature—Occult Sciences," in: *The Platonic Tradition in the Middle Ages. A Doxographic Approach*, ed. Stephen Gersh/ Maarten J.F.M. Hoenen, with the assist. of Pieter Th. van Wingerden (Filosofia Platonica 184), Berlin-New York 2002, 40–45.

Averroes' refutation of Galen's doctrine on the origin of veins is enacted by a defense of Aristotle's arguments (*rationes*) according to which the heart is the principle of the veins. The first argument (*ratio*) is that the heart is the principle of heat, digestion and food, and thus must be the principle of that which conveys the whole nourishing substances to the entire body, viz. veins. The heart is precisely this kind of body part.[97] To back up this first *ratio*, Averroes states that the principle of veins must be that which has no veins that reach it (*illud ad quod nulla uena uadit*) but only that tend toward it (*et ad quam omnes tendunt debet esse principium uenarum*).[98] This is true for all veins except the vena cava inferior, which reaches the heart from the liver. Still, Averroes considers it wrong to assume like Galen that a body part is thickest when closest to its origin and thinner when furthest from it. In his doctrine, Galen considered the vena cava inferior to originate from the liver and not from the heart, since the vena cava inferior represented only a branching out of the hepatic vein (*vena porta*).

In the first place, Averroes rules out possible objections, like Avicenna's, that it is not necessary to argue that a thing originates from another if and only if it shares a common material composition, as in the case of branches produced by the root of a plant. After all, argues Avicenna, corals (which are hard to the touch) originate from the notoriously soft sea sand. In like manner, it would not be impossible to uphold that nerves (of a hard nature) originate from the heart, which is softer. Averroes, though, discards Avicenna's objection by distinguishing the kind of generation to which corals and branches are subject. In Averroes' view, the generation of coral in the sea must be comparable to that of trees and plants on earth, rather than to the generation of branches from roots.[99] This is because Averroes believes and wants to stress that the heart is closer to the veins in its material composition as a substance than the liver is.[100] Perhaps this is the point beneath his criticism of Avicenna's metaphor.

97 Averroes, Capitulum de corde (Paris, Bibliothèque National de France, 16110, fol. 99va); Albertus Magnus, De animalibus, III tr. 1 c. 5 n. 47 (ed. Stadler, I, 297.16–21).

98 Averroes, Capitulum de corde (Paris, Bibliothèque National de France, 16110, fol. 99va); Albertus Magnus, De animalibus III tr. 1 c. 5 n. 47 (ed. Stadler, I, 297.21–25).

99 Averroes, Capitulum de corde (Paris, Bibliothèque National de France, 16110 fol. 99vb): "Et sic manifestum est quod aliter vena a radice venarum et aliter corallus ex terra molli et quia maior est assimilatio substantiae cordis quae est filosa cum venarum substantia quam epatis." Albertus Magnus, De animalibus III tr. 1 c. 5 n. 48 (ed. Stadler, I, 298.9–12).

100 Averroes, Capitulum de corde (Paris, Bibliothèque National de France, 16110, fol. 99vb); Albertus Magnus, De animalibus, III tr. 1 c. 5 n. 48 (ed. Stadler, I, 298.13–15): "Maior autem est assimilatio substantiae cordis quae filosa est cum venarum substantia, quam epatis quod est caro mollis."

Having set Avicenna's critique of Galen aside, Averroes returns to his own rejection of Galen's idea that veins originate from the liver as a branch from the roots. To upturn Galen's arguments, Averroes appeals (like Avicenna, for once) to empirical observations: by citing the case of barley, Averroes manages to reverse the Galenic correlation between material origin and thickness. Avicenna had done this by using the case of optical nerves, thinner when nearer to their origin, viz. the brain, and thicker when furthest from it (in the eye). In all cases mentioned by Averroes, branches are thicker than the roots, so one could be reasonably tempted to reverse Galen's line of reasoning and consider the heart as the origin of veins such as the vena cava inferior.[101] This is precisely the point of Averroes' argument. If veins can materially originate from the heart as thick branches can sprout from thinner roots, the next step in Averroes' *De corde* is to present the solution to this quarrel by defending Aristotle's position.

According to the Corduban thinker, it is indeed true that one may see the heart as an organ with a limited range of body parts to which it is connected through veins. As a matter of fact, the only vein which sprouts from the heart is the so-called *arteria venalis*, the artery which connects the right side of the heart to the lungs. Averroes, advocating Aristotle's point of view, tries to counter this idea by saying that the heart is the most apt among body parts to concoct and produce blood inasmuch as it presents a structure of three-chambered ventricles particularly suited to host concocted blood. Galen, though, had convincingly challenged this view by arguing that the liver is surrounded by blood vessels well-suited to wasting away the residues in the concoction of blood produced by the liver. To finally refute Galen's view, Averroes then resorted to the cardiocentric nature of the nutritive process as he had understood it from Aristotle.

Averroes bases his confirmation of Aristotle's standpoint on metaphysical grounds, heavily hinging on the concept of form in relation to the nourishment and growth of a living being. In Aristotle and in later philosophy of nature inspired by the doctrines of the Stagirite, blood plays the key role as material constituent of all body parts in a living body. The differences among these body parts depend on the degrees of concoction reached by the blood. According to Averroes, the food provided to the heart from veins such as the vena cava is still crude and receives its essential form (*suscipit formam essentialem*) from the

101 Averroes, Capitulum de corde (Paris, Bibliothèque National de France, 16110, fol. 99vb): "[…] ramus fit grossior radice […] Deinde est id quod videtur in grano ordei." Albertus Magnus, De animalibus III tr. 1 c. 5 n. 48 (ed. Stadler, I, 299.10–14).

heart, by which it constitutes the body's parts.[102] This form is received by means of the vital heat acting on the heart. By maintaining the heart as the central principle of digestion, Averroes is then able to overturn the Galenic idea that the principle of the member is located in the same place as the principle of the soul power. If nutrition is a life-giving power, and the organ that provides blood the complexion most suitable to nourish and grow body parts is the heart, then the liver comes to play a relevant role yet remains always subordinate to that of the heart.

All in all, the Averroean philosophy of nature did present Albert with a relevant array of notions for his treatment of generation. On the one hand, spirit appeared as carrier of a formative power that combines cosmic and seminal heat and does away with the Avicennian "Giver of forms" by replacing it with the more Aristotelian concept of the eduction or actualization of forms during the process of generation. Moreover, Albert found further valuable insights in Averroes' cardiocentric approach to nutrition as a key to the refutation of the Galenic doctrine on the origin of the veins. Averroes' chapter *De corde*, already known at the time of his earlier *De homine*, will prove crucial for Albert to settle the Galenic-Aristotelian quarrel over the origin of veins.

4 Albert on Generation between Theology, Zoology, and Botany

At the time when Albert composed his *De homine*, medieval theologians were debating generation-related issues from two different perspectives. The first concerned the theological and juridical determination of the rights of the embryo. Stimulated by the famous passage in the book of Exodus (21, 22–25),[103] Medieval authors had to find a solution to the thorny question about the precise chronological determination of the beginning of life and, in the wake of this, when induced abortion should be considered murder. Answers to these questions had—as nowadays—strong juridical and ethical implications. On the other hand, Medieval theologians at the time of Albert's *De homine* debated

102 Averroes, Capitulum de corde (Paris, Bibliothèque National de France, 16110, fol. 100ra): "[...] sumit formalem essenciam a corde."

103 Exodus, 22, 21–24: "If men quarrel, and one strike a woman with child, and she miscarry indeed, but live herself: he shall be answerable for so much damage as the woman's husband shall require, and as arbiters shall award. But if her death ensues thereupon, he shall render life for life. Eye for eye, tooth for tooth, hand for hand, foot for foot [...]."

the process of embryonic formation and its relation to the complex Aristotelian theory of the origin of the soul. These two aspects of the early Albert's embryology will present different solutions which, as will become evident, are clearly based on Albert's preference for Augustine as an authority in theological matters such as abortion. By contrast, Aristotle's and Galen's opinions were to be heeded as the most authoritative in matters of philosophy and medicine respectively.[104]

4.1 *Formed in 46 Days Like Jerusalem's Temple? Albert on Life, Abortion, and Murder*

When Albert discusses abortion in his *Commentary on the Sentences*, he does so following a tradition that mostly identifies the beginning of life and animation not with the motion perceived by the mother but with the complete formation of the embryo sometime around 40 days after conception. Maaike van der Luugt has highlighted this trend in theological treatises on the human soul from the 13th century.[105] This is a position that can already be found in William of Conches.[106] Moreover, theological discussions on generation in the Middle Ages tended to be flare up easily because of their connection to another very sensitive topic, viz. the transmission of original sin and the issues raised by traducianist doctrines. This is an aspect of the problem of medieval theories of generation that goes well beyond our purpose, which is humbler: namely, to establish when Albert thought the origin of the soul could occur within the theological context of discussions on abortion.

Albert takes a cue from Augustine's *De diversis quaestionibus* and equates the animation of the fetus with its complete formation. In this respect, Albert totally adheres to Augustine's position (*Dicit etiam Augustinus, quod informe puerperium non habet animam*), who symbolically equated the 46 years in which Jerusalem's temple was built to the days within which the embryo becomes fully formed. Metaphorically speaking, the human body is God's temple.

---

104 Albertus Magnus, Super II Sententiarum dist. 13 art. 2, ed. Auguste Borgnet, Paris 1894 (Ed. Paris. XXVII, 247a): "[…] sciendum, quod Augustino in his quae sunt de fide et moribus plusquam Philosophis credendum est, si dissentiunt. Sed si de medicina loqueretur, plus ego crederem Galeno, vel Hipocrati: et si de naturis rerum loquatur, credo Aristoteli plus vel alii experto in rerum naturis." This translation by Siraisi in: Nancy Siraisi, "The Medical Learning," 382.

105 Maaike van der Luugt, "L'animation de l'embryon humain et le statut de l'enfant à naître dans la pensée médiévale," in *Formation et animation de l'embryon dans l'Antiquité et au Moyen Âge*, ed. Luc Brisson/ Marie-Heléne Congourdeau/Jean-Luc Solère (Histoire de doctrines de l'Antiquité Classique 38), Paris 2008, 242–243.

106 On William's position: Van der Lugt, *L'animation de l'embryon humain*, 236–242.

For this reason, if Jerusalem's temple was completed in 46 years, the human body should be completed in 46 days. For this reason, abortion that is accidentally caused before this days cannot be condemned as murder, strictly speaking.[107]

Anthony Demaitre and Luke Travill have called the attention to the fact that the timing for the embryo's formation is borrowed by Albert from a Salernitan *florilegium*[108] and differs from the usual timing for embryonic animation, which is set around 40 days. Unlike Avicenna, Albert is not concerned here about the biological consequences of a delayed formation of the embryo, for he regards Augustine as an authority over Aristotle and Galen in matters of theology, as mentioned above. While the *Canon* and the *De animalibus* (which he largely used in the *De homine*) had stressed that the formation of an embryo in 45 days implies a delayed birth with biologically problematic consequences,[109] the occurrence of animation within this timeframe does not really complicate Albert's account. The reason behind this apparent tension in Albert's account can be sought in the main theological focus of the *quaestio*: providing a medically and physiologically accurate rendition of pregnancy with its pathological consequences is not the primary focus of the *quaestio*.

Among authors contemporary with the *De homine*, Augustine's position is endorsed by William of Auvergne in his *De anima*.[110] In a similar *quaestio* explicitly devoted to the topic (*utrum vita sit in embrione ante organizationem*), the *Summa Halensis* claims—on a similar note—that life begins when the conceived fetus is fully formed. Unlike Albert and William, the *Summa Halensis* is not specific about the precise temporal limit of when life begins.[111] By con-

107 Albertus Magnus, Super IV Sententiarum dist. 31 a. 18, ed. Auguste Borgnet, Paris 1894 (Ed. Paris. XXX, 250–251); Demaitre–Travill, "Human Embryology," 424.

108 Albertus Magnus, Super IV Sententiarum dist. 31 a. 18, ed. Auguste Borgnet, Paris 1894 (Ed. Paris. XXX, 251): "Semine conceptum sex primis rite diebus / Fit quasi lac, reliquisque novem fit sanguis: at illud/Consolidat duodena dies, bis nona deinceps/Effigiat, tempus sequens producit ad ortum." Demaitre–Travill, "Human Embryology," 424.

109 Avicenna, Canon III fen 21 c. 3, ed. Venezia 1507, fol. 361[ra]; Avicenna, Abbreviatio de animalibus IX c. 5, ed. Venezia 1508, fol. 43[rb] and the previous section of this chapter on Avicennian embryology.

110 William of Auvergne, Tractatus de anima, c. 4, pars 4 (ed. Paris 1674, fol. 109b): "Doctrina vero Christianorum habet quod quadragesimo sexto die consummatur formatio eius ad similitudinem templi illius famosissimi Hebraeorum quod 46. annis aedficatus fuit. Secundum hoc igitur anima rationalis quadragesimo sexto die infunditur fetui huiusmodi et quasi septem septimanis completur formatio eius."

111 Summa fratris Alexandri (Summa Halensis), pars 2 inq 4 tr.3 quaest. 1 tit. 2 c. 6 (ed. Quaracchi, II, 683): "Propter hoc videtur aliter dicendum. Potest ergo dici, sine praeiudicio, quod,

trast, Aquinas clearly sets the limit within which an embryo becomes a human being at 40 days.[112] These works fundamentally confirm van der Luugt's claim that Medieval discussions on induced abortion equate the beginning of life with complete embryonic formation in the second month. Whether this claim amounts to Aristotle's 40 days or Augustine's 46, the outcome is not really conflicting. As a matter of fact, more depends on the field of inquiry (theology, natural philosophy, medicine) and, as a consequence, on which authority is favored. A difference of six days seems not to be particularly significant, unless the perspective which is favored is the medical one or, at least, one which tries to give a philosophical explanation of delayed or premature births. While in thinking about abortion he may well favor Augustine, for the rest of his treatment of generation in his early works, Albert blends Aristotle, Avicenna, and Averroes although with a very skeptical attitude to the determination of the precise moment of the embryo's ensoulment.

4.2 *Albert and the Origin of the Soul throughout His Works*

The theme of the origin of the soul is a sensitive issue throughout Albert's works. It is also of paramount importance to understanding Albert's physiology of generation, firmly based on Avicennian zoology (and medicine), with relevant insights from Averroes and some skeptical additions by Albert himself. While we shall mostly focus on the physiology of Albert's theory of generation, it must not be forgotten that his theory on the origin of the soul was developed during the reworking of many sources such as Aristotle, Averroes (*De substantia orbis*), Augustine, and the pseudo-Aristotelian *De causis proprietatum elementorum*; thus he is not just interacting with Avicenna and other medico-philosophical sources, although they represent the main focus of this book.[113] Among the latest contributions, Katja Krause's article has shed some new light on some different doctrinal nuances in Albert's position that shifted over the beginning of his career that, in the end, offer a key to understanding the mature position held by Albert on the origin of the soul in humans, animals, and plants. There is some friction within his early account, since (only a) few passages seem to point towards what Krause has called "absolute psy-

---

proprie loquendo, embrio non vivit, eo modo quo illud dicitur vivere quod habet animam formam et sui perfectionem: huiusmodi autem proprie dicuntur vivere."

112 On Aquinas: Fabrizio Amerini, *Aquinas on the Beginning and End of Human Life*, trans. Mark Henninger, Cambridge 2013, 165–191.

113 On Albert's view on the origin of the soul, see also: Horst Seidl, "Zur Geistseele im menschlichen Embryo nach Aristoteles, Albert dem Großen und Thomas von Aquin. Ein Diskussionsbeitrag," *Salzburger Jahrbuch für Philosophie* 31 (1986), 37–63.

chological" creationism.[114] On the other hand, the final model developed by Albert in his early works on the origin of the human soul is that called "epigenetic creationism" by Krause, a model which is key to understanding his later views on the human soul. In this framework, generation is the result of an order of combined causes, since it combines the Aristotelian (also, Averroean) eduction of the soul from matter in the case of the lowest souls (vegetative, sensitive) with a creationist view of the rational human soul, created directly by God once the nutritive and sensitive powers have been educed from matter. Overall, Albert conceives of animal generation as the result of the interaction between three ordered causes: (1) the generative power in the male semen and the generative matter of the menstrual blood, (2) the elemental qualities of heat and moisture as active causes, and (3) the transcendental power of the heavens.[115]

To explain the origin of the animal soul, Albert appealed to the Avicennian concept of generative power, which he would call in his later works *virtus formativa*.[116] Often, Albert explains this generative power in the semen using Aristotle's famous art analogy: just as the artisan's art is found in the artefact, so is the father's soul in the male semen (*sicut artifex in artificiato*). This artefact is composed of seminal heat, cosmic heat and heat derived from the father's soul.[117] This heat serves as an instrumental and efficient cause of the formation of the embryo.[118] This doctrine is close but not identical to the *Summa Halensis*, which views generation as a cooperation of heavenly and sublunary (viz. the father's soul) virtues (*producuntur in esse a virtute generantis cooperante virtutis celesti*).[119] According to Albert, a variation in cosmic heat may determine some variation in the final result of generation, which results materially from a "kind of moisture mixed in accordance with the complexion of the semen and with nature (*materia est humidum commixtum secundum com-*

114 For an overview of these tensions, see: Krause, "Albert the Great on Animal and Human Origin," 209–221.

115 Krause, "Albert the Great on Animal and Human Origin," 221.

116 Krause, "Albert the Great on Animal and Human Origin," 223.

117 Albertus Magnus, De homine (Ed. Colon. XXVII/2, 139.30–43): "[…] dicimus quod formae substantiales sunt a qualitatibus primis agentibus et patientibus; licet enim calidum tantum sit agens, non tamen agit tantum in virtute calidi ignei, quia hoc est consumptivum, ut dicitur in libro *De substantia orbis*. Agit enim etiam cum virtute calidi caeli et cum virtute calidi animae […]".

118 Albertus Magnus, De homine (Ed. Colon. XXVII/2, 141.35–39; 141.71–142.5).

119 Summa fratris Alexandri (Summa Halensis), II inq. 4 tract. 3 q. 1 tit. 2 c. 1 (ed. Quaracchi II, 664–665).

plexionem seminis et naturam)."[120] This mixing in accordance with a suitable complexion of the semen triggers a gradual development in the embryo. After successful fertilization, embryonic development proceeds from the heart to all other body parts and presupposes the existence of a nutritive 'soul,' not in the strict sense of form but as a psychological principle of bodily organization.[121] This will also be the case with the eduction of the sensitive soul as something other than 'form', at least in the case of humans. When the embryo reaches a suitable degree of organic complexity, it receives the intellect from outside thanks to divine creation.[122] By contrast, the origin of animal soul is explained as a process of pure and gradual eduction of the form from matter through this generative power.[123]

In the early Albert, the origin of the human soul is directly attributed to God, in cooperation with a series of natural causes. This origin occurs without the mediating role of an intelligence such as Avicenna's Giver of Forms (*dator formarum*). In the Aristotelian paraphrases the model of animal psychology remains similar, even though it has been subjected to hermeneutical discussions by eminent scholars.[124] It is indeed true, as stated by Dag Hasse, that there are some passages from Albert's *Metaphysics* where Albert draws a comparison between the human rational soul and the soul of heavenly intelligences (and not of the First Cause), arguing that the former is the "image" (*imago*) of the latter.[125] Indeed, there may be a resemblance between human soul and heavenly intelligences in (rational) nature. On the other hand, though, Albert claims quite clearly in his *De natura et origine animae*, viz. the mature work in

120 Krause, "Albert the Great on Animal and Human Origin," 225–226.

121 Krause, "Albert the Great on Animal and Human Origin," 226.

122 Albertus Magnus, De homine (Ed. Colon. XXVII/2, 137.62–70): "Sine praeiudicio aliorum non discedimus a nostra sententia, quam supra posuimus in quaestione de creatione, scilicet quod formae omnes praeter animam rationalem potentia sunt in ipsa materia generabilium et corruptibilim, et educuntur per actionem calidi et frigidi et humidi et sicci commixtionem. De anima autem rationali sola dicimus quod ipsa infundendo creatur et creando infunditur a prima causa, quae est deus gloriosus et sublimis."

123 Krause, "Albert the Great on Animal and Human Origin," 226–227; Luciano Cova, "Seme e generazione umana nelle opere teologiche di Alberto Magno," in: Summa doctrina et certa experientia. *Studi su medicina e filosofia per Chiara Crisciani*, ed Gabriella Zuccolin (Micrologus Library 79), Firenze 2017, 237–256.

124 Dag N. Hasse, "Avicenna's 'Giver of Forms' in Latin Philosophy, especially in the Works of Albertus Magnus," in *The Arabic, Hebrew and Latin Reception of Avicenna's Metaphysics*, ed. Dag N. Hasse/Amos Bertolacci (Scientia Graeco-Arabica 23), Berlin-Boston 2018, 225–249.

125 Hasse, "Avicenna's 'Giver of Forms' in Latin Philosophy," 240–241.

which he addresses *ex professo* the issue of the origin of the human soul, that the human soul cannot be created from one of the cosmic intelligences but from God alone.[126]

Prima facie, this would seem to openly clash with a passage from the *De anima* in which Albert discusses briefly the origin of the human soul. There, Albert criticizes the standpoints of Alexander of Aphrodisia, who considers the soul to be completely educed from matter, and of those that consider the soul to have a completely extrinsic origin from matter, provided by something like a "Giver of Forms" (*similitudo datoris formarum*).[127] As we know, Albert upheld the view that the lowest faculties of the soul were educed from matter, while the rational soul was created by God through its light, a view that is found in his early work and that is reiterated in his mature *De natura et origine animae*.[128] Consistently with his rejection of a totally extrinsicist (favoring the Giver of Forms) or an utterly materialist explanation of the origin of the human soul, Albert rejects both opinions as false. Rather, he argues that there is an intelligence that provides the end (*intelligentia dat eum*), viz. the rational soul, to the human being, when matter has reached a suitable degree of material preparation. But what kind of intelligence is implied in this passage? Albert speaks very generically of "intelligence" here, so there is no conclusive

126 Albertus Magnus, De natura et origine animae tr. 1 c. 5 (Ed. Colon. XII, 13.9–17): "[…] dixerunt quidam animam rationalem ab intelligentia fieri; et verum dixerunt, si intelligentia dicatur intellectus primus separatus et immixtus, nulli nihil habens commune, sicut dixit Anaxagoras. Si autem intelligentia dicatur intellectus aliquis secundus, motor alicuius orbis inferioris, non potest esse verum, quod dicunt, quoniam quaecumque sit secundarum intelligentiarum, non movebit nisi in virtute primi intellectus." I thank Henryk Anzulewicz (Albertus Magnus Institut) for drawing my attention to this passage.

127 Albertus Magnus, De anima III tr. 2 c. 4 (Ed. Colon. VII/1, 183.32–44): "Similiter autem quod dicit, quod intellectualis anima educatur de semine, falsum est omnino, sed potius ipsa est ingrediens ab extrinseco a datore in materiam, et est similitudo datoris formarum, qui est intelligentia prima et non commixta corpori, sicut commiscetur virtus vel forma corporis; et ideo hoc etiam est falsum. Verum est tamen, quod ipsa est finis generationis, sed hic finis non efficitur in materia virtute qualitatum primarium transmutantium materiam, sed potius intelligentia dat eum, quando materia est disposita per principia naturalia. Et de hoc in libro *De animalibus*, ubi de generatione hominis tractabitur, erit agendum."

128 Albertus Magnus, De natura et origine animae tr. 1 c. 6 (Ed. Colon. XII, 14.19–27): "[…] tamen haec formativa non educeret eas hoc modo, prout sunt potentiae rationalis et intellectualis formae et substantiae, nisi secundum quod ipsa formativa movetur informata ab intellectu universaliter movente in opere generationis. Et ideo complementum ultimum, quod est intellectualis formae, non per instrumentum neque ex materia, sed per suam lucem influit intellectus primae causae purus et immixtus." This translation by Takahashi in: Takahashi, *Interpreting Aristotle's Cosmos*, 181.

evidence that the intelligence implied here is the analogous to the Avicennian *dator formarum*. Moreover, the link to the *De animalibus* in the end of the passage is particularly relevant insofar as it contains the possible solution to this issue. As it is well-known, the *De natura et origine animae* originally belonged to the *De animalibus* whence it was excerpted. In the *De natura et origine animae*, though, the intelligence which creates the human rational soul cannot be a second intelligence that moves planets, though, but only an utterly immaterial intelligence which Albert considers explicitly to be the First Cause (God):

> In sum, then, we say that every form of nature is a form that is induced from the intelligence (*omnis forma naturae est forma ab intelligentia* [...] *inducta*), which is the prime mover of nature. And there is no difference whether it is introduced by many or few media or by one medium (*Neque est differentia, utrum per multa aut pauca aut unum medium inducatur*), just as there is no difference in forms of art with regard to the fact that it is a form of the artist, whether it is introduced by one tool or by many, just like a medium. The introduced form differs in accordance with its proximity to matter and in accordance with its affinity to the mover and in accordance with the mediate distance and proximity to both, yet such that every form that is introduced is in the light of the intellect that moves first in nature.[129]

The prime mover of nature can hardly be a cosmic intelligence; rather, it must be identified with God. In the end, Albert seems to uphold a consistent view throughout his works on immediate origin of the rational part of the human soul, created by God, while displaying somes nuance as regards the concept of generative/formative power and the origin of the lower soul faculties (vegetative, sensitive), to which we shall now briefly turn.[130]

4.3 *Different Nuances of Formative Power and* inchoatio formae *in Albert's Mature Works*

Katja Krause has shown how the mature Albert maintains that generation results from the same combination between the work of intelligence and that

129 Albertus Magnus, De natura et origine animae tr. 1, c. 1 (Ed. Colon. XII, 4.34–46). This translation by Krause in: Krause, "Albert the Great on Animal and Human Origin," 231.

130 Therefore, the claim that Albert endorses the Avicennian doctrine of the *dator formarum* in his *De anima* (Hasse, "Avicenna's 'Giver of Forms' in Latin Philosophy," 239) should be revised.

of heavenly bodies, the formative (previously generative) power in the semen and the instrumental causes of feminine semen as matter and of heat and spirit as their shaping tool.[131] On the other hand, Adam Takahashi has shown how the material components of the heat placed in spirit, which represents the instrument of formative power, are composed by (1) seminal heat, (2) cosmic heat, and (3) the heat poured out by the father's soul.[132] This doctrine is consistent with Albert's early works.[133] The end-directedness of the action performed by formative power on embryonic matter is to be traced back directly to the action of the First Cause.[134] According to Albert, spirit is a divine body, but not exactly in the same way as proposed by Freudenthal concerning Aristotle's remarks, viz. because it enables the eternal preservation of the species. Rather, spirit is a divine body because of the divine mind at work behind it, which provides the teleology of natural processes in which spirit is involved.[135]

The gist of Albert's mature understanding of the process of embryonic formation is expressed in his theory of the *inchoatio formae* and his link to embryonic animation, studied mainly by Bruno Nardi and, more recently, Katja Krause. In particular, Nardi drew attention to some passages in the *De natura et origine animae* where Albert presents an apparently different nuance in his understanding of the process of animation and the role of formative power. To be understood properly, these passages need be considered against the background of Albert's famous theory of the *inchoatio formae*.

Nardi and, after him, Rodolfi, have shown that Albert's theory of the *inchoatio formae* pivots around the interpretation of the Aristotelian concept of privation (στερήσις, *Phys.*, I, 9), not as pure potency but as "aptitude towards the form" (*aptitudo ad formam*), and is found in his *Physics* and the *De generatione et corruptione*.[136] This interpretation is achieved through the distinction

131 Krause, "Albert the Great on Animal and Human Origin," 228–232.

132 Takahashi, "Nature, Formative Power and Intellect," 463.

133 Krause, "Albert the Great on Animal and Human Origin," 221.

134 Takahashi, "Nature, Formative Power and Intellect," 464–465.

135 Takahashi, "Nature, Formative Power and Intellect," 467: "It is thus not the spirit itself, but the formative power conveyed by the spirit that brings about the various bodily organs and members. In other words, this power is so divine because of the divine mind that is at work behind the instrument, just as the smith's mind guides the hammer in his hand in producing diverse products."

136 Bruno Nardi, "La dottrina d'Alberto Magno sull' *inchoatio formae*," in *Studi di filosofia medievale*, ed. Bruno Nardi, Roma 1960, 69–101. For a different interpretation of Albert's theory of *inchoatio formae*, see: Steven C. Snyder, "Albert the Great, *Inchoatio Formae*, and the Pure Potentiality of Matter," *American Catholic Philosophical Quarterly* 70 (1996), 63–82. Snyder argues, against Nardi's view, that Albert, like Aristotle, regarded matter as pure potentiality, even though this reading could hardly come to terms, for instance, with

between *potentia* (sheer potentiality) and *potestas* (potentiality of existence of the form in matter, although in a confused manner). Albert's doctrine of *inchoatio formae* combines his understanding of privation as aptitude for form with Augustine's theory of *rationes seminales*. This combination is witnessed in other contemporary medieval thinkers such as Bonaventure.[137] In Albert's view, the confused existence of form in matter can be causally explained as a product of the influence exerted by the heavenly spheres and by the interaction of sublunar elementary qualities.[138]

The main difference spotted by Nardi is that in Albert's zoology and *Parva Naturalia*, embryonic animation is brought about by formative power. By contrast, in the *De natura et origine animae* the action of formative power alone is necessary but not sufficient to explain the mechanism of formation and animation (*agit formando membra et animando ea et vivificando*) of the embryo. This action be effectively carried out by the nutritive and the sensitive soul, which are gradually educed from matter by the formative power and God's assent.[139] Albert's purpose seems to be to stress that the animating function of the formative power and its role in embryonic formation, much as they require the cooperation and the blending of fluids from both parents (*virtus formativa*) and divine assent, is not carried out in a totally inert matter. Rather, the embryo, which exudes a gradual organic complexity throughout the month of pregnancy, comes to represent a hylomorphic compound that undergoes a progressive development explicated in terms of progressive 'souls', not in the strict

Aquinas' criticism of the theory of *inchoatio formae* as a denial of the pure potentiality of matter. For a more recent contribution, closer to Nardi's standpoint (also, ours), see: Anna Rodolfi, *Il concetto di materia nell'opera di Alberto Magno* (Corpus philosophorum Medii Aevi. Testi e studi 18), Firenze 2004, 110–125.

137 Albertus Magnus, De natura et origine animae tr. 1 c. 6 (Ed. Colon. XII, 13.82–87): "Oportet autem scire, quod sicut in aliis ita etiam in homine inchoatio vegetativi est in materia et in esse primo substantiae animandae et inchoatio sensibilis est in vegetativo et inchoatio rationalis est in sensitivo, quia aliter homo constitutus, sicut ante probatum est, esset multa et non unum." Albertus Magnus, Physica I tr. 1 c. 3 (Ed. Colon. IV/1, 38.4–17); Nardi, "La dottrina d'Alberto Magno," 69–101.

138 For a comparison between Albert's position on *incohatio formae* and the Avicennian doctrine (which he does not make use of, in spite of some detectable doctrinal commonalities), see: Lizzini, "Flusso, preparazione appropriata e *inchoatio formae*," 142–143.

139 Bruno Nardi, "L'origine dell'anima umana secondo Dante," in *Studi di filosofia medievale*, ed. Bruno Nardi, Roma 1960, 31; see: Albertus Magnus, De animalibus XVI tr. 1 c. 4 n. 26 (ed. Stadler, II, 1073.17–21): "[...] sed virtutes calidi et humidi et aliae quae sunt in semine, sunt instrumenta illius virtutis quae in spiritum seminis influxa est ab anima quam virtutem vocamus formativam: et ipsa dirigit illas qualitates instrumentales ad finem a se intentum, hoc est ad membrorum formationem."

sense of 'forms' but, as in Aristotle and in Albert's early *De homine*, as pure capacities of the embryo to exercise certain activities.

While the stress in Albert's zoological works is strongly placed upon the physiological component responsible for animation, viz. formative power, in the *De natura et origine animae* he emphasizes the fundamental interaction between physiological and psychological causes at play in the process such as the nutritive and the sensitive souls. These 'souls', though, are not to be taken in the strict sense , but rather as a mere principle of bodily organization in the embryo, which leads to its full formation and ensoulment.[140] This understanding of the process of embryonic animation as the result of a gradual perfection of bodily structure is radically opposite to the explanation of embryonic development put forward by Aquinas, who considered the different stages in embryonic development to reflect the development and disposal (*abiectio formae*) of different 'forms' until the full formation of the embryo.[141]

Albert's mature position on formative power displays a few smaller doctrinal differences compared to his early works. In the first place, Albert makes very explicit that the semen *qua* product of the fourth digestion (as in Avicenna) needs not just to be properly concocted through the heat of the body, but also to be digested by a combination of seminal and cosmic heat.[142] In the second place, Albert develops a complex and detailed theory of rays that accounts for the diffusion of astral influences upon the sublunar world. This theory, which plays a key role in his understanding of both the cycles of generation and corruption (*De generatione et corruptione*) and climate theory (*De natura loci*), is very likely influenced by al-Kindī's *De radiis*.[143] Alessandro Palazzo has shown that, in his mature natural philosophy, Albert systematically uses rays to explain the agency of heavenly bodies and stars on the sublunar world.[144]

---

140 On the consistency of this account of embryonic development with Albert's mature account: Krause, Albert the Great on Animal and Human Origin, 226–227.

141 This assertion by Aquinas also has problematic implications in terms of personal identity: how can the form which is present at the end of generation be the same as that educed from matter after 40 days from fertilization? This problem is debated in: Amerini, *Aquinas on the Beginning and End*, 102–163.

142 Takahashi, "Nature, Formative Power and Intellect," 464. This difference will be dealt with in greater detail in section 3.4 on generation in the Aristotelian paraphrases.

143 Darrel Rutkin, "Astrology and Magic," in *A Companion to Albert the Great: Theology, Philosophy and the Sciences*, ed. Irven Resnick (Brill's Companions to Christian Tradition 38), Leiden-Boston 2013, 459–476.

144 Alessandro Palazzo, "*Philosophi aliter loquuntur de prophetia quam sancti.* Alberto il Grande e la profezia naturale," in *Immaginario e immaginazione nel Medioevo. Atti del convegno della Società italiana per lo studio del pensiero medievale* (S.I.S.P.M.), *Milano, 25–27 settembre 2008*, ed. Maria Bettetini/Francesco Paparella, with the collaboration of Roberto

Albert felt the need to introduce such mediating agents into the sublunar worlds since he opposed the idea that the sublunar world could be subject to any physical influence from the heavenly bodies without the presence of a mediating agent. This is because, according to him, every physical agent acting on a distant being needs a medium to account for its action.[145] Albert considers the amount of heat mediated through rays to depend on a number of factors such as the planet(s) or the star(s), their position with respect to earth, the presence of their rays at the time and place of impregnation, in short a number of combined and ordered causes (*implexio causarum*) which he defines as *fatum*, following the Hermetic formulation.

Alessandro Palazzo has also clarified how the scientific relevance of fate in Albert's thought provides a causal explanation of the link between the celestial and sublunar worlds, even though this explanation may be contingent and not always necessary. This is the case with many natural phenomena such as dream divination, prophecy and astrology.[146] This position on the necessary (yet contingent) role of the astral influences at play in generation is consistent with Albert's view on formative power in his mature works. On the one hand, part of the formative power is composed of seminal heat and part of it must be traced back to celestial influences acting through rays. These influences can be

---

Furlan (Féderation Internationale des Instituts d'Études Médiévales, Textes et Études du Moyen Âge 51), Louvain-la-Neuve 2009, 193.

145 Albert's criticism is formulated in the broader context of his doctrine of divinatory dreams and prophecy, see: Albertus Magnus, De divinatione. *Über die Weissagung*, Latin-German, with a transl. and intr. by Silvia Donati, (Herders Bibliothek der Philosophie des Mittelalters 48), Freiburg im Breisgau 2020, 108–109 (ed. Auguste Borgnet, vol. 9, 188b): "Quidquid autem virtutis in tali composito movente est, non potest ad locum generatorum et corruptorum pervenire nisi sit vehiculum quod vehat ipsum, sicut spiritus vehit virtutes animae per totam materiam quae subiacet moventi animae. Nec invenimus umquam probatum per philosophiam quod intelligentia simplex per se aliquid physice agat vel causet, quoniam ipsum mobile ad hoc datum est ei, ut hoc sit medium ejus quo sicut instrumento possit movere corpora, dum virtus eius per corporeum influitur instrumentum. Oportet igitur quod lumen radiale diversimode figuratum advehat nobis omnes virtutes orbis. Sic igitur nihil fit, ut videtur, per deos et inutiles sunt ad praevisionem futurorum."

146 Alessandro Palazzo, "The Scientific Significance of Fate and Celestial Influences in Some Mature Works by Albert the Great: *De fato, De somno et vigilia, De intellectu et intelligibili, Mineralia*," in Per Perscrutationem Philosophicam. *Neue Perspektiven der mittelalterlichen Forschung. Loris Sturlese zum 60. Geburtstag gewidmet*, ed. Alessandra Beccarisi/Ruedi Imbach/Pasquale Porro, Hamburg 2008, 56; Henryk Anzulewicz, "Der Einfluß der Gestirne auf die sublunare Welt und die menschliche Willensfreiheit nach Albertus Magnus," in *Actes de la Vème Conference Annuelle de la SEAC, Gdansk 1997* (Swiatowit. Supplement [Series H: Anthropology] 2), Warszawa 1999, 263–277.

positive or negative to the extent that they can even determine periods during which birth is impossible because of certain astral conjunctions. This is the case with Jupiter and Mars, for instance.[147] While the negative action of heavenly influences through rays is typical of Albert's mature production, it is not utterly absent in his early production. For instance, his *Commentary on the Sentences* displays early traces of astrological explanations of monstrosities based on Avicenna's *De animalibus*, crucial to his early theory of generation.[148]

4.4 *The Early Albert's Physiology of Generation*

In the *De homine*, Albert's take on the causal role of fluids at play in generation speak clearly in favor of a close adherence to the Avicennian model. As shown in the first chapter, according to Albert the masculine semen maintained its exclusive role of formal and efficient cause of the embryo, while what Avicenna called 'feminine semen' was deemed as matter for the embryo while the menses act as food for the embryo.[149] In the *De homine*, there is

147 Albertus Magnus, Mineralia II tr. 3 c. 2 ed. Auguste Borgnet, Paris 1890 (Ed. Paris. V, 49b): "Non igitur ignoramus quod sunt quaedam loca in caelo, in quibus cum luminaria convenerint, impediunt etiam in propria et efficaci materia figuram humanam generari: et materia tunc concrescit in horribile monstrum."; Albertus Magnus, De animalibus XVIII tr. 1 c. 6 n. 47 (ed. Stadler, II, 1214.34–36): "Ex virtute caelesti quidem quoniam quaedam loca caeli sunt in quibus quando congregantur luminaria caeli quae precipue movent ad generationem, non potest esse generatio hominis. De hiis autem in alia scientia erit opportunus dicendum." On this passage, see: Zuccolin, *I gemelli nel Medioevo*, 153–154.

148 Albertus Magnus, Super II Sententiarum dist. 7 art. 9, ed. Auguste Borgnet, Paris 1894 (Ed. Paris. XXVII, 157b): "Videmus nativitates variari secundum substantiam et secundum operationes consequentes a constellationibus, sicut dicit Ptolemaus *Quadripartito*, quod sole existente in quadam parte et minuto Arietis, non fit generatio humana: et si cadat tunc semen in matricem, monstrum nascetur. Et ut credatur, ego probavi experimento hoc in duabus matronis probis et bonis, a quibus ergo percepi quod monstra peperunt, et quaerens tempus ab eis et aequans stellas, inveni quod sole existente circa eumdem gradum et minutum secundum suas aestimationes conceperant. Item, Avicenna dicit in suo libro *de Animalibus*, quod causa monstruosorum partuum in dissimilibus seminibus non potest esse natura seminis: quoniam dissimilia semina, sicut porci et hominis, et hominis et canis, et canis et porci, non invicem adiuvant, sed corrumpunt. Et dicit, quod nulla causa est, quod partus habeat caput canis, vel homo pedem vel caput porci, nisi adspectus stellae ad ascendens vel faciem ascendentis in hora casus seminis in matricem: ergo videtur, quod hoc sit naturale: non differt autem scientia imaginum ab illa impressione, nisi sicut ars et natura: quia si natura tunc efficeret imaginem talem qualis fit per artem, facilius et melius haberet ista mirabilia quae imprimit adspectus stellae, quam imago facta per artem: ergo videtur, quod ibi nihil sit de opere daemonum. sed tantum opus artis et naturae."

149 Albertus Magnus, De homine (Ed. Colon. XXVII/2, 129.13–15): "[…] femina generat in se

no hint at the role of feminine semen in the typically Avicennian process of whitening (*dealbatio*) of the embryo by the feminine semen, though. As shown in many studies, Albert presented a consistent approach in his Aristotelian paraphrases, equally and strongly influenced by Avicennian zoology.[150] While Albert's discussion of formative power is already extant in his earlier works, in the pages devoted to the study of generation in the *De homine* we find scarce evidence for an explicit causal link between *spiritus* and formative power.

In her book on medieval embryology, Romana Martorelli Vico has noted that Albert did discuss the role of feminine semen in a context where he was among the first to make use of Avicenna's *De animalibus*.[151] Among contemporary treatises of a theological nature, the only one that compares to Albert's study of generation on a detailed biological basis is the *Summa Halensis*. Heavily influenced by Avicenna, the author of this second part of the *Summa* analyzes the question of sex differentiation (*Quales filios genuissent primi parentes quantum ad sexum*). In these passages, the *Summa Halensis* clearly follows the pattern of active agency assigned to male semen and passive agency to female by Avicenna.[152] This account is very close both chronologically and doctrinally to

ex semine alio, quod est effectivum, sed tamen ⟨etiam⟩ ex semine proprio, quod est materiale." Text amended by: Cova, "Seme e generazione umana," 251. See also: Albertus Magnus, De homine (Ed. Colon. XXVII/2, 126.45–57). In the *De homine*, though, there is no mention of the *dealbatio*, the typical function of the feminine semen according to Avicenna.

150 Albertus Magnus, De animalibus I tr. 1 c. 6 nn. 82–83 (ed. Stadler, I, 31.13–28): "Et sicut de substantia casei sunt duo, coagulum videlicet quod facit, et lac ex quo fit caseus: ita duo spermata sunt de substantia concepti: sperma quidem viri faciens, et sperma feminae suscipiens coagulationem et figurationem et formam. Et haec duo sunt prima principia generationis. Propter quod humidum spermaticum sufficiens generationi omnium animalium necesse est constare ex hiis duobus. In quibusdam autem vel forte in omnibus necesse est adesse tertiam humiditatem praesentem in loco in quo praedicta duo spermata sunt permixta: et haec est nutritiva concepti, ex qua non radicaliter, sed materialiter fiunt quaedam partes eius: et haec in animalibus generantibus sibi similia est sanguis menstruus vel alius humor qui est loco sanguinis menstrui: in ovantibus autem est humor citrinus qui vitellum vocatur, aut alius humor aquosus qui loco vitelli ad cibum concepti embrionis est praeparatus: in eis autem quae generant vermes, est sicut in aliis." Demaitre-Travill, "Human Embriology," 417; Cova, "Seme e generazione umana," 240–241; Zuccolin, "Il corpo sanabile e l'embrione," 126–128.

151 Romana Martorelli Vico, *Medicina e filosofia: per una storia dell'embriologia medievale nel XIII e XIV secolo* (Hippocratica Civitas 4), Milano 2002, 29.

152 Summa fratris Alexandri (Summa Halensis), II Inq. 4 tr. 3 q. 2 c. 4 (ed. Quaracchi II, 725): "quando semen viri praedominatur semini mulieris, tunc ad propriam naturam trahit et convertit [...] Ad illud quod obicitur quod 'semen virit praedominabitur semini mulieris' etc.: dicendum, quod hoc verum est. Numquam enim, proprie loquendo, semen mulieris

Albert: the *Summa* was completed by 1245, and Albert's *De homine* is dated sometime around 1241/1242.[153]

There are only three passages where Albert explicates a connection between spirits and the faculties of the soul responsible for generation. The first is the one in which he claims that the activities of the nutritive soul are rooted in the natural spirits (*fundantur in spiritibus naturalibus*). Unfortunately, this a generic claim that appeals to the activities of the nutritive soul in general and not to its generative faculty specifically.[154] The second passage results from a quotation of the pseudo-Aristotelian *De causis proprietatum elementorum*, a spurious work of the Arabic philosophical tradition, while the third and last is a quote from Alfred of Shareshell's *De motu cordis*. Both of these latter passages have in common a providentialist reading of the concept of spirit as the byproduct of a natural order guided by God. This is a view that is consistent with Albert's early concept of formative power, which, albeit devoid of the complex astrological features of his mature works and of the systematic and explicit connection between spirit and formative power, still emphasizes the causal role of God as the First Cause in the process of the generation and animation of the embryo.

Getting into the details, the passage from the *De causis proprietatum elementorum* claims that a spirit of life appears and is formed upon God's order (*fit in eo spiritus vitae iussu dei*) after fertilization and once that a small portion of the embryo (*frustum*) has acquired a suitably organized physiological structure (*formatur*). This passage is used by Albert to uphold a creationist interpretation, quite common in the Latin Middle Ages, of the Aristotelian passage according to which the rational soul comes from the outside (*ab extrinseco*) and, hence, is infused into it by God.[155] In this passage, the relevance of spirit

---

praedominatu semini viri; quandoque autem semen viri adeo debilis virtutis est quod magist est idoneum ad productionem mulieris."

153 For a brief survey on the role of the woman in the *Summa Halensis* and its comparison to Albert, see: Paul Hossfeld, "Albertus Magnus über die Frau," *Trierer Theologische Zeitschrift* 91 (1982), 235–237.

154 Albertus Magnus, De homine (Ed. Colon. XXVII/2, 96.71–76).

155 Albertus Magnus, De homine (Ed. Colon. XXVII/2, 73.11–16, 73.24–34): "Ad aliud dicendum quod dicens animam humanam in corpore creari, secundum esse tamen non dependere a corpore, non excidit a veritate, sed potius dicens oppositum. Nec Aristoteles umquam dixit animam humanam mori cum corpore nec dicit eam generari in corpore. [...] Quod autem non generetur, sed creetur in corpore secundum Aristotelem, patet in XVI *de animalibus*, ubi dicit intellectualem animam non esse ab aliquo principio efficiente, quod sit in semine. Quod autem eius principium sit deus, et quod ingreditur per creationem in corpus, patet per haec verba sua quae sunt in primo *de causis proprietatum elementorum*: 'Aqua viri quando cadit in matricem mulieris, decoquit eam matrix caliditate sua et com-

in the process and within the formulation "spirit of life" (*spiritus vitae*) is connected to the formation of the embryo: the spirit becomes life-giving when the embryo is fully formed.

The second passage that reports the role of spirit in generation from the *De motu cordis* strikes a similar chord. Here, the soul is said to be introduced by God into the embryo once membrane that contains the fluids in the uterus has hardened and the fluids within have become able to convert the vapor into spirits and—consequentially—to fashion the embryo's body. When the embryonic matter reaches this state, God infuses the soul into it.[156] In the end, both passages share a providentialistic view of the connection between spirit and God: it becomes life-giving when the embryo is fully formed and receives its soul from God. This approach to the role of spirit in generation is present to a greater and more complex extent in Albert's Aristotelian paraphrases, even though in a more zoocentric and less anthropocentric perspective.

4.5 Spiritus radicalis *and Formative Power in Albert's Mature Works*

Albert's mature account of the link between spirit and generation is found in detail in *De spiritu et respiratione*, I, 1, 5, where he devotes his efforts to a teleological explanation of the material composition and primary qualities of the spirit that is suitable for generation, defined either as "seminal spirit" (*spiritus seminalis*) or "radical spirit" (*spiritus radicalis*).[157] This definition sets apart this kind of spirit from the spirit responsible for nutrition and growth, viz. *naturalis*. To be understood properly, Albert's remarks on the nature of seminal or radical spirit must be placed against the background of his discussion of seminal heat from book IV of his *Meteora* and the second treatise of his *De nutrimento et nutrito*.

Takahashi has shown how in the *Meteora* Albert regarded purely seminal heat as insufficient to account for the formative power of semen. For this reason, Albert rejected Averroes' position in the Middle Commentary on the

---

miscetur aqua illa cum sanguine matricis, donec fit frustum; deinde formatur et fit in eo spiritus vitae iussu dei'." The same quotation from the *De causis proprietatum elementorum* in: Albertus Magnus, De homine (Ed. Colon. XXVII/2, 132.6–9).

156 Albertus Magnus, De homine (Ed. Colon. XXVII/2, 132.10–17): "Item, in libro *De motu cordis*: 'Inest semini vitae principium, quod extrinsecis quidem adminiculis trahitur ad actum. Mixta igitur ad vegetationem primum continentis beneficio exterius indurata, vaporem ingenitum exhalare non permittunt; qui si in spiritum converti sit habilis, corpus animatum necessario facit. Huiusmodi enim mixturae a summo omnium creatore statim anima infunditur'." Alfredus Anglicus, De motu cordis c. 12 n. 78 (ed. Baumker 1923, 61.27–62.6).

157 Takahashi, "Nature, Formative Power and Intellect," 465–468.

Meteorologica, according to which generation can occur through seminal heat alone.[158] This thesis clashes with Averroes' own Long Commentary on the *Metaphysics*, which was a relevant source for Albert. In his doctrine of generation, Albert regarded the heavenly component of the *virtus formativa* as essential to his account of generation, since it contributes to the digestion of seminal moisture in the testicles. Without this cosmic component of digestion, heat cannot spread into the whole seminal (or radical) moisture through the whole body of *spiritus* in the semen. As in Avicenna, the spirit found in the testicles does not distinguish between vital, natural and animal spirits but between their potencies.[159]

Albert's remarks in the *Meteora* clarify, despite possible doctrinal discrepancies between the different works, what Albert means when he argues that semen is the product of the fourth digestion. Semen is a haematogenetic product (i.e. originating in the concoction of blood). Usually, the fourfold scheme of digestion follows Avicenna's account according to which digestion occurs in the liver, in the veins, in the heart and then in the specific bodily members.[160] What Albert wants to stress in his *Meteora*, though, is that a good digestion of the nutritive substances does not occur without a suitable amount of cosmic heat. Once this cosmic heat has contributed to the semen's digestion, it starts to spread out over the seminal moisture, which Albert identifies, as shown by Chiara Crisciani and Giovanna Ferrari, and as it shall be seen in greater detail in the next chapter, with radical moisture.[161]

The result, given the nature of spirit as a product of the action of (cosmic and vital) heat over moisture,[162] is a new radical spirit derived directly both from radical moisture in the semen (*nascitur ex ipso vapore humidi radicali*) and

158 Albertus Magnus, Meteora IV tr. 1 c. 4 (Ed. Colon. VI/1, 215.20–28): "Sic enim est in eo vis caeli et motorum caelestium et vis complexionis sive formae illius naturae, quam format, sicut in calore seminis plantae vel animalis est vis, quae vocatur formativa animalis vel plantae. Et eodem modo est in mineralibus et calore eorum. Et sic opus caloris est educere formam de materia, quae potentia est in ipsa, et figurare organa et ordinare situm membrorum et partium sub forma terminante materiam." On this rejection, see: Takahashi, *Albert as a Reader of Aristotle's Cosmos*, 161.

159 Albertus Magnus, Meteora IV tr. 1 c. 13 (Ed. Colon. VI/1, 226.25–48; 226.55–69). For a detailed analysis of this passage: Takahashi, *Albert as a Reader of Aristotle's Cosmos*, 156–158.

160 Albertus Magnus, De nutrimento et nutrito tr. 2 c. 2 (Ed. Colon. VII/2a, 16.12–18).

161 Albertus Magnus, De anima II tr. 2 c. 7 (Ed. Colon. VII/1, 90.86–91.14). Cova, "Seme e generazione umana," 250; Crisciani-Ferrari, "Introduzione," 388–389.

162 Albertus Magnus, De spiritu et respiratione I tr. 1 c. 7, ed. Auguste Borgnet, Paris 1894 (Ed. Paris. IX, 224b–225a).

from nutritive moisture (*est autem vapor humidi radicalis et cibalis spiritus*).[163] This twofold nature of spirit as respects its origin, already noted by Chiara Crisciani,[164] can be temptatively explained in the following manner. On the one hand, spirit derives from the concoction of nutrimental moisture (*humidi cibalis*), which produces a substantial amount of (natural) spirits, defined as vapors produced by the action of heat over moisture in Albert's *De spiritu* itself. The spirit originating in the digestive process (natural spirits) reaches the testicles via the veins. On the other hand, vital spirits partially derive their material composition from natural spirits filtered through the cardiac septum. Being pumped from the heart into the body, vital spirits reach the testicles via the arteries. Vital and natural spirits partially originate in the smoky exhalations of food entering the testicles via veins and arteries, thus explaining how spirits would constitute the part of radical spirit termed "vapor of food moisture" (*vapor humidi cibalis*). On the one hand, a part of the spirit placed inside the semen reflects the nature of the spirit introduced into the testicles (*id quod de ipsa natura et substantia spiritus est in semine*). On the other hand, the second part of the spirit placed inside the semen is produced by the smoky vapors given off by radical moisture in the testicles. These (conjectural) conclusions may be valid when discussinge individual in its fully-formed state. If we reconsider mbryology, the intimate link between spirit and radical moisture can also be traced back at the first moment of embryonic formation. Since the spirit which forms the new body is placed in the first radical member, viz. the heart, then its correlation with radical moisture appears (though conjectural) perhaps clearer.[165]

Given the strict correlation between spirit and radical moisture, one might think that these arguments are human- or animal-oriented at best. Given

163 Albertus Magnus, De spiritu et respiratione I tr. 1 c. 3, ed. Auguste Borgnet, Paris 1894 (Ed. Paris. IX, 217b).

164 Chiara Crisciani, "Aspetti del dibattito sull'umido radicale nella cultura del tardo medioevo (sec. XIII–XIV)," *Arxiu de textos catalans antics* 23/24 (2005), 254.

165 Albertus Magnus, De spiritu et respiratione, I, tr. 1, c. 5, ed. Auguste Borgnet, Paris 1894 (Ed. Paris. IX, 222a): "[…] ex duobus substantia spiritus seminalis conficitur: quorum unum est id quod de ipsa natura et substantia spiritus est in semine, cum clauditur a generante id quod de ipsa natura et substantia est spiritus. Aliud autem quod nascitur ex ipso vapore humidi radicalis, cum calido loci et naturae excitatum vaporare incipit." De spiritu et respiratione I tract. 1 c. 3, ed. Auguste Borgnet, Paris 1894 (Ed. Paris. IX, 218a–b): "Si enim concedatur spiritum vitalem in omnibus ex humido naturali produci, cum humidum naturale commixtum sit elementis, sequitur necessario spiritum esse corpus ex elementis compositum." De spiritu et respiratione, I tr. 1 c. 9, ed. Auguste Borgnet, Paris 1894 (Ed. Paris. IX, 226b): "Cujus autem partis corporis sit ipsa substantia spiritualis, iam non est difficile determinare: quoniam licet origo eius sit ex humido radicali et cibali, tamen generatio ejus et essentia vitalis spiritus est in corde."

Albert's penchant for a universal kind of physiology, though, this is not the case, since Albert attributes internal digestion and radical moisture to plants as well, as we shall see in chapter 5.[166] Clearly, plants do not need the mixed fluids from their partner, as they are able to generate their own seeds without anything similar to human and animal intercourse. But we will come back to plant generation in the second part of this chapter. According to Albert, the *spiritus radicalis* or *seminalis* which results from this process of digestion in the semen and from the vapor of radical moisture is responsible for four kinds of operations concerning the formation of the new living being:

(1) melting the moisture and enabling the reception of the forms, in particular the crucial "form of life" (*dissolvitur humidum ut sit receptibile formarum*);
(2) pulsating strongly enough to facilitate the formation of holes (*foramina*), the principle of arteries in the embryo;
(3) carrying the *virtus formativa* (here, *virtus informans*) placed in the semen *sicut artifex in artificiato*;
(4) achieving the capacity to bring forth life potentially by pulsating (*pulsando sibi ingerat potentiam ad vitam*). Spirit pulsates in the testicles in a vessel which must be separated from that of radical moisture, though: since their operations are different, so are their instruments (blood vessels).[167]

From a materialistic point of view, Albert notably details the relationship between spirit and radical moisture, often left implicit in Avicenna. Having accomplished this task, he then proceeds to link the mechanical account of their relationship to the teleology of generation. The action of spirit on radical moisture is viewed as ultimately necessary for radical moisture to bring body parts to life. This is because moisture without a suitable amount of heat would be inert in terms of formative power. At the same time, radical (or spermatic) moisture represents the only possible kind of moisture that can generate living organs (*non magis posset de ipso fieri membrum actu vivum, quam de illo quovis humido contemperato*). In other words, the interaction with spirit is necessary for radical moisture to be productive of organs possessing life in potency. By contrast, spirit needs to acquire further formative power thanks to the vapors given off by radical moisture.

166 Marilena Panarelli, "How Do Plants Live and Grow? Radical Moisture and Digestion in Albert the Great's *De vegetabilibus*," *Quaestio* 20 (2020) 347–367.

167 Albertus Magnus, De spiritu et respiratione I tr. 1 c. 5, ed. Auguste Borgnet, Paris 1894 (Ed. Paris. IX, 222a–b).

The role played by spirit in Albert's teleology of generation is original because of its universal applicability to plants, animals and humans, the direct and explicit relation between spirit and radical moisture, and the clear link between spirit and formative power in the broader context of an astrologized version of Aristotle's philosophy of nature, strongly influenced by Avicenna, Averroes, and al-Kindī's *De radiis* on the issue of rays specifically. Compared to his early theology, the connection between spirit, formative power and generation presents significant doctrinal developments.

4.6 *Spirit, Formative Power, and Human Generation*

As we have observed already in the first chapter, Albert's use of the Avicennian scheme of fluids (masculine and feminine semen, menstrual blood) is well-established in scholarship and is consistent with his zoology.[168] Unlike in his previous theological writings, though, Albert goes beyond the two brief quotes from the *De motu cordis* and the *De causis proprietatum elementorum* and provides explicit and clearcut links between spirit as the receptacle of formative power and the elemental composition of formative power itself, as in the *De spiritu* (I, 1, 5) and in the *Meteora*. In his zoology, Albert considers spirit a divine body that informs matter without entering it, just as in Aristotle rennet informs milk.[169]

The direct upshot of Albert's position on the role of feminine semen is that the main argument adduced by Galen for the resemblance of offspring to their mother, hinging on an equal causality between masculine and feminine semen, was undermined *de facto*. By appealing to the Aristotelian principle of the *inoboedientia materiae* in the same way as Avicenna had, Albert both maintained the existence of a feminine semen and managed to causally justify the resemblance of offspring to their mother or to female ancestors from their maternal bloodline. This resemblance was explained as being due to the opposition exerted by embryonic matter to the formative action of the masculine semen.[170] In this manner, the only task feminine semen could be occupied with, independent of the successful formative action of the father's semen, was the "whitening" (*dealbatio*) of the embryo, as in Avicennian embryology.[171]

168 See chapter 1, 16 n. 20, and the bibliography listed therein.

169 Albertus Magnus, De animalibus XVI tr. 1 c. 16 n. 86 (ed. Stadler, II, 1109.23–36).

170 Albertus Magnus, De animalibus IX tr. 2 c. 2 n. 98 (ed. Stadler, I, 713.34–714.11); Danielle Jacquart-Claude Thomasset, "Albert le Grand et les problems de la sexualité," *History and Philosophy of the Life Sciences* 3 (1981/3), 73–93; Joan Cadden, *Meanings of Sex Differences in the Middle Ages. Medicine, Science and Culture*, Cambridge 1995, 110–134; Cova, "Seme e generazione umana," 240; Zuccolin, "Il corpo sanabile e l'embrione," 129–130.

171 Zuccolin, "Il corpo sanabile e l'embrione," 133.

Like Avicenna, though, Albert seems to violate the hylomorphic pattern of the Aristotelian scheme of fertilization, which categorically excludes the material participation of male semen to embryonic matter. This position, quite close to Avicenna's *Canon*, is to be found in his *Quaestiones super de animalibus* and not in the *De animalibus*, which remains adherent to the embryological scheme developed in Avicenna's *De animalibus* (not in the *Canon*). Nonetheless, some frictions can already be detected in the *De anima*, in which radical moisture is identified, as shown by Crisciani and Ferrari, with spermatic moisture.[172] If radical moisture is spermatic moisture, how is it possible that a part of the masculine semen enters into the material composition of the embryo-to-be, while another—as in the *De animalibus*—does not? Albert's tentative answer is found in his *Quaestiones super de animalibus*.

In the *solutio*, Albert distinguishes two parts of the male semen with different functionalities and relations to the embryo-to-be. On the one hand, we find residual humidity; on the other hand, the part of semen where spirit, charged with formative power, is located. To solve the problem, Albert holds that, while residual humidity enters into the material part of the conceived entity (*nihil tamen prohibet aliquam partem eius materialem cedere in materiam vel corpus concepti*), the second portion of semen, bearer of the formative power from the father's soul, does not.[173]

172 Albertus Magnus, De anima II tr. 2 c. 7 (Ed. Colon. VII/1, 90.86–91.14): "[…] oportet scire, quod duplici umido indiget omne quod nutritur. Unum quidem est seminale, quod est aliquid substantiae seminis, et illud est proprium subiectum caloris naturalis. Aliud autem est nutrimentale, quod accipitur de nutrimento, quod semper retinet aliquid de natur cibi et ideo semper aliquid habet de extraneo. Propter quod non adeo efficax est in eo calor naturalis sicut in primo umido, et quanto plus ingeritur de nutrimentali umido, quod additur radicali sive seminali, ex quo formantur membra radicalia et quod est infusum eis, tanto plus inspissantur et indurantur, et calor plus in eo debilitatur. Et ideo calor, qui primo fecit duas operationes, alimenti scilicet et extensionis, tandem propter sui debilitatem et membrorum duritiam non potest perficere nisi unam, quae est nutrire; et ideo tunc stat augmentum." Crisciani-Ferrari, "Introduzione," 385–391.

173 Albertus Magnus, Quaestiones super De animalibus XV q. 20 (Ed. Colon. XII 272.65–273.8): "Videtur tamen esse mihi dicendum, quod in spermate duo sunt, scilicet humiditas vel superfluitas ultimi cibi et virtus animae patris in quodam spiritu spumoso existens. Unde quantum ad humiditatem potest sperma esse pars concepti sicut menstruum, quantum tamen ad ipsum spiritum non, quia ipsa humiditas primo commiscetur cum menstruo, sicut vinum cum aqua vel coagulum cum lacte. Constat autem coagulum esse partem casei et aquam vini, etc. Sed spiritus movet ad completionem et ad digestionem completam menstrui per virtutem animae patris in eo existentem. Unde quia menstruum feminae per se sufficiens est ad ministrandum materiam fetus nec necessario requiritur alia pars spermatis pro substantia fetus, sed necessario requiritur virtus spermatis sive spiritus spumosus, qui est in spermate, ideo dicit Philosophus et proprie, quod sperma non est

To closer scrutiny, this appears an attempt at reconciling both Aristotle's and Avicenna's positions that does not prove particularly permeable to critique. Apparently, Albert wants to maintain that the material substance that carries the formative power, viz. spirit, stays out of the material composition of the embryo, thus avoiding that the formal cause of the embryo may become a material cause too. On the other hand, he concedes that a material part of the masculine semen, as held by some physicians (*medici*), enters the material composition of the embryo. Here, by *medici* Albert is very likely referring to Avicenna's *Canon*. The main drawback of Avicenna's position is making the formal and efficient cause of generation, viz. the male semen, also a material cause, thus producing a plain contradiction within the Aristotelian hylomorphic scheme. To deflect this objection, Albert tries to single out a portion of male semen that is not charged with formative power and which therefore can work as a material cause only, although this is a position with shortcomings of its own.

In the first place, it is extremely difficult to identify empirically the portions of semen that are charged with formative power and those that are not. In physiological terms, this would mean that the distribution of the heat of the formative power is not homogeneous in the semen. But semen, by definition, is a homeomerous and not an anhomeomerous part. For this reason, while Albert's hypothesis may be logically conceivable, it concretely clashes against the possibility of detecting thermic differences within semen itself. Moreover, the distinction between portions of semen charged with different degrees of heat would throw into question whether semen is really a homogeneous part of the body like blood. Indeed, in the *De spiritu et respiratione* (I, 1, 5) Albert explains in detail how the heat of radical spirit interacts with radical moisture in the human testicles. He maintains that, given the different functions of the fluids (radical moisture, spirit), they are found in separate blood vesicles. The part of semen place in the vesicle of spirit does not, according to Albert, enter the embryo-to-be materially (*istud sperma non vere est pars materialis membrorum*).[174] But again, it is very difficult to explain how the portion of fluids from

pars concepti materialis, quia sperma non dicitur, nisi inquantum habet virtutem activam et motivam, et movens et agens inquantum huiusmodi non est pars generati. Ideo proprie loquendo sperma non est pars concepti. Nihil tamen prohibet aliquam partem eius materialem cedere in materiam vel corpus concepti, ex quo commiscetur cum menstruo, et sic intelligunt medici. Et forte ex illo generantur membra principalia et radicalia, ex menstruo fluentia membra. Quantum ad virtutem spermatis intellexit hoc Philosophus."

174 Albertus Magnus, De spiritu et respiratione I tr. 1 c. 5, ed. Auguste Borgnet, Paris 1894 (Ed. Paris. IX, 222a): "[…] quia scilicet impossibile est, quod ab eodem et proximo principio fluat spiritus et humor: quia unum impediret alterum in fluxu suo: et hac eadem de causa

the vesicle of spirit in the testicles should remain separate from radical moisture found in its own vesicle once both fluids are emitted during intercourse and could become mingled in the feminine uterus.

This attempt at harmonizing Aristotle with Avicenna's *Canon* proves, in the end, impossible to achieve. But it can be achieved at the cost of admitting that part of the formal cause acts as a material cause. While this constitutes a patent infringement of Aristotle's hylomorphic scheme of generation, it highlights on the other hand the relevance attributed by Albert to the concept of radical moisture, fundamental to a physiology of the lifecycle of generation-birth-growth-decay-death. However, we shall observe the relationship between spirit and radical moisture in chapter 4, devoted to the relation between spirits and longevity. For now, suffice it here to say that Albert is not able to overcome the aporia raised by the Avicennian concept of radical moisture in generation.

In his embryology, Albert follows Avicenna on a number of other doctrinal points such as the earlier formation of masculine fetuses (*citius completur masculus et tardius femina*)[175] and the introduction of the *secundina* as the enveloping membrane of the fetus. On the other hand, though, Albert never mentions the other enveloping membranes of the Avicennian account. The Dominican master emphasizes the role of the *secundina* in the transmission of spirit and nutrition from the mother to the embryo in a way absent in Avicenna, and he adds the existence of a second membrane outside the amnion (*extra secundinam*) to which the embryo adheres (*continuatur*) during its development. This second membrane is anchored to the uterus through the cotyledons and is separated from the *secundina* by a layer of moisture.[176]

On the other hand, Albert's position on the developmental stages of the fetus seems to be much more skeptical about the data provided by the medical tradition. Even though he first mentions the classical Avicennian timeline for formation, motion and birth (40, 120, 280 days) in the embryo,[177] Demaitre and Travill have higlighted how Albert is doubtful about the truth of the medical data (*haec autem quae sic dicunt medici, non sunt certe probata, et ideo non semper inveniuntur vera*).[178] According to him, the ensoulment of the embryo

non potest spiritus per eamdem viam fluere cum humido radicali: ideo oportet quod via spiritus ubique juxta vias humidi radicalis penetret […]."

175 Albertus Magnus, De animalibus IX tr. 2 c. 5 n. 120 (ed. Stadler, I, 723.23–24).

176 Albertus Magnus, De animalibus IX tr. 1 c. 7 n. 68 (ed. Stadler, I, 700.32–701.12).

177 Albertus Magnus, De animalibus XVI tr. 1 c. 7 n. 41 (ed. Stadler, II, 1081.19–25); Demaitre-Travill, "Human Embriology," 426–427.

178 Albertus Magnus, De animalibus IX tr. 2 c. 5 (ed. Stadler, I, 723–724); Demaitre-Travill, "Human Embriology," 426.

occurs when "the organs, through which the works of soul and of life are performed, are completed in shape (*in figura perfectis*)."[179] Unfortunately, he does not specify when this should occur according to him. Despite these skeptical remarks, Albert did value the Avicennian contribution to a thorny issue he had only implicitly tackled in his early theological production, i.e. the controversy (*altercatio*) between Aristotle and Galen about feminine semen. On the matter of the origin of veins, discussed at length in his *De animalibus* with significant references to his *Parva Naturalia*, especially the *De nutrimento et nutrito*, his reliance will not only be on Avicenna but also on Averroes' short but crucial *De corde*.

4.7 Certitudo anathomiae non ostendit nisi ramificationem: *Albert on the Origin of Veins*

In the *De homine*, Albert had displayed his staunch Aristotelian cardiocentrism,[180] combined with an understanding of embryology based on the understanding that sensitive and the nutritive powers were present in the human embryo purely as potencies, becoming substantially one upon the entrance of the rational soul from the outside.[181] Even though Albert did not thematize the discussion on the origin of the veins, he already showed acquaintance with the main sources for the discussion in his mature zoology, i.e. Avicenna's *De animalibus* (III, 1) and Averroes' *De corde*. In this short passage from the *De homine*, Albert already adopts the future backbone of his mature discussion on the topic in his *De animalibus*, viz. the theoretical justification of embryological cardiocentrism through the cardiocentric nature of nutrition.[182]

---

179 Albertus Magnus, De animalibus XVI tr. 1 c. 11 n. 61 (ed. Stadler, II, 1092.28–43); Demaitre-Travill, "Human Embriology," 423.

180 Albertus Magnus, De homine (Ed. Colon. XXVII/2, 49.37–43; 142.15–19).

181 Krause, "Albert the Great on Animal and Human Origin," 224–226.

182 Albertus Magnus, De homine (Ed. Colon. XXVII/2, 244.23–40): "Ad hoc autem quod obicitur in contrarium, respondendum est secundum Avicennam in libro suo *De animalibus* et secundum Averroem in *capitulo De corde*, qui determinant contrarietatem Aristotelis et Galieni dicentes quoniam in veritate quaedam est praeparatio ad digestionem nutrimenti in hepate, ut dicit Galienus et Aristoteles non contradicit, sed illa praeparatio non potest facere, ut illa materia possit esse cibus antequam recipiat formam a calore cordis, quia sicut cor formale principium est et effectivum vitae in omnibus membris, ita est etiam formale principium nutrimenti; sed in stomacho quaedam est praeparatio in hoc quod separatur succositas ptisanaria ab impuro eiciendo, quae succositas per venas quasdam, quae miseraicae dicuntur, attrahitur ad hepar et ibi accipit praeparationem ulteriorem; completam autem formam cibi accipit a calore cordis, et quoad hoc gustus et tactus dicuntur habere instrumenta sua circa cor." On Albert's knowledge of Averroes' *De corde*, see: de Vaux, "La première entrée d'Averroës," 225; Siraisi, "The Medical Learning," 393.

Years later, in his *De animalibus*, Albert set out to discuss the origin of veins (III, 1). He divided the chapter into six sections: (1) the opinions of the ancients (*de opinionibus antiquorum de origine venarum et sanguinis*) except for the detailed analysis of the positions of (2) Aristotle (3) Galen (4) Avicenna and (5) Averroes. The first treatise ends with (6) Albert's own position (*digressio declarans opinionem nostram de altercatione Aristotelis et Galieni*) on the topic. Concerning Galen's position, we must be mindful that Albert's knowledge of Galen is, as in most of his works, dependent on mediating sources such as Avicenna and Averroes' *De corde*: up to today, it is still very difficult to ascertain whether Albert had access to genuinely Galenic works.[183]

From the time of his early theological works, Albert had assumed Aristotelian cardiocentrism as a crucial tenet in his physiological explanation of generation. In his mature works, his commitment to this point of view is restated by his rejection of Galen's stance on the origin of veins from the liver, potentially endangering the unity of the substantial form.[184] By attributing a formative power to the liver, the latter would have earned a place in the human body on equal footing with the heart, a concern which sounds very much like Avicenna's and Averroes'. Having expounded all the arguments gathered from his sources, Albert builds his rejection of Galen's position upon five premises, which he argues to have already proved elsewhere (*hae omnes iam probatae sunt in antehabitis libris*). These premises will help him demonstrate that the heart is the principle of all bodily members (*cor est principium generationis et esse omnium membrorum corporis*). The five premises are the following:

(1) until its assimilation into a specific member, food possesses a form determining its motion towards the nourished member;
(2) the motion of the form of food in the heart is is like the motion of a natural form towards its natural place;
(3) the form of food is universal and not specific until food is assimilated into a specific member;
(4) this universal form is produced by an equally universal principle completing digestion;
(5) the principle imparting the universal form for the conservation of being must coincide with the principle of being itself.[185]

183 Siraisi, "The Medical Learning," 389.

184 Albertus Magnus, De spiritu et respiratione I tr. 2 c. 1, ed. Auguste Borgnet, Paris 1894 (Ed. Paris. IX, 231b): "Principium autem animalis spiritus esse animam rationalem in capite: animas tres per substantiam differentes, et spiritus sedes ipsorum. Horum autem opiniones in tantum videtur sequi Galenus [...]."

185 Albertus Magnus, De animalibus III tr. 1 c. 6 n. 56 (ed. Stadler, I, 301.29–302.7).

But how is it possible to demonstrate that the heart is the organ out of which the liver is created? Albert appeals to his handling of nutrition in the *De nutrimento et nutrito*, where he had distinguished—following Avicenna—the four classical kinds of digestion. In this context, Albert had linked the occurrence of the fourth (assimilation to the specific member) to the form-giving action of the soul upon food. In order to be such, though, food needs to acquire its suitable form to nourish a body part. Since the soul operates in the heart, the process of digestion must be completed in the heart.[186] Relying on the idea that the heart is the organ where nutrition is perfected, Albert managed to overturn the Galen's idea that the principle of the member is where the principle of the virtue is: if Galen thought of the liver as the principle of the nutritive faculty, then he was completely wrong, since the principle of that virtue was the heart.[187] Albert's solution, in the end, seems to be based in large part on the Aristotelian and Averroean cardiocentric conception of nutrition. In particular, the use of arguments borrowed from Aristotle's cardiocentric understanding of nutrition against Galen's position betrays a clear methodological approach indebted to Averroes.

Despite the doctrinal differences between Avicenna's and Averroes' refutations of Galen, there is a key feature common to their biological explanations of embryonal development, i.e. the key role played by cardiac *pneuma*. In his *De corde*, the Averroes argues that, if heart, brain and liver were to be understood on equal footing, then the heart would not be the member capable of communicating its universal, life-giving virtue through spirit. The concrete philosoph-

186 Albertus Magnus, De animalibus III tr. 1 c. 6 nn. 57–58 (ed. Stadler, I 302.8–13; 302.16–22): "His ergo sic suppositis dicimus quod id quod dat cibo formam qua complete est cibus, est cor, et illa forma facit, quod naturaliter non tractus violenter tendit ad corporis nutrimentum supra et subtus, et a latere secundum exigentiam generantis. Tale est enim generans, sicut diximus sequentes Aristotelem in libro *de Nutrimento* [...] Adhuc autem dicimus, quod cum cor sit principium facturae totius corporis sicut dicit ultima propositio, oportet quod virtus qua efficitur cibus complete in ratione cibi, sit a corde et non ab epate: quoniam epar in substantia sua incompletum est membrum valde non habens sensum: nec potest etiam esse sedes animae: propter quod peccavit Plato ponendo in eo animam concupiscibilem: et peccatum illud secutus est Galienus." Albertus Magnus, De nutrimento et nutrito tr. 1 c. 3 (Ed. Colon. VII/2a, 7.11–20). Also, the form of food becomes life-giving once it has received in the heart the *forma vitae*: on the relevance of the *forma vitae*, see the previous chapter.

187 Albertus Magnus, De animalibus III tr. 1 c. 6 n. 60 (ed. Stadler, I, 303.4–8): "Ex hoc autem sequitur necessario, quod etiam venae omnes a corde oriuntur: et vis digestiva quae est in venis, nequaquam esset in eis, nisi a corde procederent. Sapientissime enim dixit Aristoteles, quod idem esset principium venarum, quod est principium digestionis completae et sanguinis."

ical upshot of this position would have been, as in Avicenna's case, a plurality of forms derived from the plurality of seats of soulic activities in the body, firmly rejected by Averroes.[188] In his solution to the problem of the origin of veins, Albert is well-aware of Averroes' criticism of Avicenna (*sententia Avenroys tam contra Avicennam quam contra Galienum pro Aristotele prolata*), but ultimately chooses to cite them both as the main authorities in defense of Aristotle's position against Galen, ignoring Averroes' criticisms against Avicenna.[189] The bottom line of Albert's discussion on the origin of veins, relying heavily on Averroes' and Aristotle's nutritional cardiocentrism rather than on Avicenna's embryological cardiocentrism, deems medical experience (*certitudo anathomiae*) epistemologically insufficient to settle problems through an adequate demonstrative cogency (*per rationem*) and capable only of showing "branchfulness" (*non ostendit nisi ramificationem*).[190] In his *Quaestiones super de animalibus*, a reportatio by friar Conrad of Austria on the quaestiones likely discussed sometime around 1258 by Albert with his student, shows different nuances of the discussion, perhaps influenced by the quaestiones by Peter of Spain and, in the end, with a similar outcome. Miguel de Asùa has shown that in his *quaestiones* Albert seeks a conciliatory solution to the Aristotelian-Galenic quarrel: more explicitly, he admits that to save both opinions it is necessary to distinguish between the origin of the veins *prima facie* and according to the senses (*sensitive and manifestative*) on one hand, and the origin of the veins according

188 Albertus Magnus, De animalibus III tr. 1 c. 5 n. 44 (ed. Stadler, I, 296.17–26).

189 Albertus Magnus, De animalibus III tr. 1 c. 6 n. 64 (ed. Stadler, I 305.9–12): "Sunt autem multi modernorum qui consentiunt Galieno, eo quod signa quae inducit et rationis sunt sensibiles et vulgares, quamvis sint falsa omnia quae inducit, sicut etiam alii ante nos ostenderunt, Avicenna videlicet et Avenroys." See for instance Averroes' criticism of Avicenna, reported by Albert in: Albertus Magnus, De animalibus III tr. 1 c. 5 n. 54 (ed. Stadler, I, 300.25–33): "Avicenna autem tam Aristoteli quam Galieno contradicit in hoc quod dicit, quod in principio generationis virtus formativa ex una parte materiae fecit cor, et ex alia epar, et postea coniungit inter ea per venas quas fecit ex tertia parte: et haec opera facit simul et non successive: sed est in eis ordo naturae et non temporis. Si enim sic divisa sunt membra secundum distinctionem partium materiae: tunc divisae erunt operationes membrorum, et unum membrorum non indigeret alio, quod est contra Aristotelem et contra Galienum."

190 Albertus Magnus, De animalibus III tr. 1 c. 6, n. 65 (ed. Stadler, I, 305.25–33): "Et dicemus, quod certitudo anathomiae non ostendit nisi ramificationem, quia illa est sensibilis: non autem ostendit principium esse et substantiae, quia illa non per rationem accipi potest. Dicemus insuper, quod non est in venis sicut in plantis, sicut et in animali non est sicut in plantis, quia plantae inferius habent radices ori similes, et os habet animal in supremo sui: et ideo nichil prohibet quin ramositas venarum sit extensa ad cibum et ad membra, ubi non est principium earum. Neque aliquid habet necessitatis quod dicit, neque similitudinis."

to their virtues and their roots (*virtualiter and radicaliter*) on the other.[191] Since the roots of veins are where the power of nutrition is, it follows that veins originate from the heart *virtualiter* and *radicaliter*. While Galen is right in asserting that, according to anatomical observation and sensory evidence, many veins branch out from the liver, this reality does not mirror their true (*realiter*) origin according to their being and substance, which must come from the true radical member whence all powers are derived, viz. the heart.[192]

This solution seems to be closer to Avicenna's zoology, which justifies the cardiocentric origin of veins by means of the cardiocentric nature of formative power.[193] In spite of the different tones—for in the *De animalibus* the refutation of Galen's doctrine seems to be much more influenced by Averroes' cardiocentrism in matters of nutrition—the overall perspective seems to be quite consistent with the *De animalibus*: anatomical and sensory evidence can only explain the *manifestatio ramorum*, viz. that many veins branch out from the liver, but not where the true origin of veins is. From this perspective, both discussions provided seem to be complementary rather than contradictory, insisting on the key notion of Aristotelian cardiocentrism, which can be witnessed also in Albert's doctrines on generation in oviparous animals (see the next chapter). Moreover, the key link between spirit and formative power can prove a meaningful heuristic tool to investigate not just the regular products of natural teleology but also the phenomena of the so-called "errors of nature" (*peccata naturae*) or less frequent outcomes of generation such as birthing of twins.

4.8 *Errors of Nature? Twins and Monstrosities*

A recent study by Gabriella Zuccolin has unveiled how Albert's analysis of the irregularities of nature (*Physics* II, tract. 3) offers a more systematic approach

191 Albertus Magnus, Quaestiones super de animalibus III q. 4 (Ed. Colon. XII, 125.60–64): "Unde ut salvetur sententia omnium de hoc loquentium, distingui potest, quod duplex est ortus: unus virtualis et radicalis et alius est corporalis. Primo modo venae oriuntur a corde, sed secundo modo oriuntur ab hepate." De Asùa, "War and Peace," 287–288.

192 Albertus Magnus, Quaestiones super de animalibus III q. 4 (Ed. Colon. XII, 125.64–69): "Unde ab hepate incipit ramificatio venarum sensitive et manifestative, sicut manifestatio ramorum a stipite arboris, licet realiter a radice procedat, tamen manifestative nobis apparet a stipite. Ita est de corde, quod ab ipso realiter procedunt, licet manifestative ab hepate."

193 This is the solution proposed also by Peter of Spain: according to Asùa, there are some possible influences of Peter's (Madrid) commentary, other than the mere issue of the origin of the veins: Miguel de Asùa, Peter of Spain, "Albert the Great and the *Quaestiones de animalibus*," *Physis* 34 (1997) 1–30.

than Aristotelian or Avicennian zoology on the topic.[194] In her account, Albert fits well the definition of a (Medieval) theratologist, viz. a scholar who took the study of monstrosities down to the most minute physiological details. In his doctrinal study of the irregularities or *peccata naturae*, Albert defined monstrosities as "errors of that which is in nature because of something" (*monstra enim sunt peccatum illius quod in natura est propter aliquid*). Though it is not directly involved and quoted in the discussion, the concept of spirit lurks in the background of Albert's study of monstrosities and deformities. The reasons why become evident when we consider his etiology of theratological phenomena. In this case, it is useful to report below the detailed scheme provided by Zuccolin on the four main causes for monstrous births according to Albert:

1. Lack in matter (*materiae diminutio*)
    - a. Undersized organs, caused by:
        - i. Defective amount of matter in the uterus;
        - ii. Weakness in the heat of the formative power (borne by spirit);
    - b. Deficiency[195] in both size and number organs, caused by:
        - i. Defective amount of matter;
        - ii. Weakness in the formative power;
    - c. Deficiency in the number of organs due to:
        - i. Defective amount of matter;
        - ii. Weakness in the formative power;
2. Excess in matter (*materiae superfluitas*):
    - a. Oversized organs, caused by:
        - i. Overabundance of organic matter, resulting in living beings with only a few oversized organs (mostly in the upper part of the body);[196]
        - ii. Overabundance of both organic matter and formative power, resulting in living beings with all organs oversized;

194 This section is based on Zuccolin, *I gemelli nel Medioevo*, 144–157. On twins and monsters, see also: Jos M. Thijssen, "Twins as Monsters: Albertus Magnus's Theory of the Generation of Twins and its Philosophical Context," *Bulletin of the History of Medicine* 61 (1987/2), 237–246.

195 Zuccolin has noted how the concept of "deficiency" in Albert must be interpreted not only as a pure lack of organs but also as a synonym of functional deficiency. Having a non-functioning kidney, for instance, may be tantamount to having only one out of two from a functional point of view: Zuccolin, *I gemelli nel Medioevo*, 146–147.

196 Undersized organs lie for the most part in the lower part of the body, far away from heart and liver, where the *virtus formativa* is stronger. For this reason, too much matter without a suitable amount of formative power allows the formation of defective members especially in the inferior parts of the body.

b. Overabundance in the number of organs, divided into:
   i. Premature generation of complex members (teeth, hairs), caused by an excessive heat in formative power;
   ii. Overabundant members because of excessive matter;[197]
3. Disproportioned ratio between qualities and matter (*qualitatum improportionabilitas ad materiam*):
   i. Leanness (*macies*) due to excessive heat;
   ii. Fatness (*pinguedo*) due to excessive cold;[198]
4. Defect in the uterus (*malitia continentis*):
   i. Defect in the *membrana secundina*;
      1. Problem: oozing out of matter and breaking of the *membrana*. Result: loss, malformation or displacement of embryonic members;
      2. Problem: Fusion between the amount of semen from different fetuses during formation. Result: conjoined twins.[199]
   ii. Defect in a membrane receptive to heavenly influences.[200]

Twins, especially conjoined and Siamese twins, fall under the category 2.b.ii (overabundant members because of exceeding matter). The physiological explanation provided by Albert as to the difference between non-conjoined and conjoined twins is based on the Avicennian argument of the different kinds of motions occurring in the uterus. Following Avicenna, Albert argues that some women feel a particularly strong pleasure during intercourse: if this is the case, when semen enters the uterus and stimulates its most sensitive nerves, the uterus starts to move. The result of these motions is a division of the spermatic moisture mixed with the feminine semen: if this division is regular, regular

197 Albertus Magnus, Physica II tr. 3 c. 3 (Ed. Colon. IV/1, 137.15–138.5); Zuccolin, *I gemelli nel Medioevo*, 147–150.

198 In the absence of any prevailing quality, we observe causes of hermaphroditism: Albertus Magnus, Physica II tr. 3 c. 3 (Ed. Colon. IV/1, 138.5–14): "Ea autem quae fiunt propter qualitatum improportionabilitatem, sunt sicut nimia macies et nimia pinguedo, quorum primum fit ex nimio calore et secundum ex nimio frigore, et si vincant qualitates contrariae, quarum utraque fit complexionalis et terminans, tunc contingit nasci hermaphroditos; et virtus formativa satisfacit utrique sexui, tam in membris exterioribus quam in membris interioribus, sicut ad visum probavimus in quibusdam esse utrumque sexum, ita quod a sapientibus discerni non potuit, quis praevaleret." Zuccolin, *I gemelli nel Medioevo*, 150.

199 Albertus Magnus, Physica II tr. 3 c. 3 (Ed. Colon. IV/1, 136.55–61).

200 According to Albert, the embryo possesses a membrane (*primum circundans*) specifically designed for the reception of the heavenly influences partially building the *virtus formativa*. This membrane seems even to surround the *membrana secundina*: Zuccolin, *I gemelli nel Medioevo*, 151–152.

twinning occurs; if not, the non-separated parts of the sperm are reflected in the conjoined parts of the new embryo.[201] Strictly speaking, Albert is the first medieval theratologist insofar as he seeks to reorder and rearrange all possible phenomena of monstrosities and birth anomalies according to the laws of natural philosophy, as demonstrated by Zuccolin. For instance, he explains that human-headed pigs can be caused by a disordered reception of astral influences in the membrane specifically designed for the reception of astral influences. Albert's doctrine on monstrosities is quite in keeping with Avicenna's zoology.[202] There are two main and remarkable differences, though.

In the first place, Avicenna never argues for the existence of a membrane specifically devoted to the reception of astral influences. In the second place, Avicenna is much more skeptical than Albert about the possibility of detecting a causal link between monstrosities and astrological influences. By contrast, Albert was much more open to the so-called *prognostica*, a discipline incorporated by Albert in his explanations of cycles of generation and corruption (*De generatione et corruptione*) and his theory of climate and seasons (*De natura loci*).[203]

Even if *spiritus* is not directly mentioned in Albert's account of monstrosities, its intimate link thereto as carrier of the formative power chimes well with the general role it plays in Albert's account of generation. Building on Zuccolin's (also, Takahashi's) account, we can see how spirit is the crucial instrumental cause that explains, *qua* carrier of the formative power, the workings of the soul in generation-related processes. Moreover, alterations in its thermic balance due to the complexion and age of the father or because of astral influences partially explain, at least in the case of formative power, the occurrence of the different kinds of monstruous births. To exhaustively detail these anomalies, one would need to enlarge this analysis to the study of astrologically relevant

201 Albertus Magnus, Physica II tr. 3 c. 3 (Ed. Colon. IV/1, 137.60–76); Zuccolin, *I gemelli nel Medioevo*, 148–150. On twins, see also: Albertus Magnus, De animalibus IX tr. 2 c. 6 (*Et est digressio vere declarans causam geminorum et modum et tempus naturalis partus*) (ed. Stadler, I, 727–729).

202 Zuccolin, *I gemelli nel Medioevo*, 151–152.

203 In the *De motibus animalium*, the *virtus* of the *constellatio spiritualis* (heavenly bodies) explains why, according to Albert, between two twins one will possess in the right part of his body the capacity to open any possible door or lock, while his twin is able to lock anything that had been opened before: In this, Albert criticizes those who—like Avicenna and al-Ġazālī (whose Latin version of the *Maqāṣid* was misread as an Avicennian text)—think that terrestrial events depend on heavenly intelligences and not on heavenly bodies, thus making the possibility of predicting future events very difficult, see: Albertus Magnus, De motibus animalium, I tr. 1 c. 3 ed. Auguste Borgnet, Paris 1890 (Ed. Paris. IX, 262b). More on this in: Zuccolin, *I gemelli nel Medioevo*, 155 n. 57.

notions (seasons, zodiacal signs, influences of planets, theory of climates), a task far too demanding for this chapter.

For now, suffice it here to stress once again how Albert's handling of generation is firmly rooted in his concept of formative power, the cornerstone (together with the embryonic matter from the mother) of any explanation of successful, monstruous or less frequent outcomes of generation. Still, there is a question which has escaped so far our account of generation, viz. what causal factor determines the material composition of spirit in generation from the point of view of its final cause? In other words, why is spirit in humans different from the spirit meant to preserve other species of animals and plants? To answer this question, it will be necessary to understand how Albert interpreted the difference between the respective souls of humans, animals and plants in his works.

4.9 Plus quam specie videtur differre: *Albert's Essentialist View on Human Nature and Natural Teleology*

A large consensus in scholarship has determined that Albert's belief that "man as man is intellect alone" (*homo inquantum homo solus est intellectus*) is based on a strong, essentialistic interpretation of Aristotle's *Nicomachean Ethics*, x.[204] This claim goes well beyond Aristotle's own intellectualistic conception of man: according to Albert, the axiom "man as man is intellect alone", first detectable in the *De quattuor coaequaevis*,[205] holds true not only in terms of specific (viz. behavioral) differences between men and other living beings. According to Albert, behavioristic differences between humans and animals fail to capture the real and essential differences between human, animal, and vegetative souls.

Albert's recurrent idea that man as man is intellect alone finds in the *De animalibus* his most fully-fledged expression in terms of physiological and psychological consequences. The human rational soul, placed at the horizon below eternity and above time following the *De causis*, is somatically reflected in a different and more complex organization than other animals.[206] According to

204 Henryk Anzulewicz, "Anthropology: The Concept of Man in Albert the Great," in *A Companion to Albert the Great: Theology, Philosophy and the Sciences*, ed. Irven M. Resnick (Brill's Companions to Christian Tradition 38), Leiden-Boston 2013, 340–345.

205 Albertus Magnus, De quattuor coaequaevis tr. 4 q. 69 a. 3 part. 3 quaesit. 1, ed. Auguste Borgnet, Paris 1895 (Ed. Paris. XXXIV, 703a).

206 Albertus Magnus, De animalibus XXI tr. 1 c. 1 n. 4 (ed. Stadler, II, 1323.26–38): "Patet igitur quod diximus quod videlicet perfectissimum animalium est, cuius omnis animae potentia est secundum esse naturae separatae: et ideo diximus in antehabitis huius scientiae quod homo non solum specifica differentia differt ab aliis animalibus, sed etiam secundum esse generis proximi et secundum esse generis remoti. Genus enim proximum est

Albert, we must interpret Aristotle's idea that man is the most perfect among animals (*homo animal perfectissimum*) not in merely behavioristic terms but also in essentialistic ones. In his clear analysis of Albert's doctrine, Henryk Anzulewicz has pointed out that the quiddity of the rational soul, viz. its separable and immortal nature, constitutes also the *explanans* of the physiological differences between animals and humans, which do not rest on the pure difference of the possession versus lack of rational capacities.[207] Rather, this difference is rooted in the quiddity of the human soul as opposed to the quiddity of animals and plants, doomed to perish with their bodies.

The upshot for the psychophysiology of generation is considerable. In this manner, fluids at play in human generation have the most perfect complexion among all living beings since they are meant to produce a living being whose organic complexity towers over all other animals because of the separability of their rational soul from matter after the body's death. It is not only the activities performed by the human rational soul but the ontological status of the soul *qua* perfect, separable and immortal perfection of the human body that determines a higher level of organic complexity in the embryo and a different composition (complexion) in the material substances involved in generation. Having a rational soul is not the final cause explaining why spirit needs to acquire a certain material complexion to generate a human being. This may be Aristotle's view, for instance. Rather, the reason behind the material complexion of spirit must be sought in the quiddity of the human rational soul as the final cause of the body. Given its separable nature and given the partaking of the lowest faculties in it, the presence of an immortal soul and its rational activities requires a level of organic perfection superior to any living being. For this reason, humans differ "almost generically" and not just specifically from animals (*plus quam specie videtur differre*) in their essence and psychological faculties.[208]

---

sensibile, et remotum est vivum: et nos ostendimus quod tam sentiens quam vivum participant aliqualiter rationem, licet ipsae potentiae de natura sua sint irrationales. Eodem autem modo etiam vegetativum in sensibili secundum esse alterum et aliud est quam sit ex natura sua in seipso. Et ideo dictum est quod ratio animalis et uniuscuiusque generis secundum unumquodque animalium in esse accepta est altera et altera et alterius potentiae sicut determinatum est a nobis in primo *de Anima*."

207 Henryk Anzulewicz, "*Solus homo est nexus Dei et mundi*. Albertus Magnus über den Menschen," in Multifariam. *Homenaje a los profesores Anneliese Meis, Antonio Bentué y Sergio Silva*, ed. Samuel Fernández/Juan Noemi/ Rodrigo Polanco (Anales de la Facultad de Teología. Supplementos a Teología y Vida 1), Santiago 2010, 332–333.

208 Albertus Magnus, De animalibus XX tr. 2 c. 6 n. 88 (ed. Stadler, II, 1319.32–42): "Homo autem ab aliis animalibus plus quam specie videtur differre: videmus enim ea quae nec sensu nec motu faciunt differentiam, tamen specie ab invicem differre. Cum igitur homo et in participatione sensuum differat specie ab aliis sicut probatur ex figuris organorum sensus et

Spiritus acquires, on the one hand, its life-giving power from the soul through a complex emanative process. On the other hand, the different kinds of soul involved as the final cause of generation represent the final cause of the material composition of spirits at play in generation. Given the inextricable links between creatures and their Creator, this composition mirrors a combination of complex causes that lies ultimately inactive without the assent of divine will and which reflects therefore the Neoplatonic axiom "the work of nature is the work of intelligence" (*opus naturae est opus intelligentiae*).

Conclusions

We are now in a better position to appreciate the words of Albert's *De unitate intellectus*: "opus naturae est opus intelligentiae et spiritus qui est in semine hominis, est intelligentiae instrumentum, vel etiam causae primae." The teleological role of spirit *qua* vehicle of the formative power permeates all of Albert's Aristotelian paraphrases and is developed within a psychophysiology consistent with his theories on the origin of the soul and formative power (called generative power in the *De homine*). In the Aristotelian paraphrases the connection between spirit and formative power, spelled out only in two brief passages in the *De homine*, is developed within a more full-blown theory that, as we shall see, is applied to plants as well. Based on combined insights from Avicenna and Averroes, Albert's doctrine of formative power in humans and animals and his model of "epigenetic creationism" detected by Krause will comprise the backbone of his Aristotelian paraphrases.

Albert's early model of human and animal generation is staunchly based on Aristotelian cardiocentrism and interpreted through the lenses of the physiology presented by Avicenna's *De animalibus*. The only author who possesses a similar knowledge of Avicennian zoology around the time Albert wrote the *De homine* is Peter of Spain. Like Peter and other authors, among which John

---

figura totius corporis, et insuper adiciat rationis differentiam, videtur plus quam specie homo ab aliis differre animalibus. Cuius signum est, quia irrationales vires hominis persuasibiles sunt ratione sicut concupiscibile et irascibile: et hoc non est in aliis animalibus. Plus igitur quam specie differt a brutis, et videtur ad ipsa quamdam habere generis differentiam, cum ipsam animalitatem secundum differentem potestatem ab aliis animalibus participet." Albertus Magnus, De animalibus XXI tr. 1 c. 1 n. 62 (ed. Stadler, II, 1322.6–9): "Patet igitur ex omnibus praedictis non solum in adiectione rationis esse hominem animal perfectissimum, sed etiam in omnibus potentiis et modo operationis omnium potentiarum tam sensibilis quam rationalis." Anzulewicz, "Anthropology: The Concept of Man," 345–346.

of La Rochelle and the *Summa Halensis*, Albert followed Avicenna in rejecting the active agency represented by the feminine 'semen'. Introduced by Galen and reworked by Avicenna in terms of causal agency as a purely passive and material cause of generation, feminine semen represented the matter of the embryo-to-be and accounted for the generation of female (or mother-like) offspring inasmuch as it resisted the informing action of the formal and efficient cause (male semen). In his *De homine* and *Commentary on the Sentences*, evidence for the connection between spirit and formative power are rather scant and based on quotes from the pseudo-Aristotelian *De causis proprietatum elementorum* and Alfred of Shareshell's *De motu cordis*. Both quotes emphasize that spirit acquires its life-giving power when the human soul comes to exist, created by God, in the embryo.

Albert's Avicennian approach to the physiology of generation will be maintained in his mature psychophysiology and systematically connected to the notion of spirit. Yet Albert's account of generation is not exempt from serious hermeneutical troubles, which he inherited from Avicenna himself. In his *De animalibus*, Albert maintained a faithfully Aristotelian hylomorphic scheme of generation interpreted through the lenses of Avicenna's *De animalibus*. This scheme is consistent with the early *De homine* and regards masculine semen as a formal and efficient cause, while feminine semen and menstrual blood are the material causes of the embryo-to-be. By contrast, in his *Quaestiones super de animalibus*, Albert seeks to harmonize this Aristotelian-Avicennian hylomorphic scheme with the Avicennian doctrine from the *Canon*, according to which part of the seminal moisture, viz. radical moisture, becomes part of the embryo, while the portion of semen filled with spirit and formative power remains extrinsic to the embryo. While partially corroborated by the different anatomical localizations of spirits and radical moisture in the human testicles, this distinction comes eventually to clash with the empirical evidence (difficulties in distinguishing those two fluids during intercourse). This point remains a major aporia in Albert's account of generation and, more generally, in his understanding of the relationship between natural philosophy and medicine. On the one hand, Albert rejects Galen's argument about the agency of the feminine semen because it posits two formal causes in the same process; on the other hand, he breaks the classically Aristotelian hylomorphic scheme of generation by arguing that the formal and efficient cause partially becomes the material cause at the same time.

Albert's Aristotelian paraphrases display quite a few remarkable features that signal a doctrinal development from his early theology. In the first place, the material origin of spirit and its link to semen and formative power is discussed explicitly in the *De spiritu et respiratione* and is framed within a paradigm of natural teleology synthesized by the axiom *opus naturae est opus intelligentiae*. Moreover, Albert's mature works display a concept of formative power framed within a more complex and astrologized pattern of celestial causality, based on insights from Avicenna and Averroes now blended with the Kindian theory of rays. This more detailed version of the concept of formative power explains differences in offspring according to endogenous (sex, age) and exogenous (climate) factors.

Like many of his contemporaries, Albert also discussed the relationship between life, embryonic formation and induced abortion in his theological works such as the *Commentary on the Sentences*. There, he showed a clear adherence to the traditional Augustinian view that the animation of the embryo occurs according to 46 days and not the canonical Aristotelian timing (40 days). This position closely resembles that of William of Auvergne in his *De anima* and seems to be influenced by the authoritative role exerted by Augustine in theological matters, according to Albert's own words. More generally, Albert's position falls in line with contemporary doctrines equating the principle of life (hence, the charge of murder) with the complete formation of the embryo and not with its (perceived) motion. This is the case with William but also of other contemporary texts such as the *Summa Halensis* and the writings Aquinas. While medico-philosophical texts, not least the well-known Avicennian *De animalibus* used by Albert in the *De homine*, would question whether this claim would have pathological implications (delayed childbirth), Albert's *Commentary on the Sentences* does not. Perhaps, this is due to the preeminent role of Augustine as an authority in theological matters. From a purely biological, philosophical point of view we do not find in Albert's work a clear definition of the temporal origin of life in humans, as he casts serious doubts on the Aristotelian-Avicennian identification of the beginning of life as 40 days after impregnation.

In addition to the temporal origin of life, correlated to the full formation and ensoulment of the embryo, Albert discusses his own doctrine on the origin of the soul. Throughout his works, he generally maintains a model, labelled "epigenetic creationism" by Krause, that remained consistent. This model stresses the eduction of the vegetative and sensitive powers of the human soul, with the latter being created directly by God as the first intelligence. This theory was developed following insights from many authors (Aristotle, Avicenna, Averroes, the pseudo-Aristotelian *De causis proprietatum elementorum*, Augustine's theory of *rationes seminales*, etc.) and relies, at a physiological level, on the Avicennian understanding of the fluids (male and feminine semen, menstrual blood) and Averroes' doctrine of formative power in the case of the vegetative and sensitive souls. According to Albert (and Averroes), these souls are educed from matter thanks to the formative power placed in the father's semen. This formative power is a combination of seminal and cosmic heat. While being different from Aristotle's (also, Avicenna's and Averroes') own account of generation, Albert's doctrine on the origin of the human soul tries to find a synthesis between the principles of the Aristotelian philosophy of nature (hylomorphic interaction between masculine and feminine fluids in fecundation, eduction of the soul from matter) and the Christian need to posit the human rational (and immortal) soul as being directly created by God. In the construction of a physiology of generation compatible with the principles of Aristotle's philosophy of nature and Christian orthodoxy, Albert's Arabic sources, viz. Avicenna and Averroes, play a crucial role.

Albert also maintained his staunch adherence to the Aristotelian principles of natural philosophy concerning the issue of the origin of veins. Influenced by Averroes' (*De corde*) and Aristotle's cardiocentric understanding of nutrition, Albert firmly rejects in the *De animalibus* the idea that veins may originate from the liver, conceding some kind of formative power to organs in the body other than the heart. In the *Quaestiones super de animalibus*, Albert adopted a complementary solution against Galen's view, arguing that veins must originate in the heart in terms of virtue and power. This solution seems, this time, close to Avicenna's *De animalibus*, which bases its rebuttal of Galen's position on his embryological cardiocentrism. Albert's arguments against Galen seem therefore to be two sides of the same coin of cardiocentrism, either related to the process of nutrition (Averroes) or embryonic growth (Avicenna). In his *Physics* Albert analyzed in depth possible failures in embryonic formation (*peccata naturae*), presenting a detailed analysis of monstruous births, relying on the causal role of formative power and matter that involves spirit directly. This is because any alteration in the heat of the racial spirit produced by exogenous (astral) or endogenous (complexion, age) causes may affect the formative action of spirit and result in phenomena such as undersized or oversized body parts, defects or abundance in the number of body parts, among which Albert counts twins and monstrous births such as Siamese twins.

A close study of Albert's theory of generation constitutes a good philosophical exercise for understanding the complexity and depth of thought of the *Doctor Universalis*. On the one hand, it shows how Albert's philosophy of nature was deeply permeated by doctrines from Avicenna and Averroes from the onset of his writing career (*De homine*, *Commentary on the Sentences*) and that he gradually developed the details of both of his psychological (origin of the soul) and physiological (formative power) doctrines. Moreover, it shows how Albert's intellectual consideration of the nature of humans avoids any attempt to push humans, animals and plants too closely together from a psychological point of view. On the other hand, the following chapter will show us that a typical feature of the mature Albert is represented by the great pains he takes to produce a unified account of physiological causes (spirit, formative power) of life in all ensouled beings (plants, animals. humans). This unifying account is corroborated by considerable anatomical and physiological detail, to which we will now turn.

CHAPTER 4

Spirit, Generation, and Natural Teleology—Part 2: Animal, Plant, and Spontaneous Generation

In chapter 3, we tackled the causal relationship between spirit, generation and natural teleology in the case of human generation. There it became clear how spirit represents one of the instruments of divine providence at work in generation. We emphasized how the action of spirit contributes to the progressive formation and animation of the embryo-to-be without entering its material composition. This action is derived by the suitable concoction of semen from heavenly and bodily heat and is rooted in the sum of the heat poured from the father's soul and combined with seminal heat and heat caused by astral influences. In the case of humans, we highlighted how their rational soul is created by God, while the sensitive and nutritive faculties are educed from matter without acquiring the full status of 'form' every time, as for instance in Aquinas. This creation results from the process of embryonic development, which in turn rests on Albert's Avicennian interpretation of Aristotle's physiology of generation, with some insights from Averroes added in. By contrast, we have only cursorily pointed to the origin of the animal hylomorphic compound, primarily when considering the origin of the human soul. Now it is time to turn to all instances of animal generation, including viviparous generation, oviparous generation and chance phenomena such as spontaneous generation. In this chapter, we shall also focus on the role of spirit in the generation of plants, an unprecedented doctrine in the history of philosophy in the Medieval Latin world. While previous studies have focused by way of analogy on Albert's doctrine of oviparous generation in connection to human generation,[1] we shall deal with this kind of generation in more detail here, complementing this analysis with a study of plant and spontaneous generation that builds on the recent contributions of Amalia Cerrito and Marilena Panarelli.[2]

1 Demaitre-Travill, "Human Embryology," 417.

2 Marilena Panarelli, "Converting Death into Life: Spontaneous Generation from Aristotle's Biology to Albert the Great's Analysis of Plants," *Quaestio* 22 (2022) 493–508; Amalia Cerrito, "Disclosing the Hidden Life of Plants. Theories of the Vegetative Soul in Albert the Great's *De vegetabilibus et plantis*," in *Vegetative Powers. The Roots of Life in Ancient, Medieval and Early Modern Natural Philosophy*, ed. Fabrizio Baldassarri/Andreas Blank (International Archives of the History of Ideas Archives internationales d'histoire des idées 234), Cham 2021, 105–122.

 DOI:10.1163/9789004706743_006

1 *Pneuma*, Animal Generation, and Spontaneous Generation in Aristotle

In the previous chapter, we highlighted the principles of animal generation by way of comparison with the physiological and psychological principles of human generation. We have observed how spirit plays the role of carrier of vital heat in the semen and thus enables the formation of the embryo-to-be by forming embryonic matter itself. This model, particularly evident in humans, is applicable to all viviparous animals, which are by definition the result of the interaction between masculine semen (formal and efficient cause) and menstrual blood (material cause). The model of embryonic development foreseen by Aristotle is clearly cardiocentric. On the other hand, we have highlighted how Aristotle's position on the extrinsic nature of the intellect (νοῦς θύραθεν) may set apart the origin of the human soul from the animal (sensitive) soul, educed from matter when the embryo reaches a suitable level of bodily organization. At least, this is the most common interpretation of Aristotle's doctrine on the origin of the soul and, certainly, the interpretation given by adherents to adherents of Abrahamic monotheisms such as Avicenna and Albert.

This brief recapitulation of Aristotle's overall theories of *pneuma* and the origin of the soul in viviparous animals should help us better elucidate his take on oviparous generation and sterile (or wind-eggs), which we shall base on Sophia Connell's study.[3] Aristotle discusses oviparous generation in the *Historia Animalium* and *De generatione animalium*, in which the generation of the chick represents the reference for the discussion of oviparous generation. Aristotle's take on oviparous generation displays an equally cardiocentric approach to chick development as to human and viviparous generation. According to Aristotle, the chicken egg contains two material substances crucial for the chick's formation: the egg white, which comes from the father, and the yolk inherited from the mother. While the egg white plays the role of formal and efficient cause (like masculine semen), the yolk represents the material cause of the chick's generation like menstrual blood in humans.[4] Fertilization occurs via the classic hylomorphic scheme highlighted in the previous chapter, exemplified by the rennet/milk metaphor, at least in viviparous animals. Theoretically, the same should be true of egg-laying ones. In the wake of fertilization, the egg is formed in 3 days within the hen's body and already possesses its final shape and size. By contrast, the chick inside is not yet fully formed. After laying the egg, the

3 Connell, *Aristotle on Female Animals*, 252–255.
4 Aristotle, De generatione animalium III 1 (751b5–7).

hen starts the most crucial process, viz. hatching. This process goes through different stages, where the chick develops gradually within the egg and wrapped up in a thick leather-like membrane (χόριον) performing the same function as the uterus. The upper part of the chick is completed in 10 days. The formative process comes to an end after 20 days, when the egg is broken by the chick from the inside.[5]

Unlike in the case of animal and human semen, Aristotle does not mention *pneuma* in his account of oviparous generation. Nonetheless, it seems plausible—at least this will be the most common reading according to authors like Avicenna and Albert—to read the role of *pneuma* in viviparous generation into the equally hylomorphic and cardiocentric scheme of oviparous generation. Moreover, the existence of a formal and efficient cause (egg-white) and, on the other, of a material cause (yolk) in generation lends itself to a reading of the concept of spirit analogous to the case of viviparous or human generation.

A particular case of generation in egg-laying animals are wind-eggs.[6] These eggs are produced by females of birds in the absence of their male partner: because they are usually laid in periods where winds are abundant, they are called wind- or "zephirine" eggs. According to Aristotle, these eggs are mostly laid by female exemplars of birds that do not fly much, such as pawns or hens. Because these eggs contain only the material cause of the chicken and not its formal and efficient ones, matter in the egg has no causal capacity to inform the matter meant for the chick-to-be inside the egg. For this reason, sterile eggs need to be fertilized by a male of the same species in a "suitable period of time", i.e. before white and yellow are coagulated together, to acquire that formative power which leads to chick formation.[7] Without the masculine egg-white, the formal and efficient cause of the chick, the matter itself is inert and the egg remains sterile.[8] As in the case of viviparous animals, the first organ to be formed in the offspring of egg-laying animals is the heart.[9] As we can observe so far, there is no real and explicit involvement of *pneuma* in the generation of egg-laying animals. Rather, its involvement can only be analogically inferred by comparing the case of viviparous and oviparous animals as the result of analogous formal, efficient and material causes.

---

5 Aristotle, Historia Animalium VI 3 (561b28–30).
6 Aristotle, De generatione animalium III 1 (751b21–25).
7 Aristotle, Historia Animalium VI 3 (559b5–20).
8 Aristotle, De generatione animalium I 21 (730a32); II 3 (737a33); II 4 (741a15–31); III 1 (750b25, 751b25).
9 Aristotle, Historia Animalium VI 3 (561a9–17).

While the case of oviparous generation seems easier to assess, the case of fish is somewhat different. The size of the eggs that lay ashore, argues Aristotle, is not the same as their final size. It needs, unlike a chicken egg, to grow. To do so, it requires the action of the male principle (semen), as is evident from empirical observation: male fish sprinkle milt on eggs after they have been laid.[10] On the other hand, it helps Aristotle demonstrate that the male principle is both extrinsic to female processes and the only thing able to make eggs fertile and reach their perfect and final stage.[11] As in oviparous generation, the role of *pneuma* as carrier of vital heat could theoretically be applied to generation in fish, but Aristotle, once again, again remains silent on the topic. This should also be the case, for instance, of generation in lungless and/or bloodless animals, which encompasses neither viviparous nor oviparous breathing animals such as humans and chickens. Anyway, this is an admittedly conjectural and charitable interpretation of Aristotle's ideas, which is nevertheless close to Albert's (also, Avicenna's) interpretation.

The case of plants is very different. In the first place, plants do not possess either *pneuma* or connate *pneuma*. In the second place, according to Aristotle, semen and menstrual blood constitute the formal, efficient (semen) and material (menstrual blood) causes of generation. Semen and menstrual blood are separated in nature. Luciana Repici has shown how this cannot be the case of plants for two reasons. First, plants do not have two material principles of reproduction, but one, the seed. Second, plants lack the digestive process whence semen or menstrual blood are produced in animals. To solve this puzzling situation, Aristotle thinks it reasonable (εὔλογος) to assume in the first case that plant seed encloses a substance not comparable to the masculine (formal and efficient) or feminine (material) fluids at play in generation.[12] Rather, Aristotle considers plant seed to be already analogous to the first mixture (πρῶτον μίγμα) found in the fetations (κυήματα) at the moment of conception.[13] This is because it encloses in itself the formal, efficient and material principles of generation, like a fertilized egg.[14] As Aristotle puts it, animals are somehow "divided (διερεμένα)" plants because they separate the principles of generation that are found mixed in plants. The physical origin of seed in plants is different from animal semen. While animals and humans produce semen and menstrual

10 Aristotle, De generatione animalium I 21 (730a18–24).

11 Aristotle, De generatione animalium III 5 (756a24–28). On generation in fish, see: Connell, *Aristotle on Female Animals*, 246–248.

12 Repici, *Uomini capovolti*, 31.

13 Repici, *Uomini capovolti*, 31–32.

14 Repici, *Uomini capovolti*, 32.

blood as residues (περιττώματα) of digestion of substances taken in, plants produce seeds from the concoction of taken-in substances that occurs first in the earth, functionally analogous to animal stomach.[15] The main element of difference between animal and plant digestion is that the first concoction of nutrition occurs externally to the plant. In the Aristotelian tradition, especially in the pseudo-Aristotelian *De plantis*, there is talk of two digestions, one extrinsic to the plant (as in Aristotle) and one intrinsic, the latter being operationalized due to the plant's vital heat.[16] In the case of plants, though, the only instrumental cause of a plant's soul that can produce the growth of the plant is the heat intrinsic to it. *Pneuma* is clearly not involved in this generative process, and not because Aristotle's account is fragmentary, but simply because neither Aristotle nor the *De plantis* endorse the existence of *pneuma* in plants. By contrast, *pneuma* plays a very relevant role in Aristotle's doctrine of spontaneous generation.

The concept of spontaneous generation withstood the test of time for a remarkable period. This belief was definitively eradicated only in the 19th century with the studies of Luis Pasteur,[17] and has been dealt with in depth by the key study produced by James Lennox.[18] In his article, we can see that Aristotle had introduced the explanation of this particular natural phenomenon in the fourth book of his *Meteorology*, discussing the so-called generation "from putrefaction", and in his zoology. In the *Historia Animalium*, Aristotle observed that animals such as bees, worms and testacea can be generated without parents out of putrefied matter.[19] It would seem, according to Aristotle, that these animals are able to arise out of matter, avoiding the mechanism, as Lennox has called it, of 'formal replication'.[20] By 'formal replication' Lennox means that the generation of animals such as bees, worms and testacea can occur without the formal and efficient cause provided by the father. In nature, the agency of the formal and efficient cause of generation is replaced by the action of external factors which can organize the matter of the animal-to-be in a manner so complex that this body can become ensouled and a real, living animal or plant.[21] These factors are equivocal to the life-giving semen, as they produce a living body from a corpse and converting death into life. To explain the dynamics of

15 More on this in: Repici, *Uomini capovolti*, 30–35.
16 On this issue, see chapter 2.
17 Hasse, "Spontaneous Generation," 150.
18 Lennox, "Teleology, Chance," 219–238.
19 Aristotle, Historia Animalium I 1 (487a3–5); III 21 (522b10); V 1 (539a18); V 15 (548a15); V 19 (551b27); V 19 (552a8); V 32 (557b7); VI 15 (569a28–29).
20 Lennox, "Teleology, Chance," 226.
21 Aristotle, Metaphysics, VII 9 (1034a34–b1, 1034b5–b8). Lennox, Teleology, Chance, 226.

this process, viz. how matter can generate by putrefaction without being acted upon by the formal and efficient agency of a male parent, Aristotle argues that

> Animals and plants come into being in earth and in liquid because there is water in earth, and air [*pneuma*] in water, and in all air there is vital heat; so that in a sense all things are full of soul. Therefore ⟨living things⟩ form quickly whenever ⟨this air and vital heat⟩ are enclosed ⟨in anything⟩. When they are so enclosed, the corporeal liquid's being heated, there arises as it were a frothy bubble. The differentiae which determine whether the kind is more or less honorable are ⟨determined by⟩ the organization of the vital principle in the enclosure. And both the places and the enclosed material are causes of this ⟨organization⟩.[22]

According to Aristotle, spontaneous generation takes place in earth or/and water, since both elements contain a bit of moisture and *pneuma* as a life-giving principle. When matter is enclosed within a membrane and heated by cosmic heat, then a new living being can be generated. Aristotle argues therefore that there are two factors that determine the nature or degree of perfection of the future animal, one of which is the place of generation—also conditioned by its climate—and the second the enclosed material with the surrounding membrane. Therefore, *pneuma* does not directly explain the nature of the future animal, since it is not specific for a bee or a worm. Rather, it is the nature of the enclosing membrane and the quantity of heat that oozes out putrefied matter which determines the species of the new animal.[23] This approach is in keeping with his *Meteorologica*, where spontaneous generation is closely studied in connection with putrefaction. Putrefaction, according to Aristotle, is defined as the oozing out of heat and moisture from a corpse. Paradoxically, spontaneous generation proves to be exactly the opposite of generation, which usually produces more heat and moisture in the embryonic matter to bring the embryo to life. Oddly enough, the outcome is the same. In the *Meteorologica*, spontaneous generation happens most frequently when part of the remaining heat and moisture in the corpse are drawn out because of solar heat and, together

---

22 Aristotle, De generatione animalium III 11 (762a18–27). This translation by Lennox in: Lennox, "Teleology, Chance," 224.

23 Lennox, "Teleology, Chance," 225: "But the causes of the sort of vital heat that gets enclosed and produces the species characteristics are said to be two: the location of enclosure (*oi tòpoi*) and the material that becomes enclosed (*to sòmato perilambanòmeno*). Presumably, the exact amount and strength of the *pneuma*, and consequently the vital heat, that an enclosure would 'capture' would be a function of the locale where it took shape."

with the particles of matter thrown out with it, they organize a new living being. This is the case with worms, for instance.[24] Despite the clear difference in terms of causes between spontaneous and sexual generation, Aristotle emphasizes that "the parts of the embryo already exist in the matter potentially, so that when a source of change comes about, the rest follows as a connected series, just as in the case of the spontaneous marvels."[25]

The major problem with spontaneous generation is not its physiological explanation, though. In his study on spontaneous generation, Lennox has convincingly highlighted that the theoretical problems stem from the tension between the mechanical explanation and the Aristotelian teleology of nature presented in book II of *Physics*. This problem is caused by the existence of entire species as products of spontaneous generation. The existence of entire species that are the product of spontaneous generation runs counter to the defining criteria that Aristotle had set out in his *Physics* for the term of "spontaneity" (197b14–20).[26] Indeed, chance occurrences of spontaneous generation are possible in the sublunary world. On the other hand, though, if something happens 'spontaneously' or 'by chance' it does not happen 'for the most part', let alone 'always'. Unfortunately, this is the case with entire animal species generated by spontaneous generation.[27] Among the cases that break this classification of spontaneous generation as a chance event are eels, an animal species produced entirely from putrefaction and presenting no sexual differentiation.[28]

Here we can observe how *pneuma* operates as a carrier of vital heat aimed at the reproduction of the species, even though this heat does not stem from the

24 Aristotle, Meteorologica IV 1 (379b6–8).

25 Aristotle, De generatione animalium II 5 (741b7–9). Aristotle further compares the process of generation to "spontaneous marvels" in Aristotle, De generatione animalium II 1 (734b9–17); II 5 (741b9–15). For further comparisons between spontaneous and sexual generation, see: Lennox, "Teleology, Chance," 233–234.

26 Aristotle, Physics II 6 (197b14–20). Lennox, "Teleology, Chance," 232.

27 Lennox, "Teleology, Chance," 238: "Thus, the central outstanding problem with Aristotle's doctrine of spontaneity in nature is not that it violates his overall theory of causality, nor that the chance production of organisms by nonreproductive means would not be appropriately termed 'spontaneous' according to his general doctrine. The basic problem is that to hold that every member of many species of insect and testacea are so produced would seem to be open to the following critique: 'The things mentioned, and all things which are due to nature, come to be as they do always or for the most part, and nothing which is the outcome of luck or an automatic outcome does that'."

28 Aristotle, Historia Animalium IV 10 (538a2–538a13); VI 16 (570a4–570a7). See also the recent contribution by Emily Kress, "Aristotle on Spontaneous Generation, Spontaneity, and Natural Processes," *Oxford Studies of Ancient Philosophy* 58 (2020), 157–204.

father's semen. Rather, *pneuma* substitutes the semen as a formal and efficient cause of the embryo-to-be. In addition to the lack of semen or an analogous substance, a further and lacking explanandum is the father's soul as a teleological guide for the motion of *pneuma*. In spontaneous generation, pneumatic motion is the result of a combination of material factors such as the sunny, environmental heat which takes on the role of formal and efficient cause, and a suitably organized membrane, the material substrate of generation. According to Aristotle, the frequency of spontaneous generation is higher among plants and bloodless animals, endowed with a less complex physiological structure and smaller amount of vital heat than blooded animals and humans, which both need lungs to cool down this heat.[29] Usually, the specificity of *pneuma* for the formation of a given animal rests not on the principle of synonymy, viz. by the parents' belonging to a certain species and on the elemental composition of semen (hotter and drier) and menses (colder and moister), but by other causes such as climate, the heat oozing from putrefied matter and the enclosing membrane. These equivocal factors substitute the agency of the father's soul and semen in determining the species of the animal-to-be. Strictly speaking, though, spontaneous generation does not occur to preserve a species and is not guided by an artisan like the father's semen. Spontaneous generation just occurs because of causal agents that can play a role analogous to semen and produce an animal or a plant out of putrefied matter. Still, the main problem rests with the categorization of these phenomena as 'spontaneous', when entire animal species seems to be generated out of it.[30]

Aristotle's doctrine of spontaneous generation is framed within a picture of natural philosophy that views this process as the non-teleological, chance result of the interaction between heavenly and sublunary agents. In late antiquity this picture will become more and more subject to a providentialist reading begun by Alexander and clearly detectable in later authors like Avicenna.

2 Avicenna on Spirit, Animal Generation, and (Human?) Spontaneous Generation

The previous chapter shed light on the connection between spirit, formative power and the origin of the human being in Avicenna's zoology and medicine.

---

29 Aristotle, De anima II 8 (420b20–27). On the role of lungs, see also: Aristotle, De iuventute et senectute 19 (477a14–19) and, in general, the remarks in chapter 2.

30 On spontaneous generation in Aristotle, see also: Connell, *Aristotle on Female Animals*, 255–264.

By now it should be clear how Avicenna conceives of spirit as a bridge between the causality of the heavens, channeling the heat of astral influences and combining it with the seminal heat belonging to the sublunar world. The interaction of this heat, placed into the spirit in the masculine semen, contributes to the formation of embryonic matter, whose growth occurs in a staunchly Aristotelian, cardiocentric manner. The process of fertilization is not exempt from hermeneutic tensions and contradictions, as in the case of Avicenna's *Canon*, which breaks the Aristotelian hylomorphic scheme of generation. Nonetheless, the pattern of animal generation as the result of the interaction of formal and efficient (masculine semen) and material (feminine semen, menstrual blood) causes, followed by the emanation of the soul into the body from the "Giver of Forms", represents a consistent psychophysiological model which Avicenna employs in explaining oviparous generation too.

In egg-laying animals, the fluids at play are divided into two egg whites and a yolk. Aristotle, by contrast, only speaks of one egg white. On closer inspection, the reasons for these doctrinal differences with Aristotle can be well compared to Avicenna's doctrine on human and viviparous generation. The first white has a formative function (like masculine semen) through its "formative power" (*virtus formativa*/*quwwa muṣawwira*).[31] The second white, provided by the female, though, is functionally analogous to feminine semen, viz. material cause of the chick. Likewise, the yolk provides food to the chick just like menstrual blood did with the embryo.[32] Avicenna's description of the different stages in the chick's development are basically consistent with Aristotle's: eggs are formed in 3 days, the chick acquires an upright position in ten days,[33] and after 20 days of hatching the egg mother eventually breaks the egg and sets the chick free from its egg.[34] According to Avicenna, the formation of the new chick begins really when the yolk is attracted into the second white of the egg by the action of the formative power, i.e. within the first three days. This growth follows a cardiocentric model running from top to bottom and consistent with Avicenna's cardiocentrism in humans and vivipara.[35] Even though it is not mentioned

31 Avicenna, Abbreviatio de animalibus XVI c. 1, ed. Venezia 1508, fols. 61^{va-vb} [ed. Cairo 1970, 406.6–407.9].

32 Avicenna, Abbreviatio de animalibus VI c. 1, ed. Venezia 1508, fol. 35rb [ed. Cairo 1970, 80.6–7]: "[…] membra (*aʿḍāʾ*) sunt ex albumine (*min al-bayāḍ*) et nutriuntur ex vitello (*wa-yakūnu ġidāʾuhu min al-ṣafra*)."

33 Avicenna, Abbreviatio de animalibus VI c. 1, ed. Venezia 1508, fol. 35rb [ed. Cairo 1970, 79.10–11]: "[…] post 10 dies apparet rectificata creatura."

34 Avicenna, Abbreviatio de animalibus VI c. 1, ed. Venezia 1508, fol. 35rb [ed. Cairo 1970, 82.2].

35 Avicenna, Abbreviatio de animalibus XVII c. 1, ed. Venezia 1508, fol. 62rb [ed. Cairo 1970, 416.14]: "quando attrahitur intra albumen pars vitelli fit pullus (Ar. *ǧanīn*, 'embryo')."

here, the "Giver of Forms" lurks behind Avicenna's theory of the animation in egg-laying animals, as all sublunary souls are eventually emanated into matter from its action.

Like Aristotle, Avicenna also discussed oviparous generation of sterile or wind-eggs (*ova venti/bayḍ al-rīḥ*). Consistent with Aristotelian zoology, Avicenna maintained that these eggs are usually laid by female birds that seldom fly, such as hens, doves and pawns and not by predatory birds like eagles. Since the former (doves, etc.) are birds that fly less and produce more residues, and these residues usually burn up in flight, their laying more wind eggs is a natural consequence of their life habits.[36] These sterile eggs may even be the result of mutual rubbing between females in absence of males. For this reason, they lack the formative, active principle provided by the first masculine egg white.[37] The life potentially existing in these eggs needs to be actualized either by warm winds or the masculine sperm of a bird of the same species. Fertilization by another male needs to occur in a time during which the white and the yolk have not yet been coagulated, for otherwise the process of fecundation would be ineffective.[38]

Avicenna goes on to parallel sterile eggs of birds to fish eggs, since both lack the formative principle of masculine sperm in their material composition.[39] As can be observed, Avicenna's handling of oviparous generation is pretty much in keeping with Aristotle's. There are some irreducible and innovative elements introduced by Avicenna, though. The first is represented by the tripartition of fluids (two whites and the yolk) in the reproduction of fish. The second concerns the composition of formative power as a result of combined heavenly and seminal heat. The result is a universe ruled by God's providence through the intermediaries like the "Giver of Forms" and instrumental causes such as spirit. This complex teleological system can be put to a litmus test by considering apparently non-teleological outcomes of generation such as spontaneous generation.

---

36 Avicenna, Abbreviatio de animalibus XVII c. 1, ed. Venezia 1508, fol. 62rb [ed. Cairo 1970, 415.7–9]: "Sed ova venti non sunt nisi in avibus facientibus multa ova propter multam materiam et multam superfluitatem in eis [...]."

37 Avicenna, Abbreviatio de animalibus VI c. 1, ed. Venezia 1508, fol. 35ra [ed. Cairo 1970, 79.13–16]: "Et dico ego quod non fit osculum sed potius confricatio quedam dilectionis, licet nesciat illud vere: et dicit Aristoteles quod istud accidit semper ante luxuriam ut in pluribus, et quando deficiunt masculi confricat se femina cum femina."

38 Avicenna, Abbreviatio de animalibus VI c. 1, ed. Venezia 1508, fols. 35rb–36va [ed. Cairo 1970, 79.2–4].

39 Avicenna, Abbreviatio de animalibus VI c. 1, ed. Venezia 1508, fol. 35va [ed. Cairo 1970, 75].

As observed in the first section of the previous chapter, spirit plays a crucial role in the Aristotelian doctrine of spontaneous generation, since the amount of heat enclosed within it is meant to supply the heat provided by the father's semen. Aristotle had argued that the organic complexity of the animal-to-be had to be determined both by climatic conditions (the amount of heat acting on putrefying matter) and the membrane containing the animal-to-be. The Aristotelian account strove to find a rationale account for the different kinds of animals ensuing from processes of spontaneous generation, failing to draw a clear theoretical line between animals that can be spontaneously generated and those that cannot, including humans.

This famous discussion would feature in a section of Avicenna's *Meteorology* (*Maʿdin wa-aṯar al-ʿulwiyya*, II, 6), translated into Latin by Alfred of Shareshell with the title *De diluviis in Thimaeum Platonis*, and in Avicenna's *De animalibus* analyzed by the studies of Dag Hasse and Amos Bertolacci we shall rely on here.[40] The conceptual framework within which Avicenna develops his theory of spontaneous generation is very different from Aristotle's, though. In the first place, he endorses what Aristotle was fighting in his account of spontaneous generation, viz. an extrinsic principle for the origin of the soul, as opposed to his doctrine on the origin of the soul as educed from matter. By contrast, according to Avicenna, souls emanate into matter from the "Giver of Forms". A second relevant element that sets Avicenna apart from Aristotle and stressed by Bertolacci is the Avicennian conception of time. Blending together the typically Platonic idea that catastrophic events occur cyclically (*Tim.*, 22c–23b, *Leg.*, III 677a–b) and the Aristotelian theory of the eternity of the species. Avicenna conceives of spontaneous generation as a natural process through which every species *could* be restored on earth in the wake of catastrophic events.[41] As a species of living beings, humans are not excluded from such a possibility,[42] although Avicenna has to account for this hypothetical argument.

40 Hasse, "Spontaneous Generation," 155–158; Amos Bertolacci, "Averroes against Avicenna on Human Spontaneous Generation. The Starting-Point of a Lasting Debate," in *Renaissance Averroism and Its Aftermath: Arabic Philosophy in the Early Modern Europe*, ed. Anna Akasoy/ Guido Giglioni (Archives internationales d'histoire des idées 21), Dordrecht 2013, 37–54.

41 Bertolacci, "Averroes against Avicenna," 45.

42 Avicenna, De diluviis, ed. Alonso Alonso 1949, 307.9–18 [II, 6, ed. Cairo 1965, 77.4–10]: "Et quando haec gignitio abscinditur, in multis annis possibile est ut veniat secundum quosdam constellatio et aliqua praeparatio elementorum quae faciat ea generari. Immo dicimus quod quaelibet species quae sit per complexionem elementorum in quantitatibus scitis dum elementa fuerunt, et sua divisio secundum illas quantitates et congregatio earum fuerit possibilis. Et si prima complexio (*mizāǧ*) non sufficit, haec generantur ex

While for Aristotle the fact that "man begets man and the sun" was granted by the action of the climate of a certain place and the membrane enclosing the matter subject to spontaneous generation, Avicenna's theory of spontaneous generation pivots on the crucial notion of the preparedness of matter (*praeparatio appropriate/ istiʿdād ḫāṣṣ*). This degree of material preparedness is achieved, given the absence of any seminal heat, through the cosmic heat that partially makes up formative power.[43] Hasse has stressed that, according to Avicenna, the preparedness of matter and the complexity of that which is spontaneously generated are directly linked to the theory of first and second complexion: the higher the level of complexity achieved, the more complex the generated animal.[44] Theoretically, if the level of complexity achieved possessed a sufficient amount of heat and moisture (viz. the highest possible), human beings could ensue from earth. Spontaneous human generation would be possible without the need for a uterus, which functions to defend the embryo from external threats. Since these threats would be no longer extant in case of catastrophic events such as those described in Avicenna's *Meteorology*, Bertolacci concludes, spontaneous human generation from earth could remain a possibility. This is just a theoretical possibility, though, not an opinion openly endorsed by Avicenna.[45]

Avicenna's stance on spontaneous human generation in his *De animalibus* is less direct. In these passages, studied by Amos Bertolacci, Avicenna alludes only generically to the possibility that the human species might be restored through spontaneous generation in the event of its extinction. Avicenna paraphrases Aristotle's own statement (*De generatione animalium* III 11, 762b27–32) that, should human beings and quadrupeds be generated from earth, their generation would occur either as from larvae or eggs. Again, Bertolacci has noted how this formulation is purely hypothetical.[46]

secunda et tertia. Quoniam, quemadmodum generatur animal ex complexione humorum post complexionem elementorum, non est inopinabile quod fiat complexio secunda sine semine et spermate."

43 Hasse, "Spontaneous Generation," 157–158: "Avicenna thinks it likely that the preparation to receive a particular form is itself due to an emanation, not from the giver of forms, but from the celestial bodies. Avicenna thus acknowledges that the stars have a certain influence on the process of generation, but he remains very critical of astrology in general, because he is convinced that celestial influences cannot be detected by us."

44 Hasse, "Spontaneous Generation," 156.

45 Bertolacci, "Averroes against Avicenna," 42; see also: Remke Kruk, "A Frothy Bubble: Spontaneous Generation in the Medieval Islamic Tradition," *Journal of Semitic Studies* 35 (1990), 274.

46 Avicenna, Abbreviatio de animalibus XVII, c. un., ed. Venezia 1508, fol. 62vb [ed. Cairo 1970, 419.9–10]; Bertolacci, "Averroes against Avicenna," 44.

Apart from this famous discussion, Avicenna's doctrine on spontaneous generation offers some innovative insights compared to Aristotle's remarks. Avicenna's theory encompasses also mice, snakes, scorpions, frogs and also blooded animals, something unprecedented in Aristotle.[47] The complexity of spontaneously generated animals in Avicenna seems therefore superior to that found in the zoology of the Stagirite and depends on the medical notion of complexion. Avicenna's doctrine of spontaneous generation presents another relevant case, viz. the larvae-like animals which transition from one species to another, such as butterflies. In his study of worms, Avicenna argues that the modality of their reproduction can be sexual generation, but not such that produces worms of the same species or reproduces by spontaneous generation.[48] Isabelle Draelants has shown that the generation of butterflies, silkworms and other similar insects occurs in successive stages. In the first place, putrefaction of a dead worm generates an egg, which in turn produces an incomplete larva. The incompleteness of the animal-to-be is caused by a lack in the strength of the formative power, unable to produce a fully formed living being. The future animal-to-be will then develop from the larva-like state to its fully formed state thanks to favorable environmental conditions and due to the fact that "nature does nothing in vain." This may be similar to the wind eggs, which become fertile once fertilized by winds.[49]

---

47 Avicenna, De diluviis, ed. Alonso Alonso 1949, 307.3–9 [II, 6, ed. Cairo 1965, 76.18–77.4]: "Et non est inopinabile secundum quosdam quod tu scis, quod corrumpantur plantae et multa genera eorum et post generentur per generationem, non gignitionem, quoniam multa alia fiunt per generatione et gignitionem. Et similiter plante. Et ex capilli fiunt serpentes, ex ficubus scorpiones et mus de terra et ranae de pluvia. Et omnia hiusmodi gignuntur." Kruk, "Ibn Sīnā *On Animals*," 337: "Ibn Sīnā here follows the widely accepted opinion that many "blooded" animals also could be generated spontaneously [...] Aristotle mostly restricts spontaneous generation to insects-like creature and fishes [...]"; Avicenna, Abbreviatio de animalibus XV c. 1, ed. Venezia 1508, fol. 59[va] [ed. Cairo 1970, 385.17–386.5]; on this passage, see: Bertolacci, "Averroes against Avicenna," 44.

48 Avicenna, Abbreviatio de animalibus, XV c. 1, ed. Venezia 1508, fols. 59[rb–va] [ed. Cairo 1970, 384.10–13]: "[...] anulosa iam generant se per putrefactionem, et iam erit masculinum et femininum, et erit inter illa luxuriatio (*sifād*), sed non pariunt sibi similes gusanes (ar. *dūd*, "worm"), et si parerent sibi similes essent de numero generantium sibi simile et quando generant rem non sui generis, ista generatio non vadit ad terminum ut fiat mutatio et non salvatio in genere [...]." See also: Kruk, "Ibn Sīnā *On Animals*," 334–338.

49 Avicenna, Abbreviatio de animalibus XV c. 1, ed. Venezia 1508, fol. 59[va] [ed. Cairo 1970, 384.15–385.6]: "[...] et tunc erunt quaedam animalia que generant generatione completa et quaedam incompleta: et istud est in ovantibus et generantibus gusanes. Et ovantium quedam generant ova completa sicut aves et quedam non completa sicut pisces, quoniam piscium ova crescunt postquam exeunt. Et eodem modo credo quod anulosum animal quod generatura per putrefactionem in terra non generat gusanes generatione sed ova

In his zoology, Avicenna seems to be trying to mitigate the problems caused by Aristotle's account of spontaneous generation, which categorized as 'spontaneous' the generation of entire species. We have see why this is a problematic stance according to the principles of Aristotle's philosophy of nature, since 'spontaneous' may be synonymous with 'by chance' but certainly is not with 'always' or 'for the most part', as in Aristotle.

These tensions, however, can be solved by appealing to the end-directedness of the process causing spontaneous generation, since the agency of celestial bodies is certainly God-directed. While the motions of celestial bodies may sometimes produce effects that do *not* come about via spontaneous generation (as opposed to usual generation), it is still possible. In Avicenna, this non-contradiction can be explained by his equation between the notion of 'being caused' with 'necessary'. In Avicenna, all beings other than God are considered possible in themselves but necessary (viz. caused) through something else, namely a chain of causes which goes back to the Necessary Existent, viz. God. In this sense, the existence of spontaneously generated animals can be defined as necessary *qua* caused by God, even though they represent a less usual causal effect of the actions of the celestial bodies on the sublunary world.[50]

All in all, the role of spirit in Avicenna's account of spontaneous generation seems to be rather different from Aristotle, whether in its function (restoration of entire species), its material composition (cosmic and seminal heat) or its final product (blooded and bloodless animals). In Avicenna's doctrine of generation, spirit emerges as a crucial instrumental and efficient cause at the service of the semen (or egg-white) ensuring the correct formation an embryo or a chick of the same species, just like the instruments represent a crucial tool for the artisan to bring out the desired form from matter. Spirit itself provides a fully independent physical account of generation (as in Aristotle) when considered as an interaction of material fluids. Unlike Aristotle, though, for Avicenna spirit is the receptacle of heat produced by a combined causality (seminal and cosmic) reflected in the composition of the *virtus formativa* (*quwwa*

---

quae post mutantur in gusanes et non est remotum quin etiam ille gusanes revertantur ad naturam illius a quo generatae sunt et generant ipsum: et volo dicere quod non stat sua generatio in gusanes secundum modum quo potest facere rem suae speciei [...]." On Avicenna's position, see: Isabelle Draelants, "La reproduction imparfaite: les *gusanes* et l'état larvaire des insectes chez Albert le Grand," in *Fragmented Nature: Medieval Latinate Reasoning on the Natural World and Its Order*, ed. Mattia Cipriani/Nicola Polloni, London-New York 2022, 159–163.

50 On God's will and determinism in Avicenna, see: Catarina Belo, *Chance and Determinism in Avicenna and Averroes* (Islamic Philosophy, Theology and Science 69), Leiden-Boston 2017, 91–120.

muṣawwira), which is also key to understanding deformities, such as the case of Siamese twins in chicks.

On the other hand, spirit alone is a notion unable to explain why the interaction of these material fluids produces a specific body with a determinate set of organs. Spirit can explain why a determinate set of organs are formed in an embryo or in a chick if and only if it is considered as an instrument of the father's semen that provides the mechanism of formal replication. In spontaneous generation, the role of formal and efficient cause is trusted to the heat of formative power coming from the heavens.

In the end, the metaphysical understanding of the mechanism of formal replication (in other words, the origin of the soul), or the so-called 'principle of synonymy', has two completely different meanings for Aristotle and Avicenna (*eductio formae* vs. *dator formarum*) and will be targeted by Avicenna's most fierce philosophical opponent, Averroes. Avicenna's philosophical horizon is considerably broader than Aristotle's, since he does frame the phenomenon of generation within a Neoplatonic scheme of intelligences with God at the top. It is precisely this scheme that is targeted in Averroes' criticism of Avicenna's views on the origin of soul when Averroes discusses spontaneous generation.

3 Averroes on Spirit and Spontaneous Generation

In his Long Commentary on the *Metaphysics*, Averroes considers formative power as being built up by both cosmic and seminal heat concentrated in the heart and borne by spirit in the masculine semen.[51] In the context of spontaneous generation, we have seen how Averroes discusses spontaneous generation and rejects Avicenna's doctrine of the Giver of Forms. For Averroes, the soul is actualized and not created from an extrinsic, Platonic principle as in Themistius or Avicenna. To cause the actualization of the soul in spontaneous generation, Averroes posits that the heat of the sun, mixed with that of the stars, is concentrated in certain portions of water or earth and limited by a membrane.[52] The heat enclosed within the spirit becomes the formative power that

51 Hasse, "Spontaneous Generation and the Ontology of Forms," 160.

52 Averroes, In Aristotelis Metaphysicorum libros XIIII commentarius (Aristotelis Opera cum Averrois Commentariis, vol. 8), XII comm. 18, ed. Venezia 1562, fol. 299 I–K [ed. Bouyges 1973, 1501.17–1502.7]: "Unde Aristoteles dixit quod homo generatur ex homine et sole. Et factus est ille calor in terra et aqua ex calore solis admixtum cum calore aliarum stellarum. Et ideo sol et aliae stellae sunt principium vivi cuiuslibet in natura. Calor igitur solis et stellarum divisus in aqua et terra generat animalia per putrefactionem nata. Et

shapes the future animal-to-be and brings its body to a level of organization that causes the soul to be actualized. Yet the criterion according to which certain kinds of animals can be produced by spontaneous generation and others cannot is surrounded by uncertainty. According to Averroes, humans cannot even hypothetically be generated from the earth. This is because the earth is not suitable matter for human generation. Amos Bertolacci, though, has clearly shown how Avicenna does not endorse this point of view. The gist of Avicennian theory is clear: if earth should be shattered by catastrophic events, spontaneous generation could work as a tool to restore the species. Given the absence of any external threat to humans, which is precisely that for the sake of which the uterus is conceived, the spontaneous generation of humans might hypothetically occur in the earth (*fī arḍ*). Note that the Arabic is clear enough on this point: in earth, not from earth (*min arḍ*).[53] Averroes' critique, then, takes the cue from what is a misrepresentation of Avicenna's point of view, meant to fully discard any possibility of human spontaneous generation.

Having ruled out hypothetical spontaneous human generation, Averroes argues that the complexity of the spontaneously generated animal is determined by the influence of the different kinds of constellations at play at the moment of generation.[54] Still, there is no clear physiological (complexion) and psychological (perfect or imperfect animals) criterion according to which Averroes distinguishes animals that can be spontaneously generated and those that cannot, or which can explain why animals can be spontaneously generated and humans cannot. Dag Hasse has noted how this remains an open question within Averroes' doctrine on spontaneous generation.[55] The Avicennian-Averroean dispute will be taken up by Albert, who valued both authors, as we have seen in chapter 3 and so far, in his account of formative power and generation.

4 Animal, Plant, and Spontaneous Generation in Albert the Great

In chapter 3, we shed light upon Albert's doctrinal debt to his Arabic sources on the topic of human generation. Following the lead of Krause, Takahashi

universaliter omnia quae fiunt ex non semine. Non quod illic sit anima in actu inventa in orbe declivi et a sole sicut narrat Themistius. Et omnia ista sunt declarata in libris *De animalibus*." Other relevant passages from Averroes' zoology in: Freudenthal, "The Medieval Astrologization," 121–123.

53 Bertolacci, "Averroes against Avicenna," 44–45.

54 Hasse, "Spontaneous Generation," 161.

55 Hasse, "Spontaneous Generation," 161–162.

and Nardi, we assessed the epigenetic model that underlies Albert's doctrine of the actualization of vegetative and sensitive souls and its connection to formative power. This model, following insights from Avicenna and Averroes and based on an ordered set of causes, proves crucial to understanding all of Albert's mature writings also in connection to the generation of plants, egg-laying animals and spontaneous generation. In this section, our main focus will revolve around the categories of animals that are not analogous to humans in their modes of generation, viz. egg-laying animals and plants, and on spontaneous generation. The presence of plant generation as the causal product of the action of spirit represents the gist of Albert's mature doctrine on generation. While in his early theology he used spirit and formative (or generative) power to explain the generation of animals, here the concepts of spirit and formative power are translated at a botanical level to account for the generation of plants.

In the previous chapter, we have highlighted how formative (or generative) power represents the cause by which the embryo-to-be transitions from life-in-potentiality into life-in-actuality. The threefold component of vital heat (cosmic, paternal, seminal) placed in the spirit of the semen represents the formal and efficient cause directing the interaction of the two kinds of semen and menstrual blood towards the fully-formed animal. The life-giving action of formative power runs parallel to the progressive eduction of the vegetative and sensitive soul. The potential existence and subsequent actualization of these souls (although not in the strict sense) in matter is explained by Albert according to his doctrine of *inchoatio formae*. The nutritive or sensitive soul exist in the future plant- or animal-to-be not as forms in a strict sense, but as initial principles of the organization of bodily matter. The actualization of these forms in plants or animals will be completed when these living beings reach their completed forms. None of the vegetative or sensitive souls, though, are created by God. Rather, they are educed from matter as a result of a natural process based on an ordered set of causes. In the case of vivipara, the formal and efficient causes are represented by the masculine semen, and the material causes by the feminine semen and menstrual blood. In the case of oviparous generation, Albert shows once again a clear adherence to the Avicennian model detected by Demaitre and Travill in their study of human generation, which draws some brief comparisons between the causal principles of human and oviparous generation.[56] We shall expand on their extremely brief and comparative remarks to discern the role of spirit in oviparous generation in the Aristotelian paraphrases, as there is no trace of this in the *De homine*.

56 Demaitre-Travill, "Human Embryology," 417.

4.1 *Oviparous Generation*

In his zoology, Albert produces a detailed account of the chick's developmental stages inside the egg, which follow closely the Avicennian doctrine. According to Albert, the generation of the chick must serve as the main reference for oviparous generation. This is because the formal, efficient and material principles at play in chick generation are analogous to other birds large and small (*secundum unum modum in omnibus avibus magnis et parvis*).[57] Albert adopts the Avicennian scheme of causes: the first egg-white (*albugo*) comes from the male and is functionally analogous to semen in vivipara. The external egg-white is analogous to the masculine since it derives and possesses the formative power (*in albo spisso viscoso est virtus galli*) placed in the radical or seminal spirit. While it is indeed contained in the egg, it remains external to the material composition of the second egg-white, called *albedo*. Thus, the first egg-white plays the role of formal and efficient cause in generation. By contrast, the *albedo* is analogous to feminine semen and represents the matter out of which the chick is formed (*semen feminae est albumen ovi in quo formentur membra pulli*). The yolk (*citrinum* or *vitellum*) supplies the necessary amount of food to the chick like menstrual blood does (*vitellum autem citrinum praeparatum est in supplementum et in cibum pulli*).[58] This is the gist of the comparative remarks by Demaitre and Travill, accurate and fairly limited to the roles of the main causal principles such as semen/eggwhite, feminine semen and menses/second eggwhite and yolk according to the Avicennian scheme.

---

57 Albertus Magnus, De animalibus VI tr. 1 c. 4 n. 27 (ed. Stadler, I, 452.21–24); XVII tr. 1 c. 4 n. 29 (ed. Stadler, II, 1160.13–15).

58 The study of oviparous generation is conducted analogously to plant generation: Albertus Magnus, De vegetabilibus I tr. 1 c. 7 (ed. Meyer-Jessen, 26–27): "Si autem mixtionem sexuum dare volumus plantis, per alium modum est hoc imaginandum. Sexus enim, secundum quod est in substantiis individuis completis, nullo modo est in plantis aut separatus aut mixtus. Sed sexuum vires sunt in seminibus embryis sive impraegnationibus. Embrya autem sive impraegnationes dico concepta semina. In illis enim vis masculi est formans et agens; semen autem feminae illi permixtum est formatum et patiens. Et semen quidem masculi est sicut artifex, semen vero feminae sicut artificiatum, in quod forma producitur artificis. Et sanguis menstruus est sicut cibus attractus in supplementum materiae, in qua formetur partus. Sicut videmus in uno ovo, quod ovatum a gallina vel ab alia ave, in quo in albo spisso viscoso est virtus galli; et semen feminae est albumen ovi, in quo formentur membra pulli; vitellum autem citrinum praeparatum est in supplementum et in cibum pulli, quousque egrediatur de testa. In embryis igitur omnibus virtus est utriusque sexus per modum agentis et per modum patientis. Et per hunc modum imaginabimur, sexuum virtutes esse in plantis permixtas, quia non dubitamus, in eis esse agens et formans, et aliquid esse materiam, quae actionem suscipit et formationem."

Albert does not restrain himself to a blind adherence to Avicennian doctrines of oviparous generation, though, but injected some medical comparisons into his account. In his view, the structure of the fertilized egg resembles the human head and brain, at least to some extent. He compares the membranes that protect each of the two egg-whites with the role of meningeal membranes in the brain. Thus, the membrane protecting the first egg-white (*albugo*) is analogous to the *dura mater*, since the *albugo* is the external one of all fluids involved, just as the *dura mater* is the external thing among meningeal membranes. On a similar note, the membrane protecting the *albedo* (feminine egg white) is analogous to the *pia mater*, the internal meninx. Given that Albert knew (via Avicenna) of the existence only of a *pia* and *dura mater* and not the intermediate meningeal membrane known today as arachnois, he was not able to identify a functional analogy in the brain to the membrane protecting the yolk.[59]

Albert's anatomy of the fertilized egg brought the comparison between oviparous generation and human generation to a detailed embryological level. According to Albert, the membrane that protects against the transmission of moisture, matter and food into the material component (*albedo*) of the chick is analogous to the *secundina* (*mašīma*) membrane, which envelops the embryo according to Avicenna. In the end, both membranes protect the body of the embryo or of the chick. As has emerged from this brief account, Albert clearly depends on the Avicennian doctrine of oviparous generation and its assessment of the causal role of fluids at play. On the other hand, Albert does present original comparisons of functional analogies borrowed both from brain anatomy and embryology.

In Albert, the progression of the chick's developmental stages follows Avicenna, who in turn follows Aristotle in the timing of the process but not in identifying the material substances involved. The progression is the usual one: the egg appears when the first white of the egg, endowed with formative power (*in quo magis potuit virtus formativa*),[60] attracts the nutritive matter of the yolk into the second white of the egg.[61] To occur, though, this attraction of nutritive matter must be facilitated by a three-day hatching. After six days, the egg displays the appearance of a blood-like droplet, which constitutes the material substrate for the formation of heart and cardiac *pneuma*. Cardiac spirit, which plays the role of vehicle of formative power, shapes all body parts from top to

59 Albertus Magnus, De animalibus VI tr. 1 c. 4 n. 27 (ed. Stadler, I, 452.25–31).

60 Albertus Magnus, De animalibus VI tr. 1 c. 4 n. 31 (ed. Stadler, I, 454.2–5).

61 Albertus Magnus, De animalibus VI tr. 1 c. 4 n. 28 (ed. Stadler, I, 452.32–453.4).

bottom and from the heart through the whole body.[62] The heart is the principle of two blood vessels originating in the liver and the lungs, just like the heart is the principle of veins in humans and animals. At the end of this 20-days long process, the chick comes out of the egg when the latter gets broken, either from the inside or from the outside by the mother.[63] In Albert's view, the length of the developmental process can also be influenced by external factors such as young age of the parents (as in humans), hot climate and hot food.[64] The process of growth relies, as shall be seen briefly in the next chapter, on Albert's interpretation of the Avicennian concept of radical moisture. This concept, which appears once in Avicenna's theory of oviparous generation, is introduced by Albert as well.[65]

The location of radical moisture in the generation of the chick appears to be in the second, passive white *qua* provider of matter for the chick's body. Albert's account of generation in egg-laying animals features the concept of radical moisture, which is equated with the egg white provided by the mother.[66] In the generation of oviparous animals, spirit plays the key role of instrumental cause *qua* vehicle of the heat inherent in the formal and efficient cause of generation, the male semen, and connects cosmic heat with seminal heat. Unlike with human generation and any kind of generation involving semen, Albert

62 Albertus Magnus, De animalibus VI tr. 1 c. 4 n. 29 (ed. Stadler, I, 453.4–20): "Sic autem ascendente citrino ad acumen ovi continue ascendit tractum ad illum locum ovi, ubi apparet fissura in testa, quando egreditur pullus: haec enim fissura fit in superiori parte ovi versus acumen, ubi formatur caput et rostrum avis: et ibi postquam ascendit citrinum, apparet in albo ovi quasi gutta sanguinis: et illa gutta primitus inter membra formatur in cor avis, quod statim ut formatur, spiritu plenum est ex calore digerente et formante humidum, in quo spiritu vehit se virtus formativa in formationem aliorum membrorum ex corde: qui exitus spiritus a corde non fit nisi per motum cordis secundum systolen et dyastolen, et ideo cor tunc movetur: sed a principio est lentus motus eius propter multum humidum, quod est in ipso, quod cito a calore vinci non potest. Motus autem hic relinquit supponens animam sensibilem: licet non tunc insit ut actus corporis organici physici, sed sicut artifex in artificiato faciens et instituens organa, in quibus expleat vitae potentias."

63 Albertus Magnus, De animalibus VI tr. 1 c. 4 n. 35 (ed. Stadler, I, 455.36–37).

64 Albertus Magnus, De animalibus, VI tr. 1 c. 4 n. 24 (ed. Stadler, I, 451.16–20).

65 Albertus Magnus, De animalibus VI tr. 1 c. 4 n. 58 (ed. Stadler, I, 465.17–22): "Alia autem, quae est sub ista sicut pia mater, habet intra se albam rem humidam, ex qua fiunt substantiae radicales membrorum pulli: et ipsa in generatione est quasi secundina continens totum pullum."

66 Albertus Magnus, De animalibus VI tr. 1 c. 2 n. 10 (ed. Stadler, I, 445.1–6): "[…] quando crescit ovum in gallina, in principio quidem incrementi totum non apparet nisi citrinum, ac si sit vitellum quoddam, et cum inceperit inspissari substantia ovi et compleri, dividitur substantia alba a citrina, et ponitur citrinum in medio in cibum, et album in circuitu, ut ex ipso fiat substantia membrorum radicalium […]."

never discusses explicitly the entrance of seminal fluids from the father into the chick. As we have observed, Albert's take on generation in the *De animalibus* follows the interpretation of Aristotle (and Galen) given by Avicennian zoology. This scheme seems compatible with the Aristotelian principles of generation. At least, it does not explicitly break the Aristotelian hylomorphic scheme of generation as the *Quaestiones super de animalibus* do.

The physiological model of oviparous generation laid out so far is fully comparable with human generation and follows closely the account of Avicennian zoology. Both cases of live-birthing and egg-laying generation possess two passive and material causal principles: the yolk and the second egg-white in egg-laying animals, menstrual blood and feminine semen in live-birthing ones. These material, passive principles are informed by the action of one active, formal and efficient cause: the first egg-white in oviparous animals, masculine semen in viviparous ones. In this context, spirit plays the causal role of instrument of the soul and of the First Cause *qua* vehicle of formative power, which contributes to the shaping of embryonic matter. The agency of spirit becomes such only in connection with suitable material causes (heat of formative power) and, more importantly, with the assent of the First Cause, the guiding principle of formative power.

Among oviparous animals, the Avicennian analogy of functions between whites of the egg and yolk holds only for the cases of birds and fish, though. In their case, it is impossible to tell apart the different fluids of the yolk (*citrinum*) and the second egg-white (*in ovis piscium secundum visum non sit nisi albus non humor et nullum citrinum appareat in eis*). Despite this empirical difficulty, Albert concludes that the conceptual structure that applies to generation in fish is the same as in egg-laying animals. On the one hand, there is the humor (humor) that is the material cause of the embryo and out of which radical members are formed. In viviparous animals, this humor is the menstrual blood, while for chickens it is the second egg-white, provided by the female. According to Albert, marine eggs contain the principle for the nutrition and growth of the fish-to-be, just like in chicken eggs the yolk supplies the chick-to-be with a suitable amount of nutrition. The eggs of fish are completed and the soul of the new fish is actualized when the male semen, which stays out of the material composition of the future fish, informs the matter through its heat. From this point of view, the mechanism accounting for the fecundation of marine eggs can well be compared to the so-called sterile or "wind eggs" (*ova venti*).[67]

67 Albertus Magnus, De animalibus VI tr. 2 c. 1 n. 61 (ed. Stadler, I, 466.25–31): "Adhuc autem

4.2 *An Unaccomplished Case of Generation?* Ova venti *in Albert's Zoology*

Among all cases of generation, wind eggs stand out because of the absence of a life-giving principle, viz. the father's semen. As in Avicenna's and Aristotle's versions of *De animalibus*, wind-eggs (*ova venti*) or sterile eggs are produced out of the superfluous moisture secreted by female birds like pawns and doves. Since these birds do not fly much, they retain much of their moist residues. These moist residues serve naturally as matter for the future egg and bird-to-be. Species of predatory birds such as eagles, by contrast, burn up most of their residue when hunting and therefore produce very little *ova venti* or even none at all.[68]

The laying of sterile eggs happens mostly when female pawns or doves happen to have a strong desire for intercourse in the absence of a male partner.[69] According to Albert, as for Aristotle and Avicenna, the frequency with which these eggs are laid is inversely proportional to the amount of motive activity performed. The more they fly, the fewer eggs they lay. To continue with the parallel with human-generation, Albert considers the fluid materially contained in the *ova venti* to be like menstrual blood and feminine semen *qua* derived from the mother. The *ova venti* provide only a potential principle for life, which requires the male counterpart (semen), guided by the formal cause (the father's soul), to be actualized.[70] The actualizing principle is, according to Albert, the heat of the formative power placed in the spirit.[71]

As has now emerged, oviparous generation as explained by Albert closely follows Avicenna's exegesis of Aristotelian zoology. Albert did not limit himself to injecting additional observations on the functions of the membranes protecting the different fluids in the eggs, though. He contrasted and compared these data with other anatomical and embryological analogies of the differ-

---

pullorum piscium in ovis formandorum cibus est in ventre matrum sicut et in avibus: et ideo oportet ovum piscis esse compositum ex humore duplici, quorum unus est, ex quo fiunt radicalia membra, et secundus, ex quo sicut ex cibo fiunt supplementa, quibus implentur vacuitates radicalium membrorum, licet in ovis piscium secundum visum non sit nisi albus humor et nullum citrinum appareat in eis."

68 Albertus Magnus, De animalibus, XVII tr. 1 c. 1 n. 3 (ed. Stadler, II, 1146.15–22); XVII tr. 1 c. 2 n. 8 (ed. Stadler, II, 1149.17–27).

69 Albertus Magnus, De animalibus X tr. 1 c. 3 n. 26 (ed. Stadler, I, 741.10–16); VI tr. 1 c. 4 n. 33 (ed. Stadler, I, 451.35–452.7).

70 Albertus Magnus, De animalibus XV tr. 2 c. 11 n. 145 (ed. Stadler, II, 1056.36–1057.9); De animalibus, VI tr. 1 c. 2 n. 14 (ed. Stadler, I, 446.20–29).

71 Albertus Magnus, De animalibus VI tr. 1 c. 3, n. 20 (ed. Stadler, I, 449.14–16): "Aliquando etiam ova, quae ut ova venti concipiuntur, aliquando pullificant, quando post conceptum tactum fuerit a semine galli sive masculi."

ent membranes protecting the different fluids. Like Aristotle, Albert employed the physiological model of the fertilized egg to causally explain the coexistence between formal, efficient and material causes in the seed of plants. This time, though, this coexistence also implied the presence of spirit in plants.

4.3 *Spirit and the Generation of Plants*

The second chapter of this book explained the doctrinal rationale behind Albert's peculiar introduction of *spiritus* into plants and hence into the whole realm of ensouled beings.[72] This introduction does not come without some empirical difficulties, though. As has been shown, the different material and organic composition of plants, geared toward bodily simplicity and uniformity (*similitudo corporis*),[73] makes it difficult to single out the material substances at play in these processes, viz. the different kinds of spirit and organs that are visible and separated in animals and humans. On the one hand, the predominance of homoeomerous parts in the body (*similitudo corporis*) makes it easier for plants to restore the destroyed parts through the intake of nutrimental moisture.[74] For Albert, this is principally the case with plants and, to a lesser extent, of bloodless animals such as crabs, snakes, lizards and worms.[75]

The organic simplicity of plants makes it impossible to locate precisely the *spiritus radicalis* or *seminalis* found in human and animal testicles or reproductive organs. Moreover, plants do not seem to have, according to Albert, an organ specifically devoted to the production of semen and one for its reception, like the feminine uterus. Rather, the power to form seeds appears to be spread all over the parts of the plant (*ubique virtutem habet seminis*).[76] These seeds contain in themselves the formative power. Despite the irreducible differences in physiological structure, Albert tries to draw a cogent parallel between the spermatic location and formative function of the spirit involved in plant

---

72 See chapter 2, 89–105.

73 Albertus Magnus, De spiritu et respiratione I tr. 2 c. 3, ed. Auguste Borgnet, Paris 1894 (Ed. Paris. IX, 236a).

74 Stefano Perfetti, "Rigenerazione degli animali? Alberto Magno tra *Parva Naturalia* e *De animalibus*," in Vita longa. *Vecchiaia e durata della vita nella tradizione medica e aristotelica antica e medievale*, ed. Chiara Crisciani/Luciana Repici/Pietro B. Rossi (Micrologus Library 33), Firenze 2009, 165–166.

75 Albertus Magnus, De aetate sive de iuventute et senectute, tr. 2 c. 1, ed. Auguste Borgnet, Paris 1890 (Ed. Paris. IX, 316b–317a). On regeneration of these animals: Albertus Magnus, De animalibus VII tr. 3 c. 4 n. 168 (ed. Stadler, I, 567.12–19). For a study of these passages, see: Perfetti, "Rigenerazione degli animali?," 160. More on Albert's doctrine of radical moisture in chapter 4.

76 Albertus Magnus, De nutrimento et nutrito tr. 2 c. 2 (Ed. Colon. VII/2a, 17.36–39); Albertus Magnus, De anima II tr. 2 c. 9 (Ed. Colon. VII/1, 94.49–63).

generation and that in animal generation (*sicut in semine animalium spiritus, qui vehit virtutem formantem, intra viscositatem seminis continetur*) in his *De vegetabilibus*.[77] The analogy between animal and plant generation holds in the case of coexistence of formal, efficient and material causes altogether, and in case of seed as the residue of the fourth digestion, as in animals.[78]

Despite this strong attempt at a unification of the discourse on animal and plant generation, Albert must concede some differences at the physiological level. In the first place, the material composition of fluids at play is by no means identical. The material composition of plant seed consists of an oat-like substance (*substantia farinalis*).[79] To be clear, this substance, being the material cause of generation, must be compared to the feminine semen. This comparison, though, holds only from a causal point of view and not in relation to its material composition, as it is obvious that the physiology of plants is different from that of animals.

To some extent, the model of plant generation rooted in the key role of spirit and formative power strains the similarities between plants and animals, especially those that do have something analogous to semen but no blood (like plants, which have sap). As in Aristotle's account, formal, efficient and material causes of generation coexist within the plant's seed as they do in the fertilized egg. The main difference lies in a simple detail: the fertilized egg is such after intercourse between hen and cock, plant seeds are so by nature. All in all, Albert develops his doctrine on plant generation by welding together the Avicennian scheme of three fluids (first egg-white, second egg-white, yolk) with the Aristotelian comparison between oviparous and plant generation.

Unlike animals, plants do not need sexual intercourse to produce offspring.[80] Amalia Cerrito has shown how in his botany Albert distinguishes between sex (*sexus*) as the capacity to produce a sex (*virtus sexus*) and a result or imitation of this capacity (*resultatio quaedam et imitatio virtutis sexus et non vera virtus ejus*). The first case is determined by the interaction between

77 Albertus Magnus, De vegetabilibus II tr. 1 c. 3 (ed. Meyer-Jessen, 121).

78 Albertus Magnus, De nutrimento et nutrito tr. 1 c. 3 (Ed. Colon. VII/2a, 6.59–7.10).

79 Albertus Magnus, De vegetabilibus VII tr. 1, c. 9 (ed. Meyer-Jessen, 618): "Dicamus autem nunc de seminatione. De seminis autem natura et virtute iam in praecedentibus dictum est. Sed quod hic attendendum est, hoc est, quod seminis substantia duo continet. Quorum unum est virtus formativa, quam habet e coelo cum calore et spiritu—qui instrumentaliter formativae deserviunt, calor quidem digerendo et segregando et subtiliando, spiritus autem vehendo virtutem. Alterum autem, quod habet semen, est substantia farinalis, quae, immixto sibi humido, suscipit formationem et figurationem in plantam et plantae organa."

80 Cerrito, "Disclosing the Hidden Life of Plants," 112–114.

two material substances in intercourse (*sexus est in substantiis generantibus per coitum*).[81] By contrast, the lack of this process leads Albert to consider sex in plants only as a 'result' or 'imitation' of the capacity to produce offspring with a sex. For this reason, a proper discernment (*discretio*) of sex in plants is impossible, just as with eels, animal deprived of sexual differentiation.[82]

According to Albert, sex differentiation in plants can be traced back to a hotter or colder complexion of the semen. The complexion of the seed is influenced by two factors: the heat of the sun and a *calor naturae communis* present in the semen itself. This heat is carried by the spirit in plants, although Albert is not explicit on this point. Moreover, Albert considers earth as analogous to the animal uterus (*locus terrae, in quo oritur plantae, est sicut matrix in animalibus*): the analogy with animal generation has to stop here, since there cannot be a feminine sex in plants *stricto sensu*.[83] Broadly speaking, though, there are empirical criteria for distinguishing sexual differentiation in plants: female plants produce a greater amount of softer and bigger fruit due to their moister complexion, while male plants produce rougher and drier fruit.[84]

As we hope to have shown in this section, Albert produces an original account of spirit in plant generation, trying to build his theory in doctrinal continuity with his theories of human and animal generation, based on his Aristotelian and Arabic sources. Produced by analogous causal agents (spirit, for-

81 Albertus Magnus, De vegetabilibus I tr. 1 c. 12 (ed. Meyer-Jessen, 47); Albertus Magnus, De nutrimento et nutrito tr. 2 c. 2 (Ed. Colon. VII/2a, 17.30–35): "In plantis autem omnino nulla est discretio sexus, sed ubique impraegnant et ubique impraegnantur. Et huius causa est quoniam ultimum animae vegetabilis est generativa et ab ipsa diffinitur."

82 Albertus Magnus, De nutrimento et nutrito tr. 2 c. 2 (Ed. Colon. VII/2a, 17.19–21); Albertus Magnus, De homine (Ed. Colon. XXVII/2, 129.9–12).

83 Albertus Magnus, De vegetabilibus II tr. 1 c. 1 (ed. Meyer-Jessen, 105): "Conveniunt autem plantae secundum quod planta est, vegetari tantum, et exercere actum vitae secundum augmentum et alimentum et generationem tantum. Secundum causam autem agentem plantae convenit, a sola virtute coelesti et elementorum commixtione pullulare. Forma enim prima vitae operationem habens invenitur in planta; et haec non habet magnam suarum potestatum diversitatem, propter quod etiam ex parte corporis plantae non requirit magnam suarum partium diversitatem in complexione. Et cum complexio tanto sit facilior, quanto est similior, sufficit in productione ejus agens unum commune, quod est virtus coelestis cum materia convenienter commixta. Et ideo planta omnis caret sexu, quia sexus femineus est propter principii generationis susceptionem et praeparationem. Et haec potestas est in terra convenienter commixta et praeparata; et ideo locus terrae, in quo oritur planta, est sicut matrix in animalibus; et succus sive humor in tali loco praeparatus et attractus est sicut sanguis menstruus in matricibus animalium."

84 This position is further confirmed in the chapter that Albert explicitly dedicates to sex determination in plants: see: Albertus Magnus, De vegetabilibus I tr. 1 c. 7 (ed. Meyer-Jessen, 23–24).

mative power, vital heat, matter) coexisting within the seed, plants are subject to a process of generation much more similar to human and animal generation than in Aristotle or Avicenna. On the other hand, plant generation retains some irreducible features of differentiation rooted in the nature of plants themselves, such as the lack of strict sexual differentiation. Still, Albert's strong emphasis on a 'universal physiology' of generation emerges clearly from his botany and his doctrine on generation in plants. His effort to build a unifying account of material causes at play in the phenomena of generation emerges also from his account of spontaneous generation, explained univocally through the notion of spirit in the case of both animals and plants.

4.5 *Spirit and Spontaneous Generation*

Albert's discussion on spontaneous generation is carried out in a more full-blown manner in his Aristotelian paraphrases. Still, some doctrines found in the early *De quattuor coaequaevis* and *Commentary on the Sentences* constitute relevant tools for understanding his mature position. For instance, in the *De quattuor coaequaevis* Albert presents some brief remarks on spontaneous generation, taking a cue from the curious *quaestio* on whether angelic bodies are subject to putrefaction.

Albert's answer is obviously negative, since he considers the angelic body to be of an airy nature. Nonetheless, Albert introduces a short but insightful discussion about spontaneous generation based on the key concept of formative power (*virtus formativa*) to explain spontaneous generation from putrefaction in natural bodies. The *Doctor Universalis* argues that the cosmic component of formative power is able to substitute the informing action of the father's soul and the efficient causality of masculine semen. Maaike van der Lugt has noted how this understanding of formative power mirrors a peculiar and common Medieval interpretation of the cosmological component of the principle of synonymy. In the sentence "man begets man and the sun", the phrase "and the sun" became not only integral to the physiological explanation of sexual generation but could autonomously explain the etiology of chance natural phenomena as spontaneous generation *qua* things caused by celestial virtues alone (*a virtute coelesti tantum*). The motion of heavenly bodies produces the cosmic component of formative power, which alone—that is, without the heat provided by the animal (semen, egg whites) or plant (seeds) material principles of generation—causes the spontaneous generation of plants, animals and minerals.[85]

85 Maaike Van der Luugt, *Le ver, le démon et la vierge. Les théories médiévales de la génération*

According to Albert, the heat of the formative power elicits part of the heat and moisture in the corpse (or in air and water) and organizes them. It is according to this pattern of celestial causality that Aristotle's process of *generatio ex putrefactione* as caused by extrinsic heat (*aliena caliditate*),[86] viz. the heat of the surrounding environment (*caliditas in continente localiter vincens*), must be understood.[87] In his early theological works, Albert does not develop his doctrine of spontaneous generation any further. In the *Commentary on the Sentences*, he just refers to a few cases of spontaneous generation in worms and eels.[88]

In his mature works, Albert resumed his study of spontaneous generation in the *Meteora, De vegetabilibus, De animalibus* and *Metaphysics*. In these works, the role of spirit as a vehicle of formative power becomes explicit and applies to

---

extraordinaire: une étude sur les rapports entre théologie, philosophie naturelle et médecine (L'ane d'or), Paris 2004, 142. Albertus Magnus, De quattuor coaequaevis, tr. 1 q. 1 a. 6 ed. Auguste Borgnet, Paris 1895 (Ed. Paris. XXXIV, 315b): "Sed opinio Aristotelis est alia, scilicet quod formae naturales sunt generantes formas, quae sunt in materia: semina autem sunt, quae dant formas generatis per formas quas accipiunt a generantibus primis, scilicet semen hominis, a quo descinditur, a virtute coelesti accipit virtutem formativam ad producendum formam consimilem super materiam. In his autem quae non generantur ex semine, sicut illa quae generantur ex putrefactione vel commixtione elementorum sola, sicut est generatio mineralium, est virtus formativa respectu formae substantialis a virtute coelesti tantum. Generatio autem elementorum est a motu coeli cum virtute coelesti: quia est respectu formae substantialis elementi."

86 Aristotle, Meteorologica IV 1 (379a17–19).

87 Albertus Magnus, De quattuor coaequaevis tr. 4 q. 61 a. 2, ed. Auguste Borgnet, Paris 1895 (Ed. Paris. XXXIV, 649b): "Quod autem dicit, Nec per mortem dissolvuntur, non intelligitur de morte quae est divisio spiritus moventis a corpore moto, sed de morte quae est corruptio secundum naturam: corruptio autem secundum naturam proprie putrefactio est: putrefactio autem proprie non convenit nisi corpori mixto, in quo dominatur naturae inferiorum elementorum, scilicet aquae et terrae. Et hoc patet per diffinitionem putrefactionis in IV *Meteororum*, ubi dicitur: 'Putrefactio est corruptio in unoquoque humido proprie, et secundum naturam caliditatis ab aliena caliditate': aliena autem caliditas est caliditas continentis secundum Philosophum: continens autem in medio est aer, in quo est caliditas continens secundum circulum quod est ignis, et continens secundum circulationem quod est sol: quae caliditas in continente localiter vincens calidum conservans locatum, educit ipsum, et illud eductum secum trahit humidum, et inducitur putrefactio et resolutio corporum mixtorum in elementa componentia. Sed cum corpora Angelorum sint aerea, non sic vincuntur ab aliena caliditate: et ita non inducitur putrefactio in eis. Et sic etiam intelligitur quod dicit subsequenter, quod tam ad agendum quam ad patiendum aptiora sunt aer et ignis: ad patiendum autem magis quam ad agendum humor et humus. Hoc enim est verum in mixtis et in via putrefactionis, ut dictum est."

88 Albertus Magnus, Super I Sententiarum (dist. 1–25) dist. 13 a. 9, ed. Auguste Borgnet, Paris 1893 (Ed. Paris. XXV, 385a); Super I Sententiarum (dist. 1–25) dist. 19 a. 10, ed. Auguste Borgnet, Paris 1893 (Ed. Paris. XXV, 526a).

the spontaneous generation not only of animals but also of plants, as shown in a recent contribution by Marilena Panarelli.[89] In his mature works, the explanation of spontaneous generation appears to be strictly linked to Albert's key doctrine of *inchoatio formae*, as evidenced by Albert's commentary on book Z from Aristotle's *Metaphysics*.

There, Albert argues against those who think that spontaneous generation is a valid counterargument against the Aristotelian theory of the origin of the soul as educed from matter, as opposed to an extrinsic origin of the soul from a "Giver of Forms". In these passages, Albert here presents his interpretation of the Aristotelian principle of synonymy, viz. that "man generates man and the sun". According to Albert, the principle of synonymy is granted by the cooperation between matter, the cosmic component of formative power and the potential existence of forms in matter (*inchoatio formae*). The principle of synonymy (*convenientia*) is guaranteed not in terms of being but in terms of virtue, potency, and confused essence (*in virtute et actu et essentia confusa*) by the cosmic component of formative power. In a first moment, the form that exists in the corpse (or in air and water), suitable for the future animal-to-be is such only according to an inchoative state, the so-called *inchoatio formae*. This confused and enmattered existence, according to Albert's doctrine, is actualized by the active and formal agency of the heavenly component of the formative power carried by the rays. This is the case with bees, which are spontaneously generated out of a putrefied corpse.[90] The main and innovative feature introduced by Albert's account of *Metaphysics* is—as noted by both Takahashi and Panarelli—the consideration of spontaneous generation as the product of univocal and not equivocal causes, viz. putrefied matter and embryonic matter.[91]

89 Panarelli, "Converting Death into Life," 493–508.

90 Albertus Magnus, Metaphysica XI tr. 1 c. 8 (Ed. Colon. XVI/2, 471.45–64): "Ex his igitur plane colligitur, qualiter sola opinio illa quae de potentia dicit educi actum, vera est et omnes aliae simile aliquid habent veritatis, sed non sunt verae. Ad ea autem quae inducta sunt contra positionem istam, quod omne generatum generatur ab univoco, facile est respondere per ea quae dicta sunt a nobis in locis pluribus. Non enim dicitur hic univocum pure univocum, sed id quod in forma convenit, licet non conveniat in esse illius formae; sicut forma, quae est in semine, convenit cum forma generati non in esse, sed in virtute et actu et essentia confusa, et sicut forma in anima artificis convenit cum illa quae est extra. Et hoc modo motus ut causa caloris convenit cum calore, quia motus hoc modo est dissolvens et rarificans et trahens potentiam caloris ad actum. Et hoc modo carnes vaccarum sementinam in se habent inchoationem formae apum, et commixtum sperma asini cum menstruo equae sementinam habet potentiam muli. Et in omnibus aliis facillimum est hoc modo determinare."

91 Albertus Magnus, Metaphysica XI tr. 1 c. 7 (Ed. Colon. XVI/2, 467.1–14): "Ea vero quae fiunt automato, hoc est *per se*, sunt sicut *animalia generata per putrefactionem*, et de his iam in

Apparently, these are two obverse sides of the same coin of generation and not two equivocal causes, as Aristotle himself thought. According to Albert, spontaneous generation is caused by univocal agents and is, as in the case of Averroes, a natural process rather than exception to natural, teleological process.[92]

Apart from the mention of rays and the doctrine of *inchoatio formae*, none of which is found Albert's early production, this explanation is in keeping with his approach in the *De quattuor coaequaevis*.[93] This doctrinal point chimes with Averroes, who posits the eduction (or actualization) of animal and plant souls from matter in spontaneous generation because of the cosmic component of formative power. Still, Albert shares the same tension presented in Averroes' account, viz. the identification of a clear criterion for distinguishing what animals can be subject to spontaneous generation and which cannot. According to Albert, the cosmic heat of generative power can indeed determine the production of animals as complex as mice, snakes, bees, and also plants, from putrefaction, for the humor which originates putrefied animals is perfectly comparable to the humor of the human semen. By introducing spontaneous generation in blooded animals, Albert follows Avicenna's zoology,[94] but pushes forward on the univocal agency at play in spontaneous generation, a typical trait of Averroes' philosophy. On the other hand, Albert does not accept the possibility of any extrinsic origin of sublunar souls in the case of spontaneous generation, nor that humans can be spontaneously generated, even hypothetically, as Avicenna had held.

In his discussion of the possibility of spontaneous human generation, Albert rejects Avicenna's thesis on spontaneous human generation. To do so, he follows in the footsteps of Averroes' Long Commentary on the *Metaphysics*. Despite his adherence to the theories of Averroes, Albert shows himself not com-

---

libris physicis in pluribus locis ostendimus, quod umor putrefaciens in eis est sicut umor seminis animalium, et calor, qui est in eis, et spiritus formantur virtute stellarum et caeli et loci, in quo sunt anguli incidentium radiorum, et per hanc virtutem habet virtutem formativam, quae est similis formato, sicut forma artis similis est artificiato. Haec igitur etiam fiunt a sibi univoco, et sic inductione probatum est, quod omnia quae fiunt, fiunt a sibi univoco. Ex ista propositione conclusum est, quod nihil fit nisi compositum, et non forma vel materia." Panarelli, "Converting Death into Life," 498–499; Takahashi, *Interpreting Aristotle's Cosmos*, 183.

92 Panarelli, "Converting Death into Life," 498.

93 Albertus Magnus, Metaphysica XI tr. 1 c. 8 (Ed. Colon. XVI/2, 471.45–64).

94 Albertus Magnus, Metaphysica XI tr. 1 c. 8 (Ed. Colon. XVI/2, 468.51–58): "Generantur etiam multa ex putrefactione, ut mures ex terra et apes ex carnibus vaccarum et vespae magnae citrinae ex carnibus equorum et serpentes parvi de capillis mulierum et animalia multa ex stellis. Et sol dicitur esse pater vegetabilium, et sol cum circulo declivi dicitur esse causa generationis universorum; et nihil horum videtur esse univocum […]."

pletely satisfied (*non conveniens est generaliter*) with his argument against human spontaneous generation. According to him, Averroes' argument that spontaneous generation takes place because of the *eductio formae* achieved by the action of the heat of heavenly spheres and not of an extrinsic *dator formarum* is not enough to establish a criterion for invalidating spontaneous human generation.[95] Unfortunately, Dag Hasse has shown how Albert failed to provide, unlike his pupil Thomas Aquinas, a distinctive criterion to identify those animals that can be spontaneously generated and those that cannot. By contrast, Aquinas drew a clear line: imperfect animals could be subject to spontaneous generation, while perfect animals could not.[96]

The second place where Albert takes on the doctrine of spontaneous generation is in his paraphrase of the *Meteora*, taking his cuse from the case of generation from putrefaction (*ex putrefactione*) analyzed by Takahashi.[97] Putrefaction, viz. the elicitation of heat and moisture from a corpse, is brought about by the action of extrinsic heat found in the surrounding air (*aliena caliditas continentis loci*). The result of the action of this heat over the moisture found in the body is spirit, which Albert had defined in the *De spiritu et respiratione* as a mixed body produced by the mastering action of heat over moisture (*humidum autem semper cum calido est vaporativum et convertibile in spiritum*).[98] In the *De spiritu et respiratione*, Albert had equated the spirit's pulsation with its capacity to produce life (*pulsando sibi ingerat potentiam ad vitam*):[99] this seems to be precisely the case.

The process of spontaneous generation is therefore triggered by the action of the cosmic heat of the stars and of the heavenly bodies. While describing this, Albert explicitly refers to Averroes's position.[100] The crucial doctrinal point put

95 Hasse, "Spontaneous Generation," 163: "Averroes's criticism of Avicenna, says Albertus, is not entirely appropriate (*non convenientem generaliter*), because if the disposition and property of the semen can be fully acquired from the stars, then there is no reason why a human being cannot be produced without parents. [...] For Aristotle, it depends upon the matter whether spontaneous generation is possible or not. For Averroes, the power in matter results from the motions of the sun and the stars. If Averroes's position on celestial influence is accepted, it is reasonable to conclude that it is possible to generate any natural kind from putrescent matter."

96 Hasse, "Spontaneous Generation," 163–166.

97 Takahashi, *Interpreting Aristotle's Cosmos*, 147–164.

98 Albertus Magnus, De spiritu et respiratione I tr. 1 c. 7, ed. Auguste Borgnet, Paris 1894 (Ed. Paris. IX, 224b–225a).

99 Albertus Magnus, De spiritu et respiratione I tr. 1 c. 5, ed. Auguste Borgnet, Paris 1894 (Ed. Paris. IX, 222a).

100 Albertus Magnus, Meteora IV tr. 1 c. 10 (Ed. Colon. VI/1, 223.41–43): "Et ad hoc dicit Averroes quod animalia haec innascuntur ex virtute stellarum et non ex vi caloris naturalis [...]."

forward by the *Meteora* is to emphasize how the biological explanation of spontaneous generation from putrefaction rests on the combination of two factors: the residual heat of the formative power in the dead corpse and cosmic heat.[101] The interaction between the two, usually more productive in summer than winter, produces different animals according to the different kinds of complexions. According to Albert, choleric complexion produces long and tight worms, phlegmatic complexion produces short and broad animals, while sanguineous complexion produces medium-sized animals, and melancholy short and little animals.[102] Among all living beings, plants are the easiest to be generated from putrefaction. This is because of the relatively small amount of heat placed in them and required for a plant to live.[103] For more details on spontaneous generation in plants, we need to turn to Albert's *De vegetabilibus*.

4.6 *Spirit, Formative Power, and Spontaneous Generation in Plants*

In his botany, Albert dealt with the spontaneous generation of plants. As Albert understood it, spontaneous generation from putrefaction of a corpse belonging to a certain genus usually produces living beings of the same genus. Putrefied plants only generate plants, while putrefied animals generate animals. In plants, much of those that are produced *ex putrefactione* are usually of the simpler kinds, often lacking roots and stems. This is the case with greenish leaves, generated from the putrefaction of water in the marshes (*lens paludum*), mushrooms, moss and lichen. Other cases of spontaneous generation concern forms of vegetal life sprouting from the branches of trees as woolen wires (*filariter et ut lana*) or found on the branches of long-lived plants (*illa, quae invenitur in antiquitate plantarum pullulare ex ramis*). Tellingly, Albert compares the instances of plants, especially mushrooms, which are spontaneously generated, to imperfect animals incapable of sexual reproduction. Albert argues that the most perfect plants are able to reproduce by themselves, while the less complex ones seem to be unable to do so. In this passage, Albert seems to suggest that this is the case with imperfect animals, which are often reproduced out of

101 Albertus Magnus, Meteora IV tr. 1 c. 11 (Ed. Colon. VI/1, 224.11–17): "Sicut enim in semine animalis est virtus formativa animalis secundum naturam illius seminis, quae vis est in spiritu spumoso et naturali seminis et operatur per calidum naturale, ita vis in genere animalis imperfecti remanente in subtiliori humido putrefacti, quod primo extra constat in pelle per frigidum partes faciens constare."

102 Albertus Magnus, Meteora IV tr. 1 c. 11 (Ed. Colon. VI/1, 224.42).

103 Albertus Magnus, Meteora IV tr. 1 c. 11 (Ed. Colon. VI/1, 223.72–74): "Cum enim facilioris sit generationis planta quam animal, videtur forte alicui potius ex putrefactis debere generari plantas quam animalia."

spontaneous generation.[104] Panarelli has shown that Albert does not restrict spontaneous generation to the cases of less complex plants, though. At least theoretically, he admits that also the most perfect plants can be spontaneously generated.[105] As a rule of thumb, it would seem that plants generated by putrefaction generate other plants only. There are two exceptions to this general rule, though: the thistle (*cardus/virga pastoris*), whose humoral putrefaction produces worms (*vermiculi ex putrefactione humoris eius nati*),[106] and some kinds of wasps and flies, which are produced from the putrefaction of glutinous trees (*ex putrefactione arborum glutinosarum*).[107] In her article on Albert's theory of spontaneous generation, Marilena Panarelli has also convincingly clarified that it is precisely the borderline nature of these creatures that enables the generation *per dissimilia* (animals from putrefied plants) and not *per similia* (animals from putrefied animals). The difference of animals like those which are unable to move, which live clinging to stones, and that are not divided into male and female, is quite similar to trees and fruits.[108] Albert is able to retain his argu-

104 Albertus Magnus, De vegetabilibus IV tr. 2 c. 3 (ed. Meyer-Jessen, 242–243); Albertus Magnus, De vegetabilibus V tr. 1 c. 2 (ed. Meyer-Jessen, 294): "Sed plantarum prima differentia est, quod quaedam videtur perfecta in organis et viribus vegetabilibus, quaedam autem imperfecta, sicut et animalium quoddam est imperfectum in organis et viribus sensibilibus, et quoddam perfectum: quoniam planta, quae generat ex se sibi in specie similem, videtur esse perfecta, eo quod ultima virium vegetabilis est generativa; et quae hanc generativam non habet, videtur esse imperfecta, sicut fungorum genera et illae, quae super alias plantas ex plantarum aliqua putrefactione generantur—sicut illa, quae filariter et ut lana quaedam egreditur de truncis plantarum, et illa, quae invenitur in antiquitate plantarum pullulare ex ramis, quae habet humorem viscosum, de qua etiam supra fecimus mentionem."

105 Albertus Magnus, De vegetabilibus IV tr. 3 c. 1 (ed. Meyer-Jessen, 255–257); Panarelli, "Converting Death into Life," 502.

106 Albertus Magnus, De vegetabilibus VI tr. 2 c. 19 (ed. Meyer-Jessen, 577).

107 Albertus Magnus, De animalibus XV tr. 1 c. 1 n. 6 (ed. Stadler, II, 992.33–993.2). Wasps are sometimes generated from the putrefaction of earth after rain (*post longam pluviam*): Albertus Magnus, De animalibus VIII tr. 4 c. 6 n. 191 (ed. Stadler, I, 651.5–7): "Aliquotiens etiam apparuerunt plurimi duces sive mutiri cum vespis volantes: et hoc erat post longam pluviam, ex cuius putrefactione nati fuerunt."

108 Albertus Magnus, De animalibus XV tr. 1 c. 1 n. 5 (ed. Stadler, II, 992.20–31): "Animalia vero immobilia secundum locum sicut genus durae testae, et universaliter ea quae vivunt lapidibus applicata, non habent maris et feminae distinctionem: haec enim vitae arborum assimilantur nec animalia nisi per quamdam similitudinem dicuntur et non diversantur ab arboribus nisi parum motu videlicet contractionis et dilatationis. In generatione autem huiusmodi videntur similia arboribus: quaedam enim arbores ex seminibus suis producunt fructum sibi similem secundum esse et speciem : quaedam autem non faciunt fructum similem in specie, sed qui habet quamdam convenientiam cum eis et per gradus quosdam redit ad speciem illarum arborum, quarum est fructus primus : sicut ficus

ment for univocity in the following manner: since trees are the most perfect among plants and worms are the less perfect among animals, they must share some common features and therefore allow for spontaneous generation of the latter (worms) from the latter (plants).[109]

4.7 *Accidental Species? Spontaneous Generation in Animals*

The biological explanation of spontaneous generation in animals is based, as in the case of spontaneous plant generation, on the concept of spirit *qua* carrier of the vital, life-giving formative power. This life-giving power is acquired from the power of the stars (*spiritus est vehiculum vitae infusae substantiae a stellis*), a position that chimes with the general principles of spontaneous generation both from a physiological (*Meteora, De vegetabilibus*) and a metaphysical (*Metaphysics*) point of view.[110]

The cases of spontaneous generation mentioned by Albert in his zoology concern two categories, viz. fish (testacea, Lat. *malachie*) and worms, as in Aristotle. In the *Metaphysics*, Albert had added other categories of blooded animals such as mice, borrowed from Avicenna's explanation of spontaneous generation. Albert follows Aristotle in arguing that some of the species of "frothy" (*spumosa*) fish are never generated sexually but only through putrefaction (*numquam generare, sed per putredinem nasci videtur*), viz. through the heating of water and putrefied corpses. Among these cases, the most famous are eels, which Albert had already mentioned in the *Commentary on the Sentences* as products of spontaneous generation.[111]

Albert analyses spontaneous generation occurring in other animal species, such as mutating, larvae-like animals such as butterflies or silkworms, which drastically change their morphologies. Drawing on Avicenna's study of the spontaneous generation of worms from putrefaction, Albert argues that worms

---

agrestes faciunt fructus qui per culturam iuvantur ut redeant ad generationem ficuum bonarum."

109 Panarelli, "Converting Death into Life," 499.

110 Albertus Magnus, De animalibus XIX tract. un. c. 7 n. 34 (ed. Stadler, II, 1264.4–10). On the role of spirit as *vehiculum vitae infusae substantiae a stellis*: Albertus Magnus, De animalibus VI, tr. 3 c. 3 n. 123 (ed. Stadler, I, 494.36–495.2): "Est autem differentia istorum ad ea, quae simpliciter ex putrefactione generantur: quia ista non generantur ex putrefactione substantiae putrescentis, sed potius ex ovis non putridis ex substantiali humido substantiae corruptae formatis. Illa autem, quae simpliciter ex putrefactione nascuntur, generantur ex humido ad superficiem exalante, cuius viscosum exterius frigiditate aeris duratur in pellem, intra quam calor evaporans ex putrescente continetur, qui spirare facit humidum intra pellem conclusum, quod spirans spiritualiter pulsare incipit: qui spiritus est vehiculum vitae infusae substantiae a stellis, quarum virtus est in ipso."

111 Albertus Magnus, De animalibus VI tr. 2 c. 2 n. 87 (ed. Stadler, I, 477.33–478.14).

of a certain species can be often produced (*ut frequenter*) from the putrefaction of a worm of the same species or by the mating of the most complex of these worms. Isabelle Draelants has recently addressed Albert's entomology and its study of insect generation. In her paper, she has convincingly shown how Albert's theory of generation of larvae seems to follow Avicenna's, even though with some critical remarks. Like Avicenna, Albert maintains that the generation of butterflies, silkworms and other similar insects occurs in successive stages. In the first place, putrefaction of a dead worm generates an egg, which in turn produces an incomplete larva. As in the case of Avicenna, the incompleteness of the animal-to-be is caused by a lack in the strength of the formative power, unable to produce a fully-formed living being. The future animal-to-be will then develop from the larva-like state into its complete state thanks to favorable environmental conditions and due to the fact that "nature does nothing in vain", as Aristotle teaches.[112]

The real problem with the doctrine of spontaneous generation stems from Aristotle's teaching itself and concerns the consistency between the spontaneous generation of entire biological species and the classification of this phenomenon as spontaneous. A spontaneous or chance occurrence can hardly be taken as a synonym of 'for the most part' or 'always'. This problem is a common one, shared also by Avicenna, which can be mitigated by appealing to his Neoplatonic structure of the cosmos.

In his *De natura et origine animae* and his commentary on the *Physics*, Albert gives very clear indications about the ontological status of all natural forms. In the *De natura et origine animae*, the Dominican master holds that all

112 Albertus Magnus, De animalibus, XV tr. 1 c. 1, n. 4 (ed. Stadler, II, 991.31–992.10): "Animalia autem quae sunt sine differentia perfecta sexus et sine coitu, generantia ex suis corporibus putrefactis, generant quidem etiam sicut alia, sed primo generant rem alterius generis: et in illis non est mas et femina secundum perfectam maris et feminae distinctionem: et hoc modo generant quaedam anulosorum. Quaedam etiam habent distinctum sexum et coeunt. Rationabiliter autem hoc accidit. Si enim haec secundum sua genera sibi similia generarent, necessario foret generatio eorum continua per successionem et propagationem unius ex alio: hoc autem non convenit nisi perfectis animalibus sibi similia generantibus: dicta autem animalia non sunt talia eo quod ex putrefactione sui aut ex aliis superfluitatibus putrefactis generentur. In talibus igitur non generatur simile generanti, sed potius gusanes qui dissimiles sunt generanti in figura et specie, sicut videtur: et in illis stat generatio quia illa dissimilia non alia ex se sibi et parentibus dissimilia generantur. Si enim semper per dissimilia procederet generatio, procederet in infinitum dissimilitudo: et hoc natura respuit eo quod infinitum imperfectum est: natura autem perfici et perfectum semper desiderat. Generato tamen gusane in dissimili figura, gusanes est imperfectum animal sicut et ovum et pro tempore perficitur et tunc redit ad parentis sui similitudinem." Draelants, "La reproduction imparfaite," 155–163.

natural forms are produced by the causal action of the Intelligence, viz. God (*omnis forma naturae est forma ab intelligentia, quae primus naturae motor est, inducta*).[113] Likewise, in the *Physics* he argues that, when God created the world, he placed inside prime matter the forms of generable things. Among generable things we can include not only the products of the usual generation of living beings but also the generation of minerals, and spontaneous generation too.[114] In this manner, while the categorization of the generation of mushrooms, *malachye* and other animals as 'spontaneous' remains problematic when referring to the Aristotelian understanding of 'spontaneous', if we appeal to Albert's theory of the origin of natural forms we can find a solution to the question about the nature of these creatures. As in Augustine's theory of *rationes seminales*, the future existence of natural forms is already placed in prime matter at the moment of creation, granting that all natural processes are end-directed and ultimately rooted in the agency of the First Cause, viz. God. In Albert's eyes, this means that all events, even chance ones such as spontaneous generation, are the less common product of the work of the ultimate Intelligence, viz. God.

In this we can see how Albert blends together insights from Avicenna's and Averroes' doctrines of spontaneous generation to develop his own theory. Far from solving the internal contradiction between the categorization of the generation of entire species as 'spontaneous', his theory still provides a clear teleological framework where the action of spirits represents a less frequent effect in nature of the providential action of the Divine Intelligence (*opus naturae est opus intelligentiae*).

Conclusions

Albert's doctrine of generation in his Aristotelian paraphrases is developed drawing strong analogies between human, animal and plant generation in terms of the material substances involved and their causal roles. This represents a relevant doctrinal development from Aristotle, Avicenna and Averroes, none of whom employed the concept of spirit to account for generation in

113 Albertus Magnus, De natura et origine animae tr. 1 c. 1 (Ed. Colon. XII, 4.34–46).

114 Albertus Magnus, Physica VIII tr. 1 c. 13 (Ed. Colon. IV/2, 575.40–44): "His autem suppositis dico, quod deus ex nihilo fecit mundum, quia fecit mobile et motum primum et materiam primam et incorporavit ei omnium generabilium formas et fecit, ut illae per generationem continuam ex ipsa producerentur." I would like to thank Henryk Anzulewicz (Albertus-Magnus Institut) for drawing my attention to this passage.

plants. Albert systematically connects the concept of spirit to his interpretation of formative power, heavily based on his Arabic sources (Avicenna, Averroes). On the other hand, Albert has to admit the existence of certain physiological differences in the material composition of the fluids involved (semen, egg-whites, oat-like substance etc.) as well as differences in the mechanisms of generation between different kinds of living beings. While animals produce their offspring through mixing the male and female contributions, plants do not. Rather, they resemble the case of an already fertilized egg.

These obvious material differences in the material composition (viz. complexion) of the spirit meant for generation (*spiritus radicalis/seminalis*) require a key explanatory principle, viz. the soul. The soul *qua* final cause of the organization of a new plant-, animal- or human-to-be represents that for the sake of which a certain material composition of spirit generates a plant and not, for instance, a horse. This composition can be usually produced by a combination of seminal heat, cosmic heat and heat from the father's soul. These fluids have to come from parents of the same species according to the principle of synonymy "man begets man", or at least from two parents of compatible species which may result in sterile hybrids (e.g. mules). In chance occurrences of generation such as spontaneous generation, the cosmic component of formative power can replace the role of the father's semen *qua* formal and efficient cause of the future animal- or plant-to-be. In cases of both sexual and spontaneous generation, the role of spirit as carrier of formative power is framed within and reflects a teleological and Neoplatonic structure of the cosmos with God at the top. Until now we have analyzed the causal connection between spirit and the most basic conditions for the subsistence of life and generation: now it is time to turn to the causal correlation between spirit and length or shortness of life.

CHAPTER 5

Spirit, Radical Moisture, and Longevity in Living Beings

The study of the duration of lifespan introduces into our study two of the most plastic examples of non-teleological natural phenomena, viz. aging and death, the inverses of growth and birth. For a scholar of natural philosophy like Aristotle, aging and death seemed to be fatally linked to the material nature of hylomorphic compounds, unavoidably doomed to perish. For scholars of natural of philosophy who adhered to monotheistic religions, such as Avicenna and Albert, the matter needed to be revisited. While Aristotle's natural philosophy left no room for speculation on the fate and post-mortem status of the human soul, Neoplatonic, Muslim and Christian reflections on the teleology of an apparently non-teleological natural process such as death came to embrace a further perspective by seeking a rational foundation for affirming the immortality of the human soul.

This is the case, for instance, with the opening of Albert's *De morte et vita*, a paraphrase of Aristotle's *De longitudine et brevitate vitae*. In the digression which prefaces his paraphrase of Aristotle's work, Albert makes clear that human life subsists after the body's death. The justification of this post-mortem survival is achieved by distinguishing three different levels of life (in the soul, *in fluxu ab anima*, in the body) tacitly implying that Albert's study of longevity in living beings does not affect the existence of a further and teleological dimension of death (for the human soul alone, though). Albert's biological explanation of longevity, strongly based on a blending of the doctrines of James of Venice's *translatio vetus* of the Aristotelian text and the Avicennian theory of radical moisture, has its roots already in his early works such as the *De homine*. In this work, Albert had deployed the Avicennian concept of radical moisture, usually introduced in contemporary theological treatises to explain the prelapsarian state and the *veritas humanae naturae*, viz. the material constitution of the body, which will resurrect, to account for human longevity in a purely natural and biological context.[1]

1 On the *veritas humanae naturae* and the prelapsarian state: Reynolds, *Food and the Body*, 50–66 (on the *veritas humanae naturae*), 105–119 (on radical moisture), 149–214 (on the theological debates at Paris about the *veritas humanae naturae*).

 | DOI:10.1163/9789004706743_007

As it has been ascertained by the studies of Chiara Crisciani, Albert was the first to integrate the medical, Avicennian concept of radical moisture into the template of Aristotelian philosophy of nature. This operation of "epistemic translation", as defined by Crisciani herself, is typical of his mature Aristotelian paraphrases.[2] Building on Crisciani's (also, Reynolds') studies, which themselves stress the necessity of a study of the connection between spirit and radical moisture,[3] this chapter will show how Albert drew on the chapter on the complexion of ages from Avicenna's *Canon* to shape his work on youth and old age (*De aetate sive de iuventute et senectute*), meant to supply Aristotle's unavailable work. To be understood fully, Albert's theory of length and shortness of life will be contextualized against the background of the Aristotelian and Avicennian theories of longevity, construed in terms the key interaction between heat (carried by spirit) and moisture (in Avicenna and in the medical tradition, radical moisture). Our main goal in this chapter will be to show the connection between spirit, radical moisture, and the chemistry of growth and decay, length and shortness of life, youth and old age. To do so, we shall rely not only on the key, above-mentioned works by Crisciani (and Reynolds), but also on the more recent work by Marilena Panarelli on Albert's use of the concept of radical moisture in plants.[4]

The purpose of this chapter is to show the degree to which the Avicennian medico-philosophical framework of radical moisture influenced Albert's take on length and shortness of life, youth and old age, and growth and decay. This influence was, on the one hand, doctrinally compatible with the principles of Aristotelian philosophy of nature. On the other hand, Avicenna's psychophysiology of length and shortness of life served him the purpose of supplying textual lacunas (as in the case of the *De iuventute et senectute*) in Aristotle's doctrine of youth and old age. Moreover, Avicenna's psychology provided Albert with a teleological (which is also implicitly theological) view on death as a moment of the Neoplatonic egress-regress, although Albert's position on death in relation to this classic Neoplatonic scheme seems rather to be indebted, as we will see and as established by scholarship, more on the *corpus Dyonisianum* and other sources. Given the key role assigned to the interaction between heat

2 Crisciani, "Aspetti del dibattito sull'umido radicale," 350. Peter of Spain, who wrote a *Tractatus de longitudine et brevitate vitae*, did not technically write either a paraphrase (like Albert) or a commentary (like Peter of Ireland, Adam of Buckfield or Peter of Auvergne) on Aristotle's *De longitudine et brevitate*. On the explanation of Peter's doctrine of radical moisture, we rely on: Crisciani, "Aspetti del dibattito sull'umido radicale," 345–349.

3 Crisciani, "Aspetti del dibattito sull'umido radicale," 350–354; Crisciani–Ferrari, "Introduzione," 385–396.

4 Panarelli, "How Do Plants Live and Grow?," 347–367.

and moisture in the definition of life's longevity, this chapter will explore possible connections between spirit *qua* vehicle of vital heat and radical moisture *qua* key material cause for the duration of life.[5]

1 Vital Heat, Moisture, and Longevity in Aristotle

In his *De longitudine et brevitudine vitae*, Aristotle set out to define longevity in living beings in terms of primary qualities. To be fully and properly understood, though, its account of length of life needs to be connected with his account of youth and old age. Our account will be loosely based on the study of Richard King,[6] while maintain the same reading of the concept of pneuma adopted by Freudenthal. According to Aristotle, the chemistry of the interaction between primary qualities explains the differences in length and shortness of life among species (humans are longer-lived than horses), between sexes (men are longer-lived than women) and among individuals of the same species. For instance, longevity may be due to exogenous factors such as living in different climates (hot and moist places are the best suited for a long life, for instance) or endogenous factors such as sex. Moreover, Aristotle also lists diseases as a possible cause of length and shortness of life: this is because some diseases could be fatal, while others (such as chronic illnesses) do not necessarily impact duration of life.[7] Aristotle observes, though, that the above-mentioned difference apply not 'always' but only 'for the most part' to the natural occurrences of life and death. For instance, there are inhabitants of the same place who live lives that are longer or shorter than expected, men who are shorter-lived than women, etc. Nonetheless, Aristotle sought to establish a criterion which had to come to terms with the suppositional necessity (*Physics*, II, 9) typical of natural philosophy, viz. given a certain amount and ratio of primary qualities in a living being, it follows that its life is necessarily shorter or longer. In Aristotle's account, there are two key elements within this calculus: vital heat and moisture.

Aristotle clearly states that vital heat is fundamental for defining the lifespan of a living being since "an animal is by nature humid and warm and to live

5 The second (and corollary) part of an inquiry into the relationship between spirit and radical moisture should be devoted to the relation between spirit, radical moisture and growth. This chapter will only be concerned with the relation between spirit and radical moisture in the context of longevity. For this reason, the relationship between spirit, growth and radical moisture remains open for further research.

6 Richard A.H. King, *Aristotle on Life and Death*, London 2001.

7 Aristotle, De longitudine et brevitate vitae 1 (465a4–12).

is to be of such constitution, while old age is dry and cold, and so is a corpse."[8] Vital heat, though, cannot subsist in a body without a proper substrate of inherence: as he claims in the *Metaphysics*, "heat itself is generated from the moist and kept alive by it."[9] For this reason, Aristotle's study of length and brevity of life considers heat and moisture as equally relevant principles: vital heat as the primary quality strictly and directly correlated to the existence and subsistence of life, moisture as its material substrate of inherence and subsistence. The interaction between vital heat and moisture becomes gradually clearer once Aristotle proceeds with his explanation of the longest-lived among living beings throughout his *De longitudine et brevitate*.

The first criterion for determining the lifespan of living beings is a link of direct proportionality between the amount of vital heat and the longevity of the living beings. This fact would place the living beings possessing the highest quantity of vital heat, which is directly proportional not only to the need for breathing (as observed in chapter 2) but also to the highest cognitive activities, upright position and size,[10] at the top of the *scala naturae*. This criterion would place animals like elephants right at the top. Unfortunately, this seems not to be the case. Humans, for instance, are not longer-lived because they possess a quantity of vital heat which exceeds that needed to keep an elephant alive. Rather, their longevity must be explained according to a better ratio or balance between amount of vital heat and body size, a ratio superior to that found in elephants.[11] Yet again, the assumption that vital heat alone and its ratio to body size in a living being could serve as the sole causal explanation for life expectancy is doomed to fail the litmus text constituted by plants.

Albeit possessing the smallest amount of vital heat and representing an upside-down version of humans, plants are the longest-lived among living beings. According to Aristotle, this is because they are of a most viscous moisture, and viscosity of moisture, in Aristotle's view, is directly proportional to the capacity of a living being to preserve its vital heat. This is the reason why

8 Aristotle, De longitudine et brevitate 5 (466a18–19). This translation by George R.T. Ross in: Aristotle, *The Complete Works of Aristotle*, ed. Barnes, vol. 2.1, 5.

9 Aristotle, Metaphysics I 3 (983b23): "the nutriment of all things is moist, and that heat itself is generated from the moist and kept alive by it (and that from which they come to be is a principle of all things)." This translation by William D. Ross in: Aristotle, *The Complete Works of Aristotle: the Revised Oxford Translation*, ed. Jonathan Barnes, Princeton 1984, vol. 2.2, 7.

10 For the correlation among qualitatively positive vital heat, concoction of blood and cognitive activities, see: Freudenthal, *Aristotle's Theory of Material Substance*, 62–65.

11 Aristotle, De longitudine et brevitate 5 (466a125–466b2).

plants are the longest-lived among living beings.[12] Moisture is therefore crucial not only to account for the lifespan but also for the subsistence of living beings: when living beings are not fed, they use up the body's moisture as a bigger flame consumes a smaller one by burning up its material substrate and eventually putting its fire out.[13] The antidote to the process of consumption of vital heat is found therefore in the viscosity of bodily moisture and the intake of food.

According to the criteria which combine, on the one hand, the ratio between amount of vital heat and body size and, on the other, amount and viscosity of moisture, Aristotle is able to chart an organism's life expectancy in nature. Plants top Aristotle's ranking of longest-lived beings, followed by humans, blooded land animals, oviparous animals and at the bottom, watery bloodless animals.[14] The latter are such because their moisture, extremely watery, is very prone to freezing and, thus, death.[15] Having explained the differences between plants, animals and humans, and having also singled out interspecific differences between animals, Aristotle can account for intraspecific differences in life expectancy within humans especially. In particular, inhabitants of warmer regions appear to be longer-lived than those of colder ones, while men are longer-lived than women. At the end of the *De longitudine vitae*, Aristotle concludes that the longest-lived among living beings are dwarf-like, such as trees among plants, humans and, among animals, blooded land animals.[16]

Despite these general and essential criteria for the determination of relative length of life, Aristotle individuates a number of accidental causes, which explains why, according to some accidents, living beings which should be longer-lived than others are, as a matter of fact, not so. This is the case with excessive sexual activity or toil among humans. By means of an excessive consumption of heat during these activities, life expectancy turns out to be shorter in men compared to women. This fact is not human-specific, though, but can be observed in other animals such as cock-sparrows, whose females are longer-lived than their male mates, and in mules, who are notoriously sterile animals and therefore incapable of altering their longevity because of frequent sexual activity.[17] Similarly, the exercise of sense perception—like motion—is directly

12 Aristotle, De longitudine et brevitate 6 (467a6–9).
13 Aristotle, De longitudine et brevitate, 5 (466b29–32).
14 Aristotle, De longitudine et brevitate 4 (466a9–16).
15 Aristotle, De longitudine et brevitate 5 (466b31–67).
16 Aristotle, De longitudine et brevitate 6 (467a30–467b3).
17 On length and shortness of life in men and women: Aristotle, De longitudine et brevitate 5 (466b8–17).

correlated to the transmission of heat to the outer limbs through *pneuma* embedded in the blood. Given that bloodless animals also possess connate *pneuma* as a source of strength necessary for sense perception and motion,[18] it can be concluded, as a charitable explanation of Aristotelian psychophysiology, that *pneuma* might be regarded as an accidental cause of length and brevity of life in every animal, but not in plants. Given its role as carrier of vital heat at play in generation, sensation and motion, it may be reasonable to assume that, *qua* instrument of the soul and its activities, *pneuma* may indirectly produce a decrease in life expectancy. At least, this would be the rationale behind Albert's reading of the Aristotelian text, not Aristotle's own doctrine. While Albert knew an Aristotelian work about youth and old age could exist, he did not have access to it, because the Latin version of the Parva Naturalia he had access to lacked it.[19] To better understand how Albert, in his effort to supply the unavailable Aristotelian *De iuventute et senectute*, rather relied on Avicenna (who did not know the Aristotelian *De iuventute* either), let us now recall the main features of the Aristotelian treatise here.

In the *De iuventute et senectute*, Aristotle divided possible causes of death into two kinds: exhaustion (μαράνσις) and extinction (σβήσις) of vital heat.[20] Exhaustion may occur because of an excess of internal heat: this is what happens when people drown, choke because they are strangled, or when fish suffocate. The cause of this kind of death is lack of air or water: without air in respiration or water as a principle of cooling down the heart's innate heat (for Aristotle, fish do not breathe), excessive heat ends up destroying the heart's vital heat.[21] Aristotle argues that this happens as though a bigger flame were thrown on a smaller one, the bigger flame eventually burning up itself and quenching the smaller flame by consuming the material substrate of moisture attained from food digestion.[22] This case of death by suffocation is clearly linked to breathing animals: plants and insects seem to be out of the picture here.

---

18 Aristotle, De somno 2 (456a5–25).

19 James of Venice's version of the *De longitudine et brevitate* stops at chapter 6 (467a26). Therefore, the version of the sixth chapter available to Albert does not mention: (a) the end of the discussion on the importance of cuttings in extending the lifespan of a plant, (b) the argument that males are longer-lived than females since they are hotter and more dwarf-like (467a30–b3) and (c) the mention of an existing work *On plants* (467b4). Aristotle, De longitudine et brevitate vitae [Translatio vetus], ed. Manuel Alonso, Pedro Hispano: Obras filosóficas, III (Instituto de Filosofía "Luis Vives", Serie A Núm. 4), Madrid 1952, 403–411.

20 Aristotle, De iuventute et senectute 7 (1) (469b21–30).

21 Aristotle, De iuventute et senectute, 23 (18) (479b10–13).

22 Aristotle, De iuventute et senectute, 8 (2) (470b30–470a4).

Unlike exhaustion, extinction is caused by the action of a massive quantity of cold upon the living being or when the amont of cold becomes dispersed. Extinction of vital heat occurs either whenever a living being is cut to pieces or freezes to death: in both cases, the excessive cold affecting the living being causes the failure of the nutritive faculty's action and, eventually, death.[23] Acting as a blocker of the transmission of vital heat, one might think that cold prevents the motion of *pneuma* in the blood and blocks its transmission of vital heat into the living being. When vital heat is totally overwhelmed by cold, then, the living being dies. These two kinds of death by exhaustion, though, are violent deaths: for Aristotle, natural death occurs because of the extinction of vital heat over time.

In breathing animals, age is causally linked to a hardening of the lungs. The more the lungs are rigid, the more they fail to cool down the heart's vital heat, so they end up drying out the moist substrate in which vital heat lives.[24] From this point of view, the *De iuventute et senectute* seems to be very much concerned with the determination of the process of aging through the analysis of the cooling function of the lungs and their progressive deterioration over time, thus insisting on animals as a specific category of analysis and letting the others categories, alongside plants, gradually fade into the background. The emphasis of the *De iuventute* on the hardening of the members chimes well with Aristotle's remarks in his zoology, where old age is characterized by pathologies connected to stiffening up of members such as the cornea (causing glaucoma),[25] or excessive dryness in the body's make-up, which produces loss of memory, since excess dryness is causally correlated to cases of forgetfulness.[26] A further and final aspect of old age is the growing of white hair, produced by the loss of heat and moisture in a manner comparable to putrefaction.[27]

In Aristotle's *De iuventute et senectute*, old age is considered from a perspective mainly centered on breathing animals, with particular attention to the hardening of the lungs. The weakening of the breathing function has a direct bearing on the workings of *pneuma*, which is formed in the heart and which becomes progressively less able to cool down the heart's vital heat. The outcome is a vicious circle ultimately leading to natural death. Even though most

23 Aristotle, De iuventute et senectute, 14 (8) (474b12–24).

24 Aristotle, De iuventute et senectute, 24 (18) (479a32–b5).

25 Aristotle, De generatione animalium, v 1 (780a17–b29).

26 The correlation between moisture and retention of images/sensible species emerges in Aristotle, De memoria 1 (450a25–b11); De memoria 2 (453a31–b7) is common also in medical literature.

27 Aristotle, De generatione animalium v 3 (782a11–16); De generatione animalium v 1 (780b 6–7).

living beings *qua* compounds of elemental mixture seem unable to escape the linear progression from generation to corruption, plants and bloodless animals (insects, for the most part) constitute a valuable exception, for they do not breathe.

For this reason, the only analogous phenomenon both plants and non-breathing animals may be subject to is the hardening and stiffening of body parts over time, which blocks, in the end, the transmission of heat through the body. This is an aspect of Aristotle's theory of aging and decay, however, which is not dealt with *ex professo*. So all that can be done here is to give a charitable explanation of it. Unlike bloodless, non-breathing animals, plants possess the capacity to reverse the aging process, growing branches anew, especially through the artificial practice of grafting.[28] In the end, in his handling of youth and old age Aristotle seems to be primarily concerned with the physiological mechanism of aging and decay over time because of failure in respiration. He does not explain, though, how the process of gradual decay and death connects to youth and growth, a connection mostly developed in Galen's famous system of the four complexions for the four ages.

Deprived of Aristotle's *De iuventute et senectute*, Albert would have to resort to the typical medical doctrine of radical moisture (*humidum radicale*) to explain the connection between youth-and growth, strictly connected to Aristotle's discussion on replaceable and irreplaceable members in relation to his doctrine of homeomerous (e.g. flesh, fat tissues) parts. While this kind discussion exceeds the limits of the present book, it still constitutes the second and integral part of the theory of radical moisture worth being investigated in future research. This concept was supposed to account both for (1) the differentiation between replaceable and non-replaceable members, homeomerous or anhomeomerous substances, and (2) longevity.

2 The Lamp Metaphor, Spirit, and Radical Moisture in Avicenna

Unlike Aristotle, who had dealt *ex professo* with length and shortness of life in his *Parva Naturalia*, Avicenna addressed this topic in the *Canon* and the *De animalibus* and not in his psychology. In the latter, though, the Persian philoso-

28 Aristotle, De longitudine et brevitate 6 (476a10–18). On this point, see also: Luciana Repici, "Tutto invecchia per opera del tempo. Senilità e senescenza in Aristotele," in: Vita longa. *Vecchiaia e durata della vita nella tradizione medica e aristotelica antica e medievale*, ed. Chiara Crisciani/Luciana Repici/Pietro B. Rossi (Micrologus Library 33), Firenze 2009, 28–29.

pher and physician did address some *Parva Naturalia*-related topics such as sight (book III) and the other external senses (II, 3–8). The topic of length and shortness of life, even though contained in the Arabic adaptation of Aristotle's *Parva Naturalia* (*Kitāb al-Ḥiss wa-l maḥsūs*), with which Avicenna was acquainted,[29] did not feature in Avicenna's study of the soul and its powers. Rather, his approach to the topic seems much more indebted to authors of the medical tradition like Yuhannā ibn Sarabiʿūn (Serapion for the Latins), ʿAlī ibn ʿAbbās al-Maǧūsī, author of the famous "Complete book of medical art" (*Kāmil al-ṣināʿat al-ṭibbiyya*), and Isaac Israeli's *Liber febrium*.[30]

As shown by the pioneering study of Thomas Hall, Ibn Sarabiʿūn, al-Maǧūsī and Isaac Israeli had drawn on and reworked the Galenic theory of secondary moistures to explain length and shortness of life—especially in humans—and to determine the replaceable and irreplaceable members through the concept of 'radical moisture'. To detail their answer to such questions, the starting point of these physicians was Galen's theory of secondary moistures, a doctrine that explained the assimilation of food to homeomerous parts and tissues. In his account of digestion, Galen distinguished three different stages of food assimilation, linking these stages to the production of what he called 'secondary humors', namely the byproducts of the digestive process upon the humors. These stages corresponded to three different locations of food in the body: in the veins (first secondary humor), in the spaces among the tissues (second secondary humor) and within the tissue themselves (third).[31] Although Galen did speak of a "proper" or "resident" humidity (οἰκεῖα ἰκμάς) peculiar to each mem-

29 The Arabic version of the *Parva Naturalia* was composed of three sections (*maqālāt*): (1) a first section "On sense and sensation" (*Kitāb al-ḥiss*, hence the name of the work), a second treatise in two parts: (2. 1) "On memory and reminiscence" (*fī ḏikr wa-taḏakkur*) and (2.2) a section "On sleep and wakefulness" (*Bāb al-nawm wa-l-yaqaẓa*), roughly corresponding to a reworking of Aristotle's *De somno*, *De insomniis* and *De divinatione per somnum*, and (3) a last section "On length and shortness of life" (*fī ṭūl al-ʿumr wa-qiṣarihi*). The prologue is not transmitted in the Latin translation of the *Epitome*, which begins from page 4, line 3 of the Arabic text. On Averroes' paraphrase of the *De longitudine*, see: Rotraud E. Hansberger, "Length and Shortness of Life between Philosophy and Medicine: the Arabic Aristotle and his Medical Readers," in *Philosophy and Medicine in the Formative Period of Islam*, ed. Peter Adamson/Peter Pormann (The Warburg Institute Colloquia), London 2018, 56–62. For a more general overview on the Arabic adaptation of the *Parva Naturalia*, see: Rotraud E. Hansberger, "*Kitāb al-Hiss wa-l-mahsūs*. Aristotle's *Parva naturalia* in Arabic Guise," in *Les* Parva naturalia *d'Aristote. Fortune antique et médiévale*, ed. Christophe Grellard/Pierre-Marie Morel, Paris 2010, 143–162.

30 For the influence of the previous medical tradition on Avicenna, see: Thomas Hall, "Life, Death and Radical Moisture," *Clio Medica* 6 (1971) 8–9.

31 Hall, "Life, Death and Radical Moisture," 6–7.

ber of the body, and mentioned an innate or connatural moisture (ἔμφυτον ὑγρόν/σύμφυτος ὑγρότης), he never considered the latter a fourth and distinct (radical) kind of moisture. Rather, he identified radical moisture with the second, dew-like (δροσοειδώς) secondary humor.[32]

By contrast, the medical tradition of the Islamic World produced a new taxonomy of moistures, maintaining the previous threefold Galenic classification of secondary humors, with the remarkable addition of a fourth kind of 'radical' secondary humor. In the medical tradition of the Islamic World, the second, third and fourth moisture were directly associated to the so-called hectic fevers, which Galen had fragmentarily discussed in the broader context of the notion of 'withering', i.e. the drying out of moisture, considered the material substrate of inherence for vital heat. According to Galen, the process of withering could be caused either by the consuming action of vital heat on moisture, or by harmful, extrinsic heat, such as the heat of fevers, each named differently according to the fluids subject to its attack. The crucial study of McVaugh, key for any approach to the topic of radical moisture (and our case is no exception), has laid out the general context for the discussion of the causal role of radical moisture in medical theory, viz. the case of fevers. Physicians divided fevers into three kinds: fevers that affected humors were called putrid, fevers affecting spirits were dubbed ephemeral, and fevers that attacked the second, third and fourth secondary moisture were called hectic.[33] In the history of medicine, Serapion was the first physician to link this new taxonomy of secondary humors to fevers directly. In his works, he argued that, while the first and second kind of hectic fever were somehow curable, the third kind, which attacked radical moisture, was incurable and led unavoidably to death. The lethality of this last kind of fever was caused, according to him, by its direct attack to the crucial (radical) members of the body such as the heart in the first place.

Isaac Israeli posited a similar connection between hectic fevers and secondary moisture in his *Liber febrium*. According to him, hectic fevers could originate from unnatural heat intrinsic to the body having the most diverse causes, such as aging, emotional disturbance or occurring as consequence of other fevers. In this manner, consumption of radical moisture seems to be due to natural and innatural factors. Isaac regarded this consumption as natural when caused by aging and as unnatural when caused by fevers. Both aging and fevers were associated with the consumption (*phtysis*) of radical moisture. For their part, hectic fevers constituted only a subcase of *phtysis*. More specifically,

32 Hall, "Life, Death and Radical Moisture," 7–8.

33 Michael R. McVaugh, "The '*Humidum Radicale*' in 13th-century Medicine," *Traditio* 30 (1974), 260.

hectic fevers encompassed the case of consumption of radical moisture that was due to unnatural (extrinsic) causes.[34] In the medical field, the relationship between *phthysis* and hectic fever will be then systematically taken up and developed by Avicenna.[35]

On the one hand, Avicenna was heir to a taxonomy of humors and discussions on its link between radical moisture that stemmed directly from the medical tradition of the Islamic World. On the other hand, though, Avicenna introduced elements of innovation in his account of radical moisture. Besides the usual account of the relationship between secondary humors and hectic fever developed in book IV of the *Canon*, Avicenna discussed the relationship between radical moisture and hygiene, and the topic of longevity more generally. But let us not jump the gun. In his account of hectic fevers and second humors, closely adhering to the standards of the medical tradition (Serapion, Isaac, al-Mağūsī's *Practica* from the *Pantegni*), Avicenna already displayed some elements of originality by employing the lamp metaphor to link the three stages of hectic fevers to the three moistures in the lamp. In his article on the lamp metaphor in different authors, Paul Niebyl has highlighted the taxonomy of the different moisture provided by the *Canon*: the oil imbibed into the lamp had to correspond with the second, dew-like moisture, the oily moisture imbibing the wick with the third secondary moisture and the moisture intrinsic to the wick with radical moisture itself. In Avicenna's reading, if this last moisture is subject to an attack, its drying out would cause the disruption of the whole lamp, which would substantially crumble down as a body crumbles down if its key members like the heart are subject to irreversible damage.[36]

---

34 McVaugh, "The '*Humidum Radicale*'," 260–261.

35 Note that the other main authority on the topic and a well-known medical source of Albert, al-Mağūsī, developed a taxonomy of secondary humors and hectic fevers in his *Pantegni* that was self-contradictory. In his *Practica Pantegni* he maintained a classification of secondary humors and hectic fevers consistent with Isaac, while in the *Theorica* he used the term *febris ethica* to encompass both natural (aging) and unnatural (fevers) causes for the consumption of radical moisture: McVaugh, "The '*Humidum Radicale*'," 262: "It is perhaps no wonder that the *Pantegni* never became a standard authority on this subject, when its terms were so easily open to misinterpretation. In one and the same work, *phthysis*, *ethica febris* and even *ros*, refer now to an entity and now to another. While Constantine's translations did agree broadly on the different stages in hectic fever and communicated this understanding to the West, they did not succeed in passing on a consistent terminology."

36 Avicenna, Canon IV fen 1 doctr. 3 c. 1, ed. Venezia 1507, fol. 413va [ed. New Delhi 1981–1996, vol. 4, 84.26–85.4]: "Et similitudo humiditatum primarium est oleum lucernae infusum in lucerna; et similitude secundarum est oleum imbibitum in corpore lychinii; et similitude tertie humiditatis est humiditas cum qua continuantur partes cotti de quo factus est

While Avicenna's use of the lamp metaphor follows closely the standard approach of the medical tradition to hectic fevers and radical moisture, his use of the same metaphor in connection to the role of radical moisture in hygiene (i.e. food, diets etc.) is particularly innovative.[37] In book I of the *Canon* the lamp metaphor is used to distinguish the extrinsic and intrinsic moisture of the body. While the former corresponds to nutrimental moisture, the latter equates to radical moisture. In Avicenna's view, nutrimental moisture—produced by the digestion of food—is able to partially restore the loss of intrinsic moisture. Unfortunately, the restoring action of nutrimental moisture ends up in watering down and eventually putting out the heat intrinsic to the body, proving as damaging as the extrinsic moisture in the lamp. In the lamp metaphor, nutrimental moisture can make up for the loss of intrinsic moisture in terms of matter, but not of power. Radical moisture is, like the intrinsic moisture of a lamp, crucial since it grants longevity to the living being because of its cohesive, gluing power over the radical members. When this cohesive power of radical moisture becomes gradually watered down by radical moisture, radical members of the body (heart, liver etc.) start to crumble down just like the material substance of a lamp.[38] The originality of Avicenna's intuition lies, as shown by Niebyl, in having provided a clear link between nutrition as necessary and sufficient condition for life and the topic of longevity. Avicenna will also exploit this connection between nutrition and longevity in his account of youth and old age.

2.1 *Avicenna on Youth and Old Age*

Avicenna's account of youth and old age encompasses two key yet interdependent medico-philosophical questions, viz. a suitable causal explanation of longevity and the individuation of a defining criterion to distinguish replace-

lychinius. Et cum inflammantur membra radicalia (*aʿḍāʾ aṣliyya*) et proprie cor est illud ipsa aegritudo quae est ethica secundum quod scivisti." Paul H. Niebyl, "Old Age, Fever, and the Lamp Metaphor," *Journal of the History of Medicine* 26 (1971/4), 358–359.

37 Avicenna's main source on the topic of hygiene was Galen's *De sanitate tuenda*: Hansberger, "Length and Shortness of Life," 49–53.

38 Avicenna, Canon I fen 3 c. singulare, ed. Venezia 1507, fol. 53[ra] [ed. New Delhi 1981–1996, vol. 1, 236.27–237.2]: "Augmentatur igitur caloris debilitas, propterea quod siccitas in substantia membrorum dominatur: et propter naturalis humiditatis (*bi-l-ruṭūba ġarīziyya*) diminutionem, que est sicut materia, et sicut oleum flammae lampadis (*kā-dahn li-l-sirāḫ*). Flamma namque duas habet humiditates, aquam et oleum, cum una quarum consistit, et cum altera extinguitur. Similiter calor innatus cum humiditate consistit naturali, et suffocatur seu extinguitur cum extranea, et cum augmento extraneae humiditatis, que debilitate digerendi: et est sicut aquea humiditas flammae lampadis. Et cum siccitas ad complementum accedit, calor innatus extinguitur et mors subsequitur naturalis."

able and irreplaceable members. Both questions pivot around the notion of radical moisture and connect, though indirectly, to the role of spirit. In this chapter, we shall focus, on the one hand, on the connection between radical moisture and life expectancy and, on the other hand, on the causal correlation among radical moisture, youth and old age. The topic of youth is handled both in the *Canon* (I, I, 3, 2) and in his zoology (*De animalibus*, XII, 3), which copies directly from his medical handbook. Some slight differences in doctrine between the two texts are crucial for singling out the possible sources of Albert's *De aetate sive de iuventute et senectute*. In both his medical and zoological accounts of youth and old age, Avicenna views aging and death as slow processes characterized by a gradual decrease in the amount of radical moisture and vital heat over time. This loss is understood according to the classic medical theory of the complexion of ages, divided into four. Upon a comparison, the denomination of these ages does not always seem entirely identical between the two works, but they are basically consistent:

1. First age "age of growth" (*aetas adolendi* or *crementalis/ sinn al-numūw*), 0–30 years;[39]
2. "Mature age" (*aetas status* or *standi/wuqūf* or *šabāb*), 30–35/40 years;[40]
3. "Middle age" (*sinn al-muktahilīn*) or "age of diminution without loss of virtue" (*aetas minuendi cum virtus/ʿinḥiṭāṭ maʿ al quwwa*), 40–60 years;

39 Consistency in the length of the first age occurs only in the Arabic *Qānūn* and *Ḥayawān*; unfortunately, the Latin translation of the *De animalibus* mistranslates thirty with twenty-five, an error that is hard to explain: Avicenna, Abbreviatio de animalibus XII c. 3, ed. Venezia 1508, fol. 45[ra]: "Etates sunt quattuor universaliter: etas crementalis, et vocatur anni puerorum; et durat usque ad 25 annos." The Arabic reads: Aviceenna, Ḥayawān XII c. 4, ed. Cairo 1970, 200.4–5: "There are four ages (*asnān*) in general: the age of growth (*sinn al-numūw*) is called puerile age (*sinn al-fityān*), and it [lasts] approximately until 30 years [of age] (*ilà qarīb min ṯalāṯin sanna*)." Translation is mine.

40 *Wuqūf* is translated *aetas standi* in the *De animalibus*, *aetas consistendi* in the *Canon*. The major ambiguity of the Canon's translation is the rendering of *šabāb* (youth) as *aetas pulchritudinis* and not *iuventus* in the opening lines of the chapter. This ambiguity is solved when Avicenna discusses the relationship between the heat of the first age and of the second age (*šabāb*): "Medici autem antique inter se disceptaverunt de duabus caliditatibus pueritiae et iuventutis" (Avicenna, Canon I fen 1 doctr. 3 c. 3, ed. Venezia 1507, fol. 3[vb] [ed. New Delhi 1981–1996, vol. 1, 44.29]). As for the Latin version of the *De animalibus*, it puts the end of the second age around 50 years, an addition not found in the Arabic. Avicenna, Abbreviatio de animalibus, XII c. 3, ed. Venezia 1508, fol. 45[rb]: "Et post etas diminutionis, et est cum remansione vigoris, et dicitur etas virilis usque ad 50 annos vel prope. Et aetas cum diminutione vigoris a 60 usque in finem." Avicenna, Ḥayawān XII c. 4, ed. Cairo 1970, 200.6–7: "[…] and the age of decay with the retention of virtue (*sinn al-ʿinḥiṭāṭ maʿ al quwwa*), i.e. middle age, lasts until sixty years (*wa-huwa sinn al-muktahilīna wa-huwa ilà naḥw sittīna sanna* [see above])."

4. Senile age (*aetas decrepita / sinn al-šuyūḫ ilà-l-al-aḫar al-ʿumr*), 60 years-death.

The four ages are characterized by the transition from the hot and moist age of growth to the hot and dry age of stability, then to a cold and dry complexion (old age) and, eventually, to a dry and accidentally moist (senile age) kind of complexion. In senile age, the accidentally moist complexion is caused by moist residues. These undigested residues of nutrimental substances are abundant in older people because of the poor amount of heat they possess. In turn, this scarce amount of vital heat directly affects the physical capacity of elderly people to concoct and digest food properly.[41]

According to Avicenna, death (except violent death or death by poisoning, etc.) occurs because of intrinsic or extrinsic causes. While the latter (intrinsic) are caused by putrefaction or the dessication of bodily moisture because of the surrounding air, the former are caused by the exhaustion of the radical moisture. This exhaustion can occur because of the drying out produced by the action of vital heat and, indirectly, of spirit as the vehicle for vital heat. In the *Canon* (and in the corresponding section of the *De animalibus*) Avicenna devotes particular attention not only to the causes of death but also to the different stages of the 'age of growth', which is tackled from a male-oriented point of view. As can be observed from the following table, the division of the first age is not entirely identical in both works:

Canon	***De animalibus***
Childhood (*aetas infantiae/ sinna ṭufūliyya*);	Early childhood (*infantia/sinn al-ṭufūla*);
Age of aptness for motion (*ad motiones et ambulandum apta/mustaʿidd li-l-ḥarakāt wa-l-nuhūḍ*);	Motive age (*motiva/sinn al-nuhūḍ*);
Age of teeth growing (*dentium plantativa/nabāt al-asnān*);	Age of teeth (*dentium plantativa/ nabāt al-asnān*);
Age of development (*aetas concussionis/sinn al-taraʿraʿ*) prior to puberty (*antequam poluantur/qabla al-marāhiqa*)	Pre-puerile age (*ante luxuriam/qabla al-marāhiqa*);

41 Avicenna, Canon I fen I doctr. 3 c. 3, ed. Venezia 1507, fol. 4[rb] [ed. New Delhi 1981–1996, vol. 1, 47–48].

(*cont.*)

Canon	*De animalibus*
Age of puberty (*aetas algulemati*/*sinn al-ġalāmiyya*) or adolescence (*rehac*/*riḥāq*);	Puberty (*noviter genita*/*sinn al-ġalāmiyya aw rihāq*).
Age of strength (*aetas fortitudinis*/*sinn al-šadda*).	/

As can be observed, the only clear difference in the account between the two sources is the mention of an "age of strength". While it appears in the *Canon*,[42] the list of ages in the *De animalibus* stops at "puberty".[43] Despite some lexical differences, Avicenna clearly presents an account of first age that is heavily and consistently based on Galenic medicine and its theory of complexion. At the same time, it is undeniable that Avicenna uses young males and not females as his sphere of reference for growth. Among the thorniest issues he was faced with was the difference in heat between the complexion of childhood and that of youth (here meant as second age). By Galen's time, this topic had already sparked a vivid debate among physicians, to such an extent that Avicenna, following Galen's account, divided physicians into two main groups and then reported, and endorsed, Galen's solution.

42 Avicenna, Canon I fen I doctr. 3 c. 2, ed. Venezia 1507, fols. 3^{va-vb} [ed New Delhi 1981–1996, vol. 1, 44.18–19, 44.22–26]: "Aetas autem adolendi que vocatur aetas adolescentie (*sinnat al-numūw*) et est fere usque ad triginta annos. […] Adolescentie vero aetas dividitur etatem infantie: et hec quidem est cum membra pueri nondum ad motiones et ambulandum sunt apta: et in etatem apta; et in etatem dentium plantativam que est post ambulationem et antequam sit fortis; et illud est cum nondum gingivae sunt omnibus dentibus replete. Postea est etas concussionis quod quidem est post vehementiam et nativitatem dentium et antequam poluantur. Deinde est aetas algulemati (Ar. *sinn al-ġalāmiyya*, "of puberty") vel alguadi (addition of the Latin) et rehac (ar. *riḥāq*, "puberty, adolescence") idest cum ab eo egreditur sperma que est quousque barbescant. Postea est etas fortitudinis que est donec sinat crescere."

43 Avicenna, Abbreviatio de animalibus XII c. 3, ed. Venezia 1508, fol. 45ra [ed. Cairo 1970, 200.4–5, 200.9–13]: "Aetates sunt quattuor universaliter: aetas crementalis, et vocatur anni puerorum, et durat usque ad 25 annos (Ar. "until approximately 30 years" [*ilā qarīb min ṯalāṯīn sanna*]) […] Sed puerilis aetas dividitur in noviter genitarum et etatem motivam et aetatem ante fortitudinem. Et est illa que est dentium plantativa. Et est etas que est ante luxuriam et aetas noviter genita […]". The Latin translation mentions only the *aetas ante luxuriam* (*qabla al-marāhiqa*) but omits the ending of the divisions of first age, reading precisely as the *Canon* does from this point: "Postea est etas concussionis quod quidem etc. […]"

As Avicenna reports, the first group of physicians argues that *pueri* (*ṣubyān*) have a qualitatively stronger (*maior/ašadd*) heat since they grow more than youngsters (*iuvenes/šabāb*). This empirical data is intuitively mirrored in the name of the second age itself (*aetas stativa, standi* or *consistendi*/*sinn al-wuqūf*), which indicates—especially the Arabic root *w-q-f*, the stopping of a process. According to Avicenna, growth appears to be primarily associated with longitude (*longitudo*/*ṭūl*) rather than with latitude (*latitudo/ʿarḍ*). This is because growth in longitude concerns members such as bones and nerves and, more generally, radical members (heart, liver, brain) that are defined as such because they are of spermatic origin and cannot be restored through extrinsic, nutrimental moisture.[44] Growth by width (*spissitudo*/*ʿamq*), by contrast, tends to apply to homeomerous parts of the body and not to its radical members.[45] Properly speaking, growth primarily relates to the age of growth and to the growth of radical members; secondarily, it can certainly apply to homoeomerous parts of the body which grow regardless of the age of the living being.

In Avicenna's view, radical members are impossible to replace since they are of spermatic origin and cannot grow anew thanks to nutrimental moisture. This fact alone, though, represents one of the major aporias of the theory of radical moisture. Often identified with seminal moisture, radical moisture is considered a side product of nutrition *qua* product of the fourth digestion. Paradoxically enough, it cannot be restored through nutrition itself. A further problematic idea in the Avicennian theory—as in the general theory—of radical moisture is this: that which in the father is a superfluous fluid, in the offspring it turns out to be nothing short of essential in the definition of its individuality; for these reasons, radical moisture has been defined as a 'heuristic principle' by Chiara Crisciani and Giovanna Ferrari, rather than a notion with an intrinsic consistency in the philosophy of nature.[46] Moreover, the fact that Avicenna considers the radical members to be of spermatic origin will open the door to a fierce debate on the possible entrance of the moisture of human

44 Avicenna, Abbreviatio de animalibus IX c. 2, ed. Venezia 1508, fol. 41rb [ed. Cairo 1970, 148.11–13]: "[...] et si essent membra generata ex sanguine, essent nervi et vene et ossa de genere carnis, et ossa si essent fracta (maqṭūʿa) redirent ad continuitatem sicut caro, sed non renascitur quia generatur ex spermate. Igitur ex spermate sunt membra radicalia." Note that Avicenna also assigned radical members to oviparous animals in: Avicenna, Abbreviatio de animalibus XVII c. 1, ed. Venezia 1508, fol. 62va [ed. Cairo 1970, 416.7]: "[...] albumen est magis conveniens ut sunt ex eo corporis radicalia (*wa-tanšū minhā al-uṣūl al-ūlà wa-l-mabādiʾ*)."

45 Avicenna, Liber de anima II c. 1, ed. Van Riet 1972, 106.43–52 [ed. Rahman 1959, 53.12–19].

46 Crisciani-Ferrari, "Introduzione," 340–344.

semen into the material composition of the embryo, an object of discussion for Albert in his *Quaestiones super De animalibus*.

Given the extremely relevant position of radical moisture in the Avicennian medical doctrines, one can more clearly see how the determination of the relationship between the heat of first and second age is essentially tied to Avicenna's understanding of growth in a strict sense (growth of radical organs) and broader sense (growth as growth of flesh and fatty tissues). Growth in the first age is associated with a moister complexion, closer to birth and derived from embryonic matter (also derived from radical moisture?) and observable through the growth of radical members. Growth in mature age, by contrast, applies only to a limited extent to the cases of flesh and fatty tissues.

In the light of these considerations, the reader will now be in a better position to understand the Galenic discussion on the relationship between the complexions of the first two ages. The arguments put forward by the first group of physicians, who supported the view that heat in the first age is qualitatively better than in the second, is based on the empirical evidence that digestion in children occurs more frequently (*maiores/akṯar*) and is longer (*diutius/adum*) than in youngsters. According to this first sect of physicians, higher frequency and longer duration must be causally explained in terms of a higher amount of heat in children. Since they were born more recently (*minoris temporis/ahdaṯ*) and are therefore temporally closer to the seminal heat ultimately responsible for the complete growth of the living being, they must be qualitatively hotter (viz. they have a higher temperature) than youngsters.[47]

By contrast, the second group of physicians replied that this was not the case, since heat in mature age is quantitatively greater than in the age of growth. The empirical evidence adduced by them for their statement is founded upon the frequent occurrence of nose-bleeds (epistaxis) in youngsters. According to them, these phenomena are indicators of a greater abundance of blood and stronger sanguineous flow in people of more mature age. Given that blood is, according to humoral theory, hot and wet, this points to a hotter and wetter complexion in those of mature age compared to children. Further empirical clues adduced by the second sect of physicians to support their claim consist in the colder nature of fevers (among which the tertian fever, [*tertiana/ḥummā al-ġabb*]) affecting children and a better nature of digestion in the mature age. The fact that children are more prone to vomiting than young people would seem to point to a more phlegmatic (colder and moister) complexion of the

47 Avicenna, Canon I fen I doctr. 3, c. 2, ed. Venezia 1507, fol. 3[vb] [New Delhi 1981–1996, 45.1–3]; Avicenna, Abbreviatio de animalibus XII c. 3, ed. Venezia 1508, fols. 45[ra–rb] [ed. Cairo 1970, 201.3–5].

first age compared to the second. Moreover, a further argument adduced in favor of a quantitatively lesser amount of heat in the first age is that mature people move more and have better digestion when they follow their natural appetite (lat. *istimbra*, ar. *istimrāʾ*). These *istimbrae* are causally attributed to a temporary peak of coldness in the complexion but, given that children have less heat to digest properly, they tend to vomit more than mature people do.[48]

The solution envisaged by Avicenna closely follows Galen's position, who seeks to bring the ideas of the two groups into agreement with each other. In Galen's view, the amount of heat in growing and in mature age is essentially the same; during the first age, though, heat is more abundant in quantity than in quality, while the opposite is true of the second age. The example adduced by Galen and reported by Avicenna to argue for this solution is the following: if a subtle and hot body penetrates another body, such as water or a stone, it would produce two different effects according to the primary qualities of the body subject to its action. In water or a moist body, heat would spread in a quantitatively more abundant but qualitatively weaker manner (producing a lower temperature), while the induction of heat in a stone or in a blade would cause a qualitatively stronger effect (higher temperature).[49] Galen's argument on the essential equality of heat in the two complexions is meant to maintain that heat in the first age is quantitatively more than in the second, since the amount of moisture, intimately linked to spermatic moisture and growth, is greater than in the complexion of the second age. At the same time, Galen does not want to throw out the baby (the solutions of the two sects of physicians) with the bathwater (their empirical observations). For this reason, he stresses that the qualitatively stronger (viz. the temperature is higher) nature of heat in mature age, which chimes with remarks on the higher degree of fevers in mature people than in younger people, and in their digestive capacities. Despite this, Galen had rejected the argument that the *istimbrae* were due to coldness by attributing this habit to typically childish gluttony (*gulositas/maṭʿūm*).

In concluding this discussion on the relationship between the two ages, it must be said that second age, by its own definition (*aetas standi/sinn al-wuqūf*) marks the end of growth and the beginning of the irreversible process of aging, eventually leading to death, but it is at the same time the age of maturity, during

48 The Latin version of the *Canon* does not translate the second Arabic term (*haḍm*, digestion) which clarifies the direct connection between the term *istimrāʾ* and the digestive (*hāḍima*) faculty (*Canon*).

49 Avicenna, Canon I fen 1 doctr. 3 cap. 3, ed. Venezia 1507, fols. 3^{ra-rb} [ed. New Delhi 1981–1996, vol. 1, 45.14–46.4]; Avicenna, Abbreviatio de animalibus XII c. 3, ed. Venezia 1508, fol. 45ra [ed. Cairo 1970, 201.17– 202.18].

which humans and animals have reached the full formation of their organs for their well-being. While in Aristotle the processes of aging and death are apparently non-teleological, Avicenna's position, influenced by his Muslim faith, will be different. Before jumping to the conclusion, however, it is necessary to complete our doctrinal analysis of Avicenna's theory of the different complexion of ages.

In Avicenna's view, the third age is the age of men characterized by a diminution in complexional heat but not in the capacities (*aetas minuendi cum virtus/ʿinḥiṭāṭ maʿ al-quwwa*). By contrast, the fourth and last senile age (*aetas decrepita/sinn al-šuyūḫ*) is characterized by a loss of both virtue and substance (*aetas minuendi cum manifesta virtus debilitate/ʿinḥiṭāṭ maʿ ẓuhūr al-ḍuʿf fī-l-quwwa*). Avicenna spends little space dealing with the third age and rather focuses on the fourth age and its typical diseases due to failure in the digestive process. These pathologies are manifest in elderly people, who abound in catarrhs, accidental moisture caused by imperfect digestion, and are much more prone to obstructions (lib. I, fen I, doctr. III), viz. the blocking of the flow of spirit in blood vessels or nerves. It is worth remembering that the obstructions are mentioned in the *De anima* as a proof from medical experience (*experientia physica/taǧārub al-ṭibbiyya*) that spirit is the vehicle of the soul's bodily faculties (*pendentes ex corpore/mutaʿalliqa bi-l-ǧism*).[50] Again, this is proof that for Avicenna, medical learning has value when dealing with the physiological explanation of natural bodies within the epistemic parameters set by natural philosophy.

Consistently within his study of causes for aging in the *Canon*, Avicenna presents his remedies against the diseases of old age, based especially on the six so-called Galenic *non-naturalia*: food and drink, surrounding air, sleep and waking, rest and exercise, excretions and retentions and mental affections. In the *Canon*, he provides an extensive treatment of the topic, which goes well beyond our purposes.[51] Avicenna also maintains a common position with Aristotle's handling of old age, namely the parallel between white hair and putrefaction (*putrefactio/ʿufūna*): given the nature of old age as loss of heat and moisture, these remarks chime quite well with Avicenna's handling of radi-

---

50 Avicenna, Liber de anima V c. 8, ed. Van Riet 1972, 175.49–55 [ed. Rahman 1959, 263.9–13].

51 Avicenna, Canon I fen 3 c. 3, ed. Venezia 1507, fols. 55ra–56rb [ed. New Delhi 1981–1996, 251–254]. For a thorough analysis of Avicenna's doctrine: Paola Carusi, "Vecchiaia avanzata e qualità della vita nel Canone di Avicenna," in Vita longa. *Vecchiaia e durata della vita nella tradizione medica e aristotelica antica e medievale*, ed. Chiara Crisciani/Luciana Repici/Pietro B. Rossi (Micrologus Library 33), Firenze 2009, 41–60.

cal moisture, too.[52] Considering, then, the process of aging and death from a medical and natural philosophical perspective, it would seem that death, as in Aristotle, results from an unavoidable obedience on the part of sublunary composites to the laws of generation and corruption. In this context, spirit is not mentioned by Avicenna in connection to the key notion of radical moisture.

The situation may change, though, if one considers Avicenna's handling of pathologies, especially cognitive pathologies, in the third book of the *Canon*. These sections, already analyzed in a study by Danielle Jacquart,[53] offer some hints for understanding the problematic connections between spirit and radical moisture, cognitive activities and aging. While it is known from Avicenna's own words that moisture (very likely radical moisture) gradually fades away already from the second age (*aetas consistendi*/*sinn al-wuqūf*)[54] and that elderly people show characteristic signs of weakening or impairment in their cognitive and rational capacities, it is not quite clear how the relationship between spirit and radical moisture should be thought to work. Typically, the malfunctioning of cognitive (also, motive) faculties is causally linked to a cold complexion and, in general, cold blocks the motion of spirits in the brain.[55] In his study of cognitive impairments in the *Canon*, Avicenna distinguishes between simple and temporary alterations (*alteratio*/*taġayyur*) of cognitive and rational activities from a degenerative pathology such as dementia (*amentia*/*ḥumq*). In his view, this pathology comes about through a diminution (*diminutio*/*niqṣān*) or complete destruction (*destructio*/*buṭlān*) of the cognitive activities (*cogitatio*/*fikr*) located in the middle brain ventricle, where the imaginative-cogitative (*imaginativa*/*mutaḫayyila*) and estimative (*aestimatio*/*wahm*) powers are seated.

This diminution or destruction is produced by three causes: coldness, coldness and dryness, and phlegmatic fluids. Broadly speaking, all of these are

---

52 Avicenna, Canon I fen 2 doctr. 3 cap. 3, ed. Venezia 1507, fol. 41rb [ed. New Delhi 1981–1996, 190].

53 Danielle Jacquart, "Avicenne et la nosologie galénique: l'exemple des maladie de cerveau," in *Perspectives arabes et médiévales sur la tradition scientifique et philosophique grecque: actes du colloque de la SIHSPAI (Société internationale d'histoire des sciences et de la philosophie arabes et islamiques), Paris, 31 mars–3 avril 1993*, ed. Ahmad Hasnawi/Abdelali Elamrani Jamal/Maroun Aouad; with a preface of Roshdi Rashed (Orientalia Lovaniensia Analecta 79), Leuven-Paris 1997, 217–226.

54 Avicenna, Canon I fen 1 doctr. 3 cap. 3, ed. Venezia 1507, fol. 4[ra] [ed. New Delhi 1981–1996, vol. 1, 46.18–19]: "Preterea sciendum est quod caliditas post tempus consistendi minui incipit propterea quod aer circundans materiam eius que est humiditas exiccat."

55 Avicenna, Canon III tr. 4, c. 11, ed. Venezia 1507, fol. 186[vb] [ed. New Delhi 1981–1996, vol. 3/1, 88.7–10].

material causes traceable either to the complexion of old age (coldness, dryness) or to the incapacity of elderly people to discharge away these phlegmatic, harmful substances. As noted above, coldness in general impacts the motion of animal spirit in the brain, thus affecting the cognitive capacities. Thus, the correlation between coldness, the slow motion of the spirits and cognitive impairments becomes evident. This explanation, though, accounts for cognitive failures in elderly people in a manner that requires further and more detailed explanation. The reason should rather be sought in the interaction between heat and moisture in the digestive process, which well resembles, in Avicenna's own words, the mechanism for producing animal spirits in the brain.[56]

These animal spirits are produced by a mechanism that Avicenna likens to digestion and, as is well-known from his theory of ages, digestive processes are precisely those things which suffer the most from the passing of time, becoming gradually but irreversibly less effective. This deficiency is linked to a loss in both radical moisture and in vital heat, outweighed by a growing amount of coldness. Coldness in complexion decreases, on the one hand, the efficacy of spirits *qua* vehicles of sensation and motion, which both necessitate the occurrence of heat and spirit. Moreover, a gradually freezing complexion directly affects the humoral composition of spirits, making them thicker and less digested than they should be. According to Avicenna, the annihilation (*destructio/buṭlān*) of the cognitive and rational activities depends on coldness and on the thickening of spirit itself, arising from moisture (*propter grossitiem spiritus ex humiditate/li-ġilaṯ al-rūḥ min al-ruṭūba*).[57]

Still, it is difficult to clarify what kind of moisture Avicenna is referring to here. If by 'moisture' is meant 'radical moisture', it is difficult to understand spatially where radical moisture should be located, for it should be everywhere in the body *qua* its cohesive principle, and especially in its radical members (heart, stomach, brain, muscles, nerves etc.). Perhaps Avicenna is referring to nutrimental moisture, out of which natural spirits are produced in the liver and which, in turn, enters the material composition of vital and animal spirits. But again, here we are entering the realm of conjecture and charitable interpretations, not of textual evidence. Moreover, it has been possible to observe how, according to Avicenna, radical moisture possesses a key gluing, cohesive func-

56 Avicenna, Canon III fen 1 c. 1, ed. Venezia 1507, fols. 165^{rb-va} [ed. New Delhi 1981–1996, vol. 3/1, 4.25–5.5].

57 Avicenna, Canon III tr. 1, c. 5, ed. Venezia 1507, fol. 166vb [ed. New Delhi 1981–1996, vol. 3/1, 8.3–7]: "Et sermo quidem uniuersalis in significationibus operationum est quod ipsarum diminutio et destructio fiunt propter frigus et grossitiem spiritus ex humiditate [...]."

tion of the radical members of living beings which gradually fades away over time. This fact could well impact the digestive capacities of key organs such as the stomach, the liver and the brain in the production of animal spirits as a result of successive concoctive processes of spirit from the stomach to the liver to the heart and, eventually, to the brain.

In the end, it is possible to argue that alterations in the digestive capacities, correlated to the decrease in the amount of radical moisture and the worsening concoction of nutrimental moisture, may affect the material complexion of animal spirits and indirectly cause senile dementia. Yet, the precise causal relation between spirit, radical, nutrimental moisture and pathologies of old age remains unspecified in Avicenna, thus presenting challenges for Albert the Great in his *De iuventute et senectute* and, more in general, in his handling of youth and old age, life and death.

Avicenna's approach to death from a purely materialistic perspective (death as the disaggregation of the body) is detectable in the *Canon* and *De animalibus*; by contrast, in the *De anima* he had demonstrated that, by not subsisting as form in matter, the human soul survives its severance from the body after death in view of Judgement Day (*yawm al-dīn/yawm al-qiyāma*). According to Avicenna, there are two different, coexisting perspectives on death which are compatible rather than at odds with each other, viz. death from the perspective of the body and death from the perspective of the soul. In humans, the former represents indeed a non-teleological process following the same natural rules as in animals and plants. In the case of humans, though, there is a further, eschatological dimension that is dealt with properly in psychology, which concerns death from the point of view of the soul. There, the self-subsistent and separable nature of the human rational soul, though, introduces an eschatological dimension of death, firmly rooted in the use of the Neoplatonic scheme of the procession and return (*mabdāʾ-maʿād*) of the soul to its Creator.

All in all, Avicennian *rūḥ* plays a twofold role here in the case of longevity. When accounting for length and shortness of life, its motion as a vehicle for nutrition, sense perception and motion in the body can be conjecturally thought to act as a dryer of the radical moisture so crucial for the longevity of the living being. On the other hand, spirit operates for man's well-being by forming all the radical members and providing the fullest set of organs for the exercise of rational nature. Spirit explains aging and death from a purely mechanical point of view, accounting both for failures in the functioning of certain capacities of the soul (sensation, motion), but does not exhaust the teleological explanation of aging and death. In the case of animals and plants, both subject to the rules of generation and corruption, death is just their natural destiny. For humans, though, death represents the moment when the rational part of the soul is severed from the body and returns to its origin, God. This is a conclusion Albert the Great himself would have agreed with, albeit with some differences on the parts of the soul which survive the body's death.

3 Spirit, Radical Moisture, and Longevity in Albert the Great

3.1 *The First Entry of Avicenna's* Canon*: Radical Moisture in Albert's Early Theological Works*

In his years of lecturing as a bachelor in theology (*baccalaureus*) at the University of Paris, Albert had proved quite receptive to many medico-philosophical and zoological doctrines such as those of spirit and formative power. The case of radical moisture did not escape Albert's omnivorous intellectual curiosity. Quite often, this notion had been employed by theologians, as shown by Crisciani and Reynolds, to discuss two main theological topics, the *veritas humanae naturae* and the prelapsarian state, viz. the physiological state of Adam and Eve before Original Sin.

The topic of the *veritas humanae naturae*, extensively presented by Peter Lombard's *Sentences* and traceable to Anselm of Canterbury,[58] represented a typically Medieval thematic grouping of issues concerning original sin, resurrection and personal identity within a peculiar intermingling of medicine, philosophy and theology. It revolves fundamentally around the question of individuality, the true nature of man, his material constituents and their relation to the glorious body, viz. the body that will resurrect. To determine the nature of the glorious body, theologians at Paris started to make use of the theory of radical moisture, likely circulating at that time in the faculty of medicine.[59]

In his book, *Food and the Body*, Philip Reynolds has brought to light how Albert—following the theological tradition of his time—inquires after the nature of moistures and their link to the substantial and true nature of humans (*veritas humanae naturae*).[60] The answer to this inquiry is found in his *De resurrectione* and is quite straightforward: the *veritas humanae naturae* equates with both seminal and nutrimental moisture, thus allowing for the inclusion of nutrimental moisture within the *veritas humanae naturae*. This is because

58 On the theme of *veritas humanae naturae* in 13th century theology, see: Reynolds, *Food and the Body*, 50–67; Crisciani-Ferrari, "Introduzione," 245–252, and bibliography therein.

59 On previous readings of the concept of *veritas humanae* naturae at Paris, see: Reynolds, *Food and the Body*, 149–214.

60 On the topic of the *veritas humanae naturae* in Albert, see: Reynolds, *Food and the Body*, 281–310.

these two moistures bring the body's members to their perfect and definite state (*ad debitam quantitatem*). In turn, they allow the soul's faculties to carry out the most perfect actions according to its nature (*ad actus perfectos secundum suam naturam*), which will be rewarded in the afterlife.[61] A third and final kind of moisture, defined as restoring moisture (*humidum restaurativum*), is identified by Albert as the pure provider of moisture necessary for delaying the consumption of natural or radical moisture (*humidum naturale aut radicale*). For these reasons, Albert concludes that the *veritas humanae naturae* is represented essentially by the first two moistures, thereby excluding the last one.[62]

The second case, viz. the relationship between radical moisture and the survival of Adam and Eve in the prelapsarian state (*in statu innocentiae*), is discussed in the *Commentary on the Sentences* and in three *quaestiones* (75, 76 and 77) from the *De homine* and has been discussed in a recent contribution by Evelina Miteva.[63] Miteva shows how in the *De homine* Albert proceeds to inquire into the causes of Adam and Eve's non-mortality (*potest non mori*) in Terrestrial Paradise. This is a very traditional topic in theological treatises at Paris, which dates back as early as Philip of Chancellor's *Summa de bono*.[64] According to Albert, who follows a decided trend in the so-called 'medicalization' of the discourse on theological topics,[65] the deciding factor between life and death in the Edenic condition, which balances the mortal (complexion) and immortal (God's grace) aspects of the human being, is will. Obedience to God's command ("you must not eat from the tree of the knowledge of good and evil" [*Gen.*, II, 17]) grants Adam and Eve that neither of them is subject to

61 Albertus Magnus, De resurrectione tr. 1 q. 6 a. 9 (Ed. Colon. XXVI, 255.15–19, 255.43–52): "Solutio: Distinguimus triplex humidum, ex quo sumitur substantia humani corporis. Primum est seminale, ex quo fit substantia prima membrorum, scilicet substantia ossis et carnis et manus et pedis et huiusmodi. [...] Secundum humidum est nutrimentale, quod per conversionem additum primo deducit organa corporis ad debitam quantitatem. Huius autem necessitas est, quia anima non solum habet potentias ad actus, sed etiam ad actus perfectos secundum suam naturam. Ergo oportet organum non solum proportionari actibus animae in se, sed etiam actibus perfectis. Illa autem non sunt organa diminuta, quae fiunt ex humido seminali, sed completa per augmentum et nutrimentum in consolidatione et quantitate."

62 Albertus Magnus, De resurrectione tr. 1 c. 6 (Ed. Colon. XXVI, 255.67–70; 255.75–256.10).

63 Evelina Miteva, "The Bodies of Adam and Eve: The (Missed?) Encounter of Natural Philosophy and Theology in the Account of Albert the Great," *Syzetesis* 7 (2019/2), 393–404.

64 Philippus Cancellarius, Summa de bono, De bono nature quod est diminuibile per malum culpe pars 1 q. 4 (ed. Wicki 1985, vol. 1, 310–314).

65 On the term "medicalization", see: Joseph Ziegler, "Medicine and Immortality in Terrestrial Paradise," in *Religion and Medicine in the Middle Ages*, ed. Peter Biller/Joseph Ziegler (York Studies in Medieval Theology 3), York 2001, 224.

the usual physical laws of life and death.[66] In these *quaestiones*, though, there is no trace of the concept of radical moisture. In the *Commentary on the Sentences*, though, the situation is a bit different. Here, the criteria envisaged for the survival of Adam and Eve are more complex and go beyond simple human will. Albert's explanation relies on five factors: (1) drinking from a perpetual source of life; (2) the specific *habitus* of Adam's body, able to keep the its bodily proportions, composition and complexion free from alteration; (3) eating from the tree of life (*lignum vitae*), which removes signs of old age, stiffness of limbs and decay of the humors; (4) eating from other trees, which supply the loss of *humidum radicale* and (5) God's grace. Even though four out of five elements in this list are all necessary for non-mortality, the fifth factor, necessary and sufficient, is God's Grace: without it, no avoidance of the physical rules of life and death is possible.[67]

In this context, the function of radical moisture (*humidum radicale*) is not to grant duration of life, as one would expect, but rather to avoid anxiety, sickness or old age and is not connected to what is usually called the "tree of life" (*lignum vitae*).[68] The association between radical moisture, longevity in the prelapsarian state and *lignum vitae* can be found, as shown by Chiara Crisciani, in the contemporary *Summa Halensis*.[69] The theological roots of the correlation between immortality and eating of the tree of life go back to Augustine's interpretation of Genesis and can already be spotted in the above noted *quaestio* from Philip's *Summa de bono*. Albert, though, while using the same medical concept (radical moisture), shifts slightly away from his main theological authority, viz. Augustine. By choosing to link the restoration of radical moisture with eating from the tree of life, he does not correlate the restoration of radical moisture with Adam's non-mortality, but rather to the avoidance of other physiological phenomena correlated to a loss of radical moisture, such as wrinkling and aging.

The previous analysis of Albert's use of the concept of radical moisture helps the reader understand that, on the one hand, he was following in the footsteps of canonical debates (*veritas humanae naturae*, prelapsarian state). On

66 Albertus Magnus, De homine (Ed. Colon. XXVII/2, 561.55–61); Miteva, "The Bodies of Adam and Eve," 399.

67 Albertus Magnus, Super II Sententiarum dist. 19 a. 4, ed. Auguste Borgnet, Paris 1894 (Ed. Paris. XXVII, 334a–b); Miteva, "The Bodies of Adam and Eve," 399.

68 Miteva, "The Bodies of Adam and Eve," 401.

69 Chiara Crisciani, "Premesse e promesse di lunga vita: tra teologia e pratica terapeutica (secolo XIII)," in Vita longa. *Vecchiaia e durata della vita nella tradizione medica e aristotelica antica e medievale*, ed. Chiara Crisciani/Luciana Repici/Pietro B. Rossi (Micrologus Library 33), Firenze 2009, 68.

the other hand, his use of the concept of radical moisture in the *De homine* in the sections devoted to the nutritive powers of the soul is much less obvious and shows his interest in an in-depth analysis of psychophysiological processes, a typical feature of the *De homine* compared to contemporary theological treatises on the human soul. His main source on the topic is the *Canon*,[70] which is quoted anonymously when discussing whether all parts of the body can be subject to the loss and restoration of material substance (*utrum in omnibus partibus materiae possit fieri deperditio et restauratio vel in quibusdam tantum*). In the question devoted to this topic, Albert draws on, as noted by Reynolds, the famous Avicennian lamp metaphor.[71]

In this passage, the threefold distinction of moisture into radical moisture and two distinct kinds of nutrimental moisture, from book IV of the *Canon*, is employed by Albert to argue that radical moisture can never be completely restored through food, even though radical or spermatic moisture is basically of spermatic, and therefore nutrimental, origin.[72] His position chimes well with Avicenna's. In Albert's *De homine*, consumption of radical moisture is directly proportional to length and shortness of life, since radical moisture acts as a converting principle of food, fundamental for the persistence of the activity of the nutritive faculty.[73] Further, if it is true that radical members can grow (as shall be seen, during the growing age), the loss of those members cannot be replaced by nutrimental moisture in humans; in the case of other animals, such as crabs and the like, this is possible.[74]

In the *De homine* and in the *Commentary on the Sentences*, radical members are primarily associated with the formative power, since those organs cannot be replaced because of their spermatic origin, unlike other parts of the body such as flesh which, though generated in spermatic moisture, can be replaced by nutrimental moisture. Radical moisture is crucial since it is both

70 The only explicit reference to Avicenna's medical work, named *anatomia*, is: Albertus Magnus, De homine (Ed. Colon. XXVII/2, 203.56–59): "Ad aliud dicendum quod in veritate septem paria diffundentia sensus per totum corpus oriuntur ab anteriori cella capitis, quae sunt multarum divisionum, ut dicit Avicenna in sua anatomia." Siraisi, "The Medical Learning," 393.

71 Albertus Magnus, De homine (Ed. Colon. XXVII/2, 110.45–63); *Reynolds, Food and the Body*, 234–237.

72 Albertus Magnus, De homine (Ed. Colon. XXVII/2, 110.18–45).

73 Albertus Magnus, De homine (Ed. Colon. XXVII/2, 111.1–7): "Unde sicut destructa radice in planta non fit postea incrementum, quia non est quod convertat nutrimentum ad naturam plantae sic destructo radicali, quod diffusum est per organa, non erit vegetatio membrorum. Radicale enim est convertens et activum per virtutem animae, quae est in ipso, sed nutrimentale est convertibile tantum."

74 Albertus Magnus, De homine (Ed. Colon. XXVII/2, 111.14–32).

able to attract menstrual blood (*tractivum sanguinis menstrui*) out of which (with feminine semen) shapes the embryo, and at the same time can restore the loss of nutrimental moisture in organs (*restaurant deperditum in organis factis*).[75] This is the short treatment given by Albert to the notion of *humidum radicale* on purely philosophical grounds in his early theological works. The handling of radical moisture as a notion connected to the theme of growth and replacement of members, life and death, and not just the prelapsarian state or the *veritas humanae naturae* does not find relevant parallels, to our present knowledge, in authors chronologically close to Albert and authors of theological treatises on the human soul. One of the few parallels can be observed in John of La Rochelle's *Summa de anima*, where heart, brain, liver, and testicles are called "radical members" (*radicalia*).[76] No treatment of the concept of radical moisture as a naturalistic notion, though, is provided in his works. At this moment, therefore, Albert represents the first theologian—at least, among those who we know (until now) to be active in Paris—to develop the use of radical moisture within a naturalistic framework tied to the determination of lifespan and the distinction between replaceable and non-replaceable members in a theological work. By contrast, he is not the only one at his time to make use of the concept within an Aristotelian philosophy of nature. This is the case, for instance, with Peter of Spain's *Tractatus de longitudine*. Because of its naturalistic framework permeated by Avicennian medico-philosophical doctrines, this work can well-be compared to Albert's mature *De morte et vita*.

---

75 Albertus Magnus, De homine (Ed. Colon. XXVII/2, 142.50–58): "Sed non intendit dicere quod eadem anima sit quae facit corpus et regit ipsum, sed quod sit eadem virtus informativa faciens quodlibet organum et continens ipsum factum. Virtus enim informativa organo facto manet in humido radicali, quod est in organis, et sicut prius erat tractivum sanguinis menstrui, ex quo multiplicato humido seminali fecit organum, ita postea per attractum humidi nutrimentalis restaurat deperditum in organis factis." The same connection between radical moisture and *virtus formativa* in: Albertus Magnus, Super III Sententiarum dist. 3 a. 26, ed. Auguste Borgnet, Paris 1894 (Ed. Paris. XXVIII, 65b): "Contra: Philosophus in II *de Anima* dicit, quod omnium natura constantium terminus est et ratio magnitudinis et augmenti: hic autem terminus et ratio sumitur a duobus: a quantitate humidi radicalis, quod est materia prima generationis, et a vigore virtutis formativae, et informatione, sicut innuit Galenus, qui dicit quorumdam defectum membrorum causari ex defectu materiae, quorumdam autem ex defectu virtutis igitur illa materia proportionata est tanto corpori in tanta quantitate [...]".

76 Johannes de Rupella, Summa de anima c. 39 (ed. Bougerol 1995, 121.58–59): "[....] fontes autem dicuntur principalia et radicalia membra, cor scilicet, cerebrum et epar [...]"; Summa de anima c. 82 [16] (ed. Bougerol 1995, 222.39–40): "Hec enim est que, ut artifex, membra plasmat ex semine; primo radicalia ut cor, cerebrum , epar, testiculos [...]."

Albert's handling of the topic keeps the notions of radical moisture and spirit apart and associates a reflection on the concept from a primarily human point of view, despite some references to particular cases of animals. In conclusion, in his early works Albert introduced the concept of radical moisture to tackle both theological (*veritas humanae naturae*, prelapsarian state) and philosophical (life and death, replaceable vs. replaceable members) topics, though never connecting explicitly the *humidum radicale* to spirit. The causal connection between spirit and lifespan will separate Albert's doctrine from that of contemporary commentaries on Aristotle's *De longitudine et brevitate* (by Adam of Buckfield, Peter of Ireland), while providing interesting comparisons with his contemporary Peter of Spain, author of the *Tractatus de longitudine*. The connection between spirit and radical moisture will appear in his own *De morte et vita* (a paraphrase of Aristotle's *De longitudine et brevitate*) and the *De aetate sive de iuventute et senectute*, meant to supply Aristotle's unavailable work. Translated by James of Venice together with the *De respiratione et inspiratione*, this translation is extant only in four manuscripts.[77]

3.2 **Desiccativum est humidi radicalis*: Spirit and Radical Moisture in the* Parva Naturalia**

The *humidum radicale* plays a relevant role in Albert's definition of length and brevity of life. In contrast to his earlier works, though, this concept is extended not only to humans and animals (as in Avicenna) but, as Marilena Panarelli has recently shown, also to plants[78] and minerals.[79] Albert's approach to radical moisture is pretty much in keeping with Albert's unifying use of concepts such as spirit and formative power observed so far.

While the concept of radical moisture in minerals falls out of the scope of the present book, a few remarks on the concept of radical moisture in plants (based on Panarelli's work) seem in order here. Usually, the concept of radical member (and radical moisture) is primarily associated in animals and humans to the heart *qua* organ from which all organs are originated in the embryo, and to other non-replaceable parts which are considered as such—in plants, they are called *partes integrales essentiales*—because of their spermatic origin. An organ analogous to the heart can be found, as shown in the second chapter, in

77 De Leemans, "*Secundum viae naturae et doctrinae*," 203.

78 Panarelli, "How Do Plants Live and Grow?," 347–367.

79 Albertus Magnus, De vegetabilibus IV tr. 2 c. 2 (ed. Meyer-Jessen, 235): "Planta enim in generatione sua duobus indiget, quorum unum est materia, ex qua fit; et alterum est locus suae generationi conveniens. Locus enim principium est generationis, quemadmodum et pater; et materiam quidem impedit loci salsugo, eo quod exsiccat humidum radicale

the roots. Moreover, plants of a more complex nature have something like veins (another typically non-replaceable, radical member), which serve as a channel for the distribution of the spirit in the plants or the marrow, both qualified by Albert as "radical parts", for they supply a non-replaceable function, i.e. the diffusion of the spirit in the plant.[80] Strictly speaking, plants with roots do not have internally digestive organs but do have, as Albert had highlighted in the *De nutrimento et nutrito*, four digestions, just like animals, which prepare the seed to produce new bodily organs. In the *De vegetabilibus*, by contrast, these digestions appear to be only two: one in earth and one inside the plant, more specifically in the marrow. As argued above following Panarelli's interpretation, these statements are not so much contradictory insofar as these four digestions occur (1) in earth (*in terra*), (2) in the roots (*in radice*), (3) and (4) in each particular organ. While distinguishing the three different moments of digestion inside a plant, both the *De nutrimento* and the *De vegetabilibus* maintain that the concoction and digestion of nutritive substances begins in earth and is fully actualized within the plant itself, although in not a completely identical manner.[81] As we have noted in chapter 3, Albert identified radical moisture with seminal moisture and regarded this as the material cause of non-replaceable, radical members.[82]

In Albert's account of human and animal generation, this system of digestions produces the formative power of the semen, carried by what Albert calls *spiritus radicalis*, i.e. the spirit placed inside the semen. Unlike generation in animals, though, the origin of radical moisture in plants seems to be univocal: it comes from the seed, no matter the sex. This is because in the seed, as in the fertilized egg, formal, efficient and material cause are already mixed together. By contrast, in animals the situation is completely different: as Albert had argued in the *De anima*, radical moisture is of spermatic origin, and therefore comes

---

plantarum. Cujus signum est, quod terrae sale commixtae vel sale seminatae steriles efficiuntur."

80 On marrow, see: Albertus Magnus, De vegetabilibus II tr. 1 c. 3 (ed. Meyer-Jessen, 122): "Medullae autem plantarum videntur esse vicarii virtutis radicalis, sicut nucha in animalibus est vicarius cerebri; et per medullam arboris decurrit et pulsat plus de spiritu plantae, quam per aliquam aliarum viarum; et in ea est vigor spiritualis plantae, quia aliter partes distantes nimium a radice, non bene vivificarentur ad formam speciei plantae." On the definition of veins as radical members: Albertus Magnus, De vegetabilibus VII tr. 1 c. 10 (ed. Meyer-Jessen, 623).

81 The *De vegetabilibus* argues that the second digestion occurs in the marrow, while the physiology of digestion of the *De nutrimento et nutrito* distinguishes three different places of internal digestion: Panarelli, "How do Plants Live and Grow?," 355.

82 Albertus Magnus, De anima II tr. 2 c. 7 (Ed. Colon. VII/1, 90.86–91.14). See above (chapter 3), 172 n. 172.

from the male alone. In the *De vegetabilibus*, we can find two notable examples of Albert's application of the concept of radical moisture: the first concerns the already mentioned case of "essential and integral parts" (*partes integrales essentiales*), while the second concerns fruits, which Albert dubs "accidentally essential" (*partes essentiales accidentales*) parts. This is because, while the former are essential for the survival of the living being, the latter are not, being only essential for the sake of the perpetuation of the species. The first case can include members such as the roots, the vein-like channels, the knots, the marrow and the bark, as opposed to other parts, such as the wooden parts between the bark and the vein-like channels that: as shown by Gila Wöllmer, these parts correspond to flesh, and therefore to a replaceable part.[83] The most explicit and intriguing example, though, concerns perhaps Albert's study of fruits: according to Albert, the inner humor, of seminal origin, represents the radical part of the fruit, while the second constitutes its fleshy, replaceable part.[84] This also applies if we think of fruits in view of their reproductive function: if their seed is sown, they can generate new plants, while the second part is not essential for the fruit (in this case a cotyledon) to generate new plants.

Having elucidated Albert's innovation in terms of the scope of the notion of radical moisture, attention can now be diverted to fleshing out Albert's approach to the incidence of spirit on length and shortness of life. Contextually, we shall try to highlight its role in the processes of growth, aging and death. To do so, it is necessary to resume to Albert's understanding of vital heat and radical moisture in the *De morte et vita*, relying on both the aforementioned studied by Crisciani and Panarelli, although neither of those studies has dealt with the connection among spirit, radical moisture, and Albert's theory of length and shortness of life and ages, our main goal here. In his paraphrastic reworking of Aristotle's *De longitudine et brevitate*, Albert upheld the (Aristotelian) view that length and shortness of life are determined by the quantity and quality of heat in moisture (*quantitas et qualitas calidi in humidum*) and vice-versa (*quantitas virtutis humidi in calido*). In addition to the two classic Aristotelian criteria, he adds the quality of members in which heat can operate (*qualitas membrorum in quibus operatur*) and (4) the disposition of the vessels through which

83 Wöllmer, "St. Albert and his Botany," 236.

84 Albertus Magnus, De vegetabilibus III tr. 1 c. 2 (ed. Meyer-Jessen, 169): "Duo autem dicti humores, unus videlicet seminum, et alter carnis circumpositae seminibus, sunt sicut duo humores in animalibus, materialis videlicet—qui nutrimentalis vocatur—et formalis—qui est radicalis, ex quo fit prima rei formatio. Sic etiam humor ex radice prodiens circumfunditur humido nutrimentali, qui virtutem generativam nullo modo habet."

moisture is concocted (*dispositio vasorum in quibus decoquitur*).[85] These criteria provided a rationale not only for understanding the Aristotelian ranking of longest- and shortest-lived beings in the *scala naturae* but also for explaining the bearing of non-teleological outcomes (deformities) of generation or of non-teleological processes (aging) on the longevity of the living being.

According to Albert, the capacity of moisture to preserve vital heat explains how plants are for the most part (*ut in pluribus*) the longest-lived among living beings, followed by blooded and bloodless animals.[86] This is because, despite their relatively small amount of vital heat, their moisture is so viscous, as in Aristotle, that it prevents an early loss and extinction of vital heat. These considerations on the longevity of animals are, as in Aristotelian psychology, true only for the most part. For instance, bees are longer-lived than some blooded animals because they have a better qualitative ratio of heat in moisture and moisture in heat.[87] As in Aristotle, the size of an animal is directly proportional to its vital heat and to a longer life only for the most part, since the amount of vital heat must be supported by a quantitatively and qualitatively amount of heat and moisture.

Faithful to the Aristotelian approach, Albert argues that men are essentially (*naturaliter*) longer-lived than women, but there are a number of causes, such as toil or frequent sexual intercourse, which make men accidentally shorter-lived than women.[88] Moreover, Albert adds to the Aristotelian examples of the male cock-sparrows (shorter-lived because of their frequent semen emission) and mules (longer-lived than their parents for a contrary reason) the examples of hens frequently laying eggs (*gallinae multum ovantes*) and women who conceive and give birth very often (*mulieres multum concipientes et parentes*).[89]

According to Albert, these criteria are conditioned by the so-called *non naturalia* (food and drink, surrounding air, sleep and waking, rest and exercise,

85 Albertus Magnus, De morte et vita tr. 2 c. 7, ed. Auguste Borgnet, Paris 1890 (Ed. Paris. IX, 362a).

86 Albertus Magnus, De morte et vita, tr. 2 c. 5, ed. Auguste Borgnet, Paris 1890 (Ed. Paris. IX, 358b–359a).

87 Albertus Magnus, De morte et vita tr. 2 c. 10 ed. Auguste Borgnet, Paris 1890 (Ed. Paris. IX, 367b).

88 Albertus Magnus, De morte et vita tr. 2 c. 8, ed. Auguste Borgnet, Paris 1890 (Ed. Paris. IX, 365a): "Naturaliter autem loquendo in hominibus mares sunt longioris vitae quam foeminae, nisi per accidens aliquod exercitii vel luxuriae et hujusmodi abbreviet eorum vitam."

89 Albertus Magnus, De morte et vita tr. 2 c. 8 ed. Auguste Borgnet, Paris 1890 (Ed. Paris. IX, 364b–365a). On this passage, see Tilke Nelis, "A Jumble of Writings: Commentaries on Aristotle's *De Longitudine et Brevitate Vitae* Attributed to Adam of Buckfield," *Early Science and Medicine* 27 (2022/1) 36–37.

excretions and retentions and mental affections), relevant also in Avicenna's medical approach to aging. Albert makes it clear, though, that the study of *non naturalia* for medical purposes exceeds the scope of natural philosophy. By contrast, sleep, climate and nutrition are actually studied by Albert, albeit with a philosophical and not a medical purpose.[90] To explain how length and shortness of life are conditioned by the material composition of the living being, Albert addresses systematically the corruption of the ratio (*proportio*) between active and passive qualities. By not discussing the incidence of the physiological state of the members on length of life, Albert implicitly assumes that the primary defining factors for length and shortness of life include a ratio (*proportio*) between active and passive qualities, while the latter two criteria are to be discussed—if anything—elsewhere or very synthetically. This is the case, for instance, of his doctrinal study of the *peccata naturae* presented in the *Physics*, which depends heavily on the ratio between heat and moisture in the formative power and accounts for deformities. In the end, Albert does present some remarks on the pathologies of old age in the *De iuventute et senectute*, but does not explicate any possible link between deformities and life expectancy. Perhaps, this would have been a more suitable matter for physicians to discuss.

Albert's handling of the topic proceeds according to what, from Aristotle onwards, were considered the essential causes of length and shortness of life, viz. heat and moisture. Both clearly linked to the notion of spirit, they are given an entire digression (*digressio declarans calidum et humidum secundum quod sunt principia vitae*). In this digression, Albert distinguishes the qualitative corruption of this ratio (*proportio*) into "with" or "without matter" (*cum materia/sine materia*), viz. extrinsic or intrinsic causes of this corruption. The starting point is represented by the decay of heat. A first corrupting agent may be represented by extrinsic heat, which is, according to Albert and medical literature, represented by fevers: in this case, non-lethal, putrid fevers attacking humors.[91] On the other hand, the intrinsic cause of the qualitative corruption of heat is represented by lack or excess in heat (*excedit in qualitate caloris vel*

90 Albertus Magnus, De morte et vita tr. 2 c. 6, ed. Auguste Borgnet, Paris 1890 (Ed. Paris. IX, 361b): "Causae autem horum omnium peccatorum in calido et humido sunt valde multiplices. Aliquando ex membris quidem corporis existentis: aliquando autem ex locis, vel temporibus, vel exercitiis, aut cibis et potibus, aut ex accidentibus animae quae sunt gaudium, et tristitia, et hujusmodi: sed de his in medicina intenditur, et non spectat ad scientiam universalem naturalem."

91 Albertus Magnus, De morte et vita tr. 2 c. 6, ed. Auguste Borgnet, Paris 1890 (Ed. Paris. IX, 361a): "[…] et fit calor immoderatus, qui consumit humidum, sicut est in febribus putridis."

in remissione), causing either the premature fading away of heat or the burning up of all the available moist substrate.[92]

Albert also studies the causes of the quantitative and qualitative corruption of the ratio between heat and moisture from the point of view of moisture, dividing again the causes for the corruption of this balance into intrinsic and extrinsic. The extrinsic (*sine materia*), quantitative cause of this corruption is represented by an excess of quantitative amount of moisture, which ends up suffocating the vital heat as excessive smokes quenches the fire of a lamp.[93] The intrinsic cause affecting the quality of moisture is represented by a lack of sufficient heat, unable to convert food into nutrimental moisture and avoiding the persistence of residues, as typical of the older ages, especially the *aetas decrepita*.[94] The extrinsic, quantitative cause of corruption of the proportion between heat and moisture can be caused by the mixture of other substances, which must be eliminated by the means of purgation (*oportet quod educatur per purgationem*).[95] Until now, Albert, has spoken very generally of 'moisture' in his account of longevity: to understand his account properly, one needs to understand what kind of moisture Albert is here referring to.

When discussing the essential features of length and shortness of life, Albert cites the Avicennian lamp metaphor. In this manner, he makes clear that the general kind of 'humidity' he has talked about so far when explaining Aristotle's text amounts to the Avicennian concepts of nutrimental and radical moisture. According to Albert, the key material cause that helps define the longevity in living beings is represented by the radical moisture, whose watering down and consumption appears closely linked to its interaction with nutrimental moisture. Albert adds, in keeping with his position in the *De homine*, that the first moisture cannot be restored through the second and third moistures (*humidum nutrimentale* and *humidum restaurativum*), thus maintaining

92 Albertus Magnus, De morte et vita tr. 2 c. 6, ed. Auguste Borgnet, Paris 1890 (Ed. Paris. IX, 361a).

93 Albertus Magnus, De morte et vita tr. 2 c. 6, ed. Auguste Borgnet, Paris 1890 (Ed. Paris. IX, 361a).

94 Albertus Magnus, De morte et vita tr. 2 c. 6, ed. Auguste Borgnet, Paris 1890 (Ed. Paris. IX, 361a).

95 Albertus Magnus, De morte et vita tr. 2 c. 7, ed. Auguste Borgnet, Paris 1890 (Ed. Paris. IX, 362a): "Quoniam hic aliquando est paucus, aliquando vero multus, aliquando autem grossus, aliquando subtilis ultra aequalitatem debitae complexionis. Et quando quidem est paucus, tunc debilitate abbreviat vitam, sicut est in melancholicis. Quando autem multus, tunc exsufflando exstinguit calorem, sicut ventus exstinguit lampadem. Quando autem grossus, tunc impedit operationes virtutum vitalium et naturalium, et praecipue animalium. Et quando est nimis exilis, tunc de facili exspirat, et mortem inducit."

the aporia of a fluid generated through nutrition but which, at the same time, is irreplaceable by it.[96] Radical moisture, whose loss can be partially compensated for by the intake of nutrimental moisture, represents the material substrate of inherence of heat during the entire lifespan of the individual. Vital heat will consume it, slowly but relentlessly, unto death. This is also the position, as highlighted by Chiara Crisciani and Giovanna Ferrari, presented by Peter of Spain in his comment to the *Parva Naturalia*.[97] Radical moisture, linked too by Peter to seminal moisture in his *Tractatus de longitudine*,[98] is key to defining the lifespan of living beings. To explain exactly how it is and what material substances are involved in the process of nutrition, growth, aging and death, Peter resorts, like Albert, to the Avicennian lamp metaphor.[99]

The same Avicennian lamp metaphor is also employed in the commentary by Peter of Ireland, even though he does not develop a theory of radical moisture such as that of Albert. As a matter of fact, in his commentary Peter speaks instead of substantial humidity (*humiditas substancialis*) rather than radical moisture itself.[100] The use of the lamp metaphor, though, appears to be quite common in authors working on the earliest versions of Aristotle's *De longitu-*

96 Albertus Magnus, De morte et vita tr. 2 c. 7, ed. Auguste Borgnet, Paris 1890 (Ed. Paris. IX, 363a). The lamp metaphor is quoted extensively in Albertus Magnus, De morte et vita tr. 2 c. 6, ed. Auguste Borgnet, Paris 1890 (Ed. Paris. IX, 360a–b): "Primum autem istorum humidorum est simile illi radicali humido in quo sicut in subjecto proprio haeret calor naturalis, qui dispersus est in membris radicalibus, ex calore seminali procreatus. Secundum autem simile est illi parti nutrimentalis humidi, quod jam assimilatum est in membris ad cibum imbibitum et attractum. Tertium est simile illi quod nondum plene assimilatum est, sed est adhuc transfusum membris in vasis nutrimento deservientibus, quod tamen per calorem digestivum in continuo tempore assimilatur." Note that Averroes' third section of the *Parva Naturalia* on length and shortness of life, does not present a theory of radical moisture but rather relies solely on Aristotelian elemental theory: Hansberger, "Length and Shortness of Life," 63–67.

97 Crisciani-Ferrari, "Introduzione," 374–375.

98 Peter of Spain, Tractatus de longitudine et brevitate I c. 2 (ed. Alonso 1952, 429.3); "humidum radicale ex primo spermate procedit."

99 The lamp metaphor is presented extensively in: Peter of Spain, Tractatus de longitudine et brevitate, I c. 2 (ed. Alonso 1952, 427–428).

100 Petrus de Hibernia, Expositio et quaestiones in Aristotelis librum de longitudine et brevitate vitae, ed. Michael Dunne (Philosophes médiévaux 30), Louvain-la-Neuve/Paris, 1993, 76.116–124: "Et quia inpossibile est consumpcionem huius humidi ex toto prohibere, propter hoc inpossibile est uita perpetuari; tamen eius consumpcionem retrdari conueniat per humidum nutrimentale, sicut retardatur consumpcio humidi substancialis licinii per humidum oleoginosum, tamen necessarium ad ultimum finitur. Quanto ergo humidum tale est uiscosum, tanto magis resisti consumpcioni, et quanto plus est in quantitate, tanto in longiori tempore consumitur, et quanto minus in in minori uel breuiori, et sic accidit quedam esse naturaliter brevi suite et quedam longioris."

dine et veritate to account for the progressive depletion of moisture derived from the semen over time and its relation to moisture derived from nutrition. The penetration of Avicennian medical learning, though, has different outcomes in the authors considered here.

If we go back to Albert's paraphrase of Aristotle's text, we can observe that a natural consequence of the great pains Albert takes to detail the reciprocal interaction between the active quality par excellence, viz. heat, and its substrate of inherence, moisture, leads naturally to a question about the role of spirit in longevity because of its nature as carrier of vital heat. Moreover, it leads to the very sensitive question—left open by Avicenna—about the relationship between spirit and radical moisture. In his *De morte et vita*, Albert presents an entire digression to the topic. In order to give a conclusive evaluation of the role of spirit as an accidental cause for length and shortness of life, though, it is necessary to contextualize the mechanical explanation for youth and old age, growth and decay in Albert's *De iuventute et senectute*.

In the *De iuventute et senectute*, Albert explicitly follows the classical medical division of ages into growing age (*aetas puerilis* or *augens*), mature age (*aetas stativa/standi/virilis*), old age (*senectus*) and senile age (*senium sive aetas decrepita*), which he mutuates from Avicenna's *Canon* (I, I, 3, 2) and *De animalibus* (XII, 3).[101] According to Albert, age (*aetas*) is "a period of lifetime during which the virtue of life can act in living beings—provided that no accidental cause intervene—according to the material conditions it meets at that stage of life (*spatium temporis in vivente relictum, in quo virtus vitae omnem perficit suae mensurae modum, nisi per accidens impediatur*)".[102]

According to the medical theory of complexions, the first age (*aetas augens*) is characterized by abundance of heat and (soft) moisture that, if solidified by the action of heat, helps bodily organs grow until their completion.[103] Following in Avicenna's footsteps, Albert argues that if the living being possesses a sufficient amount of heat and (tender) moisture, the growing faculty is able to shape the radical members until the beginning of the second age, where growth stops (*stat augmentum*).[104] The second age is named *aetas stativa* to mark the end of the process of growth, a process which, according to Albert concerns pri-

---

101 Albertus Magnus, De morte et vita tr. 2 c. 7, ed. Auguste Borgnet, Paris 1890 (Ed. Paris. IX, 363a).

102 Albertus Magnus, De aetate sive de iuventute et senectute tr. 1 c. 1, ed. Auguste Borgnet, Paris 1890 (Ed. Paris. IX, 306a).

103 Albertus Magnus, De nutrimento et nutrito tr. 2 c. 1 (Ed. Colon. VII/2a, 15.13–15).

104 Albertus Magnus, De aetate sive de iuventute et senectute tr. 1 c. 3, ed. Auguste Borgnet, Paris 1890 (Ed. Paris. IX, 309b): "Haec autem aetas secundum omnem sui differentiam convenit virtuti augmentativae, secundum scilicet quod procedit in tempore, ita debilitatur

marily the development of the members according to length (*longitudo*) and not width (*latitudo*) and breadth (*spissitudo*), closely following to Avicenna's position.

Since radical members grow primarily according to their *longitudo*, and growth is primarily (*precipue*) associated with length, growth appears to be primarily associated with radical members. Radical members do not exhaust the list of growing body members, though, since homeomerous, non-radical members such as flesh and fatty tissue (*caro et adeps*) are involved in the process of growth beyond the first age. It is an empirical datum that people fatten up and slim down over the course of years and not necessarily during their first 30 years only. For this reason, the size of their flesh and fatty tissues can vary over the course of life, whereas the physical shape of radical members such as the heart is life-long (*manentia per vitam*).[105] For this reason, growth is a kind of nutrition which shares the same physiological process, but with the different purpose of either perfecting the radical members or increasing the size of the homeomerous members: if the former function lasts only for the length of the first age, the second persists for the whole of life: this distinction between the two kinds of growth, inspired by Avicenna, appears already well-rooted in the *De homine*.[106]

If we now consider spirit (more specifically, *spiritus naturalis*) in connection with growth, we can see how its action as a vehicle of the growing faculty acquires two purposes, precisely as the faculty of growth: on the one hand, it enables the spread and transmission of heat, causing the formation of radical members. On the other hand, its action as carrier of food helps it produce other,

---

augmentum et minoratur. Causa autem hujus est, quoniam molle et humidum facilius extenditur quam siccius et durius."

105 Albertus Magnus, De nutrimento et nutrito tr. 2 c. 1 (Ed. Colon. VII/2a, 14.29–31, 14.38–50, 14.55–58): "[…] non omnis additio quanti facit augmentum prout augmentum est motus ad perfectam quantitatem animatorum. […] Augmentum autem animatorum precipue secundum longitudinem determinatur, eo quo dilla fit in partibus radicalibus proportionaliter. Augmentum autem per latitudinem et spissitudinem fit per totam vitam animatorum. Et fit magis in supplementis, quae sunt precipue caro et adeps, eit illa sunt partes maxime in animatis secundum materiam determinate influentes et effluentes continue. Radicales autem magis sunt in ossibus et nervis et arteriis et chordis et ligaturis et cor et hepate et pulmone et cerebro et testiculis et aliis huiusmodi partibus, in quibus et per quae anima exercet vitae operationes sicut per organa. Horum autem augmentum secundum longitudinem determinatur […] Augmenta autem quae ista recipiunt proportionaliter sunt et manentia per vitam, licet non continue augeantur, sed tempore iuventutis." By *iuventus* Albert means the fifth stage of the *aetas augens*: Albertus Magnus, De aetate sive de iuventute et senectute tr. 1 c. 3 ed. Auguste Borgnet, Paris 1890 (Ed. Paris. IX, 309a).

106 Albertus Magnus, De homine (Ed. Colon. XXVII/2, 124.51–62).

supplementary (*supplementi*) members such as flesh and fatty tissues. While in the formation of radical members is for the well-being of the plant, animal or human, the case of excessive growth in fleshy or fatty tissues can, by contrast, produce a physiologically problematic situation which is not beneficial neither for the plant's being nor well-being. In his *De iuventute et senectute*, Albert presents a list of ages which is, on the one hand, clearly inspired by Avicenna male-oriented classification of ages and, on the other hand, strictly compares the ages of man to ages in plants:

Humans	Plants
Infantia (0–1 years)	Age during which plants are extremely soft and tender (*primo sunt valde molle et tenerrima*);
Plantativa (1–7): growing of teeth;	Growth of small branches, leaves and roots (*plantantur ramusculi et folia plura et radices*)
Aetas concussiva (7–12), stronger action of heat over moisture (viz. in the growth of organs);	Stronger action of heat over moisture (further formation of organic parts);
Age of puberty (12–16);	Hardening and oozing out of superfluous moisture (*durescere incipiunt resudantes humidum superfluum*);
Age of beard growing and semen ripening (*iuventus sive crinita*, 16–24).	Ultimate completion (*ad statum perveniunt suae quantitatis*).

After presenting and explaining this fivefold division of the first age,[107] Albert takes on the study of second age (*aetas stativa*). To do so, he devotes a large part of his chapter on the relationship between the heat of the *aetas augens* and the *aetas stativa*. The problem, vivaciously debated in scholastic literature, concerned primarily the quantity and quality of heat in the two different ages and their relation to possible signs indicating differences in complexion, such as fevers and digestive processes. When discussing this controversy among the different groups of physicians (heat is more in the first or in the second age),

107 All ages listed above in: Albertus Magnus, De aetate sive de iuventute et senectute tr. 1 c. 3, ed. Auguste Borgnet, Paris 1890 (Ed. Paris. IX, 307b–308b).

Albert reports faithfully the arguments presented by Avicenna, carefully distinguishing between the theoretical rationale (*rationes*) and empirical evidence (*signa*) adduced by each sect to argue that heat is greater in the first or in the second age. Albert's main difference with the Avicennian account is that Albert reverses the order of the sects: the first one is that which argues that heat is greater in the second age, the second is that arguing that it is greater in the first.

An additional and remarkable feature of Albert's account is that it paraphrastically expands on Avicenna to give a clearer understanding of both the theoretical and the empirical background of the thesis. Unlike Avicenna, he does not identify the adherents of these sects as mere "physicians" (*medici*), but also as "ancients" (*antiqui*) and "Peripatetics" (*Peripatetici*). To be clear, the identification of such philosophical authorities—if any, besides those implied in Avicenna's account—advocating for the two opposite theses will need a proper ascertainment achievable only through a critical edition. What may be interesting to note here, though, is that a typically medical problem, viz. the relationship between the complexion of ages, becomes in Albert's eyes a problem of natural philosophy. Those who discussed it can be, for this reason, identified with the term *Peripatetici*.

In the case of the second sect listed by Avicenna, Albert clarifies how the rationale behind their position is rooted in the idea that heat never reaches the same degree of temperature in simple, moist substances as it does in drier ones (*calidum in simplicibus non est adeo intensum cum humido sicut cum sicco*). Then, Albert goes on to faithfully report the list of empirical observations (children have colder fevers than mature youngsters, etc.) upon which they based their claims.[108] In this manner, the argumentation of this sect of physicians becomes theoretically clearer and more articulated. In a symmetrical manner, Albert reports the rationale behind the idea of the first sect that the amount of heat is major in the first age because of a chronological proximity to the moment of birth. By the same token, they argue, all of the following ages must have a smaller amount of heat than the first one.[109] Having reported their arguments (longer digestions, more frequent appetite), Albert goes on to briefly assess the role of digestion in children or *istimbrae*, a nearly *verbatim* transliteration of the Arabic *istimrāʾ*.

108 Albertus Magnus, De aetate sive de iuventute et senectute, tr. 1 c. 4, ed. Auguste Borgnet, Paris 1890 (Ed. Paris. IX, 311b).

109 Albertus Magnus, De aetate sive de iuventute et senectute, tr. 1 c. 4, ed. Auguste Borgnet, Paris 1890 (Ed. Paris. IX, 311b–312a).

Those phenomena that are defined as quick appetites with a good digestion are not to be causally traced back to a colder complexion in children but, Albert argues, to the presence of superfluous humors and food (*propter superfluitatem humoris et nutrimenti*).[110] As was claimed in the first chapter, the mention of *istimbrae* appears only in the *Canon* and not in the *De animalibus*. For this reason, it can be assumed that Albert was drawing at least on both works in his *De iuventute et senectute*.[111]

The solution to this medico-philosophical dilemma is, in the end, based on the Galenic solution transmitted by Avicenna in his *Canon* and *De animalibus*. For Albert, heat is essentially the same, but its quantity is stronger in the first age and its quality in the second age. Albert then reports the Galenic example of the object endowed with the same essential heat which, induced into water or moisture, produces a quantitatively stronger heat (so a lower temperature), while, when induced into a stone or a blade, produces a qualitatively stronger heat (higher temperature).[112] In Albert's view, Galen's solution is fundamentally right (*vera est valde*), although it presents a difficulty: how is it possible that heat in children does not evaporate at all (*nihil de calore expiraverit*)? Is it therefore possible that heat does not undergo any quantitative or qualitative change from the first to the second age? Albert argues that if one considers Galen's solution accurately (*si quis eam subtiliter considerat*), even though it appears *prima facie* as partially false (*aliquid falsitatis continere*), there is a solution at least to some extent to rectify Galen's position. As Galen himself had claimed, if the persistence of the same essential amount of heat is granted by the capacity of the body to carry out a perfect assimilation of food to body, then

110 Albertus Magnus, De aetate sive de iuventute et senectute tr. 1 c. 4 ed. Auguste Borgnet, Paris 1890 (Ed. Paris. IX, 312a).

111 As stated previously, this claim could be better substantiated or, by contrast, revised, upon the appearance of a critical edition of Avicenna's *De animalibus* (in preparation by Aafke van Oppenraaij) and of Albert's *De aetate sive de iuventute et senectute* (in preparation by Silvia Donati). See above, chapter 1, 29 n. 60.

112 Albertus Magnus, De aetate sive de iuventute et senectute tr. 1 c. 4 ed. Auguste Borgnet, Paris 1890 (Ed. Paris. IX, 312a–b): "Post ambas autem istas sectas venit Galenus, et dixit quod istae aetates sunt aequaliter caldae: sed prima aetas excedit in quantitate caloris, et secunda in qualitate [...] Et hujus exemplum esse dicit, quod si nos imaginamur aliquid calidum vel corpus aliquod calidum et subtile primo ingredi in corpus humidum sicut in aquam: et postea tantum quantum quod est illud, ingredi in corpus siccum spissum minus in quantitate quam sit humidum, sicut in lapidem vel ferrum: tunc enim non est differentia in essentia corporis calidi. Hoc est unum et idem in utroque: et tamen differentia est in qualitate calidi, quia acutius est in sicco quam in humido: et differentia est in quantitate, quia magis dilatatur calor in humidum quam in sicco [...]."

the heat taken in through nutrition can replace all the heat eventually lost during the first age.[113]

Having settled this quarrel between Galen and the two sects of physicians following Avicenna, Albert moves then on to analyze the third age (*senectus* or *diminutiva virtutis*), a time in the life of humans during which failures in the motive capacities or signs of aging (skin wrinkling, hunching of the back) begin to manifest.[114] To explain how the *diminutio* concerns only the embodied virtues of the soul and not the substance of the body, Albert compares the process of aging to a lamp giving off vapor. The fact alone that the lamp gives off vapor does not necessarily imply that this lamp is using up its substance. Rather, it is when the lamp gives off too much vapor that one might start to assume a loss in substance.[115] The deficiency *in virtute* is caused by a drier complexion, implying the loss of moisture, the material substrate of spirit (*materia spiritus*).[116]

---

113 Albertus Magnus, De aetate sive de iuventute et senectute tr. 1 c. 5, ed. Auguste Borgnet, Paris 1890 (Ed. Paris. IX, 312b): "Ista est sententia Galeni: et si quis eam subtiliter considerat, vera est valde, licet secundum superficiem videatur aliquid falsitatis continere. Videtur enim secundum superficiem omnino impossibile esse, quod in corporibus tenellis in prima aetate nihil de calore exspiraverit: et cum nihil sit de calore in secunda aetate nisi quod transmissum est ab aetate prima, videtur calor etiam in essentia esse diminutus in secunda aetate, et major esse in prima. Sed Galenus reliquit, quod secundum digestionis bonitatem primae aetatis digestio quae vocatur epsesis, perfecte assimilat nutrimentum ei quod nutritur [...]."

114 Albertus Magnus, De aetate sive de iuventute et senectute tr. 1 c. 5, ed. Auguste Borgnet, Paris 1890 (Ed. Paris. IX, 312a): "Sed cum nervi non adhuc contrahantur, neque corrugetur pellis, neque incurventur spondiles dorsi, neque aliquod aliorum membrorum diminuatur vel curvetur, dicitur non deficere substantia."

115 Albertus Magnus, De aetate sive de iuventute et senectute tr. 1 c. 5 ed. Auguste Borgnet, Paris 1890 (Ed. Paris. IX, 313a–b): "Cum autem dicitur quod est sine defectu substantiae, debet intelligi quod est sine defectu substantiae membrorum quia nec in qualitate, nec in quantitate sentitur defectus membrorum similium, sive etiam heterogeneorum: sed non potest esse sine diminutione substantiae humiditatis naturalis: quia si illa maneret, maneret etiam calidum: quia hoc est proprium subiectum caloris naturalis, et quoad illam in aliquo deficit substantia. [...] Contingit autem hoc propterea quod calor antecedentium aetatum optime restituit deperdita, quae adhuc manent, ita quod non potest deprehendi in membris similibus si quid est diminutum: virtus autem remissior est, ut diximus: et hoc percipitur propter exsiccationem frigidam in humido radicali: huius enim diminutio non statim sentitur in membris, licet subtrahitur in seipsa. Cuius exemplum est, quod oleum in lampade et imbibitum in licinio, et videtur evaporare de licinio: non tamen sentitur defectus eius in lampade, set tunc primum sentitur, cum multum fuerit evaporatum per ignem."

116 Albertus Magnus, De morte et vita tr. 2 c. 4, ed. Auguste Borgnet, Paris 1890 (Ed. Paris. IX, 357a–b): "Humidum autem cum sit materia spiritus in animatis quae sunt sensibilia, excellit post calorem."

This decline in the amount of heat and moisture becomes particularly evident in the last, senile age (*aetas senium sive decrepita*),[117] which is linked to a deficiency in both virtue and substance. Compared to the previous age, senility is drier on the inside (*interius siccior*) and moister on the outside because of the presence of unnatural, corrupted moistures and humors produced by failures in the digestive process.[118] The natural decay of the complexion toward coldness and dryness affects directly the activities carried out by spirit, especially the psychic, mental activities. Albert argues that the spirit's thickening induces older people to act childishly (*incipiunt tandem puerilia amare et facere*). Worse, by being unable to pass on to the middle ventricle of the brain (*non penetrant ad cellam logisticam*), their exercise of reason is inevitably flawed, making them subject to delusion, like a lyre out of tune (*sunt deliri, eo quod distemperata est harmonia virtutum animalium eorum, sicut lyra dissona*). Also typical of senile age is forgetfulness caused by excessive dryness, usually associated with the capacity of retention and thus with memory.[119] Albert divides this last, fourth age into three parts: a first part, colder and dryer, closer in complexion to *senectus* or *diminutiva virtutis*, a middle part characterized by the abundance of moist fluids such as catharrs in the lungs and a last part where body parts

117 Albertus Magnus, De aetate sive de iuventute et senectute, tr. 1 c. 5 ed. Auguste Borgnet, Paris 1890 (Ed. Paris. IX, 313a).

118 Albertus Magnus, De aetate sive de iuventute et senectute tr. 1 c. 6, ed. Auguste Borgnet, Paris 1890 (Ed. Paris. IX, 314b); Albertus Magnus, De nutrimento et nutrito tr. 1 c. 5 (Ed. Colon. VII/2a, 11.57–61): "Econtra in secundum naturam corruptis, sicut in senibus et in decrepitiis, propter debile calidum et siccatum humidum radicale quod erat in eis parum vel nihil de cibo sumpto adducitur ad membra."

119 Albertus Magnus, De aetate sive de iuventute et senectute tr. 1 c. 6, ed. Auguste Borgnet, Paris 1890 (Ed. Paris. IX, 314b): "Cum autem comparata fuerit haec aetas ad senectutem praecedentem, tunc est interius siccior, et exterius humidior in membris humiditate innaturali: et haec est causa, quod multas corruptas humiditates ejiciunt senes, et sunt multae egestionis et multe comestionis: et egestio multa eorum causatur ex multa comestione eorum indigesta. Comestio autem eorum multa causatur ex frigido et pondere cibi, quod semper tendit ad fundum per portanarium stomachi ad secessum, sicut de melancholicis diximus in libro *de Somno et Vigilia*: et sunt eorum egestiones quasi fluxus lienterici, et sunt malarum operationum animalium, ita quod incipiunt tandem puerilia amare et facere quod contingit propter spirituum animalium grossitiem et pigritiam, qui non penetrant ad cellam logisticam, sed remanent in interiori capitis cum defectu: et ideo non prohibetur apprehensio phantastica in eis per actum rationis: et tunc sunt deliri, eo quod distemperata est harmonia virtutum animalium eorum, sicut lyra dissona. Sunt etiam tunc memoriae eorum malae ad recipiendum, propter frigus et interiorem siccitatem. Haec autem, sicut diximus, in homine magis apparent: sed imaginari nos oportet, quod proportionaliter sint in omnibus aliis vivis. In omni enim eo in quo agit calidum et humidum, tandem necesse est induci talem exitum vitae et induci corruptionem."

are stiff in with *rigor mortis*. This stiffness in the member can be found, Albert argues, also in the terminal stage of life in plants as an effect of the lack of assimilation of food and the passing of time.[120] For Albert, death is not something tragic, but, as remarked by Crisciani and Ferrari as well, occurs when the members are almost numbed by the passing of time.[121] At the end of the study of Albert's process of life and death, youth and old age, there are two key questions which emerge. The first concerns the causal relationship between spirit and longevity, while the second concerns the relationship between spirit and radical moisture in the case of cognitive diseases. Let us begin with the first question.

As stated above, in the *De morte et vita*, Albert devotes an entire section of his own (*digressio*) to the reason why, in keeping with Aristotle (also, Avicenna) spirit is not an essential cause in length and shortness of life (*digressio declarans, quod spiritus non est causa longioris vitae per se*). This is because the essential causes of relative length of life are heat and moisture, out of which spirit is generated (*spiritus materialiter generatur ex humido, et active generatur ex calido naturali*).[122] Spirit, on the other hand, can only be regarded as an accidental cause (*per accidens*) of longevity or shortness of life inasmuch as it consumes radical moisture *qua* carrier of vital heat (*desiccativum est humidi radicalis*).[123] This happens through the diverse activities carried out by the spir-

120 Albertus Magnus, De nutrimento et nutrito tr. 1 c. 5 (Ed. Colon. VII/2a, 12.8–12).

121 Albertus Magnus, De morte et vita tr. 2 c. 7, ed. Auguste Borgnet, Paris 1890 (Ed. Paris. IX, 363a): "Haec autem quae dicta sunt probant, quod primum humidum radicale numquam accipit restaurum per omnia simile, et praecipue in senibus. Et ideo succedunt sibi aetates in his qui vivunt et sic post calidum humidum, calidum siccum, et post hoc frigidum siccum, et tandem intus siccum et extra alienam humiditatem in senio putrescente. Haec etiam est causa, quare non amara est mors senum, cum calidum pungens in eis sit quasi nullum: et quod est, est ab alieno humido suffocatum, et ideo impotens ad pungendum, et membra sunt jam quasi putrida et insensibilia." Crisciani-Ferrari, "Introduzione," 389.

122 Albertus Magnus, De morte et vita tr. 2 c. 7, ed. Auguste Borgnet, Paris 1890 (Ed. Paris. IX, 362a–b): "Secundum Aristotelem subtilem intellectum, spiritus nequaquam est principium vitae, neque etiam longioris aut brevioris vitae causa, nisi per accidens: cujus probatio est quod supra diximus, quod spiritus materialiter generatur ex humido, et active generatur ex calido naturali. Et ideo paucitas et multitudo spiritus, similiter autem et caeterae dispositiones ejus sunt causatae aut ex calido, aut ex humido, aut ex utroque. Et ideo causa prima et per se vitae in calido est et humido, et non in spiritu."

123 Albertus Magnus De nutrimento et nutrito tr. 1 c. 4 (Ed. Colon. VII/2a, 9.57–59): "Calor enim qui agit et spiritus in omnibus animatis perforativa sunt partium animati et exsiccativa humidi radicalis quod est in eis [...]"; Albertus Magnus, De aetate sive de iuventute et senectute, ed. Auguste Borgnet, Paris 1890 (Ed. Paris. IX, 309a): "[...] omnia viva ex humido terrestri permixta habent generationem per calidum, quod est humidi radicalis formativum et terminativum et exsiccativum."

its: as a vehicle of the nutritive faculty, it helps nutrimental moisture water down radical moisture. As substrate of the inherence of the heat responsible for generation, sensation and motion, all influential factors in the determination of lifespan, it contributes to the dessication of radical moisture. From this perspective, spirit represents an accidental cause for length/shortness of life, but in a way which seems to be much more connected to animals and humans. It is not quite clear, by contrast, how it could do so in plants. Among the earliest commentators on Aristotle's *De longitudine et brevitate*, Albert is the only author who explicates the role of spirit as an accidental cause in the determination of longevity, since it dries out radical moisture. By contrast, Tilke Nelis has demonstrated that contemporary authors such as Adam of Buckfield—who does not adopt the theory of radical moisture—and Peter of Ireland claimed that vital heat could affect the lifespan of living beings in the case of excessive toil and sexual intercourse.[124] This is also the case with Peter of Spain's *Tractatus de longitudine*.[125] The peculiarity of Peter, though, is that, like Albert, he tries to make sense of the connection between spirit, radical moisture and longevity. He does not go as far as Albert to say that spirit dehydrates radical moisture. Rather, he considers spirit to be an accidental cause of length of life in a different manner, arguing that the strength (*fortitudo*), abundance (*multitudo*) and natural disposition of spirit is causally correlated to a long life, the opposite to a short one.[126]

Coming to the second question, viz. the relationship between spirit and radical moisture over time, it is crucial to recall, as in Avicenna's case, the anatomical scheme for the production of animal spirits. Albert, as Avicenna, maintains the correlation between the lapsing of time and pathologies, i.e. thickening of spirits and cognitive impairments. According to Albert, though, dementia and other pathologies of old age such as loss of memory have to be understood analogically in plants and animals.[127] Clearly, this claim must not be considered in terms of clinical analogy with human pathologies, for plants do not suffer of

124 Nelis, "A Jumble of Writings," 33–39.

125 On the bearing of sex, toil, motion and sensation on life expectancy, see: Peter of Spain, Tractatus de longitudine III c. 2 (ed. Alonso 1952, 467.3–8).

126 Peter of Spain, Tractatus de longitudine III c. 2 (ed. Alonso 1952, 466.19–20): "Spirituum multitudo, fortitudo, ac dispositio naturalis ad longam operantur vitam, contraria ad brevem."

127 Albertus Magnus, De aetate sive de iuventute et senectute tr. 1 c. 6, ed. Auguste Borgnet, Paris 1890 (Ed. Paris. IX, 314b): "Haec autem, sicut diximus, in homine magis apparent: sed imaginari nos oportet, quod proportionaliter sint in omnibus aliis vivis. In omni enim eo in quo agit calidum et humidum, tandem necesse est induci talem exitum vitae et induci corruptionem."

maladies affecting their (inexistent) mnemonic capacities. Rather, these claims must be understood in biological terms of analogous biological decay, viz. of spirit thickening over time. In turn, the thickening of spirit makes it impossible for it to flow through a body, be it that of a plant, of an animal or of a human. This thickening of spirit either blocks or impairs the exercise of the soul faculties involved by cutting out or slowing down the flow of spirits to the organs of certain faculties. If all living beings have a radical moisture and a spirit, if this radical moisture fades away over time and, likewise, spirits tend to thicken during the fourth age causing dementia and other pathologies, there is an apparently implicit correlation between loss in radical moisture, spirit thickening and pathologies of old age. The overly thick complexion of spirits is, according to Albert, also the cause for diminished cognitive capacities in the other age that is characterized by abundant moisture in the complexion, viz. childhood. This in keeping with the findings of Stefano Perfetti, who has higlighted that Albert, following Aristotle's remarks, compares childhood *qua* age with diminished cognitive capacities to the cognitive capacities of non-rational animals.[128] This is a position which finds clear parallels in the *Tractatus de longitudine* by Peter of Spain, who correlates loss in radical moisture, digestive capacities, thickening of spirits and defects in cognitive capacities.[129]

Still, as in Avicenna, while it clear that the activities of the soul's faculties are weakened over time—in this case, in all living beings and not just humans—there is a problem in determining the causal connection between radical/nutrimental moisture and spirit. This correlation, absent in the *De iuventute et senectute*, was made explicit in a passage of Albert's *De spiritu et respiratione*, where spirit is said to be produced from radical and nutrimental moisture.[130] While the determination of the precise localization of radical moisture is, as we have shown, more difficult, the connection between spirit and nutrimental moisture seems to offer a possible hint toward a better understanding of the process of aging and its pathologies. If we assume, as Albert did in his *De animalibus*, the Avicennian scheme of the relationship between natural, vital and animal spirits, then failures in the digestion of nutrimental moisture can be viewed as the material cause impacting the complexion of vital and animal spirits. The final result would be a complexion of animal spirits which would make them

128 Stefano Perfetti, *Nature imperfette. Umano, subumano ed animale nel pensiero di Alberto Magno. Con un saggio di Amalia Cerrito* (Philosophica, serie viola 226), Pisa 2020, 30–35.

129 Peter of Spain, Tractatus de longitudine I c. 2 (ed. Alonso 1952, 430–431).

130 Albertus Magnus, De spiritu et respiratione I, tr. 1, c. 9, ed. Auguste Borgnet, Paris 1894 (Ed. Paris. IX, 226b): "Licet origo ejus sit ex humido radicali et cibali tamen generatio eius et essentia vitalis spiritus est in corde."

unsuitable to flow through the human brain (causing dementia) or the body of animals and plants. Moreover, the thickening of spirit over time can be related to the concept of radical moisture. If, as Albert argues in the *De animalibus*, radical moisture has a gluing, cohesive power on the radical member,[131] its loss would trigger a loss in the concoctive and digestive power of organs like the stomach and liver. The more nutrimental moisture outweighs radical moisture in a living being, the weaker its digestive capacities and, in the end, the thicker the material composition of its animal spirit.

In this respect, Albert's theory of the relationship between radical moisture and spirit has some flaws, first and foremost due the fact that radical moisture is, as noted by Chiara Crisciani and Giovanna Ferrari, more of a heuristic principle to define the lifespan of living beings than a consistent medico-philosophical notion.[132] To his credit, it must be said that Albert is the first author to have made explicit and brought into discussion, with all its possible limits, the link between spirit and radical moisture implicit in Avicenna's fragmentary doctrine of radical moisture. Still, the relationship between the two concepts, except for an intuitive yet implicit correlation between decrease in radical moisture, aging, spirit thickening and pathologies of old age, remains difficult to ascertain.[133]

3.3 *What Is Death for? Albert between Natural Philosophy and Theological Eschatology*

The workings of spirit within generation offer an explanation of natural phenomena in terms of material, efficient causality but not of final causality. As we stressed in the introduction, it is hardly a coincidence that Albert opens his *De morte et vita* with a long digression on life as an act of the soul which, in the case of humans, subsists after the soul's post-mortem severance from the body. Dying, therefore, acquires a twofold dimension: from the point of view of material necessity and laws of natural philosophy, it is just the necessary outcome of the reciprocal interaction between active and passive qualities which tend to destroy each other. From the point of view of the soul, though, death is not really the end of the journey for humans.

131 Crisciani, "Aspetti del dibattito sull'umido radicale," 350–352.

132 On these doctrinal tensions within the concept of radical moisture itself: Crisciani-Ferrari, "Introduzione," 340–344.

133 Another recent and remarkable contribution on Albert's theory of radical moisture is: Gabriella Zuccolin, "L'umidità come principio di vita. Trasformazioni di un tema classico tra medicina e filosofia," in: *Fonti, flussi, onde: l'acqua tra realtà e metafora nel pensiero antico, medievale e moderno*, ed. Massimiliano Lenzi/Olga L. Lizzini/Pina Totaro/Luisa Valente (Micrologus Library 111), Firenze 2022, 27–52.

According to Albert's typically Neoplatonic psychology, death represented the principle of the soul's return (*reductio*) to its Creator, God. In his early works, Albert had built up a system of natural teleology ruled by God, conceived as both first efficient and final cause of every living being (*primum ut origo/primum ut finis*). For this reason, life was understood as a procession of every living being from the First Cause (*vita est processio procedens ex ente primo quieto, semper aeterno, et primus motus*).[134] In the case of the human rational soul, though, this procession had its aim in the return to God and the Beatific Vision at Judgement Day.[135] As noted by Krause, this process of regression is grounded by Albert in the substantial union of the rational soul with the human body, which acquires a disposition for the Resurrection when the God-created soul enters into the body after complete embryonic formation.[136] Further, man *qua* body-soul union returns to God not only in his rational part but—unlike in Avicenna's eschatology—in his psychosomatic integrity to enjoy the Beatific Vision and affective union with God.[137] Death in Albert's early theology is interpreted in a manner that is very similar, but not identical to—especially concerning the survival of the lowest psychic faculties—the psychology of Avicenna, for it relied—as shown by the studies of Anzulewicz—on the *corpus Dyonisiacum* and on the *Liber de causis* rather than on Avicenna.[138] Albert's insistence on the regress to God as a condition of the survival of the soul in the aftermath of death is typical of Albert's first theological production but is also consistent with his mature views, in particular with his *De causis et processu universitatis*. Here, Albert emphasizes that the reductio to God *qua* ultimate goal of our existence (also, happiness) is a constitutive and essential part of the human being.[139]

---

134 Anzulewicz, "Zur Theorie des Menschlichen Lebens," 38.

135 Anzulewicz, "Zur Theorie des Menschlichen Lebens," 38–39; Krause, "Albert the Great on Animal and Human Origin," 212–213.

136 Krause, "Albert the Great on Animal and Human Origin," 213.

137 Krause, "Albert the Great on Animal and Human Origin," 212–213.

138 Henryk Anzulewicz, "Die Denkstruktur des Albertus Magnus. Ihre Dekodierung und ihre Relevanz für die Begrifflichkeit und Terminologie," in *L'élaboration du vocabulaire philosophique au Moyen Âge. Actes du Colloque international de Louvain-la-Neuve et Leuven 12–14 septembre 1998 organisé par la Société Internationale pour l'Étude de la Philosophie Médiévale*, ed. Jacqueline Hamesse/Carlos Steel (Rencontres de Philosophie Médiévale 8), Turnhout 2000, 369–396.

139 Albertus Magnus, De causis et processu universitatis a prima causa I tr. 1 c. 11 (Ed. Colon. XVII/2, 24.91–25.8): "Secundum exitum ergo rerum in esse primum erit principium ut artifex, secundum autem reductionem rerum ad perfectum primum erit principium ut finis. Propter quod a quibusdam theologis non inconvenienter ut principium et finis laudantur. Omnia enim ab ipso sunt et propter ipsum et ad ipsum. Ab ipso enim non

The metaphysical and psychological perspective of the Aristotelian paraphrases, which often display a strong distinction in principles and methodology between philosophy and theology, is compatible with a philosophical justification of the rational soul's continued existence after the death of the body. Albert's position throughout these works seems to undergo a change from an initial skepticism to a subsequent optimism. In his first commentary on the *Nicomachean Ethics*, *Super Ethica*, he is dubious of a possible philosophical explanation of the soul's post-mortem state.[140] This initial skepticism is overcome in his later *De anima*, the *De natura et origine animae* and in the second commentary on the *Nicomachean Ethics*, *Ethica*. In the first two works, especially *De natura et origine animae*, Albert discusses at least three relevant topics: (1) the modality of knowledge proper to the human soul after its survival of the body and in comparison to intellectual activity during life, (2) the relation of the soul to previously acquired knowledge, and (3) the knowledge of separate substances.[141] This position does not mirror the position of the *Super Ethica*, because in these later works Albert does provide a study of the post-mortem knowledge of the rational soul from a philosophical perspective. In *Ethica*, for instance, Albert emphasizes, in reliance on al-Farābī's conception of acquired intellect (*intellectus adeptus*/*ʿaql mustafād*) and other philosophical doctrines, that intellectual perfection lays the ground for post-mortem intellectual happiness.[142] This view is consistent with Albert's *De natura et origine animae*, where intellectual perfection and happiness in this life (*beatitudo quae est viae*) are taken to lay the grounds for post-mortem happiness (but also

propter aliud sunt quam propter seipsum, eo quod extra seipsum nihil habet quod consequi intendat vel consequi intendere possit per opus." On Albert's position, see: Henryk Anzulewicz–Caterina Rigo, "*Reductio ad esse divinum*. Zur Vollendung des Menschen nach Albertus Magnus," in *Ende und Vollendung. Eschatologische Perspektiven im Mittelalter*, ed. Jan A. Artsen/Martin Pickavé (Miscellanea Medievalia 29), Berlin 2002, 389.

140 Albertus Magnus, Super Ethica I, lect. 13 (Ed. Colon. XIV/1, 71.73–79): "Dicendum, quod hoc quod animae defunctorum remaneant post mortem, non potest per philosophiam sufficienter sciri. Et suppositio, quod remaneant, de statu earum et qualiter se habeant ad ea quae circa nos fiunt, omnino nihil sciri per philosophiam potest, sed haec cognoscuntur altiori lumine infuso non naturali, quod est habitus fidei."

141 For a thorough discussions of the passages from the *De anima* and *De natura et origine animae*, see: Henryk Anzulewicz–Katja Krause, "Albert der Große und sein holistisches Konzept menschlicher Erkenntnis," in Veritas et subtilitas. *Truth and Subtlety in the History of Philosophy. Essays in Memory of Burkhard Mojsisch (1944–2015)*, ed. Tengiz Iremadze/Udo Reinhold Jeck (Bochumer Studien zur Philosophie 59), Amsterdam 2018, 157–194.

142 More on this in: Anzulewicz–Rigo, "*Reductio ad esse divinum*," 411–412.

for unhappiness).[143] In short, Albert's mature doctrinal position discusses the post-mortem state of the rational soul, justifying its survival and post-mortem knowledge from a philosophical point of view, but his position over the years developed changed from an explicit explanation provided in his early theology, to a more skeptical stance in the first commentary on the *Ethics* and, finally, into a more optimistic one.[144] In none of these cases, though, does death take on the role of a pointless phenomenon devoid of any teleology. Indeed, it is a necessary event due to the natural course of life and incidentally caused by spirit through its dehydration of radical moisture. On the other hand, though, death reflects the teleological order of the cosmos, bringing back man to his Creator.

Conclusions

Albert's psychophysiology of longevity builds upon a traditional concept, viz. radical moisture, mostly employed in theological discussions on the *veritas humanae naturae* and the prelapsarian state at the time of his *De homine*. In his early career days, Albert started to develop a detailed naturalistic and physiological analysis of the process of longevity, which he would later expand on in his *De morte et vita*, his paraphrase of the Aristotelian *De longitudine et brevitate*, and based on Avicenna's medico-biological theories. Albert made use of the concept of radical moisture to account for the lifespan of living beings and to differentiate between replaceable and irreplaceable members in a manner that finds few parallels among his contemporaries, especially in his botany. He can only be compared to Peter of Spain's work on the *Parva Naturalia* and his *Tractatus de longitudine*. Later on, Albert would employ the concept of radical moisture to explain the psychophysiology of length and shortness of life in humans, animals and plants.

In his mature theory of radical moisture—at least in the works considered in this book (*Parva*, zoology, botany)—Albert developed an explicit but yet problematic connection between the workings of spirit and radical moisture, the cornerstone of his psychophysiology of longevity. On the one hand, spirit was regarded as an accidental cause of life-shortening *qua* vehicle of certain activities (sex, sense perception, motion) that can dehydrate the living beings. These

143 Anzulewicz-Krause, "Albert der Große und sein holistisches Konzept," 178–179.

144 For a comparison between the above-mentioned from the *Super Ethica* and Albert's *Ethica*, *De intellectu et intelligibili* and *De natura et origine animae*, see: Anzulewicz-Rigo, "*Reductio ad esse divinum*," 410–411.

considerations on the spirit as an accidental cause of lifespan, addressed in a digression in the *De morte et vita*, apply most to humans and animals, for plants do not experience sensation, motion or sexual intercourse. On the other hand, the material composition of spirit appears to be explicitly and directly related to radical and nutrimental moisture. This is because, in the Galenic system of spirits, mutuated mostly from Avicenna, the quality of the digestive processes impacts directly the material formation of natural, vital and animal spirits. The quality and amount of radical moisture left in the individual and its cohesive functions on the radical organs is key for a correct functioning of the radical parts (heart, stomach, liver, veins, arteries) involved in the material formation spirits.

The gradual decay of the agential power exerted by spirits over time proves their nature as material substance subject to the laws of material necessity leading to death. Vital spirits will be less vital because of fading vital heat, animal spirits will be thicker because of the presence of corrupted humors in the body and will thus lead to impairments in cognitive and rational activities. Albert's position on the correlation between the decay in the power of radical moisture and in the power of spirits finds very interesting parallels in Peter of Spain's *Tractatus de longitudine*, influenced by the Avicennian theory of radical moisture as well, though there these notions are not so detailed or systematically extended to plants, as in Albert's work.

With these latter considerations we have ended our long journey on the relationship between spirit and life, its definition, its origin at the moment of generation and the criteria for defining the length of it. From now on, we shall turn to the role of spirit as a vehicle of the sensitive faculties, in particular of the five external senses and common sense, and of the so-called inner sense. By so doing, we shall abandon the field where Albert introduced his main elements of doctrinal originality in psychophysiology, viz. botany, and turn to human and animal psychophysiology.

PART 2

Spirits of Perception

CHAPTER 6

Spirit and *potentiae apprehensivae deforis*: the Psychophysiology of Sense Perception and Common Sense

This chapter will begin to take us through the second part of this book, viz. the study of the 'spirits of perception'. The journey will consist in two logical steps: the first will investigate the role of spirit in sense perception (viz. the five senses) and the powers of the soul responsible for the monitoring, coordination and unifying of sensory stimuli, viz. common sense (chapter 6). The second step will revolve around the connection to perception-related activities and the corresponding psychic powers such as imagination, estimation and memory (chapter 7). In the case of imagination-related activities such as dreaming, it will be necessary to recall the role of common sense, crucial to explain perceptive awareness during the dreaming process (chapter 8). In Medieval Latin philosophical psychology, the five senses plus common sense were together named "externally apprehensive powers" (*potentiae apprehensivae deforis*). By contrast, faculties such as imagination, estimation or cogitation and memory took on the description of "internally apprehensive powers" (*potentiae apprehensivae deintus*). In this chapter, we shall start with the study of the role of spirit in the former category (*apprehensivae deforis*).

Medieval Latin debates on sense perception, often sparked by Aristotle's psychology (*De anima, De sensu et sensato*),[1] presented at least two kinds of difficulties to exegetes and scholars. The first issue was represented by the controversial theory of sensible species and the discussion on the mode of existence of these species. The main problem to account for was how the soul could perceive an immaterial, sensible form which existed first in the object, then in the medium—in the case of distal senses such as sight, hearing and smell—and then in the body of the sentient, living being. From Albert to William of Ockham, the relation of the ontological status of the sensible species during the different stages (in the object, in the medium, in the sentient subject) of the process of sense perception sparked many scholarly discussions across European universities. In Albert's works, this discussion involved treatment of the

1 Sten Ebbesen, Véronique Décaix, Christina Thomsen-Thörnqvist, "Questions on *De sensu et sensato, De memoria and De somno et vigilia*," *Bulletin de Philosophie Médiévale* 57 (2015), 59–115.

 | DOI:10.1163/9789004706743_008

role of spirit in the changing modes of existence among sensible species during sense perception, which underwent a doctrinal change over the years.

The discussion of the psychological and metaphysical categories needed to explain the actualization of sense perception and common sense was only part of the problem facing theories of sense perception in the 13th century, though. A second problem was represented by the compatibility of the new medico-philosophical explanations of sense perception inherited from Galenic physiology, especially from the Arabic medico-philosophical tradition. The central role attributed to the brain and its seven pairs of cranial nerves not only corrects the fundamental flaws of Aristotelian physiology of sense perception, which totally ignores the role of nerves; it also requires a detailed and accurate explanation of the relationship between the powers of the sensitive soul, seated in the brain, and the central role of the heart, cause and principle of life. Mostly under the lead of Arabic authors like Avicenna, Medieval Latin scholars discussed the relationship between the unity of the soul and the seat of its powers to reconcile Aristotelian psychology and Galenic medicine. In this framework, we shall try to observe how common sense represents a key faculty for a number of activities exerted during the waking state such as the coordination and unification of different sensory stimuli and second-level perceptive awareness (e.g. we see that we see when we see).

Moreover, this chapter will investigate the teleology of sense perception and common sense. On the one hand, it will be necessary to scrutinize the role of spirit according to the two different finalities of sense perception laid out by Aristotle himself (*De anima*, II 415b20–21), i.e. that "of which" (τό οὗ) sensitive faculties represent the finality (proper and common sensibles) and that "for which" (τό ᾧ) these proper and common sensibles are meant to be beneficial and in what manner.

To discuss these three aspects of the externally apprehensive powers, this chapter will first tackle the general psychological framework of the relationship between the soul, the heart and the powers of the sensitive soul. At the same time, the investigation will show how the role and nature of spirit relates to the connection between heart/brain, sensation and the unity of the soul so as to provide a safe background to understand the psychophysiology and teleology of distal (sight, hearing, smell) and contact (touch and taste) sensation and common sense. While a full and detailed account of Albert's psychophysiology of sense perception has been already laid out by the works of Anzulewicz, Steneck, Dewan and Hasse,[2] our task in this chapter will be to sketch a uni-

2 Nicholas H. Steneck, "Albert on the Psychology of Sense Perception," in *Albertus Magnus and*

tary account of Albert's physiology of the *potentiae apprehensivae deforis* (sense perception and common sense)[3] pivoting around the role of *spiritus animalis* throughout his works, still a desideratum in the current state of research. To explain Albert's psychophysiology of sensation, it will be necessary to start our journey with Aristotle's own one.

1 Heat, *pneuma*, and Sensation in Aristotle

As we observed in chapter 3, Aristotle places great emphasis on the cardiocentric origin of the soul's powers, as do Avicenna, Averroes and Albert. According to the Stagirite, the heart represents the unitary center that diffuses vital heat through the whole body. The cardiocentric nature of psychic powers, as stressed by Charles Kahn and Gad Freudenthal, applies to sensation as well.[4]

Despite their difference in terms of activities (ἐνέργειαι) or capacities (δύναμεις), the organs of sense perception are not viewed as single, independent or isolated sources of information. Rather, they are considered integral parts of a unified physiological system with a single chief organ, the heart. In his work, Pavel Gregoric has shown that the faculty responsible for the coordination and monitoring of sensation, not to mention the capacity of discriminating heterogeneous sensibles (e.g. sweet and white) and compounding the different (specific, common and accidental) sensibles into a single image, is common sense, as we shall see in greater detail.[5] On the one hand, common sense repre-

---

the Sciences, Commemorative Essays, ed. James A. Weisheipl (Studies and Texts 49), Toronto 1980, 263–290; Nicholas H. Steneck, "Albert on the Classification and Localization of Internal Senses," *Isis* 65 (1974/2), 193–211; Lawrence Dewan, "St. Albert, the Sensibles, and Spiritual Being," in *Albertus Magnus and the Sciences. Commemorative Essays 1980*, ed. James A. Weisheipl (Studies and Texts 49), Toronto 1980, 291–320; Henryk Anzulewicz, "Konzeptionen und Perspektiven der Sinneswahrnehmung im System Alberts des Großen" (Micrologus Library 10), Firenze 2002, 199–238; Hasse, *Avicenna's* De Anima, 92–107 (on shellfish and nerves and the Latin reception of Avicenna's doctrines), 107–127 (on Avicenna's theory of vision and its Latin reception).

3 Note that in this chapter common sense shall be dealt with only in relation to the five senses. By contrast, the role of common sense in activities such as sleep and dreaming, and its relation to the different classification of internal senses and degrees of abstraction presented in Avicenna's and Albert's writings, will be studied in the next two chapters (7 and 8).

4 Freudenthal, *Aristotle's Theory of Material Substance*, 131: "Because sense perception is a single soul-faculty, it must be centred in one bodily part, the heart." Charles Kahn, "Sensation and Consciousness in Aristotle' Psychology," in *Articles on Aristotle. Vol. IV. Psychology and Aesthetics*, ed. Jonathan Barnes/Malcholm Schofield/Richard Sorabji, London 1979, 20.

5 Pavel Gregoric, *Aristotle on the Common Sense*, Oxford, 2007.

sents the faculty that avoids the risk of a fragmentary understanding of sense perception at a psychological level by being primarily seated and rooted in the seat of the soul, viz. the heart. A whole different question, though, is represented by the unity of the sensitive soul as respects the psychological structures of animals and humans. Given that these living beings are endowed with other faculties or souls (nutritive, rational ones in the case of humans), one may wonder about the relationship between the different localizations of these psychic powers in the body and the cardiocentric localization of the soul. A tentative answer can be found in Aristotle's insistence on the unity of perception, which appears to be strictly correlated to his physiological cardiocentrism: if the soul is one and it is the one principle of all bodily activities, then it must have a single organ which is the source for his nutritive, perceptive and motive activities, viz. the heart.

Aristotle's consideration of sense perception is twofold, accounting for the psychological and the physiological, as with any embodied faculty of the soul. Psychologically or metaphysically, sense perception is conceived, given the immaterial nature of the sensitive faculty, as an affection (πάθος) or a motion (κίνησις) produced by an immaterial sensible species (color, sound, etc.), which leaves an impression on the psychic faculty in the same manner that wax receives the seal of a ring without absorbing its matter.[6] Physiologically, the sense organ is said to be affected by the sensible species that does not have matter in it but the qualities related to the species (e.g. hot, cold to the touch). To perceive properly, these organs need to possess a suitable amount of vital heat, which turns their potential capacity for sensation into actual sense perception. To be fully actualized, the sensible forms of sense perception need to reach the heart.[7] Gad Freudenthal has shown that in the case of sight, hearing and smell, Aristotle hints to the fact that *pneuma*, the only material substance conducive to sensation, fills pores (πόροι) branching out from the blood vessels into the sense organs. The filling of sense organs with heat and *pneuma* enables the first moment of sensation, viz. perception in the sense organ itself.[8]

6 Aristotle, De anima II 12 (424a17–24).

7 Kahn, "Sensation and Consciousness," 26: "This fact of centring of the soul is normally expressed by the characteristically Aristotelian term ἀρχή [...]: it is the 'source' or 'principle' of sensation which is located in the heart. Sensation proper occurs only when the stimulus reaches this centre; and unless it does so, the eye cannot see nor the ear hear."

8 Freudenthal, *Aristotle's Theory of Material Substance*, 130–134. By contrast, blood seems not to be per se conducive to sensation: Aristotle, Historia Animalium III 19 (520b12–17); De partibus animalium II 10 (656b19–22); De partibus animalium III 4 (666a16–18), at least according to Freudenthal's reading. Gregoric (*Aristotle on Common Sense*, especially 45 n. 17) gives a different account for the transmission of sensibles: *pneuma* grants that sensible forms are received

To be fully actualized, though, sense perception needs the perceived sensible species to reach the seat of the soul where sensation is completed, viz. the heart. Yet Aristotle does not clarify how the physiological flow of blood and spirit into the sense organs relates to the complete actualization of sense capacities. Even worse, the remarks on the role of *pneuma* are left completely isolated in the Aristotelian psychophysiology of sense perception, as they connect only to one of the two kinds of Aristotelian sensation, namely distal (sight, hearing, smell), and not contact (touch, smell), sensation. In the wake of this, it appears that a study of the evidence connecting Aristotelian *pneuma* and sense perception must be largely conjectural,[9] for it is supported by textual evidence only in the case of distal senses.

The major problem with sight, hearing and smell in Aristotle's theory of sense perception was represented by the question about the transmission of the sensible species from the sensed object through the medium to the sentient subject. According to Aristotle, sensibles are of three kinds: specific (colors, sound, smells, flavors, tangibles), common (size, shape, rest and motion, number) and accidental (e.g. somebody being Diares' son). While accidental sensibles will be dealt with in the next chapter, as they represented one of the major issues in Medieval debates on human rationality and animals' reason-like capacities, it is necessary to say a few words about common sensibles before plunging into the problems of Aristotelian theory of distal perception. The perception of common sensibles still puzzles scholars by means of a simple yet intricate question: are they perceived by common sense—this would be, as we shall see, Averroes' interpretation—or by a shared ability of the senses which is causally correlated to the existence of the common sense?

Common sensibles, although not defining the single senses, are still considered an essential part of perceptual content, since they can be perceived if and only if sensation occurs and, when it occurs, they are perceived together with the different proper sensibles.[10] Sight, for instance, perceives much more common sensibles than does smell. Common sense, for its part, must not be considered an over-and-above sense with respect to the individual senses. Rather,

in the sense organs, but the transmission of these sensibles to the heart and the full actualization of sense perception is ascribed by Gregoric to the blood, which he considers conducive to sensation. The backflow of blood would therefore connect sense organs to the heart via blood vessels. *Pneuma* is therefore active only in a part of the process, viz. in the sense organs. In short, the disagreement between Freudenthal and Gregoric rests on the role of blood and the subsequent considerations on the role of the *pneuma*.

9 Freudenthal, *Aristotle's Theory of Material Substance*, 133.

10 Kjeller-Johansen, *The Powers of Aristotle's Soul*, 183. For a more detailed explanation of the perception of common sensibles, see: Gregoric, *Aristotle on the Common Sense*, 193–199.

it is a shared ability between the senses that varies according to the nature of senses. Still, this element remains at times both challenging and puzzling. We will come back to this issue in due course, when analyzing the psychophysiology of common sense.

As we pointed out in the beginning, a major issue with Aristotelian sense perception was represented by the modality according to which the sensed object was able to actualize the capacity of sense perception in the subject. But this is less of a problem if we consider the mode of being of the sensible species in contact sensation, for the sensible species directly affects its sense organ. In the case of distal sensation, this poses major problems. For instance, it becomes difficult to understand how we can perceive different colors—even contrary colors—in the same medium. If these sensible species were present in the medium according to a material being, then they would violate the Aristotelian tenet that there cannot be more than one physical object in the same material thing. If they do not exist in the medium according to a material being, then they should exist in an immaterial being. If so, how does this immateriality of the sensible species relate to the object it exists with (the object of sensation), the subject of sensation and its sensitive faculties? This is the major question left open by Aristotle, one that will cause major hermeneutical problems within the Medieval Arabic and Latin philosophical traditions on the topic of distal senses in particular. On the other hand, scholar authors of this age had also serious trouble dealing with the contact sense that Aristotle considered the most basic among living beings; that is to say, touch.[11]

1.1 *Touch*

Aristotle's study of touch represents perhaps one of thorniest issues in psychophysiology. On the one hand, his remarks in the *De generatione et corruptione* (II, 2) on the object of touch (tangibles) and their nature (hot-cold, wet-dry, heavy-light, viscous-brittle, coarse-fine) provides a relatively straightforward explanation of how they are derived—except for heaviness and lightness—from the opposition between active (hot, cold) and passive (wet, dry) primary qualities.[12] According to Aristotle, viscousness, subtleness and softness derive from moisture, while brittleness, thickness and hardness derive from dryness.[13] The main issue in the psychophysiology of touch, though, was represented by the individuation of the sense organ of touch.

11 The following section on Aristotle's theories of sense perception is mostly based on: Kjeller-Johansen, *The Powers of Aristotle's Soul*, 146–169.

12 Aristotle, De generatione et corruptione, II 2 (329b26–34).

13 Aristotle, De generatione et corruptione, II 2 (329b35–330a12).

Concerning this question, Aristotle seems to present a somewhat contradictory position. On the one hand, if we consider that the organ of a certain sense needs to possess the capacity of being actualized by the qualities of the sensed object, then it is reasonable to assume that flesh, *qua* compound of elemental mixture, is a suitable candidate for being the organ of touch.[14] This claim would be in keeping with the doctrines of the *De generatione et corruptione* on the nature of tangibles, essentially derived from primary qualities. In a nearby part of his corpus, though, Aristotle contradicts himself, arguing that the organ of touch is located inside the body, near the heart, and that the flesh acts as its medium.[15] But why does Aristotle feel so compelled to argue that touch (also, taste) has a medium even if it is a contact sense? In his study on Avicennian psychology, Dag Hasse had pointed out that the Aristotelian localization of the organ of touch inside (*intus*/ἐντός) the body[16] or near the heart (πρός τῇ καρδίᾳ) may be introduced, in keeping with his cardiocentric approach to sense perception, for reasons of philosophical consistency with his account of perception at a distance (hence, through media).[17] The result is that Aristotle's locating the organ of touch near the heart seems, on the one hand, to stress the cardiocentric unity of the sense of touch and make it more similar to the other distal senses, able to transmit the specific qualities of the senses to their sense organs, and such is the case with this reading of touch. By contrast, the definition of flesh as the medium and not the organ of touch makes the notion of "medium" in perception equivocal: usually, the medium is placed outside the body; in the case of touch, it is the fleshy surface of the body itself.

In the case of touch, the perception of the tangibles is triggered anyway by direct contact of the tangible with the animal's flesh, which in turn allows for the immaterial, sensible species placed on the surface of the sensed object to be transmitted. Theoretically, this process should lead the sensible species from flesh to the heart via the blood vessels, thus actualizing the individual capacities of perceiving tangibles thanks to the mediating role of *pneuma*, which is conducive to sensation. Unfortunately, this is an aspect of Aristotle's psychophysiology of sense perception that can only be inferred and is not based

---

14 Aristotle, De partibus animalium II 1 (647a19); II 8 (653b25).

15 Aristotle, De partibus animalium, II 10 (656b34).

16 Aristotle, De anima II 11 (423b22–23); Aristoteles Latinus, De anima [A.L. XII.1], ed. Decorte–Brams (online): "Aut et manifestum est quod intus sit possibilis tangi sentitivum."

17 Aristotle, De sensu et sensato 2 (439a2–3); Aristoteles Latinus, De sensu et sensato [A.L. XIII.1], ed Peeters (online): "Et ideo iuxta cor instrumentum eorum, gustus et tactus." Hasse, *Avicenna's* De anima, 99.

on hard textual evidence. A similar problem concerning the missing role of *pneuma* in the physiology of sense perception also occurs in the second contact sense: taste.

1.2 *Taste*

According to Aristotle, taste is a "certain kind of touch" (ἁφή τις),[18] since it perceives the flavors directly on the surface of its sense organ, the tongue, without any medium. The distinction between touch and taste might seem problematic, since we touch and taste food at the same time and with the same organ (the tongue). On the other hand, Kjeller-Johansen has proposed to solve this tension in this manner: since the objects of the sense (primary elemental oppositions and flavors) are different, they affect not a single, polyfunctional sense-organ, but rather two different parts of the same organ.[19] A further, essential element of difference between taste and touch is that taste takes place only in part of the body (the tongue) and that this part of the tongue needs to be moistened by saliva, clearly not the case with touch.[20] Moreover, saliva moisture is not considered by Aristotle to be a medium for taste but rather a sheer material substance facilitating the mechanism of taste.

On the one hand, taste shares the nature of contact sense with touch, but on the other displays many similarities with smell. As a matter of fact, most of the adjectives employed to describe smells derive from flavors because of their more perspicuous nature (at least in humans): as Aristotle argues in the *De sensu et sensato*, the sensibles of taste range from sweetness—associated with moisture—to saltiness or bitterness, associated with dryness.[21]

1.3 *Smell*

The case of smell enables us to transition from the more conjectural realm of senses where the presence of *pneuma* can only be inferred on grounds of correlation between heat, *pneuma* and sensation, to the senses where this doctrine can actually be corroborated by actual (though scant) Aristotelian statements. According to Aristotle, odors are of a twofold species in humans and are only

---

18 Aristotle, De anima II 10 (422a8).

19 Kjeller-Johansen, *The Powers of Aristotle's Soul*, 173.

20 A further way to distinguish the object of taste in: Thomas Kjeller-Johansen, "Aristotle on the Sense of Smell," *Phronesis* 41 (1999/1), 11: "In actual tasting there is a liquefaction of the organ of the tongue (*De anima*, II, 10, 422b1–11). That is why we do not taste anything if there is already too much liquid on the tongue. For it is the flavor of the liquid on the tongue that we taste rather than the flavor of what is put on the tongue."

21 Aristotle, De sensu et sensato 4 (442a18–28).

flavor-related in animals.[22] This is because humans have developed a sense not only for flavor-related odors perceived both by humans and animals, but also for human-specific odors that can be judged as pleasant or unpleasant with respect to the brain's health.[23]

Among animals, humans do not have the sharpest sense of smell. Their sense only enables them to detect the strongest smells, just as hard-eyed animals see only the most evident colors.[24] In his study of odors, Aristotle takes issue with Plato's idea (*Timaeus* 66b) that odors are produced by smoky vapors, arguing that odors are actually an immaterial sensible species propagated through air and water.[25] After its detection, the immaterial species of smell is imprinted on the body of the animal that perceives it. Though smell is primarily associated with breathing in breathing animals, which inhale through the nostrils, Aristotle uses empirical arguments to prove that bloodless and non-breathing animals have the same sense, too. Proceeding from close scrutiny of their behavior, Aristotle argues that turning away from or coming near to food by animals such as fish or bees indicates that they actually possess a sense like smell.[26] A further empirical argument for the existence of smell in all animals is that bloodless animals are killed by the same lethal smoky exhalations that kill humans, such as that emitted by sulphur and bitumen: if they were not to inhale them, they would not be killed.[27] Notwithstanding the physiological differences between bloodless and blooded animals, their psychological endowments in terms of faculties seem to be similar, even though not every ani-

22 Aristotle, De sensu et sensato 5 (443b19–444a8).

23 Aristotle, De sensu et sensato 5 (443b27–444a18); Aristoteles Latinus, De sensu et sensato [A.L. XIII.1], ed Peeters (online): "Et illis quidem, quia secundum accidens habent delectabile, dividuntur species secundum sapores, illius autem non, propter naturam ipsius esse secundum ipsam delectabilem aut tristem. Causa autem ut proprium esse hominis talem odorem, propter frigus quod circa cerebrum est. [...] Proprium autem hominis nature talis odoratus genus, quia plus cerebrum et humidius habet aliis animalibus ut secundum magnitudinem; propter hoc enim quasi solum [differenter] dicere sentit aliis animalibus homo et gaudet florum et talium odoribus."

24 Aristotle, De anima II 9 (421a13–16).

25 Aristotle, De sensu et sensato 5 (443a21–b2). In another passage of De sensu et sensato, though (2, 438b20–27), Aristotle seems to endorse a position close to evaporation theory: Dag N. Hasse, "The Soul's Faculties," in *The Cambridge History of Medieval Philosophy*, ed. Robert Pasnau, Cambridge 2010, 312.

26 Aristotle, De sensu et sensato 5 (444b8–b14).

27 Aristotle, De anima II 9 (421b20–26); Aristoteles Latinus, De anima [A.L. XII.1], ed. Decorte-Brams (online): "Odorabilis enim sensus et mali odoramenti et boni odoramenti olfatus est. Amplius igitur et corrumpi videntur a fortibus odoribus ex quibus quidem homo est, ut ⟨...⟩ et sulfure et huiuscemodi." The term *bitumen* (ἄσφαλτον) is omitted in the Latin translation.

mal possesses the full set of sensitive faculties. If one were to consider the role of *pneuma* in smell, one could see that its vital action and its being conducive to sensation result in its being a carrier of different perceptual content in humans and animals, for humans perceive a second and further class of odors related to the brain's health. By contrast, the material composition or blend/mixture (μίξις) of humans and animals makes them more or less apt to possess a sharp sense of smell. As Aristotle argues, it is difficult to talk about smell, because humans have a less sharp sense of smell than animals do.[28]

1.4 *Hearing*

The case of hearing appears to be less problematic than smell when it comes to the definition of its object. Sound, be it high- or low-pitched, is defined as the species produced by a percutient body on another body and the air in the middle of it. This means that a body causes a species of sound that can be propagated through air and water and eventually reach the human or animal ear.[29] Sound, according to Aristotle, is perceived because of the presence of "natural air" in the human and animal ear, transsonant just like air or water: this fact makes the human ear receptive to the sound, just like the mixed elemental nature of flesh made it receptible to tangibles.[30] According to Aristotle, hearing does not take place in air only but also, though less effectively, in water. To successfully occur in humans or land animals, it is necessary that water does not enter the connate air of the ear or the ear itself: if this were not so, then we would be able to hear underwater as well. According to Aristotle, fish possess hearing as well, even though without—as in humans—connate air in the organs analogous to ears.[31] In the case of hearing, *pneuma* works as the receptacle of the sensible species of sound once they are imprinted in the ears, filled with blood and *pneuma*. From the sense organ, the blood transmits the sensible species to the heart so that perception thereof can move from an initial stage to full sensory actualization.

1.5 *Sight*

Among the faculties of sense perception, the nature and implications of the sight process are doubtlessly the most debated. A first matter of disagreement was represented by the general understanding of the act of seeing, viz. whether it was produced by the extramission of rays from the eye, as in Plato, or resulted

28 Aristotle, De sensu et sensato 4 (440b30–441a2).
29 Aristotle, De anima, II 9 (419a18–22).
30 Aristotle, De anima, II 9 (420a9–14).
31 Aristotle, Historia Animalium, IV 8 (533b4–534a11).

from a passive activity of perception, as in Aristotle.[32] Light, defined by Aristotle as the "color of the translucent" (χρῶμα τοῦ διαφανοῦς)[33] or "act of the transparent *qua* transparent",[34] is that which enables the observer to see color. The existence of light, according to Aristotle, is never conceived apart from the medium.[35] This is a position which sets him apart from later thinkers like Avicenna, who will argue that certain bodies (e.g. the sun) possess light as their intrinsic quality, which exists independently of the translucent medium (viz. air or water). Color, according to Aristotle, is placed on the extremity of the surface of the object delimiting the translucent body. Since its existence depends on the actualization of the transparent medium by light, color is defined as "the limit of the transparent."[36] The range of colors is divided in the *De sensu et sensato* into seven different kinds ranging from white to black, plus the color-as-sensible itself, which makes an eight color of itself. Once the transparent medium is actualized by light, color is transmitted to the eye, which receives these visible forms. Given that Aristotle assumes that *pneuma* fills in the organ of sight with blood and considers *pneuma* conducive to sensation, there is good reason to believe that this material substance could serve as an instrument for the soul to actualize its own capacity of perceiving visible species.

Up to now we have seen how *pneuma* plays a fragmentary and limited role in Aristotle's psychophysiology of sense perception. We have examined the single senses, highlighting how their cardiocentric nature vouches for the unity of sensation. This assertion, though, is not complete when considering the physiological and also the psychological power which maintains, coordinates, monitors, compares and unifies the different sense stimuli, viz. common sense, not to mention its complex relationship to common sensibles. An examination of its psychophysiology is therefore crucial here.

32 An apparent exception is represented by the passage in his *Meteorology* when, discussing rainbows as objects of vision, he apparently endorses an extramissionist theory, see: Aristotle, Meteorologica III 4 (373b1–10).

33 Aristotle, De sensu et sensato 3 (439a18–19).

34 Aristotle, De anima II 7 (418b10–11).

35 On the fact that water is more easily contained than air, Aristotle, De sensu et sensato 2 (438a15–16); Freudenthal, *Aristotle's Theory of Material Substance*, 173.

36 Aristotle, De sensu et sensato 3 (439b811–814); Aristoteles Latinus, De sensu et sensato [A.L. XIII.1], ed Peeters (online): "Quare color utique erit perspicui extremitas in corpore determinato, et ipsorum perspicuorum, velud aqua et si quid aliud tale, et in quocumque videtur color proprius inesse, secundum extremitatem similiter omnibus inest."

1.6 *Common Sense*

According to Aristotle, common sense and common sensibles (size, shape, number, motion, rest) represent an essential part of perceptual content which simply cannot be ignored. The perception of colors comes often if not always with the perception of a magnitude, a shape and the like. On the other hand, common sensibles do not define the perceptual content of the individual senses. Rather, the ability to perceive common sensibles seems a feature common to all of them. Moreover, common sense is responsible for other features of sense perception, such as: second-level perceptive awareness (we see that we see), the discrimination of different sensibles (white is not sweet), and the compounding of specific and common sensibles into a single image (*phantasma*) of the thing.[37] While the Aristotelian psychology of common sense seems on its own, as evidenced by the book of Pavel Gregoric on common sense, to prompt questions about the faculty responsible for the perception of common sensibles, Aristotelian physiology provides no evidence linking the psychological activities of common sense to their explanation based on blood, *pneuma*, and the chief role played by the heart.[38]

The main problem with this account is that is simply not extant in Aristotle, so every claim that can be made about *pneuma* as a vehicle of the cognitive content of common sense (common sensibles, etc.) can only be inferential. For instance, let us consider Aristotle's own statements that the organs of sight, hearing and smell have to be filled with blood and *pneuma* in order to be conducive to sensation, which is later actualized in the heart. Since common sensibles are perceived together with proper sensibles such as colors, there is good reason to assume that *pneuma* could be conducive both to proper and common sensibles. Given that the actualization of the perceptive capacity of common sense takes place in the heart, the perception of common sensibles can follow a similar route as that of proper sensibles. As for other activities such as second-level perception, they can be seen as correlated to the waking state: as long as there is a single sensitive faculty actualized (e.g. touch), there can be second-level, perceptive awareness, which disappears upon our falling asleep.

---

37 Kjeller-Johansen, *The Powers of Aristotle's Soul*, 183.

38 On the individuation of heart as the organ of common sense, see: Gregoric, *Aristotle on the Common Sense*, 47: "The heart is said to be 'the sense organ common to all peripheral sense organs', and 'the master sense organ to which all sense organs lead' (DSV 2 455a33–34)." The primary link between the coordinating function exerted by common sense over the individual senses and the cardiocentric nature of the powers of the sensitive soul seem to be quite in keeping with this interpretation of Aristotelian psychophysiology.

Aristotle's reflection on the correlation between the deactivation of the powers of common sense and of touch, carried out in *De somno* 2, will become a matter of discussion in Albert's psychophysiology of sleep and dreaming, tackled more extensively in chapter 8. Suffice it here to say that in this chapter, Aristotle considers common sense to be the root of all sensible powers and sleep the deactivation of all senses. Since he regards both common sense and the sense of touch as the basic foundations of the different perceptive powers, his text sparks a debate about which faculty (common sense/touch) may be considered as primarily responsible for the periodical alternation of sleep and waking in animal life and, we may add, as a primary cause of second-level perceptual awareness, which belongs to the powers of common sense. These aspects are investigated in depth by the studies of Pavel Gregoric we are here summarizing.[39]

Now we have undertaken an investigation into Aristotle's psychophysiology of sense perception and common sense which temporarily sets the issue of finality in sense perception apart. To gain a more complete grasp of the role of *pneuma*—in spite of all its problematic features within this theory of Aristotelian sense perception—we need to recall for a moment the teleology of the senses, an explanation that accounts for sense perception at a level beyond pure psychophysiology.

1.7 *The Teleology of Sense Perception*

According to Aristotle, there is a double finality in sense perception: the existence (εἶναι) and the well-being (εὖ εἶναι) of the animal. If we consider the first of these as an aspect of teleology, touch (followed by taste) represents the primary sense inasmuch "these senses (sc. touch and taste) are necessary to an animal, and it is evident that an animal cannot be without touch, the other senses (sc. sight, hearing, and smell) are for the sake of doing well."[40] This statement, though, must not be taken too strictly since taste is considered by Aristotle to be also related to well-being, since it causes pleasure or pain.[41] Moreover, Aristotle adds that sight, hearing and smell exist for the animal's being (*esse*/εἶναι) as such, and not only for its exercise of cognitive capacities. This is because

39 Gregoric, *Aristotle on the Common Sense*, 174–192.

40 Aristotle, De anima III 12 (434b18–24).

41 Aristotle, De anima III 13 (435b19–25); Aristoteles Latinus, De anima [A.L. XII.1], ed. Decorte–Brams (online): "Alios autem sensus habet animal, sicut dictum est, non propter esse, sed bene; ut visum, quia in aere et aqua est, ut videat, aliter autem, quia in lucido est; gustum que propter dulce et amarum, ut sentiat in alimento et desiderio et moveatur; auditum autem, ut significet aliquid sibi ipsi; linguam vero quatinus significet aliquid alteri." Aristotle, De sensu et sensato 1 (436b13–436b17).

they can detect threats approaching or help animals endowed with locomotive capacities direct themselves towards desirable locations or to food.[42] If we consider the purpose of sense perception from the point of view of the animal's or human's ontology, the sense which is primarily and exclusively associated to it is touch, followed by taste, smell, hearing and sight. On the other hand, if we consider the well-being of the animal the order is reversed: sight, hearing, smell, taste and touch. The reason for this theoretical assumption is that sight is the major provider of perceptual content, enabling the exercise of the cognitive faculties, viz. the highest and ultimate purpose of the animal's well-being.

But there is an additional factor that must be considered in the purpose of sense perception, i.e. the possibility of impairments with its concrete implications. If we assume that sight is the first and primary sense in terms of cognition, we have to note that hearing is accidentally the most essential sense in order to develop cognitive—if not intellectual—capacities. The reason for this, Aristotle argues, is that a person born blind can become wiser compared to somebody born deaf. Compared to a deaf person, a blind person still retains the possibility of learning new cognitive or intellectual content via hearing.[43]

Inferring how the workings of *pneuma* as an instrument of the sensitive soul relate to the Aristotelian teleology of sense perception requires that one consider the Aristotelian teleology of sense perception. On the one hand, we can infer that *pneuma* provides the soul with crucial information for the preservation of the existence of the individual living being it dwells in. This information can come from all of the senses in animals, but is primarily associated to touch, the sense that differentiates plants and animals, and sight, the richest sense in cognitive content. On the other hand, *pneuma* provides the soul with cognitive content directly connected to the improvement and exercise of the cognitive faculties of the sensitive soul. An additional point to our reconstruction of the role of *pneuma* in sense perception, strongly based on Freudenthal's, is the role of *pneuma* in common sense. In the case of common sense, *pneuma* serves not only as a vehicle of common sensibles but appears to be causally correlated to the perception of proper sensibles (in distal senses) which often come with common sensibles. Moreover, the strict correlation between heat, *pneuma* and sensation points towards its causal role as the activator, together with vital heat,

42 Aristotle, De sensu et sensato 1 (436b18–437a3).

43 Aristotle, De sensu et sensato 1 (437a4–16); Aristoteles Latinus, De sensu et sensato [A.L. XIII.1], ed Peeters: "Secundum vero accidens ad sapientiam auditus multum confert; sermo enim audibilis existens causa est discipline, non secundum se sed secundum accidens; ex nominibus enim constat; nominum vero unumquodque simbolum est. Quare sapientiores a nativitate privatis ab utrolibet sensu sunt ceci mutis et surdis."

of key functions of common sense such as the discrimination, association and comparison of different sensibles, and to second-level perceptual awareness.

The main gaps left by Aristotle's physiology of sense perception concern the connection between *pneuma*, contact senses (touch and taste) and common sense. Moreover, it completely lacks any explanation on the transmission of the sense data from the sense organ to the heart in physiological terms. On the other hand, Aristotle does attempt to give a physiological explanations of certain impairment such as glaucomas and similar diseases affecting eyes, basing his etiopathology on the consumption of moisture over time, which in turn leads to a hardening in the cornea, and making sight gradually more difficult.[44] These lacunae are fulfilled in the *corpus* of the Galenic medical tradition, whose psychophysiology of sense perception revolved around the "animal spirit" (πνεῦμα ψυχικόν/*rūḥ nafsāniyyah*/*spiritus animalis*) and provided a much more comprehensive and detailed account of the physiology of sense perception than Aristotle did.

2 Animal Spirits and Sense Perception in Costa Ben Luca, Nemesius, and Other Medico-philosophical Sources

We have observed in the second chapter how Albert's minor medico-philosophical sources, among which towers Costa ben Luca's *De differentia spiritus et animae*, suffer from a weak compatibility between their psychology and their physiology, as is evident in the case of the role of vital spirits in defining life, tackled in chapter 2. By contrast, this seems to be less the case with sense perception, for Costa ben Luca does give a detailed account of sense perception based on animal spirits and, on the other hand, includes in his list of faculties the five senses. The only notable exception to Aristotelian psychology, which Costa ben Luca listed as a source in his prologue, is represented by the lack of common sense, *ḥiss muštarak*, in Arabic. Senses are therefore, in Costa ben Luca's psychology, described in detail from a physiological point of view that lacks a unifying faculty which grants, as in the case of Aristotelian common sense, that each sense does not represent an isolated source separate from the other senses. Like Galen and unlike Aristotle, though, the *De differentia* shows a clearly encephalocentric understanding of sense perception, based on the flow of animal spirits through the seven pairs of cranial nerves responsible for sensation. The physiology of the *De differentia* presented Albert with a more

44 Aristotle, De generatione animalium v 1 (780a17–b29). See chapter 4, 229 n. 25.

state-of-art anatomy of sense perception than Aristotle's physiology, which totally ignored the role of the brain and of the nerves in sense perception. Costa ben Luca and other sources like Nemesius constitute a good reference-point for his debates on the anatomy of cranial nerves and correlation to sense perception, in spite of some psychological shortcomings, which could be well supplemented by Avicennian psychology.

In spite of its psychological flaws, the advantage of the *De differentia* is, as stressed above, a quite detailed account of the anatomy of sense perception and the etiopathology of diseases affecting the different brain areas and cranial nerves. Like Galen, Costa ben Luca considered animal spirit (*spiritus animalis*/*rūḥ ḥayawāniyya*) to have its material origin in the vital spirit originating in the heart. In turn, this substance was pumped from the heart to the anterior part of the brain via the aorta superior, the carotid arteries and that complex net of blood vessels underlying the brain known as *rete mirabile*. These progressive steps led to the entrance of the vital spirit into the brain,[45] causing the vital spirit to thin out.[46] As in Galen, the subtleness and thinness of spirit appears to be directly proportional to the complexity of the activities carried out by it. This thinness is achieved when spirit enters the front brain ventricle and, from this cavity, starts to spread out into the organs of sense perception through the seven pairs of cranial nerves. Costa ben Luca's description of the different stages of purification and subtilitation of animal spirits resembles quite closely Constantine the African's *Pantegni*, another well-known source of Albert's.[47]

Costa ben Luca also provides a taxonomy of cranial nerves, which is not quite as influential as Avicenna's, but which will be still valued by Albert. According to the *De differentia*, the first nerve is responsible for vision, the second pair for the motion of the muscle around the eye, the third couple of nerves is responsible for taste, the fourth partially for taste as well,[48] the fifth for hearing and the sixth—albeit partially—for touch, while the seventh causes the tongue's

45 Costa ben Luca, De differentia spiritus et animae, ed. Wilcox 1985, 151.112–153.24 [ed. Troupeau 2011, viii.10–ix.5].

46 Costa ben Luca, De differentia spiritus et animae, ed. Wilcox 1985, 187.542–546 [ed. Troupeau 2011, xxx.4–7].

47 Constantinus Africanus, Pantegni [Theorica IV.19], ed. Burnett 1994, 114–115.

48 Note that the Latin does not translate the Arabic *"fa-yu'addī ilayhi ḥiss al-ṭa'm"* [Costa ben Luca, Risāla al-faṣl bayna al-rūh wa-l-nafs, ed. Troupeau 2011, xii.9–10], i.e. "and brings the sense of flavor to it (sc. the palate)." From this perspective, what for Albert seems to be different functions of the same nerve respectively in Qusṭā and Avicenna (who links this nerve to flavor as well) is not due to a doctrinal difference between the two but rather to an omission in the Latin translation.

motion.[49] Costa ben Luca's classification of cranial nerves, though in keeping with the crucial role attributed by Galen to them in sense perception, betrays some departures from Galen. While Costa ben Luca agrees with Galen on the function of the first, fifth and seventh nerves, he parts ways from him concerning the function of the remaining pairs of cranial nerves.[50] Among Albert's sources, Qustā's classification of the functions of seven pairs of cranial nerves is paralleled by Avicenna and, though to a minor extent, Nemesius.[51]

The remarkable feature of Nemesius and Costa ben Luca's physiology of sense perception, as will appear in Avicenna, is that it clearly excludes any possible connection of the cranial nerves to the sense of smell. A small group of secondary sources of Albert's, i.e. the *Anatomia Richardi*[52] and the two anatomies attributed to the *Magister Nicolaus Peripateticus*, will assert the contrary.[53] As in Nemesius, the manner in which the brain is affected by odors is still unclear in the *De differentia*. While Costa ben Luca explains how impairments affecting the animal spirit, such as the coagulation of water in the eye or the production of corrupted humors between the brain and the organs of sense perception, causes blindness and deafness, he does not quite explain how this would be the case with smell.[54] While the mechanism leading to sensorial

49 Costa ben Luca, De differentia spiritus et animae, ed. Wilcox 1985, 159.192–208 [ed. Troupeau 2011, xii.3–15].

50 Judith Carol Wilcox, *On the Transmission and Influence of Qusta Ibn Luqa's "On the Difference Between Spirit and the Soul"*, Unpublished Ph.D. Thesis, New York 1985, 54–55: "However, there are a few discrepancies between their respective distribution of functions to the sensory nerves. Both Galen and Qusṭā attribute to the first vision. To the second, Galen attributes the motion of the eye, and Qusṭā the movement of the eyelid. Galen attributes to the third pair sensation in the face, while Qusṭā attributes to it the sense of taste. According to Galen, the fourth pair is responsible for motion in the face, while this joins the palate according to Qusṭā. Both Qusṭā and Galen attribute to the fifth the sense of hearing. Galen says that the sixth is for motion of the tongue and pharynx, while Qusṭā has it travel down to the viscera to impart to it sensation and then back to the uvula, probably for speech. This would put them somewhat in agreement as to part of its function. On the function of the seventh pair, which is to move the tongue, they are in agreement."

51 The role of nerves in sense perception can be observed in: Nemesius, De natura hominis c. 6 (ed. Verbeke-Moncho 1975, 75.35–38); c. 7 (ed. Verbeke/Moncho 1975, 81.54–56); c. 8 (ed. Verbeke/Moncho 1975, 83.11–84.3).

52 Richardus (Anglicus), Anatomia (ed. Tarrasch 1898, 15): "alii duo (*sc.* nervi) ad caruncula narium ad odoratum, et dependent ad modum mamillarium."

53 Nicolaus Physicus, Anatomia (ed. Redeker 1917, 39.255–40.259): "Item oritur nervus a cellula phantastica qui protenditur usque ad carunculas narium, quae spongiosae sunt et formatae ad modum uberae mamillarum. Per hunc nervum defertur spiritus odorabilis ad illas carunculas."

54 Costa ben Luca, De differentia spiritus et animae, ed. Wilcox 1985, 159.208–161.219 [ed. Troupeau 2011, xiii.1–9].

impairment may be clear in the case of other senses, all endowed with a single sensory nerve of their own (except for touch), this is left in the dark as it pertains to smell.

Among other senses, Costa ben Luca mentions touch, which he often links to the physiology of motion in paralysis cases: if the flow of spirit from the brain through the nerves to the outer part of the body is blocked, the outer skin looks perfectly regular, but the nerves are unable to answer the sensory stimulus of touch or begin any kind of motive process. According to Costa ben Luca, there are only two available solutions to this impairment: either the organism overcomes this illness spontaneously, or it needs a treatment (*medicamentum*/*ʿilāǧ*) remove the physical impediment.[55] Further cases of sensory impairment are produced by exogenous factors such as excess of heat, moisture and vapors in the surrounding environment: if—for instance—a person enters a bathroom (as in a typical *ḥammām*, for instance) with too much heat and humidity, his sight would be impaired; similar circumstantial impairments would apply to other senses.[56]

The role of animal spirit *qua* instrument of the sensory faculties in many of Albert's minor sources chimes with Nemesius' *De natura hominis*, Isaac Israeli's *De definicionibus*, and Alfred of Shareshell's *De motu cordis*, all of which display the same connection between animal spirit (*spiritus animalis*) and sense perception or phantasy in the front brain ventricle, cognitive activities in the middle ventricle and memory in the posterior ventricle.[57] As we have seen,

55 Costa ben Luca, De differentia spiritus et animae, ed. Wilcox 1985, 161.220–223 [ed. Troupeau 2011, xiii.8–9]: "Cumque aperti fuerint ipsi meatus aut per medicamentum aut per directionem nature reluctante (Ar.: "or by a natural resistance to the disease" [*bi-muqāwamati-l-ṭabīʿiyya li-l-ʿilla*]), scilicet natura contra infirmitatem, redit membrum ad opus suum et efficitur sanum atque equale (ar. *fa-ṣāra ṣaḥīḥan wa-mustawiyyan*)." *Scilicet contra infermitatem* is an addition of the Latin text.

56 Costa ben Luca, De differentia spiritus et animae, ed. Wilcox 1985, 163.247–261 [ed. Troupeau 2011, xiv.11–xv.4]: "Et fortassis accidit spiritui qui est in omnibus ventriculis vel in quibusdam ex eis aliquod impedimentum ex mala complexione vel commixtione malorum vaporum et propter hoc destruuntur actus eiusdem membri. Verbi gratia: cum ille spiritus tantum qui est in anterioribus ventriculis aliquod impedimentum vel corruptionem passus fuerit, fit ex hoc destructio sensuum quemadmodum accidit ei qui ingreditur balneum et ibi moras facit: obscurantur enim oculi eius et nichil videt; aut quemadmodum accidit ei in quo excitatur colera rufa (ar. *mirra*, "bile") et eius vapor ascenderit ad caput eius iunctusque fuerit vapor spiritui qui est in cerebro anteriori: similter obscuratur visus eius et non videt aliquid, et similiter accidit auditui ceterisque sensibus."

57 Nemesius, De natura hominis c. 8 (ed. Verbeke/Moncho 1975, 71.52–54): "Organa autem eius sunt anteriores cerebri ventres et qui in eis est animalis spiritus et hi qui ex his sunt nervi, perfusi ab animali spiritu, et constructione membrorum sensum." This passage is not extant in Alfanus' version of the work. See also: Isaac Israeli, Liber De definicionibus

Costa ben Luca's detailed approach to the formation of animal spirit is comparable to the chapter on spirits from Constantine the African's *Pantegni*, a purely medical work, though the latter does not mention the role of cranial nerves at all. What is peculiar to Costa ben Luca among these authors is an emphasis—shared by the *Pantegni*, i.e. al-Maǧūsī, however—on the connection between the physiological motion of vital spirit in the body that reifies a state of purity, clarity that enables it to perform the operations of sense perception and, as we shall see in the next chapter, cognition and memory.[58]

As we have shown, the main focus of Costa ben Luca and of Albert's other minor medical sources is medical and pathological rather than philosophical: they provide an explanation of purely material and efficient causation, without relating spirit to a consistent theory of sense perception and its teleology. Plus, all these sources had neglected the role of common sense, the different typologies of perceptual content (proper, common, accidental sensibles) and the end-goal of sense perception given the medically oriented perspective. For this reason, even though texts such as the *De differentia* could provide Albert with a detailed physiological account of sense perception to replace the obsolete Aristotelian physiology (in fact, one of the most detailed after Avicenna's),[59] Albert needed to turn elsewhere to find a more full-blown and integrated discussion of the role of spirit within a (revisited) Aristotelian framework of sense perception. In Albert's view, this 'elsewhere' had a precise name: Avicenna.

3 Animal Spirits in Avicenna's Theory of Sense Perception and Common Sense

In the second and third chapter, we observed how Avicenna stresses the unity of the soul by appealing to the Neoplatonic theory of flow (*fluxus*/*fayḍ*), which enables him to distinguish ontologically the soul from its powers. This not-

(ed. Muckle 1937/8, 318.23–319.2); Alfredus Anglicus, De motu cordis c. 8 n. 2 (ed. Baumker 1923, 31.22–32.2).

58 As we shall in Albert's section on animal spirits, Qusṭā's emphasis on the purity and clarity of the animal spirit will be relevant not only in Albert's discussion of the characterization of animal spirit in the *De spiritu et respiratione*, but will also become a criterion by which to rank the different sense faculties.

59 Charles Burnett, "The Introduction of Aristotle's Natural Philosophy into Great Britain: a Preliminary Survey of the Manuscript Evidence," in *Aristotle in Britain during the Middle Ages: Proceedings of the International Conference at Cambridge, 8–11 April 1994, organized by the Société Internationale pour l'Étude de la Philosophie Médiévale*, ed. John Marenbon (Rencontres de Philosophie Médiévale 5) Turnhout 1996, 35.

unproblematic distinction is correlated to one of the major tenets of Avicennian psychophysiology, viz. the cardiocentric origin of all psychic powers. As we saw in chapter 3, the clearest exemplification of his cardiocentrism can be found in Avicenna's psychology (*De anima* V, 8) and in his discussion on the origin of veins (*De animalibus* III, 1). Both discussions are hinged on his key notion of formative power.

The obverse side of the coin, viz. the refutation of the equality attributed by Galen to heart, liver and brain, can be found in Avicenna's account of the cardiac origin of the sensory powers of the soul (*De anima* V, 8). To be sure, Avicenna did value the Galenic medical tradition and its detailed account of the role of the brain and cranial nerves in sense perception. As in the case of the liver, though, Avicenna simply could not bring himself to accept placing the heart and the brain on equal footing as ruling powers. Faithful to his Neoplatonized version of Aristotelian cardiocentrism, read through the lens of the concept of *fluxus*/*fayḍ*, Avicenna qualified the heart as the "first principle" (*principium primum*/*mabdāʾ ʿawwal*) of the soul's powers. By contrast, brain and liver are classified as "secondary" principles, defined as such because they derive their powers from the heart.[60] According to Avicenna, the cardiac origin of the nervous system demonstrate that its powers derive (*transsumere*/*istamadda*) from the heart as the unitary seat of the substantial form.[61] As we shall see, in Avicenna the question about the unity of the sensitive soul—not the

60 Avicenna, Liber de anima V c. 8, ed. Van Riet 1968, 176.72–75 [ed. Rahman 1959, 264.6–8]: "Oportet ergo ut anima principaliter pendeat ex corde; impossibile est autem prius eam pendere a corde et postea a cerebro (*laysa yağūzu an tataʿallaqa bi-l-qalb ṯumma bi-l-dimāġ*): cum enim pendet a primo membro fit corpus animal; in secundo autem non operatur nisi mediante primo (*fa-innamā tafʿalu fīhi lā mahāla bi-tawassuṭ haḏā alʾawwal*)." Avicenna, Liber de anima V c. 8, ed. Van Riet 1968, 177.78–90 [ed. Rahman 1959, 264.11–21]: "In cerebro autem perficitur complexio spiritus qui est aptus ad vehendum virtutes sensus et motus ad corpora (*allaḏī yaṣluhu li-ʿanna yakūna ḥāmilan li-l-quwà al-ḥiss wa-l-ḥaraka ilā-l-aʿḍāʾ*), ad hoc ut fiant apta ad exercendum suas actiones. Sic etiam est dispositio epatis comparatione (*bi-l-qiyās*) virtutis nutritivae, quamvis cor sit primum principium (*mabdāʾ awwal*) ex quo principaliter pendet et a quo diffunditur ad alia et fiunt actiones in reliquis membris; sicut principium sensus (*mabdāʾ al-ḥiss*), secundum eos qui dissentiunt ab hac dictione, non est nisi cerebrum, sed actiones sensus non sunt ex eo, nec in eo, sed ex aliis membris, sicut ex cute et oculo et aure, nec propter hoc necesse est cerebrum non esse principium. Similiter cor potest esse principium virtutum nutritivarum, quarum actiones sunt in epate, et virtutum imaginationis et memorialis et formalis, quarum actiones sunt in cerebro." See also: Avicenna, Liber de anima V c. 8, ed. Van Riet 1968, 177.91–178.99 [ed. Rahman 1959, 264.22–265.4].

61 Avicenna, Liber de anima V c. 8, ed. Van Riet 1968, 178.99–179.8 [ed. Rahman 1959, 265.4–11]. More on Avicenna's position and the issue of the "corporeal form" (*forma corporeitatis*) in chapter 3, especially 126–127.

soul as a unity of essence and powers—will be answered, as in Aristotle, with reference to the faculty of common sense.

Avicenna's psychophysiology of sense perception blends Aristotelian zoology and its cardiocentric background with Galenic anatomy and physiology. This fact is witnessed by the chapters on nerves (*De animalibus*, XII, 8), on sight (*De animalibus*, XII, 11) and on hearing (*De animalibus* XII, 12) in his *Book of Animals*, copied directly from the *Canon*. These chapters provide a detailed material explanation of the organs of sense perception, while the topic of the sensitive faculties is analyzed in the chapter on the faculties (*Canon* I, I, VI), more specifically in the subsection of the chapter which he devotes to the "animal" faculties charged with perception (*virtutes comprehendentes* / *quwà ḥayawāniyya al-mudrika*). Even though less systematically permeated by medical notions, the physiology of sense perception in his *De anima*, from touch to taste (II, 3), smell, (II, 4) hearing (II, 5), sight (book III) and common sense (III, 8/ IV, 1), can be easily integrated within his zoological and medical doctrines. This section of ours on Avicenna's animal spirits takes its cue from studies by Dag Hasse and Danielle Jacquart, complementing them with a few detailed observations on the implication of spirit in the so-called phenomenon of "double vision".[62] The purpose of this section will be to shed light upon how Avicenna succeeds in a major task of his philosophical project, namely in integrating the faculty theory of the Aristotelian sensitive soul (common sense included) with the Galenic anatomy of sense perception. The functional order of the couple of cranial nerves foreseen by Avicenna is the following: 1. sight; 2. eye motion; 3. facial sensation and, partially, taste (in the palate); 4. taste (on the tongue); 5. hearing; 6. sensation in the interior members and in the uvula; 7. motion of the tongue. Avicenna's anatomy of sense perception and its classification of cranial nerves will prove influential for centuries both in the East and in the West, constituting a valuable part of any psychophysiology of sense perception, most relevantly those senses strictly connected to the functioning of nerves, such as touch.

3.1 *Touch*

Avicenna's doctrine on touch was influenced by his strongly Galenic medical background. As a matter of fact, his doctrine of sense perception is rooted on the role of brain nerves. On the other hand, Avicenna does not blindly adhere to Aristotle's doctrines on the psychology of sense perception, as we shall see

62 Hasse, "The Soul's Faculties," 305–319; Danielle Jacquart, "La notion philosophico-médicale de *spiritus*," 15–33.

in the case of smell. In the case of touch itself, Avicenna presented a view—to which he was not committed—that argued that the sense of touch could be considered one in genus but fourfold in faculties because of its ability to sense objects (viz. tangible species) derived from a fourfold opposition: hot and cold, wet and dry, hard and soft, rough and smooth. This fact alone would have caused the number of senses to increase to eight instead of the usual five, but Avicenna leaves to philosophers the confirmation (*certificare*/*tahqīq*) of this matter.[63] If we leave apart the slight differences in terms of tangible species between Avicenna and Aristotle's *De generatione et corruptione* II, 2—the latter had hot-cold, wet-dry, heavy-light, viscous-brittle, coarse-fine, compared to Avicenna's hot-cold, wet-dry, hard-soft, rough-smooth—the greatest point of controversy between Avicenna and Aristotle appears to be the identification of the sense organ of touch.

Contrary to Aristotle's statements in the *De Anima* or *De sensu et sensato*,[64] Avicenna thinks touch requires a medium which he dubs nervous flesh.[65] In Avicenna's view, nervous flesh or flesh-like substances (e.g. teeth) provide the brain with the tangibles through the flow of the animal spirit (*spiritus animalis*/*rūḥ nafsāniyya*). This position was reported by al-Ġazalī in his *Intentions of the Philosophers*.[66] While Avicenna's position on the sense organ of touch

63 Hasse, "The Soul's Faculties," 308. Avicenna, Liber de anima I c. 5, ed. Van Riet 1972, 85.80–87 [ed. Rahman 1959, 42.14–43.1]; Avicenna, Canon I fen 1 doctr. 6 c. 5, ed. Venezia 1507, fol. 24[vb] [ed. New Delhi 1981–1996, vol. 1, 128.13–17]: "Et si dixerimus quod sint octo, erit huius causa quod pluribus certificatorum videtur quod tactus sit plures virtutes, scilicet quattuor et ponunt proprie umumquodque quattuor generum tactus virtutem per se [...] hoc autem certificare est philosophi." Once again, we can see Avicenna stressing that the determination of the soul's faculties belongs to philosophy and not to medicine.

64 Avicenna, Liber de anima II c. 3, ed. Van Riet 1972, 138.2–139.12 [ed. Rahman 1959, 71.20–72.6]; Hasse, Avicenna's *De Anima*, 100–101. In his definition of the faculty of touch in *De anima*, I, 5, though, Avicenna omits any mention of the medium, while clearly locating the faculty of touch in flesh and nerves: Avicenna, Liber de anima I c. 5, ed. Van Riet 1972, 84.77–85.80 [ed. Rahman 1968, 42.11–13]: "Et ex illis est tactus, qui est vis ordinata in nervis cutis totius corporis et eius carnibus, ad apprehendendum id quod tangit illum et afficit contrarietate permutante complexionem et affectionem compositionis."

65 Avicenna, Liber de anima II c. 5, ed. Van Riet 1972, 141.44–45 [ed. Rahman 1959, 73.13–14]: "istae virtutes diffusae sunt per omnia membra (*fī kulli al-ālāt*) aequaliter [...]."

66 Algazel, Metaphysica pars 2 tr. 4 c. 3, ed. St. Clair 2005, 63.77–87 [ed. Cairo 1936, 40.2–9]: "Sensus tactus manifestus est qui est virtus diffusa per omnem cutem et carnem per quam apprehenditur calor et frigiditas, humiditas et siccitas, durities et mollifies, asperitas et lenitas, gravitas et levitas. Et haec virtus pertingit ad partes carnis et cutis mediante corpore subtili quod est vehiculum eius, quod dicitur spiritus, et discurrit per compagines nervorum quibus mediantibus pertingit ad partes carnis et cutis. Et hoc corpus subtile non acquirit, neque haurit virtutem hanc, nisi a corde et cerebro, sicut postea dicetur. Nisi

is unmistakably different from Aristotle's, Avicenna's insistence on the causal role of spirit as a vehicle of sense perception can be, to some extent, brought in agreement with Aristotle's scattered remarks on the role of *pneuma* as a vehicle of sense perception *qua* carrier of vital heat. To be clear, while there may be some general compatibility between the role of spirit in Aristotelian and Avicennian natural philosophy respectively, the role of spirit is here interpreted, as in Costa ben Luca and other authors of the medical tradition, through the concept of animal spirits and Galen's theory of humors and of complexion. Unlike previous authors, though, who all presented a human-centered view of animal spirits,[67] Avicenna's framework has the merit, in the eyes of Latin authors like Albert, of extending the domain of animal spirits to animals as well.

3.2 *Taste*

As in Aristotle, taste is similar to touch *qua* contact perception: unlike touch, taste needs the presence of a salivary moisture (Lat. *humor salivalis*, "salivary humor", Ar. *ruṭūba luʿābiyya* "salivary moisture"). Yet this moisture does not constitute any medium through which flavors are perceived.[68] Rather, its function is to facilitate taste by helping the flavor come into contact with the surface of the tongue and the nerve therein. This nerve is the only organ responsible for it,[69] a position echoed by al-Gazālī.[70] In the *Canon* and in the *De animalibus* (XII, 8), Avicenna adds that flavors are transmitted to the brain not only via the fourth branch of the third pair of cranial nerves (spread out over the

autem convertatur qualitas cutis in simile apprehensi—scilicet in frigiditatem vel caliditatem vel in ceteras earum—non fiet apprehendens, et ideo non apprehendit nisi id quod est calidius eo vel frigidius eo; id autem quod est tale non agit in tangentem." Hasse, *Avicenna's* De Anima, 104 n. 130.

67 This claim is true if we consider the Latin translation of the authors known by Albert: Qusṭā, for instance, ascribed the existence of spirit to animals too: Costa ben Luca, Kitāb min ʿilal iḫtilāf al-nās fī aḫlāqihim (ed. Sbath 1941, 120.6–7): "[…] this spirit (sc. vital spirit) is shared by humans and animals (*yaštariku fī mā al-insān wa-l-bahīmah*) and they are distinguished from plants through it (*wa yanfaṣilāni bihā ʿan al-nabāt*)." Translation is mine.

68 Avicenna, Liber de anima II c. 4, ed. Van Riet 1972, 143.76–80 [ed. Rahman 1959, 75.4–7].

69 Avicenna, Liber de anima I c. 5, ed. Van Riet 1972, 84.73–77 [ed. Rahman 1959, 42.9–11]: "Et ex illis est gustus, qui est vis ordinata in nervo expanso super corpus linguae, ad apprehendendum sapores solutos ex corporibus contingentibus ipsum, cum permiscentur humori unctuoso linguae permixtione permutanti." Compare this position to Costa's position in the *De differentia spiritus et animae*: see above, 287–291.

70 Algazel, Metaphysica pars 2 tr. 4 c. 3, ed. St. Clair 2005, 65.121–125 [ed. Cairo 1936, 42.4–8]: "Gustus autem fit per virtutem attributam nervo expanso in exteriore superficie linguae, mediante humiditate suavi, quae nullius saporis est et est diffusa in superficie lingue. Haec autem apprehendit saporem saporiferi, et convertitur in ilium et coniungitur illi nervo, et sic nervus apprehendit eum per virtutem quae est in ipso."

tongue, as in the *De anima*),[71] but also through a second pair of cranial nerves (the fourth) connected to the palate.[72] The resulting picture produces a little tension between the *De anima* and the *Canon*/*De animalibus*: though maintaining tongue as the main organ of taste, there are seemingly other organic parts, such as the palate and its related nerve, that provide further perceptual content through the spirit flowing through them to the anterior brain ventricle, where these sensibles are perceived.

3.3 *Smell*

Consistently with other authors with medical backgrounds such as Costa Ben Luca and Nemesius, Avicenna does not explain smell as the transmission of the sensible entity through a pair of cranial nerves. Instead, like Plato he attributes the transmission of vapors through a couple of fleshy lumps (*duo carunculae anterioris partis cerebri*/*zāʾidatā muqaddim al-dimāġ*) placed inside the nostrils and connected to the front part of the brain near the first pair of cranial (optical) nerves.[73] This position is found in al-Ġazālī as well.[74] As noted by Hasse,

---

71 Avicenna, Canon I fen 1 doctr. 5 c. 2, ed. Venezia 1507, fol. 19rb [ed. New Delhi 1981–1996, vol. 1, 107.12–13]: "Quartus autem ramus tertii paris vadit penetrans a foramine suo quod in mandibula exit superiore et vadit ad linguam […] et tribuit ei sensum proprium et gustum." Avicenna, Abbreviatio de animalibus XII c. 8, ed. Venezia 1508, fol. 47vb [ed. Cairo 1970, 238.12–14]: "Sed ramus quartus huius paris tertii transit per unum foramen quod est in mandibula superiori et vadit ad linguam et dividitur in ramos in duplo lingue superiori et dat ei sensum proprium et gustum."

72 Avicenna, Canon I fen 1 doctr. 5 summa 3, c. 2, ed. Venezia 1507, fol. 19rb [ed. New Delhi 1981–1996, vol. 1, 107.16–17]: "Quarti vero paris origo est post secundum (Ar. *ṯāliṯ*, "third" and not second, as in the *Ḥayawān*) sed ad cerebri basim magis declinans, cum tertio quemadmodo diximus miscetur, deinde ab eo separatur et ad palatum vadit et dat ei sensum." Avicenna, Abbreviatio de animalibus XII c. 8, ed. Venezia 1508, fol. 47vb [ed. Cairo 1970, 238.17–19]: "Sed par quartus nascitur post tertium et est proprinquum medulle basis. Et miscetur cum tertio et post dividitur ab illo sicut diximus et venit ad palatum et dat sensum."

73 Avicenna, Liber de anima I c. 5, ed. Van Riet 1972, 84.69–73 [ed. Rahman 1959, 42.6–9]: "[…] olfactus, qui est vis ordinata in duabus carunculis anterioris partis cerebri similibus mamillarum capitibus, ad apprehendendum id quod offert ei aer attractus de odore qui est in vapore permixto cum aere (*fī l buḫār al-muḫālit lahu*), aut de odore impresso in illo ex permutatione quae fit ex corpore odorifero." Avicenna, Canon I fen 1 doctr. 5, c. 2, ed. Venezia 1507, fol. 19ra [ed. New Delhi 1981–1996, vol. 1, 106.4–5]: "Cerebro septem nervorum paria oriuntur: primi paris principium est ex fundo duorum ventriculorum cerebri antecedentium apud duorum additamentorum vicinitatem mammillarum extremitatibus similibus quibus fit odoratus." Avicenna, Abbreviatio de animalibus XII c. 8, ed. Venezia 1508, fol. 47va [ed. Cairo 1970, 236.15–16]: "Et primum par habet originem a duabus medullis primis anterioribus ubi nascuntur due mamille olfactus."

74 Algazel, Metaphysica pars 2 tr. 4 c. 3, ed. St. Clair 2005, 63.88–92 [ed. Cairo 1936, 40.10–14]:

Avicenna's view on odors differs notably from Aristotle's, for it does not postulate the existence of an immaterial species of different flavors in the medium but rather endorses two possible views on the transmission of smell and rejects one, which he deems untenable. According to Avicenna, the tenable explanation for the transmission of odors is that they may be small, smoky particles given off by the scented object and mixed with air. The second tenable hypothesis is that the medium (air or water) can be physically affected by the action of the odorous body. The third, and untenable, hypothesis is that the medium does not undergo any physical change but works as the medium for the transmission of the odor only: ironically enough, this was actually Aristotle's own view.[75] Avicenna's account of smell suffers from internal tensions, too, however. Avicenna produces evidence to support his view that, in order to smell, air must become filled with vaporous particles issued from the smelled object, as in the case of the apple that shrinks in the wake of its giving off odorous vapors. There are a limited number of cases which work as a litmus test for Avicenna's theory. For instance, he concedes that the case of vultures flying to distant places for prey cannot be explained primarily in terms of smell as material evaporation. Rather, given the considerable distance between the sentient being (vultures) and the source of odors (carcasses), it is reasonable to suppose that vultures individuate carcasses by sight rather than by smell.[76]

Tommaso Alpina has highlighted that, just like Aristotle, Avicenna posits the existence of senses in animals by appealing to empirical arguments.[77] For instance, a common tactic used by fishermen to capture fish consists in soaking the fishing nets with milk or roasted bait: according to Avicenna, this would prove that fish smell. Further, Avicenna seems to follow Aristotle's argument that annelids (*anulosa*) possess a particularly sensitive sense of smell and die because of the same exhalations that provoke death in humans.[78] Therefore,

---

"Olfactus fit per virtutem quae est in duabus carunculis cerebri quae sunt similes sumitatibus uberum, quae apprehendit mediante corpore quod patitur ab odoribus et retinet eos, vel partes odoriferorum permiscentur illi corpori, sicut est aer et aqua. Non sequitur autem quod partes odoriferorum permisceantur aeri."

75 Hasse, "The Soul's Faculties," 318.

76 Hasse, "The Soul's Faculties," 312–313; Avicenna, Liber de anima II c. 4, ed. Van Riet 1972, 148–154 [ed. Rahman 1959, 78–81].

77 Tommaso Alpina, "Translating Method: Inference from Behavior to Anatomy in Avicenna's Zoology," in *Premodern Experience of the Natural World in Translation*, ed. Katja Krause/Maria Auxent/Drora Weil, New York 2023, 136–152.

78 Avicenna, Abbreviatio de animalibus IV c. 2, ed. Venezia 1508, fol. 33rb [ed. Cairo 1970, 61.10; 62.7–8; 62.11–13; 63.15–16]: "et nisi olfacerent non venirent ad retem propter lac et alias res in rete olfactibiles [...] Et piscatores accipiunt afastem et assant ipsum et ponunt ipsum in rete, et statim intrant pisces rete propter odorem [...] Et animalia anulosa habent sensus

the case of annelids and fish shows that smell depends usually on breathing but is shared by most animals, especially those endowed with motive capacities. The physical modality of the transmission of odors to the brain stands out among all the other senses. If animal spirit is usually considered the vehicle of perceptual species, his role in smell is different. Given that there is no sensory nerve implied in transmission, we are left to assume that vapors affect the brain directly and that their sensible species are imprinted in the nervous terminations of the brain, filled with animal spirit, without the prior mediation of any cranial nerve.

3.4 *Hearing*

Hearing, by contrast, represents perhaps the less controversial case of distal sense perception. According to Avicenna, our sense of hearing is actualized when a physical body is hit by another, either by cutting (*scissio*/*qalʿ*) or percussion (*percussio*/*qarʿ*),[79] thus generating a wavy motion (*commotio*/*tamawwuǧ*) which, in turn, produces the immaterial species of sound, propagated through air or water until it reaches the ear.[80] From the outer surface of the ear, sound is transmitted to the air in the ear and then to the acoustic nerve,[81] i.e. the fifth

multos acutos, et olfaciunt et gustant et audiunt a remotis, et sunt quidam odores qui eis placent et interficiuntur a quibusdam odoribus, sic ab odore sulfuris et arse cutis et origani agresti […] et apes non veniunt ad locum mali odoris et non quiescent nisi in loco boni odoris et ubi est dulcedo."

79 On the definition of the faculty of hearing: Avicenna, Liber de anima I c. 5, ed. Van Riet 1972, 84.63–69 [ed. Rahman 1959, 42.1–5]; Algazel, Metaphysica pars 2 tr. 4 c. 3, ed. St. Clair 2005, 166.11–16 [ed. Cairo 1936, 40.11–15]: "Et cum motus ille pervenit ad aerem quiescentem intus inter concavitatem auris, imprimitur ille aer coniunctus nervo expanso in profundo concavitatis, quemadmodum corium extenditur in timpano. Fit igitur ibi tinnitus, et apprehendit eum virtus sita illi nervo."

80 Avicenna, Liber de anima II c. 5, ed. Van Riet 1972, 157.85–158.90 [ed. Rahman 1959, 83.12–84.4]: "Ergo sonus (*ṣawt*) accidens est quod accidit ex hoc praedicto motu quem consequitur et est cum illo. Cum autem commotio aeris aut aquae pervenit ad nervum expansum intus in aure receptibilem soni (*fa-iḏā intahā al-tamawwuǧ min al-hawāʾ aw al-māʾ ilā al-ṣimāḫ wa-hunaka taǧwīf fīhi hawāʾ iatamawwuǧu bi-tamawwuǧin-mā*), ante quem est concavitas in qua est aer quietus, movetur motu eius qui pervenit ad ipsum et post quem est quasi paries, super quem est nervus expansus sentiens sonitum, tunc sentitur sonitus."

81 Avicenna, Liber de anima I c. 5, ed. Van Riet 1972, 84.62–69 [ed. Rahman 1959, 42.1–5]: "Ex illis etiam est auditus, qui est vis ordinata in nervo expanso in superficiem nervi optici, ad apprehendendum formam eius quod sibi advenit ex commotione aeris qui constringitur inter percutiens et percussum resistens constrictione coacta, ex qua provenit sonus, et pervenit eius commotio ad aerem collectum quietum qui est in concavitate nervi optici, movens illum ad instar sui motus, et commotiones illius motus tangunt nervum."

pair of cranial nerves.[82] Once received in the acoustic nerve, sound is imprinted upon the receptacle of sense perception, animal spirit, and transmitted to the front ventricle of the brain.

3.5 *Sight*

Avicenna, in line with most of the philosophical and medical literature, devotes his most attentive analysis to sight, studied in the entirety of book III of his *De anima* and, as respects its physiological aspects, in his *De animalibus* (XII, 11). Particularly relevant for the present book is the chapter of *De anima*, III, 8. In this chapter, Avicenna also analyzes possible causes of the impairment of sight in relation to the motion of spirit in the eye, in the cranic nerves and in the brain. Before tackling Avicenna's physiology of sight, though, we need some preliminary clarification on Avicenna's theory of sight, which is quite different from Aristotle's.

Earlier in his life, as witnessed by his epistolary exchange with al-Bīrūnī,[83] Avicenna adopted the Aristotelian theory of vision. Later on, though, he changed his mind. The main object of his doubts concerned the definition of light in relation to the translucent medium. As shown by Dag Hasse, in the *De anima* Avicenna employed a distinction between "natural light" (*lux*/*ḍawʾ*) and "light" (*lumen*/*nūr*):[84] "Light", as defined in the "Easterners" (*Mašriqiyūn*), is a quality of certain bodies such as the sun and fire, whose light is just essential and non-separable from them.[85] On the other hand, "natural light" is the quality "acquired" (*istafāda*) by a natural body if and only if light hits upon it. Avicenna's conception of light moves significantly away from the Aristotelian idea of light as "color of the translucent" inasmuch light can exist independently of it.[86] While in Aristotle vision only needs light to function,

82 Avicenna, Canon I fen 1 doctr. 5 c. 2, ed. Venezia 1507, fol. 19[va] [ed. New Delhi 1981–1996, vol. 1, 107.27]: "Gustus autem in quarto creatus fuit nervorum et auditus in quinto [...]" Avicenna, Abbreviatio de animalibus XII c. 8, ed. Venezia 1508, fol. 47[vb] [ed. Cairo 1970, 239.9]: "Et nervus gustus (Ar. *ḏawq*, "taste" and not "nerve of taste") nascebatur a pari quarto et nervus auditus (Ar. *samʿ*, "hearing" and not "audible nerve") a pari quinto.' The Arabic is the same in the *Qānūn* and the *Ḥayawān*, while the Latin rendition is different: as we can see, these differences in the Latin are worth pointing out not per se but in order to highlight what material was employed by Albert at different stages of his production (*De homine*, *De animalibus* etc.)

83 Hasse, *Avicenna's* De anima, 108–109.

84 Hasse, *Avicenna's* De anima, 109.

85 Hasse, *Avicenna's* De anima, 109–112. An exception to the rule that essentially luminous bodies are colorless is represented by Mars and Saturn: both these planets have natural, non-acquired light and color at the same time.

86 Hasse, *Avicenna's* De anima, 109.

in Avicenna's account there are two different ways the translucent can be actualized. In the first case, the natural light (*lux*/ *ḍawʾ*) of the essentially luminous body produces some kind of motion and change in the translucent (*translucidum*/*šaffāf*), which causes the actualization of the medium and makes the *per se* luminous body visible. The second way in which the translucent medium is actualized is the acceptance of light (*lumen*/*nūr*) by the surface of visible objects. In his theory of vision, Avicenna shares the Aristotelian concern for an explanation of sight based on the actualization of the translucent medium. On the other hand, Avicenna considers light not as the actualized state of the translucent—as Aristotle would—but as the effect of a change in the medium itself.[87] To complete our picture of Avicenna's psychophysiology of sight, we need to turn to Avicenna's understanding of the physiology of sight.

3.5.1 Double Vision, Spirit, and Sight Impairments

As argued so far, proper and common sensibles perceived by sight are transmitted by the "psychic spirit" (*spiritus animalis*/*rūḥ ḥayawāniyya*). But we still need to provide a precise physiological explanation of this process. When the visible species gets imprinted upon the eye's crystalline humor, it reaches the anterior ventricle of the brain via the first pair of cranial nerves, viz. optical nerves.[88] In the chapter of the *De animalibus* (XII, 11) excerpted from the *Canon* (I, I, 5, 2) and specifically devoted to the anatomy of the eye, Avicenna details the motion of animal spirits from the eye, highlighting the key role of optical chiasm, the seat of the production of the final image of sight transmitted to the anterior ventricle of the brain.[89] According to Avicenna, when the two optical

87 Hasse, *Avicenna's* De Anima, 111: "What is Aristotelian […] is the emphasis on the need for actualization of the medium, but what is not Aristotelian is the fact that light, being the quality of certain bodies, exists independently of the medium. Where Aristotle says that light is the actualized state of the translucent, Avicenna relegates this actualization to a movement or change of bodies."

88 Avicenna, Canon I fen 1 doctr. 5 c. 2, ed. Venezia 1507, fol. 19[ra] [ed. New Delhi 1981–1996, vol. 1, 106.4–5]: "Cerebro septem paria oriuntur: primi paris principium est ex fundo duorum ventriculorum cerebri antecedentium apud duorum addimentroum vicinitatem […] Et ille qui a sinistra oritur dextratur. Et ille qui a dextra oritur sinistratur […]"; Avicenna, Abbreviatio de animalibus XII c. 8, ed. Venezia 1508, fol. 47[va] [ed. Cairo 1970, 236.10–12]: "Primum par habet originem a duabus medullis primis anterioribus […] et ille que venit a dextra venit ad sinixtram et e econverso […]"

89 According to Avicenna, the optical chiasm has a threefold utility: (1) providing an additional amount of spirit to the other eye when one is closed, (2) to unify the two simulacra of the visible object and (3) to substain physically the other optical nerve: Avicenna, Canon I fen 1 doctr. 5 c. 2, ed. Venezia 1507, fol. 19[ra] [ed. New Delhi 1981–1996, vol. 1, 106.10–19]

nerves meet in the optical chiasm, they produce a single image out of the two pyramidal simulacrum (*piramidis*/*maḫrūṭ*) initially imprinted upon the two crystalline humors. The pyramidal base of these images corresponds to the perceived object on the surface of the eye and its vertex in the crystalline humor. Avicenna relies to some extent on the previous medical tradition (Galen and Ḥunayn ibn Isḥāq) but also moves significantly away from it in arguing that the faculty of sight is not seated in the crystalline humor as in Galen or Ḥunayn. Rather, he locates the faculty of sight in the optical spirit (or the optical nerve) beside the optical chiasm. This is because it is there that, in his view, the two visual images from the two eyes become a unified, single one.[90]

Given the nature of the optical spirit as the material vehicle for these visible images, possible affections befalling the animal spirit can result in misperception. This can happen, for instance, when we perceive a single thing as two. According to Avicenna, there are four possible causes for this phenomenon. The first cause is the distortion of spirit (*contorsio instrumenti quod reddit simulacrum*/*intiqāl al-āla al-mu'addiya li-l-šabḥ*) as it transmits the visible image to the faculty of sight. If spirit deviates from its regular trajectory, the two pyramid-like images fail to clearly unify when they meet in the optical chiasm.[91]

The second cause is the rightward or leftward wavy motion of the spirit (*commotio ad anteriora, dextrorsum et sinistrorsum*/ *ḥarakatu al-rūḥ al-bāṣir wa-tamawwuğuhu yaminatan aw yasiratan*) as it enters the crystalline humor in one or both eyes. If spirit passes the natural end-point of its motion and tends toward one of the extremities of the crystalline humor, then one or both of the images impressed upon spirit are incompatible with each other. According to Avicenna, the result is as blurry as the reflection of the sun's image on rippled water.[92]

A third cause of double vision is the backward and forward motion of spirit after it passes the optical chiasm. According to Avicenna, this backward or forward motion is a possible side effect of the speed by which these images are transmitted through animal spirits to the front brain ventricle. This speed would cause a sort of short circuit in the motion of animal spirits. Contrary motions of this kind can occur either in the optical nerve or in the front brain ventricle, producing the apprehension of two successive image. When the first image impressed by spirit on the organ of common sense has not been deleted

---

Avicenna, Abbreviatio de animalibus XII c. 8, ed. Venezia 1508, fol. 47[va] [ed. Cairo 1970, 237.1–11].

90 Hasse, *Avicenna's* De anima, 122–123.

91 Avicenna, Liber de anima III c. 8, ed. Van Riet 1972, 273.3–14 [ed. Rahman 1959, 154.12–21].

92 Avicenna, Liber de anima III c. 8, ed. Van Riet 1972, 273.15–27 [ed. Rahman 1959, 155.1–9].

from it and stored in retentive imagination (*imaginatio*), the second image reaches the organ of common sense. The result is a double vision because of the quick succession of simulacra received by common sense.[93] To provide a further physiological example of such a short circuit, Avicenna reports the case of the observation of a point placed upon an object rotating at high speed, which seemingly draws a straight, rectilinear line in the eyes of the observer because of its high speed. Even though spirit is designed to receive quickly the visible species, the excessive speed of the rotation in the object—and, consequently, of the images imprinted upon spirit—causes an overlap in the transmitted images. This overlap of rapidly successive images makes the observer see this rotating point as though as it constituted a unique straight line and not a series of discrete points upon a rotating surface. This phenomenon could cause a dizziness comparable to vertigo.[94] Vertigo, though, has a different causal explanation: usually, it comes about because of an excess in vapors inside the brain; these vapors, in turn, impede the correct reception of visible forms.[95]

The misperception of one thing as two is further accentuated in this case by the material composition of spirit, which is very thin and quickly receives sensory stimulus: if this quickness is outstripped by excessive speed in the perceived object, then the perceived image does not mirror exactly the individual instances of perception but rather their sum, without allowing the perceiver to single out the discrete moments of perception.[96] This pertur-

---

93 Avicenna, Liber de anima III c. 8, ed. Van Riet 1972, 273.28–275.49 [ed. Rahman 1959, 155.10–156.6]. Once the visible image has been unified, so have the common sensibles embedded in it: more on this aspect of the Avicennian account of vision in Charles Lindberg, *Theories of Vision from al-Kindi to Kepler*, Chicago 1976, 43–52. Lindberg's account, though, must be read against the backdrop of the cautions expressed by: Hasse, *Avicenna's* De anima, 123, since Lindberg characterizes the Avicennian account of vision as a *tout court* defense of the Aristotelian standpoint which, as we have seen, it is not.

94 Avicenna, Liber de anima III c. 8, ed. Van Riet 1972, 275.52–276.68 [ed. Rahman 1959, 156.9–20]. On this explanation of vertigo, see also: Carla Di Martino, Ratio particularis: *La doctrine des sens internes d'Avicenne à Thomas D'Aquin. Contribution a l'etude de la tradition arabo-latine de la psychologie d'Aristote* (Études de philosophie médiévale XCIV), Paris 2008, 32.

95 Avicenna, Liber de anima IV c. 1, ed. Van Riet 1968, 3.34–41 [ed. Rahman 1959, 164.8–12]: "[…] vertigo (*dawār*) autem non est nisi causa motus vaporis qui est in cerebro et in spiritu qui ibi est, cum accidit spiritui circumvolvi; ergo virtuti quae ibi est, accidit id quod iam expedivimus. Similiter contingit hominem inspicere aliquid velociter circumvolvi, sicut praediximus; hoc autem non fit propter aliquid quod sit in aliqua parte oculi nec in spiritu ibi expanso, cuiusmodi est similitudo velocitatis mobilis punctalis quae videtur directa vel circularis, sicut iam praedictum est."

96 Avicenna, Liber de anima III c. 8, ed. Van Riet 1972, 277.82–90; 277.98–278.6 [ed. Rahman 1959, 157.9–15; 158.1–7]: "Debes autem scire quod praeter has causas est ibi alia causa mate-

bations occur especially in the case of vertigo, in the front brain ventricle: this is because the perceived images remain in the faculty of common sense as long as there is a relation (*nisba*, which the Latin translates as *comparatio*) between object, visible species and observer: when the sensed object is gone, this relation no longer exists and the image is stored in the formative faculty, otherwise known as imagination (*formativa, imaginatio/muṣawwira, ḫayāl*).

The fourth and final cause of double vision is attributed by Avicenna to the pupil's motion, which he calls "the hole of the grape-like tunic of the eye" (*foramen uveae/ ṯuqba al-ʿaynabiyya*): according to him, its constriction or dilatation makes the visible object look smaller (when constricted) or larger (when dilated).[97] If the motion of the pupil is leftward or rightward, then the perceived images are have a different angle. This is because the perspective from which they were observed differs at the time when the sensible species are imprinted upon the crystalline humor and then upon the spirit, in a manner that quite resembles the second case of double vision.[98]

So far, we have observed how Avicenna integrates spirit (*spiritus/rūḥ*) as a vehicle of proper and common sensibles into a physiological and anatomical account of sense perception that blends together the Galenic medical tradition of the Islamic World and Aristotelian psychology. The integration of the Galenic anatomy of sense perception into a theory of sense perception based on a revisited version of Aristotelian psychology is particularly observable in Avicenna's theory of "inner senses" (*sensus interiores/ḥawāss al-bāṭina*): common sense, imagination, imaginative faculty, estimation and memory, which

rialis adiuvans illas, quae est haec scilicet quod substantia spiritus est substantia ultimae subtilitatis et velocis oboedientiae ad recipiendum motum, ita ut, cum contigerit in eo causa propter quam debeat mutari simulacrum de una parte ad aliam, sequatur moveri substantia spiritus aliquo motu in directum illius partis, quamvis sit parvissimus ille motus. Causa autem huius est quoniam unaquaeque virtutum apprehendentium impellitur naturaliter ad suum comprehensibile, ita ut quasi delectetur cum illo; […] Et propter hanc causam, cum inceperit homo inspicere aliquid quod circumvolvitur, videntur ei omnia alia circumvolvi, eo quod contingit in spiritu motus circularis ex hoc quod sequitur motum simulacri; et praeter hoc, cum inspexerit quod cito movetur in directum, contingit in spiritu motus rectus e contrario illius partis; pars enim motus rei est contraria parti motus habentis simulacrum. Oportet ergo videri omnia moveri contra illam partem: simulacra enim rerum non sunt fixa."

97 Avicenna, Liber de anima III c. 8, ed. Van Riet 1972, 278.7–14 [ed. Rahman 1959, 158.8–13].

98 Avicenna, Liber de anima III c. 8, ed. Van Riet 1972, 278.14–17 [ed. Rahman 1959, 158.13–15]: "[…] aut continget cum inclinabitur (sc. foramen) ad unam partem, videri in alio loco, et continget quasi primum visum sit praeter visum secundum, praecipue cum apparuerit alia forma antequam destruatur prima."

will become clearer in the next chapters. Among these inner senses, we shall consider here the case of common sense and its relation to sense perception.

3.6 *Common Sense*

Common sense is an inner sense because it perceives common sensibles only through the proper sensibles (color, sound etc.) inasmuch as they are related to them. Indeed, the common sensibles constitute an essential part of perceptual content, but they cannot be perceived independently of proper sensibles themselves. Common sense is the sense that, according to Avicenna, allows one to hear, see, smell etc. all at once and, at the same time, works as the collector of the sensory images from the different sense organs.[99] As stressed by Dag Hasse, this central role of common sense in perception makes it the faculty that "truly senses" (*vere quae sentit*/*bi-l-ḥaqīqa allatī tuḥissu*) and unifies all of the sensorial stimuli.[100] Unlike Aristotle, common sense is not the faculty charged with second-level perception, i.e. perceptual awareness, however. Rather, in Avicennian psychology, this function is attributed to estimation.[101] Common sense works as a sort of operative center of sense perception, in a manner which Peter Pormann has compared to the Stoic concept of *egemonikòn*, connected to external senses through shoots (*rami*/*suʿb*).[102] This psychological fact is physiologically exemplified by Avicenna's conception of the flow of animal spirits, which moves from the front brain ventricle—where common sense is seated—into the sense organs, mostly through cranial nerves. This is true in all cases of sense perception except for smell, for its vapors directly affect the brain via the fleshy part in the nostrils connected to the brain or via the pores in the skin.

Jari Kaukua has recently highlighted how Avicennian common sense seems to possess an active role in structuring the content of sense perception in a manner which goes well beyond the purely passive structure of an Aristotelian psychology of sense perception. This phenomenon is illustrated by two case studies found in the *De anima* (III, 7 and III, 8), viz. the raindrop and the point rotating on an oblong surface. In both cases, the cognitive content of common sense encompasses three distinct moments in sense perception: the place at which the drop or the point was in a prior phase of the movement, the place

---

99 Avicenna, Liber de anima I c. 5, ed. Van Riet 1972, 87.19–22 [ed. Rahman 1959, 44.3–6].

100 Avicenna, Liber de anima IV c. 1, ed. Van Riet 1968, 2.56–59 [ed. Rahman 1959, 165.6–8]: "Et haec virtus est quae vocatur sensus communis, quae est centrum omnium sensuum et a qua derivantur rami et cui reddunt sensus, et ipsa est vere quae sentit." Hasse, The Soul's Faculties, 309.

101 Di Martino, Ratio Particularis, 26–27.

102 Pormann, "Avicenna on Medical Practice," 104.

at which it is now, and the extension between the former and current places. Because of the high speed of both the sensed object and the motion of the animal spirit responsible for sight, the image passed to common sense is an image composed of these three distinct aspects, as if they were sensed all at once.[103] If we consider that these three aspects of sense perception cannot be reduced to a single moment in sense perception itself and that they rather relate to a continued motion, motion itself being a measure of time as maintained by Avicenna (following Aristotle), then, concludes Kaukua, common sense gives us an experience of time.[104] As Avicenna stresses in his *Physics* when discussing the relation between temporally extended motion and its dependence on the perceiving soul, the image (such as the raindrop) is retained in the mind by the formative faculty as if the 'now' when this image was perceived was singular. When the mind considers this image stored in the imagery, though, it can actually become aware that one of the nows is prior in existence to the other.[105] In the end, time as duration comes to be in perception through common sense and through the form-bearing faculty which stores these images. The relation of these time-related perceipts to the form-bearing faculty is, unfortunately, far from clear.[106] Expanding on Kaukua's argument, it can be added that animals are able to perceive time through common sense via the workings of animal spirit. Given the high receptivity of spirits to external sense stimuli, their flow from the eyes to the brain is able to replicate in the brain an image that looks as if it was sensed in a single moment. But Avicenna's account opens the door to a further question: are all animals aware of time when they perceive? This will remain an open question, one which Albert himself will address.

To complete our study of common sense, let us briefly recall Avicenna' handling of common sense in the *Canon*. In his chapter on the faculties, he deals

103 Jari Kaukua, "Avicenna on the Soul's Activity in Perception," in *Active Perception in the History of Philosophy. From Plato to Modern Philosophy*, ed. José Filipe Silva/Mikko Yrjönsuuri (History of the Philosophy of the Mind 14), New York 2014, 104; Avicenna, Liber de anima III c. 7, ed. Van Riet 1972, 255.20–256.32 [ed. Rahman 1959, 142.15–143.5]. On the case of the point moving on the surface, see the section above.

104 Kaukua, "Avicenna on the Soul's Activity," 104.

105 Kaukua, "Avicenna on the Soul's Activity," 105: "[…] Avicenna consistently recognized the need to account for the constitution of our experience of time, and that he found the theoretical means for such an account in his theory of the internal senses." Avicenna, Sufficientia II c. 13, ed. Van Riet 1992, fol. 36va [ed. Cairo 1983, 258].

106 Kaukua, "Avicenna on the Soul's Activity," 105: "Whereas the psychological context speaks only of the common sense, here we have the two impressions (past and present) relegated to the imagery. This seeming incoherence notwithstanding, the central point remains: time as duration first comes to be in perception, whether in the common sense or in the imagery." Avicenna, Sufficientia II c. 1, ed. Van Riet 1992, fols. 23^{vb-ra} [ed. Cairo 1983, 112].

with "psychic apprehensive faculties" (*virtutes animales comprehendentes/ quwà al-nafsāniyya al-mudrika*) and classes common sense under the usual medical category—found also in Nemesius and Costa ben Luca—of the virtue of phantasy. This mention comes without any reference to the awareness of time or to the possible pathological implications of sight, which can be found elsewhere in the *Canon* in the second book, which provides a tip-to-toe discussion of illnesses in the human body. It is remarkable to remember that Avicenna adds to his own brief statements about common sense that, according to philosophy, common sense and phantasy are two distinct virtues. In this manner, he stresses once again that the determination of the number and essence of the soul's faculties or powers belongs to natural philosophy alone and not to medicine.[107]

3.7 *The Teleology of Sense Perception*

A further aspect of Avicenna's psychophysiology of sense perception that emerges only in his psychology concerns the teleology of sense perception and common sense. Yet, as proven by a comparative table of contents between Aristotle's and Avicenna's *De anima* presented in the recent book by Tommaso Alpina on Avicennian psychology, the discussion of the teleology of sense perception in Avicenna's *De anima* is much less explicit than Aristotle's and the evidence for it quite scant.[108] Like Aristotle, Avicenna regards touch as the primary sense in relation to the subsistence of the animal.[109] On the other hand, the only extensive discussion which he provides is incidentally found in his

107 Avicenna, Canon I fen 1 doctr. 6 c. 1, ed. Venezia 1507, fol. 24vb [ed. New Delhi 1981–1996, vol. 1, 128.17–20]: "Et virtus quidem comprehensiva occulte et non sentiens est sicut genus quinque virtutum. Una est virtus que vocatur sensus communis et phantasia et apud medicos quidem sunt una virtus; sed apud certificantes qui sunt ex philosophis due sunt virtutes. Sensus enim communis est illa que omnia sensu percepta recipit et propter earum formas patitur quae in ipsa copulantur." Pormann, "Avicenna on Medical Practice," 105–106.

108 Alpina, *Subject, Definition, Activity*, 9–11.

109 Avicenna, Liber de anima II c. 3, ed. Van Riet 1972, 131.87–132.1, 132.5–6 [ed. Rahman 1959, 67.12–68.2, 68.5–6]. A similar position in: Averroes, De sensu et sensato, ed. Shields/Blumberg 1949, 4.35–40 [ed. Blumberg 1972, 4.5–12]: "Tactus quidem et gustus sunt necessarii; auditus vero et visus et olfactus sunt propter melius (*min ʾaǧli al-afḍal*): per gustum enim distinguitur cibus conveniens a non conveniens (*yumayyazu al-ṭaʿam al-mulāʾim min ġayr al-mulāʾim*); et per tactum distinguuntur ea quae corrumpunt corpus ex extrinseco et ea quae conservant corpus (*wa-bi-ḥāssati al-lims yumayyazu al-ʿumūr allatī tufassidu badanahu min ḫāriǧ wa-allatī taḥfazuhu wa tanāsabahu*). Alii vero sensus non sunt innati distinguere talia; et ideo non sunt necessarii (to the animal's existence [Ar. *fī wuǧūd al-ḥayawān*, omitted in the Latin])."

study of the process of learning. In *De anima* v, 1 he remarks that voice is essential in teaching and is able to produce knowledge without the need for any other medium. Moreover, he deems the other senses such as sight as valuable insofar as they can provide additional information, as in the case of the teacher communicating to students through facial expressions. This is true, though, only in the case of humans and not of animals, which do not, strictly speaking, learn.[110]

As we have shown, expanding on previous studies, Avicenna's view of the role of spirit as a vehicle of sense perception rests on the Galenic concept of animal spirit and complexion, whose function covers all the senses except for smell and, unlike his predecessors, common sense. His most typical feature consists in the blending of the Galenic concept of animal spirit—mostly human-related—with the Aristotelian framework of sense perception, encompassing the whole of the animal kingdom. The enlargement of the notion of spirits concerns even animals with connate *pneuma*, such as insects and the like. This is not to say that in these animals we can discern a physiological system as apparent as in humans (with brain, cranial nerves etc.), but that this system is functionally analogous to that possessed by humans. Moreover, animals possess an analogous system of differentiating the material substrate of the soul's powers, viz. their different humors and complexions.[111]

---

110 Avicenna, Liber de anima v c. 1, ed. Van Riet 1968, 71.28–72.48 [ed. Rahman 1959, 203.5–204.4]: "Ex his ergo causis et aliis minus evidentibus sed pluribus numero necessarium fuit homini habere naturaliter potentiam docendi alium sibi socium quod est in anima eius signo aut opere. Ad hoc autem commodior fuit vox quae dividitur in elementa ex quibus fiunt multae compositiones sine labore corporis, et sunt tales quae non permanent et secretae ei qui non debet eas percipere. Post vocem autem est nutus qui est eiusmodi, sed vox magis indicat quam nutus: nutus enim non indicat nisi secundum hoc quod subiacet visui, quod fit ex quadam proprietate eius, quia cum quis voluerit aliquid indicare alii, necesse est movere oculum ad partem propriam multis motibus quibus nutus intelligatur; vox autem sibi sufficit nec est ei necesse ex aliqua proprietate sua indicare motibus; praeter hoc etiam non eget medio ad hoc ut apprehendatur, sicut non eget eo color, quod necessarium est nutibus; ergo natura fecit ut anima ex sonis componeret aliquid per quod posset docere alium. Reliqua quoque animalia habent sonos, quibus sciunt alia dispositionem eius quod est in cordibus eorum; ipsi autem soni non significant nisi naturaliter et confuse quod appetitur aut quod respuitur indiscrete; quod autem habet homo de hoc est ad placitum, eo quod humani appetitus quasi infiniti sunt; unde non potuit homo naturaliter habere sonos sine fine."

111 Avicenna, Liber de anima v c. 8, ed. Van Riet 1968, 175.58–176.63 [ed. Rahman 1959, 263.15–19]: "Habet autem complexionem propriam, quae mutatur etiam prout necesse est in ea esse diversitatem propter quam fit vehens virtutes diversas: complexio enim qua concupiscitur non est apta ut per eam sentiatur, nec complexio quae est apta visibili spiritui est apta spiritui moventi; si autem complexio una esset, virtutes quae sunt in spiritu essent una, et actiones earum una."

From the point of view of the teleology of sense perception, not much of the Aristotelian discussion on the primacy of senses is left. Though maintaining the primary role of touch as the first sense for the existence of the individual and the primacy of hearing over sight in the process of learning, there is no room for any classification of sense according to the criterion of being or well-being. On the other hand, Avicenna's psychology of common sense presents two elements (the role of animal spirits and the perception of time) that develop and expand on Aristotelian psychophysiology, correcting its flaws by implementing Galenic medical learning and adding, as in the case of time, some doctrines peculiar to Avicenna alone. Be that as it may, the extension of the Galenic concept of animal spirit to the whole animal kingdom, welded together with a revisitation of Aristotelian psychology of sense perception, marks a crucial step in the history of natural philosophy and zoology that is crucial for understanding Albert's approach to *spiritus* from his *De homine* onward.

4 *Spiritus animalis* and *potentiae apprehensivae deforis* in Albert's Psychophysiology

Albert's psychology and physiology of sense perception is, in keeping with his general attitude, much indebted to Avicennian psychology. Expanding on Hasse's, Anzulewicz's, Steneck's and Dewan's studies,[112] we shall try to complete the study of Albert's psychophysiology of sense perception and common sense by producing a more organic account of his notion of animal spirit (*spiritus animalis*) in his early (*De homine*) and mature works (Aristotelian paraphrases). As correctly noted by Steneck, Albert does discuss at length the role of animal spirits and, more generally, the anatomy and psychophysiology of sense perception. Contemporary treatises on the soul, are not as detailed as Albert in this respect.[113] There are only a few remarks, for instance, found in Roland of Cremona's *Summa* and William of Auvergne's *De anima* about the role of spirit in sensation. These remarks are far too scant to be comparable to the psychophysiology of sense perception laid out in the *De homine*.[114] While his

112 See 274, n. 2 at the beginning of this chapter.

113 Steneck, "Albert on the Psychology of Sense Perception," 275.

114 Rolandus Cremonensis, Summae magistri Rolandi Cremonensis, II CXVIII n. 14 (ed. Cortesi-Midali 2017, 170): "Quilibet autem istorum sensum habet proprium spiritum, qui est instrumentum eius, sive vehiculum, ut dictum est. Verbi gratia. Instrumentum visus est spiritus visibilis, qui est in pupilla oculi, sive vehiculum et spiritus ille est quoddam corpus subtile igneum, et appellatur etiam calor naturalis, quia de vapore cordis efficitur

detailed account of the role of animal spirits in sense perception sets him apart from his contemporaries, there are other aspects of Albert's theory of sense perception shared with contemporary authors which make him a child of his time. The first one is represented by the unity between the faculties of the sensitive soul and of the soul itself, while the second concerns the famous doctrine of so-called "spiritual being" (*esse spirituale*), referred to the mode of existence of the sensible species in the sensed object, in the medium of sensation—in distal senses such as sight, hearing and smell—and in the sentient subject. This doctrine was supposed to answer a straightforward yet intricate question: is the form that we receive in our senses without matter the very same form as that which exists with matter in the object perceived?

The first philosophical question, though, viz. the unity of the different faculties of the sensitive soul in relation to the soul itself and its cardiocentric localization, posed a challenge to its Medieval Latin readers because of the relationship between Aristotle's psychology and (outdated) physiology of sense perception. Aristotelian psychophysiology of sense perception simply ignored the role of nerves and downplayed that of the brain. For Medieval readers of Aristotelian psychology, the challenge was to blend the new physiology of sense perception inherited from Galenic medicine with Aristotelian psychology. Very often, this problem was solved by drawing on Avicenna, who happened to be pursuing this very task. This attempt to maintain the primacy of the heart over the other key members (brain, liver) while at the same time emphasizing the unity of the soul has been already analyzed in chapter 2 and 3 as concerns the overall structure of the soul and its relation to its powers (chapter 2), especially the powers of the nutritive soul (chapter 3). This discussion, as we have seen, could have different outcomes: on the one hand, authors who endorsed the plurality of forms (such as John of La Rochelle, the *Summa Halensis* etc.), while other authors like Albert and Peter of Spain favored a solution based more on Avicenna's own psychological ideas.

As we saw in the second chapter, Albert conceives of the soul as substantially one but manifold in its powers, which emanate from it according to the

qui est calor naturalis. Aliquando ille spiritus deficit in homine, et tunc fit homo caecus. [...] Sed, si opiletur via ne possit de cerebro venire spiritus, semper remanebit homo cecus, donec removetur illa opilatio. Similiter auditus habet suum, et sic de alii. Sicut enim motus virtutum sunt in anima, sic sunt ipsi spiritus in membris sensum. Et, cum homo dormit, trahuntur isti spiritus versus suum principium, idest versus cor, et ideo membra sensuum in dormitione remanent sensibilia." William of Auvergne, Tractatus de anima c. 1 p. 5 (ed. Paris 1674, fol. 70^{b}): "Manifestum est igitur ex omnibus his quia potentia sentiendi maxime est in primis est et maxime in nervis et in spiritibus."

Neoplatonic notion of *fluxus*. Much as Albert maintains the problematic Avicennian and Boethian distinction between the soul's essence and its powers, we have observed how Albert also tries to stress the unity of the soul, as opposed to the frequent pluralistic views circulating in Paris already at the time of the *De homine*. In keeping with this approach, Albert borrows from Avicenna (*De anima* V, 8) his explanation of the derivation of the sensitive powers seated in the brain from the heart via the cardiac *pneuma*. According to Albert and Avicenna, the cardiac origin of the powers of the sensitive soul grant unity between the powers of the sensitive soul and those of the nutritive soul. Moreover, it makes their localization in the brain not dangerous, for their localization in the brain does not put the brain on an equal psychological footing with the heart, which Albert considers the "seat of the soul" (*sedes et domicilium animae*). This definition, borrowed from Alfred of Shareshell's *De motu cordis*, emphasizes the cardiocentric nature and origin of all psychic powers, already explained in the embryological context. This view appears already in his early *De homine*.[115] While it is relatively clear that the powers of the sensitive soul seated in the brain and in the nerves should come to the brain from the heart via cardiac *pneuma*, this fact alone does not explain how Albert, like Aristotle and Avicenna, does not view the different senses as isolated sources of information. To understand how the single senses are integrated into a unitary framework of monitoring and coordination of sense perception, it will be necessary to consider, as we shall do later in this chapter, the key role of common sense.

Albert comes back to his discussion of the cardiocentric origin of the brain- and nerve-seated powers of the sensitive soul in his mature *De animalibus*, more specifically in the chapter dedicated to the resolution of the Galenic-Aristotelian quarrel over the origin of veins (III, 1, 6). When refuting Galen's position on the origin of veins, Albert states that the power of the brain is such only because it is derived from the heart.[116] More generally, we have seen how

---

115 Albertus Magnus, De homine (Ed. Colon. XXVII/2, 275.8–13): "Aliter tamen solvit Avicenna dicens quod nihil prohibet quod aliqua virtus oriatur in uno membro et perficiatur in altero et tunc completa revertatur ad ipsum. Unde ipse vult quod virtus sensitiva oriatur a corde et perficiatur in cerebro […]."

116 Albertus Magnus, De animalibus III tr. 1 c. 6 n. 63 (ed. Stadler, I, 304.29–39): "Sensus autem primo non est in cerebro, quia ipsum non sentit: nec primo est motus in ipso, cum ipsum non moveatur: sed utrumque istorum primo tempore et natura ostendemus in sexto decimo huius scientiae libro inesse cordi: propter quod aliud est principium sensus et aliud sensus virtutem determinans et distinguens et approprians. Principium autem sensus et motus est cor: approprians autem et distinguens et determinans sensus est organa determinatorum sensuum. Et quia principium utriusque istorum iam cor probatum est esse,

Albert strongly stresses the unity of the soul's dependence upon the cardiocentric seat and origin of its powers. From this perspective, his doctrine closely resembles Avicenna's and can well be harmonized with Aristotle's and Averroes' cardiocentrism.

By contrast, Albert's take on the mechanism of sense perception seems to be strongly influenced by Averroes' doctrine of "spiritual being" (*esse spirituale*). Averroes' highly influential doctrine of spiritual being is aimed at explaining how the form existing in the sensed object is able to affect the soul's perceptive capacities. Rotraud Hansberger has shown how, based to some extent on the adaptation of Aristotle's *Parva Naturalia*,[117] Averroes' doctrine claims that sensible species possess progressive degrees of spirituality, from the lower one found in the sensed object to the highest degree of spirituality found in the object of memory.[118] According to Averroes, the fact that sensible species exist in their medium and, in general, that they exist in the sense organs of sense perception, is due to their possession of a more 'spiritual' being than the form as it was found on the object itself. This higher degree of spirituality allows for the form to be gradually rinsed from its material appendages until it reaches its most spiritual degree, found in memory. The fact that a sensible species exists in the medium (air, water) corresponding to a spiritual being allows, according to Averroes, the coexistence of sensible species with those of contrary sensibles (e.g. black and white) in the same medium (e.g. air) without their canceling each other out, just like contrary notions can exist in the intellect without mutually excluding each other. Moreover, the doctrine of the 'spiritual being' of sensible species in the medium was supposed to make up for the shortcomings of a materialist explanation of sensation, which clashed against empirical examples such as vultures, able to smell dead corpses from afar.

The first reception of this doctrine in the Parisian milieu—where Albert first studied theology—is witnessed in the anonymous *De potenciis animae et obiectis* edited by Daniel Callus and dating back to the 1220s. This date places the first reception of the Averroist doctrine of spiritual being significantly earlier than Albert's *De homine* (1241/1242).[119] The reception of the doctrine of spir-

ideo necessario concluditur cor etiam esse principium nervorum qui sunt instrumenta quibus sensus et motus per corpus distribuuntur."

117 On the debt of Averroes' theory of "spirituality" to the Arabic adaptation, see: Rotraud E. Hansberger, "Averroes on the Internal Senses," in *Interpreting Averroes. Critical Essays*, ed. Peter Adamson/Matteo Di Giovanni, Cambridge 2019, 148–149.

118 Averroes, De memoria et reminiscentia, ed. Shields/Blumberg 1949, 58.6–59.2 [ed. Blumberg 1972, 42.11–43.4].

119 Rega Wood, "Spirituality and Perception in Medieval Aristotelian Natural Philosophy," in

itual being had different nuances in Medieval thought, though. On the one hand, thinkers like Alexander of Hales (*Quaestiones antequam esse fratris*), Robert Grosseteste (*Hexaëmeron, Commentary on the Posterior Analytics*) and Richard Rufus of Cornwall (*Sententia cum quaestionibus in libros De anima*) endorsed it.[120] The peculiarity of the doctrine of spiritual being professed by these authors rests in what Rega Wood has called their "light theory": light is the agent whose action abstracts the sensible species from their material substrates and gives the species their spiritual being. This abstracting action of light, though, does not apply merely to colours but to sensible species of all senses: this is the doctrinal peculiarity of their theory, which Albert was acquainted with.[121] On the other hand, authors such as Aquinas endorsed the doctrine of *esse spirituale* of the form in the medium or in the different degrees of abstraction. Renée Antoine Gauthier has emphasized that Aquinas, in so doing, deprived the Averroean concept of spiritual being of its typical features by equating 'spiritual' with 'immaterial'.[122] One of the few exceptions to this apparent consensus to the endorsement of the doctrine of spiritual being, in spite of its different interpretations, is represented by Roger Bacon. In his study, Smith higlights Bacon's straightforward rejection of the existence of the sensible species and, on the other hand, his claim that the sensible species which existed in the medium existed only in a state of corporeal existence (*esse corporeale*) since they require a material (spherical) medium within which to subsist. The light, according to Bacon, is then propagated from any luminous or illuminated object through the medium in a form of a sphere and does not need to exist in it according to spiritual being.[123]

Albert, by contrast, developed his own model of spiritual being, which can already be observed in his early works such as the *De homine* and in a manner that is quite original, to some extent, compared to his contemporaries. While

---

Medieval Perceptual Puzzles: Theories of Sense Perception in the 13. and 14. Centuries, ed. Elena Baltuta (Investigating Medieval Philosophy 13), Leiden 2020, 157 n. 12.

120 Wood, "Spirituality and Perception," 169 n. 48 (on Alexander and Grosseteste), 168–169 (on Rufus). Note that Grosseteste's theory differs in that it upholds an extramissionist theory of vision.

121 Wood, "Spirituality and Perception," 168–169: "light is responsible for the abstraction of sensible forms. But in Rufus light is responsible for the spiritual being of all sensible qualities […] all sensibles evolve from incorporated light, not just colour, but also sound and every other natural, external sensible."

122 Renée Antoine Gauthier, "Préface," in: Thomas Aquinas, Sentencia libri de anima, ed. Fratres Praedicatores (Rene-Antoine Gauthier), Roma–Paris 1984 (Ed. Leon. XLV/1), 226–227.

123 A. Mark Smith, "Perception," in *The Cambridge History of Medieval Philosophy*, ed. Robert Pasnau/Christina van Dyke, Cambridge 2010, vol. 1, 339–340.

we might well expect to find traces of the doctrine of *esse spirituale* in a study of distal senses (sight, hearing, smell), it is less intuitive for it to appear in the context of contact senses such as touch and in a manner that actively involves the concept of animal spirit as well. That it appears in such environs in Albert's work testifies to how his thought develops by following doctrines developed in his historico-philosophical thinking and putting his own spin on to them.

4.1 *Animal Spirits and Sense Perception: Touch*

Over the entire course of his intellectual production, Albert maintained Aristotle's classic approach to the tangible qualities, reduced to the oppositions among primary qualities. Concerning the understanding of the faculty of touch itself, though, Albert presented two different solutions in his early and mature works. In his *De homine*, Albert follows Avicenna in distinguishing two different aspects of it, viz. touch as form of the body (*secundum quod est forma et perfectio*) and as faculty responsible for the judgment of the tangibles (*secundum iudicium tangibilium*).[124] What Albert draws from Avicenna's remarks is that, if we follow the first criterion (*secundum quod est forma corporis*), the organ of touch is constituted by the whole animal body, primarily by flesh, thus being in agreement with Aristotle. If one considers that not every part of the body is equally receptive to sensory stimulus, though, but that its degree of sensitivity is rather dependent on the presence of nerves and animal spirit (*spiritus animalis*), then one should consider the nervous parts of the flesh as the form of the body, following Avicenna.[125] Aware of the discrepancies between Aristotle and Avicenna, Albert tries to build an agreement between the two authors by emphasizing common ground between them, namely embryological cardiocentrism. According to Albert, the cardiocentric origin of the body, if put in relation to the nature of flesh and the whole body as the organ of touch, allows us to claim that the organ of touch is primarily associated with the heart *qua* origin of the whole animal body.[126] According to the first criterion, then, the

124 Albertus Magnus, De homine (Ed. Colon. XXVII/2, 246.68–74): "Solutio: Dicimus quod tactus duo habet de ratione sui. Quorum unum est quod est iudicium tangibilium, et sic est potentia et pars animae sensibilis. Alterum est quod est esse et perfectio, a qua sumitur sermo diffinitivus animae sensibilis, et sic non est pars animae sensibilis, sed est id quod constituit ipsam et facit esse sensibilem animam [...]"; Albertus Magnus, De homine (Ed. Colon. XXVII/2, 249.77–80; 252.79–80); Dewan, St. Albert, the Sensibles, 300.

125 Albertus Magnus, De homine (Ed. Colon. XXVII/2, 252.80–253.7); Hasse, *Avicenna's* De Anima, 103; Dewan, "St. Albert, the Sensibles," 300.

126 Albertus Magnus, De homine (Ed. Colon. XXVII/2, 253.8–13): "Cum enim tactus sic sit forma animati corporis, terminat et complet totam materiam suam. Et propter hoc percipit quicquid immutat carnem et quicquid immutat cutem et quicquid immutat nervum,

organ of touch appears to be primarily seated near the heart, as in Aristotle's own doctrine. Albert remarks that one should understand Aristotle's position on flesh as a medium for touch as follows: the action of the tangible on the flesh and the nerve is simultaneous, but comes to the nerve via the flesh. This, says Albert, is what Aristotle believes.[127]

A different localization of the organ of touch emerges when considering touch according to its capacity of judging tangibles (*secundum quod est iudicium tangibilium*): in this case, Albert follows the principles of Galenic anatomy by locating this judgemental capacity in the front brain ventricle.[128] Albert refers explicitly to al-Ġāzālī's stance on the pivotal role played by the spirit as vehicle of tangibles. As it has been observed so far, al-Ġāzālī's doctrine of spirit in the *Intentions of Philosophers* mostly reports Avicennian psychological doctrines (*De anima*).[129] Dag Hasse has demonstrated that Albert's position on the role of spirit as a vehicle of tangibles is not only a blending of the positions of Avicenna and al-Ġāzālī, but also draws on the consensus of the medical literature avaible in his time (*Pantegni*, Costa Ben Luca's *De differentia spiritus et animae*, Nemesius' *De natura hominis*, and other minor sources).[130] These all localize the capacity for judging sensibles in the front brain ventricles where, as Albert may add in agreement with Avicenna, common sense, viz. the root of all senses, is seated.

Years later, when writing the *De anima*, Albert adopts a somewhat different standpoint on the role of flesh as a medium in touch. It is again Hasse who has shown how Albert, despite his awareness that anatomical evidence proved that flesh in itself is insensitive and nerves are the actual anatomical part conducive to sensation,[131] claimed that flesh is the medium for touch. He did so by "wishing both to save the truth and giving reverence to the father of the philosophers, Aristotle" (*et veritatem salvare cupientes et reverentiam exhibere*

et quoad hoc tactus principaliter est in corde, quod est principium totius corporis et per fila nervorum continet totum corpus." Hasse, *Avicenna's* De Anima, 103.

127 Dewan, "St. Albert, the Sensibles," 300.

128 Albertus Magnus, De homine (Ed. Colon. XXVII/2, 253.14–16).

129 Albertus Magnus, De homine (Ed. Colon. XXVII/2, 253.16–24): "Licet enim cerebrum propter sui frigiditatem et humiditatem nullum habeat tactum, tamen ab ipso diffunditur spiritus animalis, qui est principium sensuum, sicut dicit Algazel, quod tactus 'pertingit ad partes carnis et cutis mediante corpore subtili, quod est vehiculum eius, quod dicitur spiritus, et discurrit per compagines nervorum, quibus mediantibus pertingit ad partes carnis et cutis, et hoc corpus subtile non acquirit neque haurit virtutem hanc nisi a corde et cerebro'." Algazel, Metaphysica pars 2 tr. 4 c. 3, ed. St. Clair 2005, 63.79–84.

130 Hasse, *Avicenna's* De Anima, 104.

131 Albertus Magnus, De anima II tr. 3 c. 35 (Ed. Colon. VII/1, 147.19–23).

patri philosophorum Aristoteli).[132] While this statement may seem *prima facie* to endorse one of the two interpretations of Aristotle's position on touch—flesh actually is its medium—the meaning of this statement by Albert is blurrier than expected. This blurriness is caused by the polysemy which Albert attributes to the term "flesh" itself: according to him, flesh has three different meanings: (1) properly speaking, flesh or what is analogous to it, (2) organs with the same complexion as flesh (such as teeth or the like), and (3) organs mixed with flesh (such as nerves and nervous flesh).[133] In the end, while this position seems to preserve Aristotle's own words, it also finds a compromise with Avicenna's own views on nervous flesh as the crucial organ in touch. Despite Albert's different positions on the medium of touch, the relevance of the nervous system and of the physical instrument of the sensitive soul, viz. animal spirits, emerges consistently as the cornerstone of Albert's physiology of sense perception throughout his intellectual production.

As shown by the crucial study of Dewan on the concept of spiritual being (*esse spirituale*) throughout Albert's works, the *De homine* and Albert's mature works contain two different understandings of the causal role of spirit within sense perception. Certain conclusions about this, however, remain quite implicit, and rightfully so, because of Dewan's primary focus on the concept of spiritual being. In the *De homine*, Albert distinguishes three stages in the actualization of the faculties of the senses: the imprinting of the sensible onto the sense organ, the existence of the sensible species in the animal spirits flowing through the nerves, and the entrance of the sensible species carried by the spirit into the front brain ventricle.[134] Albert proposes a special type of being for the sensible in the medium, for all three senses have a genuine external medium. In the object, it has material being (*esse materiale*); in the sense, it has spiritual being only (*esse spirituale tantum*); but in the medium it has sensible being

132 Albertus Magnus, De anima II tr. 3 c. 35 (Ed. Colon. VII/1, 147.40–45): "Nos autem et veritatem salvare cupientes et reverentiam exhibere patri philosophorum Aristoteli dicamus carnem esse medium tactus; sed carnem generaliter vocamus id quod vere caro est vel id quod est simile carni. Simile autem carni dicimus tripliciter accipi." Hasse, *Avicenna's* De Anima, 104.

133 Albertus Magnus, De anima II tr. 3 c. 35 (Ed. Colon. VII/1, 147.45–82); Hasse, *Avicenna's* De Anima, 104.

134 Albert's position is expressed with reference to sight but can be taken as an example of the physiological processing of sense data, see: Albertus Magnus, De homine (Ed. Colon. XXVII/2, 146.60–66): "Ad aliud dicendum quod virtus visiva secundum sui perfectionem est in anteriore parte cerebri, sed secundum sui inceptionem est in humore crystallino, sed secundum progressum ad perfectionem est in nervo optico et spiritu visibili, qui decurrit in illo. Et sic nihil prohibet quod vis visiva diversimode considerata diversa habeat instrumenta, sicut etiam una vis visiva duos habet oculos."

(*in medio vero sensible*).[135] As a preliminary requirement for the tangible to be sensed, the sensible species of heat and cold (and so the other elements and their derived oppositions) are not perceived by touch when their complexion is equal to the complexion of the mixture in the flesh. These tangibles are perceived only if they exceed (*excellens*) the ratio of heat, cold, wet or dry in the mixture of the perceiving organ;[136] a similar position is expressed in the *De anima*.[137] While these preliminary requirements for the imprinting and perception of a sensible species on flesh remain consistent throughout Albert's works, the explanation of the transmission of sensible species through animal spirits displays certain doctrinal changes through the years, which relate to his changing view on the doctrine of spiritual being.

In the *De homine*, Albert carves out an account of sense perception which differentiates the causal role of animal spirits in distal (sight, hearing, smell) and contact (touch, taste) senses. In his discussion of the physiological transmission of sensibles belonging to contact senses (tangibles, flavors), Albert ascribes to animal spirit, as noted by Dewan, the capacity to abstract the sensible species and cause its existence in the sense organ according to 'spiritual being'.[138] The abstracting action of animal spirits avoids, for instance, the conclusion that hot objects may heat up the brain by acting according to their material being. This capacity for abstracting the sensible species from the object of sensation itself parallels the role of light, as we have observed in the case of his contemporaries Rufus, Alexander of Hales and Roberte Grosseteste. Dewan has higlighted, though, how Albert rejects the idea that light works as the source of spiritual being in all distal senses. To be sure, the abstracting role of light can explain why species of contrary colors (white and black) exist in the same medium (air), actualized at the same time without canceling each other out.[139] Albert, though, rejects the theory of *esse spirituale* in the

---

135 Dewan, "St. Albert, the Sensibles," 301.

136 Albertus Magnus, De homine (Ed. Colon. XXVII/2, 247.65–69).

137 Albertus Magnus, De anima III tr. 2 c. 3 (Ed. Colon. VII/1, 179.12–15): "Diximus enim superius de organo tactus, quod recipit excellens se tangibile, cum tamen habeat media in se tangibilium; medium enim ab extremo differt specie et convenit genere."

138 Albertus Magnus, De homine (Ed. Colon. XXVII/2, 262.61–67): "Duo autem sensus qui sunt per medium intra, habent sensibilia, quae actu sunt et in obiecto et in medio et organo, et sensibilia illorum esse spirituale acquirunt per spiritum animalem abstrahentem et deferentem intentiones illorum sensibilium ad cerebrum. Et hoc patet, quia intentio calidi abstracta per spiritum animalem ab organo tactus non calefacit cerebrum." Dewan, "St. Albert, the Sensibles," 301–302. Note that in the *De homine* the term *intentio* is both a synonym for *species*—as in this passage—and the Latin translation for the object of the Avicennian faculty of estimation (*maʿnan*).

139 Dewan, "St. Albert, the Sensibles," 294.

other distal senses (hearing, smell), because hearing and smelling can occur at dark. For this reason, the doctrine of spiritual being as the product of light (the causal agent of *esse spirituale*) upheld by some of his contemporaries must be dismissed.[140] The sensibles of hearing and smell do not exist in their media according to spiritual being qua product of the light's abstracting action but rather according to sensible being (*esse sensibile*), viz. a being that can actually allow the respective faculties to sense them.[141] In the case of sight, sensible being equates with spiritual being, while in hearing and smell it does not. As a last note on the *De homine*'s theory of spiritual being, it is worth adding that there seems even to be some isomorphism between abstractive light and animal spirits: both share a luminous nature and both confer spiritual being on the object of the senses through their abstractive action.[142]

In the Aristotelian paraphrases, Albert presented a more consistent, though changed, position on the doctrine of spiritual being, which will be reflected in the causal role of animal spirit. In the *De anima*, this question was assumed to be of paramount importance in sense perception, as a voucher for the actualization of the psychic powers devoted to sense perception. According to Albert, sensible species belonging to all (distal, contact) senses possess gradually more abstracted degrees of spirituality, viz. of spiritual or intentional being (*esse intentionale*). Albert carefully distinguishes two senses of *intentio*: *intentio* as sensible (which can be called also *species*), belonging to the five senses, and *intentio* as a form of the hylomorphic compound in its totality.[143] According to Albert, it is the form of the hylomorphic compound *qua* form which enables the latter (the *intentio* or *species*) to acquire the spiritual being necessary for its perception by the soul. The action of the form *qua* form produces a *multiplicatio specierum* in the medium (in distal senses), while in contact senses (touch and taste) the tangible or the flavor acts directly on the sense organ according to its material being (*esse materiale*), and in sight, hearing and smell the medium grants that the sensible reside in the medium according to a spiritual being that varies from the most spiritual (sight) to the least spiritual (smell).[144]

140 Dewan, "St. Albert, the Sensibles," 298.

141 Dewan, "St. Albert, the Sensibles," 301.

142 Dewan, "St. Albert, the Sensibles," 293–294: "Light is regarded as an agent which 'abstracts,' i.e., dissociates from its material subject, 'liberates,' one might say, the colour. [...] Averroes' term 'spiritual,' he tells us, is not to be taken as used of soul in distinction from body. It is taken from 'spirit' said of the body with which the transparent has something in common, i.e., the celestial body. He also argues that it is not colour and light which in excess directly."

143 Dewan, "St. Albert, the Sensibles," 303.

144 Albertus Magnus, De anima II tr. 3 c. 6 (Ed. Colon. VII/1, 105.85–106.1): "Esse autem intentionale et spirituale non una ratione est in sensibilibus, quia unum est multo spiritualius

As noted by Dewan, Albert's gnoseological theory of *multiplicatio specierum* is somehow reminiscent of Grosseteste's.[145] Spiritual being in Albert's Aristotelian paraphrases is, in the end, the result of this multiplicative process of form *qua* form. The upshot of this doctrinal modification is that animal spirit loses its exclusive function as source of spiritual being for tangibles and flavors. This task is then taken on by the medium which, in the case of touch, is flesh according to its many senses. As Albert claims in the *De anima*, the relationship among sensibles and the media for distal senses is isomorphic compared to the relationship between the media for touch and taste and these senses themselves. What Albert means here is that mediums of touch and taste (flesh) play an analogous role to sensible species, viz. the causal agent for the existence of the sensible species according to their spiritual or intentional ontologies. In the cause of distal sensation, the spiritual existence of a sensible species links to its existence in the medium, while spiritual existence of contact sensation links to the existence of these species in the faculty by which they are perceived.[146] While in Albert's early works it was the spirit alone that acted as a source for the spiritual existence of the sensed form in contact perception, the scant evidence we are provide here seems to point primarily toward a key role played by the flesh, which could also be aided in the process by its nervous parts filled with animal spirits. This last inference is conjectural.

4.2 *Taste*

In the *De homine*, Albert's psychology of sense perception in touch is then followed his doctrinal study on taste, thus going from touch to sight through taste, smell, and hearing. According to Albert, animal spirits produce spiritual being

alio, quia unum afficit et medium et organum secundum esse materiale agens in ipsum, sicut est in obiecti tactus; et ea quae habent idem medium sicut visibilia et audibilia et odorabilia, non habent ilud medium secundum unam et eandem naturam medii sed secundum diversas, sicut inferius ostendemus." Albertus Magnus, De anima II tr. 3 c. 27 (Ed. Colon. VII/1, 138.40–42): "[…] gustus non accipit speciem saporis denudatam a corpore saporoso, sicut faciunt tres sensus, de quibus dictum est." Dewan, "St. Albert, the Sensibles," 304.

145 Dewan (St. Albert, the Sensibles, 306) points out how Albert's doctrine of the *multiplicatio specierum* differs from Grosseteste's because of its purely gnoseological, and not ontological, implications. Moreover, it must be remembered that Grosseteste adhered to an extramissionist view of vision, which runs counter to Albert's Aristotelian doctrine of sight.

146 Albertus Magnus, De anima II tr. 4 c. 4 (Ed. Colon. VII/1, 152.62–66): "Gustabilia autem et tangibilia per materiam coniunguntur mediis, sed tamen medium non agit in organum nisi esse intentionale; et ideo videtur, quod sicut in aliis sensibus sensibilia ad medium se habent, ita in gustu et tactu se habent media ad sensum."

in the sensible species of taste thanks to their abstracting action. In the Aristotelian paraphrases, animal spirits act as the vehicles of sensitive faculties, but without any explicit attribution of an abstracting capacity. Rather, the transition of sensible species from their enmattered state (in the object of sensation) to their spiritual being in the sense faculty is granted by the physiological process of taste, strongly based on Avicenna's account.

Albert's psychophysiology of sense perception in the *De homine* shows remarkable doctrinal commonalities with Avicenna's account. Following Aristotle, Albert maintains taste to be "a certain kind of touch" (*quasi tactus particularis*), concerning only a specific part of the body (the tongue) and requiring a previous moistening action of saliva on the tongue to facilitate the perception of flavors. When confronted with the task of localizing the faculty of taste, he confronts Avicenna's *De anima* (I, 5), *De animalibus* (XII, 8) and Nemesius' *De natura hominis*. On the one hand, Avicenna's *De anima* localizes the faculty of taste in the (third) pair of cranial nerves spread out over the tongue's surface. Nemesius and Avicenna's *De animalibus*, by contrast, add that palate is partially responsible for taste too, since the fourth pair of cranial nerves vehiculates part of the flavor to the brain too. According to Albert, both the fourth and third pair of nerves contribute to taste, but the fourth pair does it best (*secundum sui optimum est in lingua vel in membro, quod est loco linguae*).[147] Albert considers taste in a twofold way, i.e. as a faculty passing judgement on flavors (*secundum quod est iudicium saporum*) and as a sense of the ingested food (*secundum quod est sensus alimenti*). In the first case, its localization is found in the brain. If it is considered a sense of the food, then the cardiocentric nature of nutrition and its instrument (vital heat) favors a localization near the heart.[148]

Drawing on Aristotelian zoology, Albert considers taste not only in humans, but also in animals, admitting that there are some animals who do not have,

---

147 Albertus Magnus, De homine (Ed. Colon. XXVII/2, 238.55–59): "Solutio: Dicendum quod in veritate gustus est iam in nervi palati, sicut obiectum est, sed secundum sui optimum est in lingua vel in membro, quod est loco linguae, et cum ab optimo sumatur diffinitio, diffinitur a nervo expanso in superficie linguae."

148 Albertus Magnus, De homine (Ed. Colon. XXVII/2, 244.11–21): "Solutio: Dicendum quod gustus duobus modis accipitur. Unum secundum quod 'iudicium saporum' est, ut dicit Aristoteles in *IV Ethicorum*, et sic nervus gustativus principiatur a cerebro et expanditur in lingua et palato, et sic non habet instrumentum circa cor. Alio modo secundum quod est sensus alimenti, et sic est sensus commixti ex calido et frigido, humido et sicco, et quoad habet instrumentum suum circa cor. Et per omnem eundem modum distinguimus in tactu, scilicet secundum quod est iudicium tangibilium, vel secundum quod ipsum tangens nutritur tangibilibus."

strictly speaking, a tongue, but something with its function (*loco linguae*).[149] Albert then describes the different shapes of tongues in animals (snakes, fish, chickens) in connection with their digestive systems and shows how certain animals possess a complete set of organs for digestion, like humans, while others do not. In this manner, Albert, unlike many of the contemporary treatises on the human soul, maintains a comparative and zoological approach to sense perception even in its most material and physiological causes.[150]

In the Aristotelian paraphrases (*De anima*), Albert's position on taste closely follows his early *De homine*, maintaining that flavors are perceived on the tongue and aided by the mediating role of the saliva.[151] While his analysis of cranial nerves in the *De animalibus* shows some commonalities with the *De homine*, it emphasizes the role of the tongue in the process of touch and its need for saliva fluid as a means to facilitate the process. In his zoology Albert tackles the issue in three different passages: a passage on the anatomy of the cranial nerves (lib. I tr. 2, c. 18), based on Avicenna's reports in the *Canon* and *De animalibus*; the chapters based on Aristotle's own remarks on the role of the tongue in what we know as *Historia Animalium* (lib. I, tr. 2, c. 10); and *De partibus animalium* (lib. XII, tr. 3, c. 5). In the first discussion on cranial nerves, based on Avicenna's medical learning, Albert seems to follow the same solution he had opted for in the *De homine*, arguing that taste should be associated—as in Avicenna—both with (the fourth branch of) the third[152] and the fourth cranial nerves,[153] joined to the tongue and to the palate respectively. In the chapter on the composition of the tongue (lib. I tr. 2, c. 10), Albert stresses again the localization of the faculty of taste in the nerve spread out onto the

149 Albertus Magnus, De homine (Ed. Colon. XXVII/2, 238.60–239.1).

150 Albertus Magnus, De homine (Ed. Colon. XXVII/2, 239.4–24).

151 Albertus Magnus, De anima II tr. 3 c. 27 (Ed. Colon. VII/1, 138.50–51): "Umor autem salivalis est medium ad bene esse gustus et non ad esse […]".

152 Albertus Magnus, De animalibus I tr. 2 c. 18 n. 363 (ed. Stadler, I, 128.27–30). Avicenna, Canon I fen 1 doctr. 5 c. 2, ed. Venezia 1507, fol. 19ra [ed. New Delhi 1981–1996, vol. 1, 107.12–13].

153 Albertus Magnus, De animalibus, I tr. 2 c. 18 n. 364 (ed. Stadler, I, 128.35–38, 129.16–22): "Quarti vero paris nervorum septem parium origo est statim post tertium vadit ad palatum et dat ei sensum […] Gustus autem in quarto fuit nervo creatus, et auditus in quinto: quoniam instrumento auditus fuit necessarium ut esset discoopertum, ne clauderetur via aeris qui pertingere debet ad ipsum. Organum gustus necessarium fuit ut esset protectum et clausuris cavernarum circundatum propter fontem salivalem. Et ideo convenienter nervi auditus facti sunt duri: et ideo ortus eorum postremo cerebri fuit propinquior." Avicenna, Canon I fen 1 doctr. 5 c. 2, ed. Venezia 1507, fol. 19va: "Gustus autem in quarto creatus fuit nervorum et auditus in quinto".

tongue's surface.[154] As noted already in Stadler's edition of the *De animalibus*, the anatomical description of the tongue and its parts is borrowed from Avicenna's *Canon* (lib. III, fen 6, doctr. 1, c. 1).

In the chapter on the teleology of the tongue (*De animalibus*, XII, 3, 5), Albert deals in greater detail with the different kinds and shapes of tongues in animals: river fish, for instance, have something analogous to a tongue (*loco linguae*). Many of them even have a palate like humans (*in pluribus piscium fluvialium erit aliquando palatum carnosum et valde molle*), while others do not, as in the case of reptiles, which possess a forked tongue. The list of animals endowed with tongue goes on to cover annelids (*anulosa*), testacea and shellfish: the latter possess an organ for taste that is hard (*durus*) to the touch, as opposed to most animals, who possess a soft (*mollis*) tongue.[155] All in all, Albert's zoology shows that the equivocal localization of taste depends on the different physiological structures and processes in various animals. For this reason, taste is localized in the tongue in animals without a palate, while animals with a palate localize taste in the nerves placed on their tongue and palate. In animals that have palate, taste is primarily connected to the tongue and the fourth pair of cranial nerves as the first organ. In all this we see that Albert endorsed Avicenna's account of the cranial nerves over other theories available to him, especially Costa Ben Luca's, because of the anatomical accuracy proffered by Avicenna.

If we want to draw some conclusions on the role of animal spirits in taste, we can argue that Albert sides from the beginning with the medical explanation, according to which flavors come to the brain via the nerve on the tongue, or something analogous to it, in less complex animals. In more complex animals such as humans, the transmission of the sensible species comes through the branch of the third pair of cranial nerves, which reach the palate. This does not change the fact that the most suitable organic means for sensing flavors is the nerve on the tongue's surface, as in all kinds of animals. While differences in the role of animal spirit in the physiology of taste are due to the organic structure of living beings, doctrinal changes in the role of animal spirit in the psychology of taste depend, as shown in the case of touch, on Albert's changing doctrine of spiritual being (*esse spirituale*). In the *De homine*, the source for the spiritual

---

154 Albertus Magnus, De animalibus I tr. 2 c. 10 n. 235 (ed. Stadler, I, 84.19–20, 84.27–29): "Et opus quidem gustus habet ex hoc quod est bene passibilis in anteriori extremitate sua, in qua est expansus nervus gustativus. [...] Cum tamen sit ubique in lingua et in ore aliquis ramusculus nervi gustativi ad adiuvandum in discernendo sapores, sed perfecte non est nisi in anteriori parte linguae."

155 Albertus Magnus, De animalibus XII tr. 3 c. 5 (ed. Stadler, II, 881–883).

existence of sensible species carried by animal spirit through the nerves is represented by light (sight) and animal spirits (touch and taste). In the *De anima* and in the Aristotelian paraphrases, that which causes the sensible species to exist in the living being according to a spiritual being does not depend anymore on the abstracting actions of animal spirits anymore, but on the form *qua* form. This changing role of animal spirits, though, applies to contact senses (touch and taste) only, as the case of distal senses (smell, hearing and sight) proves to be quite different for a number of reasons which we shall start to analyze with the case of smell.

4.3 *Smell*

The chapter on the nerves (XII, 8) from Avicenna's *De animalibus* strongly influences Albert's handling of both taste and smell in the *De homine*. His understanding of the object of smell and of its transmission clearly favors clearly Avicenna's view over Aristotle's and Averroes'. According to Albert, odors are caused by vapors affecting the medium of air and water and are produced by the detaching of smoky particles from the sensed object. This is supposed to be proved by the case of the shrinking apple mentioned by Avicenna and faithfully reported by Albert.[156] Albert also follows Avicenna in the case of vultures, arguing that their capacity to locate dead carcasses depends more on their sight than on their sense of smell.[157] While heavily indebted to Avicenna, Albert retains some genuinely Aristotelian elements in his thinking here: for example, the distinction—typical of Aristotle's *De sensu et sensato*—between odors that are pleasant in relation to food, sensed only by animals, and odors that are pleasant or unpleasant per se *qua* good or bad for the human cerebral complexion.[158] Albert, although refusing to concede that these kinds of odors are different in species, proposes a threefold distinction of odors: *per se*, further divided into pleasant per se or *qua* related to food, and a third, emotion-related kind, which triggers emotion.[159]

---

156 Dewan, "St. Albert, the Sensibles," 299.

157 Albertus Magnus, De homine (Ed. Colon. XXVII/2, 235.66–237.46); Dewan, "St. Albert, the Sensibles," 299.

158 Albertus Magnus, De homine (Ed. Colon. XXVII/2, 235.4–10): "Delectabile autem quod confert ad sanitatem cerebri, est per se in odore. Cum enim odor sit quaedam evaporatio, pertingit ad locum cerebri, ad quem non potest pertingere substantia cibi, et ideo necesse est quod hoc delectabile sit in odore per se et non in comparatione ad cibum; et hoc est praecipue, quod est in floribus et aromaticis."

159 Albertus Magnus, De homine (Ed. Colon. XXVII/2, 234.50–60): "Solutio: Dicendum quod species odoris distinguuntur secundum tres modos, scilicet in se et in comparatione ad delectationem vel tristitiam, quam infert olfactui. In se autem distinguuntur secundum

Albert's psychophysiology of smell in the *De homine* is strongly permeated by Avicennian doctrines. On the one hand, he sides with his illustrious predecessor on the material existence of odors in the medium as vapors. On the other hand, his localization of the faculty of smell in the fleshy part of the nose (*carunculae*) follows Avicennian remarks in the *De homine*,[160] blending them together with similar remarks in al-Gazālī (who depended on Avicenna),[161] Nemesius[162] and Costa ben Luca,[163] who all denied the existence of an olfactive nerve. In his *quaestio* on the individuation of the organ of smell,[164] Albert sides with Avicenna, Nemesius and Costa ben Luca (*Constabulus*), targeting the author of an 'anatomy' who asserts the existence of an olfactory nerve connected to this fleshy appendage of the brain.[165] Perhaps the author meant here wrote the *Anatomia Vivorum*, *Anatomia Richardi* or the *Quaestiones Nicolai Peripatetici*, works which all argued in favor of the existence of an olfactory nerve.[166]

As in Avicenna and the other authors of the medical tradition, smell is the only sense in which the sensible does not reach the brain via a sensory nerve filled with animal spirit: according to Albert, if spirit were to carry the odors, they would not be able to affect the brain's complexion.[167] In his study of smell,

duos modos, scilicet absolute et in comparatione ad sapores. Absolute autem non habemus propria nomina, eo quod male odoramus. Unde in illa distinctione non habemus nisi nomina excellentiarum, et unum dicimus redolens et alterum foetens, et si habemus medium, hoc erit per abnegationem extremorum et non per commixtionem, scilicet quod neque redolens est neque foetens."

160 Albertus Magnus, De homine (Ed. Colon. XXVII/2, 224.29–36): "Secunda vero scribitur in libro *de animalibus* Avicennae in capitulo de anatomia nervorum, ubi sic dicit: 'Nervus faciens sensum nascitur ab anteriori, quod est magis humidum, et faciens motum a posteriori, quod est magis siccum, et a medulla nascuntur septem paria nervorum, et primum par habet originem a primis duabus partibus medullae anterioribus, ubi nascuntur duae mamillae olfactus'."

161 Albertus Magnus, De homine (Ed. Colon. XXVII/2, 223.55–224.4).

162 Albertus Magnus, De homine (Ed. Colon. XXVII/2, 224.13–16). Nemesius, De natura hominis c. 10 (ed. Verbeke/Moncho 1975, 85.49–86.52).

163 Albertus Magnus, De homine (Ed. Colon. XXVII/2, 224.18–21).

164 Albertus Magnus, De homine (Ed. Colon. XXVII/2, 224.67–70): "Solutio: Dicendum ad praedicta quod in veritate organa olfactus sunt immediate in cerebro secundum auctores, scilicet Avicennam, Constabulum et Gregorium Nixenum."

165 Albertus Magnus, De homine (Ed. Colon. XXVII/2, 224.37–42): "Sed contra obicitur in hunc modum: (1) In anatomia quae secundum Galienum dicitur esse facta, sic dicitur: 'Duo nervi sensibiles diriguntur ad duas carunculas narium, quae dependent ad modum duarum mamillarum in naribus ad faciendum odoratum sive olfactum'."

166 On these authors, see the section above: 286, nn. 51, 52, and 53.

167 Albertus Magnus, De homine (Ed. Colon. XXVII/2, 225.24–34): "Causa autem quare immediate in cerebro est sensus olfactus, haec est, quia odor secundum suum esse naturale

Albert sketches a brief comparative picture of it, contrasting its nature as accidentally but primarily associated with breathing with Aristotle's and Avicenna's zoology (IV, 2). Both emphasize the existence of smell in non-breathing animals,[168] while maintaining the unity of the faculty of smell (*Dicendum quod olfactus non dicitur aequivoce in olfacientibus diversimode et diverso naso*) in spite of its different physiological unfolding.[169] Albert's overall approach to smell in the *De homine*, both psychologically and physiologically, follows in the footsteps of Avicenna.

Dewan has demonstrated that Albert's position on the nature of odors changes considerably In the *De anima* and the *Parva Naturalia*. In these works, he no longer supports Avicenna's—also Plato's—view that smells are produced by vapors. Rather, he favors a view inspired by Aristotle and Averroes, while maintaining his previous physiology of smell, carved out from Avicenna, Costa ben Luca, and Nemesius. As emphasized before, the major change in Albert's mature doctrine of distal senses consists in the fact that the theory of spiritual being (*esse spirituale*) has now acquired a role of paramount importance in explaining sense perception.[170] The natural upshot of his adoption of the doctrine according to which any sensible species of distal senses exists in its medium according to its spiritual being led him to reconsider and eventually reject Avicenna's position on the nature of odors as products of vapors which exist in their media according to material existence (the species of the odor is enmattered in the vapor), as opposed to the spiritual, intentional (also, immaterial) being of the sensible species as understood by Albert and Averroes.

Albert questions Avicenna's doctrine because, in his eyes, it does not explain the diffusion of particles of rarefied bodies over great distances, as in the famous case of vultures. According to Albert, this makes a materialistic account of the existence of sensible species in a medium simply untenable.[171] There-

et materiale, quod habet in vaporativo sicco, iuvativus est cerebri, et ideo oportuit quod in sua materia pertingeret ad ipsum et non tantum per speciem sensibilem separatam a materia, sicut accidit in aliis sensibus. Si autem esset vis ordinata in nervo concavo deferente spiritum odorativum, ille spiritus non reciperet impressionem nisi tantum speciei sine materia; et cum species separata a materia nec sit calida nec frigida, non conferret ad iuvamentum cerebri."

168 Albertus Magnus, De homine (Ed. Colon. XXVII/2, 237.50–65). The example put forward by Albert to demonstrate the existence of smell in non-breathing animals is fish (as in Avicenna's *De animalibus*, IV, 2).

169 Albertus Magnus, De homine (Ed. Colon. XXVII/2, 238.3–4).

170 Dewan, "St. Albert, the Sensibles," 303–307.

171 Dewan, "St. Albert, the Sensibles," 310–311.

fore, concludes Albert, the sensible species of odors cannot act on a sense organ but according to their own spiritual and immaterial existence. To be fair, though, smell enjoys the lowest degree of spirituality among the distal senses, i.e. compared to hearing and sight, as stated above. There are some cases where Albert concedes that vapors come together with odors, as Avicenna's—also, Aristotle's—remarks on the suffocation of animals (*De animalibus*, IV, 2) seem to suggest. Nonetheless, Albert argues that this phenomenon represents an accidental occurrence and not the essence of odors, understood as the alteration of the medium by the simple quality of the odor.[172]

In the *De sensu et sensato*, Albert devoted an entire section to support Aristotle's assertion that the perceptual content of smell in humans, though inferior to other animals, is not only food-related—in other words, odors are an affection of nutritional substances—but can also serve as a tool to defend the complexion of the human brain.[173] From the perspective of the nature of the sensible in its medium, Albert favors a view closer to Averroes', turning against Avicenna; on the other hand, his physiological account of smell remains faithful to the Avicennian and Galenic physiology of smell, against any role of spirit in the transmission of odors to the brain.

In the *De animalibus* (I, 2, 18) Albert replicates his physiological explanation of smell. In these works, he claims that odors affect the brain through its fleshy part, placed next to the first, cranial (optical nerve), endorsing the view that the nose, in breathing animals, may be the instrument of smell.[174] In his zoology, he enlarges his considerations on smell in animals on empirical grounds: that fishes can smell is proven by a number of examples, most of which connect to the different methods of baiting, and by the fact that they flee from other fish because of their stink (*propter fetorem*).[175] This is the same argument employed, as seen above, by Avicenna, disguised here as an empirical proof of

172 Dewan, "St. Albert, the Sensibles," 313.

173 Albertus Magnus, De sensu et sensato I tr. 2 c. 12 (Ed. Colon. VII/2a, 84–87).

174 Albertus Magnus, De animalibus I tr. 2 c. 18 n. 358 (ed. Stadler, I, 126.35–127.1); Albertus Magnus, De animalibus I tr. 2 c. 8 n. 214 (ed. Stadler, I, 77.17–29): "Nasus vero qui interiacet inter oculos et descendit ab ipsis, est instrumentum sensus odoratus: quod instrumentum sic est compositum: Extremitas quidem cerebri versus os quod dicitur coronale vel puppis capitis deficit in duas papulas, quae sunt similes papulis mamillarum: et iacet una iuxta alteram versus duos triangulos narium, et in omnibus odorantibus per inspirationem operiuntur velamine panniculari, ne laedantur obviante aere, calore vel frigore. Et ideo talia animalia non odorant, nisi cum vento inspirationis exsuflatur velamen illud: eo quod virtus olfactiva est in illis papulis sicut visus in oculo. In animalibus autem non inspirantibus sunt nudae papulae illae: quia illa habent cerebella sicciora, quae non ita facile laeduntur ab obviantibus."

175 Albertus Magnus, De animalibus IV tr. 2 c. 1 (ed. Stadler, I, 394–395).

the existence of smell in animals. Likewise, ringed animals (annelids, Lat. *anulosa*) have smell since they are seen suffocating from the same smokes (e.g.: bitumen and sulphur) that are also lethal to humans or fish, as already in Aristotle.[176]

Coming to an explanation of the psychophysiological role of animal spirits in smell, Albert provides us with scant evidence about the link between odors and spirits. Whatever the presupposed doctrine for the transmission of sensible species through a medium (spiritual being *qua* product of light/animal spirit or form *qua* form), Albert conceives of animal spirits as vehicles of the faculties of the sensitive soul and its instruments. For this reason, despite Albert's evidence linking the transport of odors to the front brain ventricle, we can make some tentative remarks on the role of animal spirits in smell. On the one hand, it is uncontroversial that animal spirits vehiculate sensible species to the brain via cranial nerves, and so is the fact that smell does not come through cranial nerves to the brain but via a fleshy part joined to it and reaching the nostrils, at least in breathing animals. Moreover, species of odor act directly on the brain, as Albert himself admitted distinguishing the different kinds of odors, not least the lethal ones and those perceived only by humans and judged to be noxious. While the function of animal spirits in olfaction seems common between humans and animals, viz. the direct imprinting of these species on the brain through the fleshy particle under it, there are two differences in the role of animal spirits in the physiology of smell worth being mentioned. The first difference occurs between animal species and concerns the modality of this imprint and transmission, determined by an animal's breathing or non-breathing nature.[177] The second difference concerns the sensory content of smell: while in animals odors are only food-related, humans sense a further class of unhealthy odors, noxious to the brain's health. Therefore, animal spirits carry two different sets of sensorial information between humans and animals, a remarkable case of distal senses.

4.4 *Hearing*

The case of hearing is somewhat less complex than smell. As we have previously shown, Albert rejected in the *De homine* the theory of spiritual existence both for smell and sound as a byproduct of the abstracting action of light. Albert's

176 Albertus Magnus, De animalibus IV tr. 2 c. 1 (ed. Stadler, I, 396.3–5, 396.20–26).

177 As Albert points out, the nose *qua* organ of smell, broadly speaking, is particularly sensitive to touch: this fact is in keeping with the peculiar exception of the elephant's trunk Albertus Magnus, De animalibus I tr. 1 c. 2 (ed. Stadler, I, 79), functionally analogous to hands (*utitur ipso elephas loco manus*).

theory on the transmission of sounds in the *De homine* hinges, therefore, on his consideration of air not only as medium but also as matter of sound: without air, sound does not exist. For this reason, Albert considers the travelling of sounds in water to be produced by air forcibly making its way into water.[178] According to this peculiar understanding of the role of air in hearing, Albert explains that sounds propagate through air as their matter—not only as their medium—right up to the hearer. Unlike sight, where contrary colors coexist in the same medium if actualized at the same time, sounds can cancel each other out.[179]

In the *De homine*, Albert relies on Avicenna and Costa Ben Luca[180] for their localization of the faculty of hearing in the fifth cranial nerve, linked straight up to the front brain ventricle. A similar emphasis on the role of the nerves is shared by Nemesius' *De natura hominis*, who only speaks about the role of 'nerves' in the brain.[181] In the *solutio* to the *quaestio* on the organ of hearing, Albert substantially endorsed Avicenna's localization of the faculty of hearing in the fifth cranial nerve.[182] In the *De homine*, Albert undertakes a remarkable approach based on the comparison between the human-oriented description of the role of spirit and nerves in hearing (as it was for smell) and animal physiology of sense perception relying on passages from Avicennian and Aristotelian zoology. For instance, Albert points out that fish, though lacking ears, do possess an analogous organ connected to the brain through a set of nerves, as witnessed by their behavior.[183]

---

178 Dewan, "St. Albert, the Sensibles," 299.

179 Dewan, "St. Albert, the Sensibles," 297–298.

180 Costa ben Luca's list of cranial nerves is quoted in: Albertus Magnus, De homine (Ed. Colon. XXVII/2, 202.59–203.2).

181 Nemesius, De natura hominis c. 10 (ed. Verbeke/Moncho 1975, 85.38–39).

182 Albertus Magnus, De homine (Ed. Colon. XXVII/2, 203.41–49): "Solutio: (1) Dicendum est ad primum quod nervus opticus sive nervus posticus, qui est nervus audibilis, ut dicit Avicenna in suo libro *de Animalibus*, 'in veritate nascitur ab anteriori parte medullae' cerebri et 'vadit ad telam auricularem', quae dicitur tympanum, 'et in ea dispergitur'. Et quia uterque ramus eius dividitur in duo, maior pars illius rami diffunditur in tympano auris, et 'minor pars exit per foramen ossis petrosi, quod dicitur foramen caecum'." Avicenna, Abbreviatio de animalibus, XII c. 8, ed. Venezia 1508, fol. 47vb.

183 Albertus Magnus, De homine (Ed. Colon. XXVII/2, 203.67–204.2): "Ad aliud dicendum quod in veritate, ut dicitur in libro *de animalibus*, pisces non habent vias auditus apertas in capite, quia cum elementum aquae sit fluxibile, subintraret ad medullam cerebri et infrigidando illam exstingueret naturam. Et propter hoc tympanum piscium est tela in piscibus cooperta, in qua expanditur nervus audibilis, qui immutatur a commotione soni, qui subintrat per aquam, et hoc propter connaturalitatem, quam habet nervus ille ad aërem commotum, et propter connaturalitatem spiritus audibilis, qui est in ipso, in quo imprimuntur species sonorum; et ille spiritus est eis pro aëre quieto."

In his Aristotelian paraphrases such as the *De anima*, Albert has changed his mind about the doctrine of the presence of spirit in hearing. As shown by Dewan, in his mature works Albert assumes a logical distinction on the nature of air as both matter and medium of sound (as in the *De homine*) and as medium alone.[184] As a consequence, one would expect Albert to provide an example of how sounds can exist in air *qua* their medium alone rather than only as their matter *and* medium. As noted by Dewan, the only example put forward by Albert as respects the nature of air as pure medium for hearing relates to a particular case, viz. echo. When describing this natural phenomenon, Albert points out that every kind of sound takes place through reflexion. Given that the echo is a phenomenon where reflexion is indisputable, it also poses the problem of how two sounds from two opposite directions can coexist in the same physical place. Albert solves this conundrum by appealing to the doctrine of the spiritual existence of the sound in the medium, arguing that the opposite waves cancel each other out just in a single point, as in the case of the two crossing arches of a circumference diametrically opposed to each other. Since the other parts of the waves do not cancel each other out, they coexist in the medium like opposite colors coexist in air.[185] In conclusion, sounds seem to enjoy an existence in air as in their medium according to "spiritual being" (*esse spirituale*), but the evidence for this claims is restricted to the case of echo alone and does not comprehend the more general class of sounds, as one would expect.

While displaying noteworthy doctrinal changes in its psychological framework, viz. the doctrine of spiritual ontology, Albert's psychophysiology of hearing maintains a consistent anatomical localization (*De animalibus*, I, 2, 18) of the faculty of hearing in the fifth pair of cranial nerves, following the insights of medico-philosophical sources, above all Avicenna.[186] In his zoology, Albert

---

184 Dewan, "St. Albert, the Sensibles," 310: "Sound is no longer thought to be found in air only as in a matter which simultaneously has the role of medium. It is also found in air merely according to spiritual being. In this latter situation it is comparable to the coloured thing, which produces its *intentio* alone in the air, though we are reminded once more that sound's spiritual being is not as spiritual as that of colour". Albertus Magnus, De anima II tr. 3 c. 18 (Ed. Colon. VII/1, 126.11–21).

185 Dewan, "St. Albert, the Sensibles," 310–311; Albertus Magnus, De anima II tr. 3 c. 19 (Ed. Colon. VII/1, 127.88–128.1): "[…] videtur, quod circuli lati ex opposito venientes, se tangunt in parte, quae est quasi in aiis partibus se nec tangunt nec se confringunt, et in illis est soni figuratio, et ab illis iterum fit in aëre vicino, ubi obviante opposite circulo confracuts fuit. Fit autem non per iterationem sonans, sed per reflexionem, ut supra diximus; est enim facilis generatio soni, eo quod in aëre aliquo non est nisi per intentionem […]".

186 Albertus Magnus, De animalibus I tr. 2 c. 18 n. 365 (ed. Stadler, I, 129.16–22).

detects, following Aristotle and Avicenna (*De animalibus* IV, 2), the existence of the sense of hearing in animals, especially in fish.[187] The inferential arguments used to demonstrate the existence of hearing are the following: (1) fish start to move suddenly when awoken and (2) they almost faint when fishermen use a high-pitched sound produced from bells to stun them.[188] According to Albert, fish and snakes, like chicken, have audible pathways (*viae auditus*) rather than ears, less evident than in humans because of the hardness of their skin (*duritia cutis*).[189] In hearing, the role of animal spirits is not clearly specified—perhaps this is trivial for Albert here, as in other cases—but amounts to the transmission of the sensible species from the fifth pair of cranial nerves to the front brain ventricle.

4.5 *Sight*

In the *De homine*, Albert took on the study of sight, basing his observations on Aristotle's, Avicenna's and Averroes' intramissionist theories. His starting point contained therefore a rejection of the extramissionist theory of Plato, Galen, Euclides (*De visu*) and al-Kindī (*De aspectibus lunae*). Albert's position on sight is remarkable for more than its mere historiographical importance. Among his contemporaries, he was the only one not to misinterpret Avicenna's theory of light as a quality of a body existing independently of the medium of air.[190] As shown by Dag Hasse in his *Avicenna's* De anima *in the Latin West*, the major obstacle to a correct understanding of Avicennian doctrine was the problematic Latin translation of the oft-quoted short definition of the faculty of sight (*De anima* I, 5). The translators (Gundissalinus and Avendauth) rendered this passage as if the process of vision occurs when the likeness of the colored thing becomes visible through the actualization of rays (*corpora radiosa*) to the surface of polished bodies such as the crystalline humor.

The major issue with this translation is that it ascribes the visibility of the likenesses of colored bodies (*similitude corporum coloratorum*) to the actualization of rays. In Avicenna's own theory, this actualization of the likenesses (sensible species) has nothing to do with the actualization of rays but, rather, with the actualization of the translucent medium itself. This depends, in turn, on the fact that the Latin renders the Arabic *šaffāf* (translucent) with *radio-*

187 Albertus Magnus, De animalibus IV tr. 2 c. 1 n. 79 (ed. Stadler, I, 393.17–26).

188 Albertus Magnus, De animalibus, IV tr. 2 c. 1 n. 82, (ed. Stadler, I, 394.25–28).

189 Albertus Magnus, De animalibus XIV tr. 2 c. 5 n. 55 (ed. Stadler, II, 976.4–8). More general considerations on the mechanism of hearing in: Albertus Magnus, De anima II tr. 3 c. 20 (Ed. Colon. VII/1, 128.9–73).

190 Hasse, *Avicenna's* De Anima, 126–127.

sus.[191] Albert was able to dodge a misinterpretation of Avicenna's doctrine by contrasting and comparing other passages, where the Arabic *šaffāf* (the medium for vision) is rendered with *translucens* or *pervius*. In the *De homine* and in other works—such as the commentary on the *De divinis nominibus*—Albert was therefore able to grasp the gist of Avicenna's theory of vision correctly and spell it out explicitly, pointing out the difference between *lucidus* (Ar. *muḍīʾ*, "with natural light" [*ḍawʾ*]) as referring to essentially luminous bodies like the Sun[192] and the adjective *illuminatus*, denoting a body with acquired light.

A good understanding of Avicenna's theory of vision did not imply that Albert subscribed to Avicenna's own views on light, though. In the *De homine*, when discussing the nature of light, Albert forcibly asserted that all natural philosophers agree on the fact that light is a *habitus* of the transparent. This is something that Avicenna would have denied, for he considered some light (viz. natural light) to exist independently of its medium (translucent air).[193] Light, as we observed previously, is the abstracting agent that gives the visible form its spiritual being (*esse spirituale*) in its medium. In sum, while accepting Avicenna's distinction of objects as endowed with either essential or acquired light, Albert favored a view on light mutuated from Averroes, viz. light as *habitus* of the medium. As we have seen, Albert's early ideas on light shares some things with his contemporaries like Rufus, Grosseteste and Alexander of Hales on the abstracting action of light itself, which causes the sensible species to acquire "spiritual being" (*esse spirituale*). According to Albert, this case can apply to sight alone and not to all distal senses.

While his discussion on the nature of light disagrees with Avicenna's position, Albert's own physiology of sight is widely and deeply influenced by Avi-

191 Avicenna, Liber de anima I c. 5, ed. Van Riet 1972, 83.58–84.62 [ed. Rahman 1959, 41.19–42.1]: "Ex quibus est visus, qui est vis ordinata in nervo concavo (*quwwa murattaba fī ʿasbati al-maǧūfa*), ad apprehendendum formam eius quod formatur in humorem crystallinum ex similitudinibus (*ašbāḥ*) corporum habentium colorem quae veniunt per corpora radiosa (*fī al-aǧsām al-šaffāfa*) in effectu ad superficies corporum tersorum." This definition, very well-known among Medieval Latin Scholars, is also very misleading: on the reasons for such a phenomenon: Hasse, *Avicenna's* De Anima, 123–124.

192 Albert also distinguished, trying to avoid terminological confusion, the Avicennian sense of *lucidus* from the Graeco-Latin translation of *lucidus* for the Aristotelian "translucent", see: Hasse, *Avicenna's* De Anima, 118.

193 Albertus Magnus, De homine (Ed. Colon. XXVII/2, 160.11–14): "Et super illud dicit Averroes: 'Ex hoc patet quod lux est habitus diaphani et quod colores non acquirunt habitum ex luce'. Hoc etiam videtur dicere omnes philosophi naturales." Hasse, *Avicenna's* De Anima, 118. Averroes' position in: Averroes, Commentarium magnum in Aristotelis De anima libros, II, comm. 69, ed. Crawford 1953, 236.31–33.

cenna on many levels. In terms of the localization of the sense organ, Albert confronts the different localizations of sight presented in Avicenna's writings. Not content with Avicenna's summarized definition of the faculty of sight in *De anima* I, 5, Albert goes on to explain that the faculty of sight can be perfected at three different stages: in the crystalline humor according to its principle (*secundum sui inceptionem*), in the visible nerve according to its progressive unfolding (*secundum progressum*) and in the front brain ventricle according to its completion (*secundum sui perfectionem*).[194] In the physiological explanation of sight proposed by Albert in the *De homine*, there is an additional factor that must be added to the painstaking localization of the different faculties of sense perception pointed out by Steneck, i.e. Albert's particular attention to impairments of sensitive faculties, especially sight impairments.

In the *De homine*, Albert emphasizes the crucial role of the complexional balance in sense perception: if the soul and the body are in a proper state of harmony, the soul's disposition in the acquisition of sensibles is effective (*dispositio fit per spiritum animalem et calorem naturalem et per harmoniam organi*).[195] If this harmony is lacking due to physiological causes, then misperception ensues. In the case of the spirit responsible for sight (*spiritus visibilis*), causal agents which may disrupt this harmony are represented by drunkenness—an argument inspired by Aristotle's *De sensu et sensato*—and rage.[196] In both cases, the outcome is an excessive production of vapors, which end up impairing the receptive capacity of the *spiritus animalis* and, in the wake of this, damage capacity for sight. Albert also reports the four Avicennian arguments (*De anima*, III, 8) for double vision: while the first, i.e. the distortion of the trajectory of the spirit's flow, is directly associated to the pressure of an object on

194 Albertus Magnus, De homine (Ed. Colon. XXVII/2, 146.60–64): "Ad aliud dicendum quod virtus visiva secundum sui perfectionem est in anteriore parte cerebri, sed secundum sui inceptionem est in humore crystallino, sed secundum progressum ad perfectionem est in nervo optico et spiritu visibili, qui decurrit in illo."

195 Albertus Magnus, De homine (Ed. Colon. XXVII/2, 258.4–9): "Quando enim potentia animae coniungitur cum organo, est potentia prima ad sensum; quando vero dispositio fit per spiritum animalem et calorem naturalem et harmoniam organi, tunc est potentia sensus in secundo gradu; quando vero habet speciem sensibilis, tunc est potentia completa per actum." Steneck, "Albert on the Psychology of Sense Perception," 271.

196 On drunkenness, see: Albertus Magnus, De homine (Ed. Colon. XXVII/2, 150.31–39): "[…] Item, secunda (*sc.* ratio) est per experimentum sumptum in ebriis, in quibus vinum calidum superat cerebrum frigidum et calore resolvit et facit fluere humidum oculi frigido coagulatum, et ex vaporibus calidis et grossis diminuit ipsum in parte et sibilem. Hoc autem non accideret, si dominaretur in oculo ignis; calidum enim a calido non diminueretur, sed augeretur." Albertus Magnus, De homine (Ed. Colon. XXVII/2, 227.21–23, 28–36). On rage, see: Albertus Magnus, De homine (Ed. Colon. XXVII/2, 152.2–16).

the eye,[197] the other three report the Avicennian doctrine faithfully.[198] Albert also tries to broaden the scope of his doctrine of impairment in sight, relating them generally to analogous phenomena in animals.[199] These passages reflect a typical attitude of Albert's, which never loses an occasion to compare human and animal sensation, combining at the same time various sources such as Avicenna's *De anima*, Aristotle's *De sensu et sensato* and the medico-philosophical notion of *spiritus*.

Sight, in Albert's *De homine*, represents the only sense whose sensible species exists in its medium (air or water) according to its spiritual being. The noteworthy difference, as we have observed, is given by the source of spiritual being in the medium: light in the *De homine*, the form *qua* form in the *De anima*. Here Albert—as noted by Hasse—first introduces a concept of light which comes close to Avicenna's definition of acquired light (*lumen/nūr*) but ends up sticking to the Averroean stance of acquired light as *habitus in natura diaphani*.[200] In his mature works, Albert maintains that the faculty of sight depends on the transmission of sensible images from the crystalline humor through the first pair of cranial nerves to the front brain ventricle in humans.[201] The eyes, defined as *flores animae*,[202] are connected to the brain—or whatever is analogous to it—through a couple of optical nerves, since every animal possesses two eyes but not everyone makes use of them (moles, for example, do not).[203]

In his Aristotelian paraphrases, Albert insists on some pathological phenomena connected to sight, such as double vision. In the *De sensu et sensato* he devotes an entire *digressio* to the case of double vision, relying heavily on Avicenna (*De anima*, III, 8) and displaying a position very close to his early *De homine*. In this section, Albert repeats the same arguments of Avicenna on the distortion of the trajectory of spirit on its way to the optical chiasm, the motion of spirit in the eye which goes beyond its natural endpoint, and the

197 Albertus Magnus, De homine (Ed. Colon. XXVII/2, 201.15–33) and apparatus.

198 Albertus Magnus, De homine (Ed. Colon. XXVII/2, 202.34–74) and apparatus.

199 Albertus Magnus, De homine (Ed. Colon. XXVII/2, 152.29–34): "Ad primum ergo dicatur quod quaedam animalia sunt, quae debilem habent visum et non bene coadunatum, et illa claritate diei vincuntur. Et propter hoc in die quiescunt et de nocte moventur, quando lumen lunae et stellarum est in aëre, quod debilius est lumine solis."

200 Albertus Magnus, De anima II tr. 3 c. 8 (Ed. Colon. VII/1, 110.68–74). Hasse, *Avicenna's* De Anima, 118–119.

201 Albertus Magnus, De animalibus, I tr. 2 c. 18 (ed. Stadler, I, 126–127). In these pages, Albert paraphrases Avicenna's description of the first pair of cranial nerves.

202 Albertus Magnus, De animalibus I tr. 2 c. 3 n. 142 (ed. Stadler, I, 51.32–33).

203 Albertus Magnus, De animalibus IV tr. 2 c. 1 n. 76 (ed. Stadler, I, 392.16–27).

backward-forward motion of the spirit in the optical nerves. This last cause for double vision is connected by Albert to drunkenness and, as in Avicenna, to the observation of a point upon a rotating surface. Albert then mentions the fourth cause for double vision in Avicenna's account, viz. the pupil's constriction or dilation,[204] and completes his account of misperception with further examples of optical pathologies. For instance, myopia (shortsightedness) appears to be caused by a retracted position of the crystalline humor in the eye, which in turn impacts the nature of the image transmitted by the spirit to the brain.[205] By contrast, long-sightedness is due to the drying out of the action of spirit and is connected to aging. Hypermetropia—long-sightedness—consists in the inability of the spirit to receive the impression from the closer objects of vision, while, by contrast, animal spirit is able to see afar.[206]

In his explanation of sleep as an incapacitation of the sensitive powers, Albert expounds other examples of pathologies, among which the case of epilepsy looms large. According to Albert's *De somno et vigilia*, excesses in moist exhalations and vapors produce epileptic shock: this physiological explanation is shared with another phenomenon, e.g. vertigo, which, following Avicenna, is caused by excessive vapors arising in the brain as well. For this reason, Albert places so much emphasis in the *De animalibus* on the cooling and balancing role of the cerebral complexion, which cools down the excess in heat produced by the motion of animal spirits or the heat of the vital spirit originated in the heart and entering the brain; otherwise, the operations of the faculties become ineffective (*commixtae et confusae*). According to Albert, seizures can begin during sleep because of the great amount of moisture that come from digestive vapors, the material cause of sleep itself.[207]

---

204 Albertus Magnus, De sensu et sensato I tr. 1 c. 11 (Ed. Colon. VII/2a, 45.35–46.20) and critical apparatus.

205 Albertus Magnus, De sensu et sensato I tr. 1 c. 14 (Ed. Colon. VII/2a, 52.17–25): "Ex diversitate autem horum spirituum et ex diverso situ humoris oculi diversitas magna accidit circa visum, quoniam habens humorem crystallinum non in centro, sed versus interiorem superficiem oculi non videt nisi de propinquo."

206 Albertus Magnus, De sensu et sensato I tr. 1 c. 14 (Ed. Colon. VII/2a, 52.26–36): "Si autem solum spiritum debilitari contingat, sicut fit in senibus et exsiccatis ex vigiliis et studio, tales non vident prope iuxta oculos sed eminus ad distantiam aliquam. Cuius causa est quia quando visibile est iuxta oculos, tunc ad directum oculi non diriguntur nisi paucae partes eius, et quaedam sunt super oculum et quae dam infra et quaedam a latere divertentes a directo oculi; et quia spiritus debilis est, tales partes non percipiuntur. Cum autem removeatur magis ab oculo visibile, tunc magis ex directo percipiuntur, eo quod magis diriguntur lineae quae sunt inter visum et visibile."

207 Albertus Magnus, De animalibus I tr. 3 c. 1 (ed. Stadler, I, 185). For a similar explanation of

In humans as well as in animals, the brain seems to be endowed with enough coolness to avoid vertigo, double vision or epilepsy *qua* products of excessive vapors affecting the spirit. Usually, vapors derive from the heating of moist liquids, so a better hot-cold balance in body temperature avoids this kind of pathological phenomena. Here we cannot but observe how Albert's painstaking depiction of the physiology of sight and its pathological explanations emerges as peculiar among his contemporaries when writing the *De homine* and is even expanded in his Aristotelian paraphrases, where he knits together examples from medical and philosophical literature.

4.6 *Common Sense*

In her book *Ratio particularis*, Carla di Martino has shown how Albert's localization and understanding of common sense owes a lot of its physiological perspective to Avicenna, even while he retains meaningful elements of Aristotle's thought and remains indebted to Averroes. In the *De homine*, Albert places common sense in the front brain ventricle and sees it, like Avicenna and Averroes, as the source and root power for the other senses both at a psychological and at physiological level, for it is from the front brain ventricle, where common sense is seated, that animal spirits flow into the different pairs of cranial nerves.[208]

The psychological functions that Albert ascribes to common sense result from a synthesis of Aristotle's and Averroes' positions. In the *De homine*, Albert considers common sense as a faculty charged with the perception of common sensory objects, a recurrent position in Aristotelian exegesis at the time.[209] Common sense in Albert's early theological writings is quite different from Aristotle's and Avicenna's and is responsible for three main activities. Unlike Aristotle, the perception of common sensibles belongs to common sense as its proper object. Like Aristotle and unlike Avicenna, it accounts for perceptual awareness, which the Persian philosopher and doctor would have assigned to estimation.[210] Like both Avicenna and Averroes, though, Albert considers com-

---

the cooling role of the cerebral complexion on animal spirits: Avicenna, Canon III fen 1 tr. 1, ed. Venezia 1507, fol. 165ra.

208 Albertus Magnus, De homine (Ed. Colon. XXVII/2, 273.39–42): "Prima enim concavitas cerebri est locus, in quo prima est congregatio spiritus animalis, qui per distributionem ad organa specialia operatur quinque sensus."

209 Di Martino, Ratio particularis, 72. On Averroes' position, see: Averroes, Commentarium magnum in Aristotelis De anima libros, II, comm. 65, ed. Crawford 1953, 228.30–32: "Et etiam sensibilia communia, ut declarabitur, sunt propria sensui communi (quemadmodum ista sunt propria unicuique sensuum) […]".

210 Albertus Magnus, De homine (Ed. Colon. XXVII/2, 279.38–49): "Solutio: Dicendum quod

mon sense the faculty that is able to compound common and special sensibles into a single whole stored in the form-bearing faculty and, moreover, to reciprocally contrast, compare and discriminate proper sensibles (e.g. sweet is not white etc.).[211]

In his early theology, Albert showed a good degree of acquaintance with Avicenna's psychology and psychophysiology, observable in his reporting the example of the raindrop as a proof that common sense allows us to experience time. In the *De quattuor coaequaevis*, Albert quotes the Avicennian raindrop example comparing it to our vision of the circular motion of a stick. To our sight, the circularly moving stick appears to be here and there (*accipit hic et ibi*), while our common sense perceives these two distinct moments of sense perception as if the object was one, i.e. a circle.[212] From a certain perspective, Albert seems to follow Avicenna in pointing out that common sense perceives continuity of motion as if it were the product of a single moment of sensation. Unfortunately, Albert does not develop the point further, and neither does he connect it to the perception of time. In his later *De sensu et sensato*, Albert replicates the Avicennian example of the raindrop and of the rotating stick he had mentioned in the *De quattuor coaequaevis*, and that the quickness of the

tres actus sunt sensus communis, ut probatum est per rationes Philosophi, sed non eodem modo conveniunt sensui communi. Apprehensio enim sensati communis est actus suus per se et per illum diffinitur. Apprehensio autem sensati proprii convenit ei per posterius, scilicet inquantum sensatum proprium est in sensato communi ut in subiecto; sed apprehensio actuum convenit ei per accidens, scilicet inquantum unusquisque propriorum reducitur in ipsum. Sicut enim dicit Avicenna, omnes sensus proprii oriuntur a communi et revertuntur ad ipsum cum lucro." Di Martino, Ratio particularis, 72–73; Anzulewicz, "Konzeptionen und Perspektiven," 215–217.

211 Albertus Magnus, De homine (Ed. Colon. XXVII/2, 276.46–48): "Solutio: Dicendum secundum Alfarabium et Avicennam quod sensus communis ponitur propter compositionem et distinctionem sensibilium propriorum [...]" On the reasons for this misattribution: Helmut Gätje, "Der *Liber de sensu et sensato* von Alfārābī bei Albertus Magnus," *Oriens Christianus* 48 (1964), 107–116.

212 Albertus Magnus, De quattuor coaequaevis tr. 4 q. 69 a. 3, part. 3, quaesit. 2, ed. Auguste Borgnet, Paris 1895 (Ed. Paris. XXXIV, 710a): "Solutio: dicendum, quod virtutes sensibiles animae quaedam sunt apprehensivae, et quaedam motivae. Apprehensivarum autem quaedam sunt apprehensivae de intus, et quaedam de foris, sicut sensus particulares qui particularia sensata apprehendunt praesente materia: ab intus autem sicut sensus communis, qui componit sensata particularia cum magnitudine continua et caeteris sensatis communibus. Cujus exemplum est in gutta cadente, quam accipit sensus particularis in loco hoc et illo divisim: sensus autem communis continuat locum ad locum, et apprehendit in similitudinem lineae. Aliud exemplum ejusdem est in virgula in una parte immota, et in alia parte circumducta, quam sensus particularis accipit hic et ibi: sed sensus communis apprehendit continuitatem per modum circuli."

motion of spirits (*frequentia motus spiritus*) in the brain produces vertigo and causes somebody to fall down.[213]

Unlike in Avicenna, we do not find reflections on the correlation between common sense, unified images from sense perception and time awareness. However, Albert does draw a line between humans and animals in terms of perception of time in his account of memory. As shown recently by Veronique Décaix, in the *De memoria* Albert argues that humans rationally apprehend time as "number of motion" (*secundum quod est numerus motus*), while animals only obscurely sense the passing of time.[214] Here, the psychological framework is different from Avicenna's, since the different perception of time relates certainly to the past, but is triggered by the activity of memory (devoid of past-directedness in Avicenna) and not of common sense and the form-bearing faculty.

Albert's doctrine of common sense undergoes a relevant doctrinal development in his mature psychology, expounded in the *De anima*. On the one hand, he retains some of the most basic and Aristotelian features of common sense already rooted in the *De homine*, such as the capacity of common sense to compare and discern different kinds of sensibles and to exercises a second-level perceptive awareness.[215] By contrast, he goes back to a more Aristotelian position by depriving common sense of the perception of common sensibles and assigning it to the individual senses.[216] The danger of making common sensibles the proper object sof common sense would have led, as noted by Carla Di Martino, to a consideration of the sensible forms perceived by common sense as accidental sensibles. This change of mind may be due, adds Di Martino, to Avicenna's (perhaps, also Aristotle's) influence. The Persian author held that common sense did not perceive common sensibles as if they were objects of

213 Albertus Magnus, De sensu et sensato I tr. 1 c. 11 (Ed. Colon. VII/2a, 46.1–16).

214 Albertus Magnus, De memoria et reminiscentia tr. 1 c. 2 (Ed. Colon. VII/2a, 116.63–117.5). Veronique Décaix, "What is Memory of? Albert the Great on the Proper Object of Memory," in *Memory and Recollection in the Aristotelian Tradition: Essays on the Reception of Aristotle's* De memoria et reminiscentia, ed. Véronique Decaix/Christina Thomsen Thörnqvist (Studia Artistarum 47), Turnhout 2021, 165 n. 48.

215 Albertus Magnus, De anima II tr. 4 c. 7 (Ed. Colon. VII/1, 157.1–5; 157.8–13): "Inquantum autem communis est, habet duo, sine quibus sensibilis cognitio non perficitur. Quorum unum est iudicium sensibilis operationis, sicut nos apprehendimus nos videre, quando videmus, et audire, quando audimus, et sic de aliis [...] Secundum autem est comparare sensata diversorum sensuum per hoc quod invenit ea in uno communi sensato coniuncta vel disiuncta; hoc enim facere non potest aliquis sensuum propriorum, quia comparatio inter plura est, et oportet comparantem plura simul habere apud se."

216 Albertus Magnus, De anima II tr. 4 c. 6 (Ed. Colon. VII/1, 155.18–22; 155.59–67; 155.72–74).

an over and above sense, and so did Albert in his mature works.[217] All in all, when Albert introduces his rejection of the perception of common sensibles by common sense, he refers to supporters of the opposite theory: while behind this *aliquis* is very likely Averroes, yet this list could feature the early Albert himself.[218]

While Albert's psychology of common sense undergoes some relevant modifications throughout the course of his intellectual production, his physiology of common sense is consistently derived from Avicenna. In the Aristotelian paraphrases he makes use of the *Canon*, while the *De animalibus* influences the earlier *De homine*. In both cases, common sense is thought to be, at a physiological level, the source for sense perception, because it is from the organ of common sense (the front brain ventricle) through cranial nerves and the nervous system in general that animal spirit flows into sense organs.[219] As we shall see, the role of common sense as the root of the powers of the five senses and as the center for the distribution of animal spirit into the members is crucial when considering the physiology of sleep and dreaming. For now, suffice it to say that the psychological differences between *De homine* and *De anima* bear physiological implications. In the *De homine* the perception of common sensibles is assigned to common sense which is localized in the front brain ventricle, while in the *De anima* this perception is assigned to the individual sense, whose powers are localized in different nerves and, in the case of smell, fleshy particle of the nose, but certainly not in the front brain ventricle where common sense is seated.

As in the case of sight, Albert takes the study of common sense also to the level of analyzing possible pathological implication in its malfunctioning. Ani-

---

217 Di Martino, Ratio particularis, 81.

218 Albertus Magnus, De anima II tr. 4 c. 6 (Ed. Colon. VII/1, 154.78–155.4): "Forte *autem* dicet aliquis, quod organum *aliquod proprium quo sentimus, est* eorum sensibilium quae supra vocavimus sensata *communia*; et sic ille sensus qui est illius organi, erit sextus sensus proprius, ita quod sensata communia sint sibi propria. Sed si hoc esset verum, tunc sensata communia *sentiremus unoquoque* sensu *secundum accidens* et non essentialiter."

219 Albertus Magnus, De homine (Ed. Colon. XXVII/2, 274.70–275.2): "Solutio: Sicut dicit Avicenna, 'dispositio sensus communis est talis quod ex ipso est principium cuiuslibet virtutis sensibilis, et postea redit ad ipsum cum lucro'. Cum ergo nervi sentiendi nascantur ex prima parte cerebri, eo quod prima pars mollior est, et mollities facit multum ad impressiones sensibilium, dicimus quod organum sensus communis est in prima concavitate cerebri, quae plena est spiritu sensibili." Albertus Magnus, De anima II tr. 4 c. 7 (Ed. Colon. VII/1, 156.84–90): "Dicamus igitur, quod in omni natura, quae est pluribus communis, oporte esse unum fontem, ex quo oriatur illa communitas. Sensibilis autem cognitio est communicata quinque sensibus, et ideo oportet esse unum fontem, ex quo omnis sensus oriatur et ad quem omnis motus sensibilium referatur sicut ad ultimum finem."

mal spirits are the closest among all spirits to the soul's workings (*plus accedit ad opera animae*) because of their nature.[220] If they are subtle and bright, they are well-suited to receive sensible forms. By contrast, their being cold and moist qualifies them for a good retention of these forms.[221] These qualities, most notably the capacity to quickly receive the sensible forms, can turn against a correct perception, though. In the case of double-sight and other phenomena produced by the heat caused by excessive quickness in the motion of spirits, we have seen how the cold complexion of the human brain plays a key role in avoiding any such heat-caused harm.[222]

Albert, like Avicenna, associates the excess in vapors with other pathological phenomena, such as vertigo. According to the Dominican master, vertigo is produced by two contrary motions of spirit in the pulse, i.e. *pulsus* and *tractus*. These motions resemble dilation (diastole) and constriction (systole) in the heart: the *pulsus* is represented by the motion of the form from common sense to the form-bearing faculty stored in imagery that travels from it to the anterior part of the brain, where common sense is seated. The *tractus*, on the other hand, is caused by the natural motion of forms perceived by common sense. Like any other perceived form, they go backwards towards imagination to be stored. According to Albert, it is the contrast between these two contrary motions of animal spirit that is responsible for vertigo.[223] Among other pathologies produced by vapors affecting the animal spirit features epilepsy, produced by vapors, triggered by the action of common sense, and blocking

220 Albertus Magnus, De spiritu et respiratione I tr. 2 c. 4, ed. Auguste Borgnet, Paris 1894 (Ed. Paris. IX, 236b).

221 Albertus Magnus, De spiritu et respiratione I tr. 1 c. 4, ed. Auguste Borgnet, Paris 1894 (Ed. Paris. IX, 220a): "Aquae quidem habet proprietatem in humido aliquantulum spisso, ut et bene formas recipiat et convenienter teneat: haec enim causa est, quia anterior pars cerebri in qua sigillantur formae sensuum, medullosa et humida facta est: et tunc a simili probatur, quod idem oportet esse in spiritu animali."

222 Albertus Magnus, De spiritu et respiratione I tr. 2, c. 6, ed. Auguste Borgnet, Paris 1894 (Ed. Paris. IX, 240a–b).

223 Albertus Magnus, De homine (Ed. Colon. XXVII/2, 372.26–36): "Ad id quod quaeritur de motu, dicendum quod est motus vertiginis, qui componitur ex pulsu et tractu; et pulsus est ab imaginativa ad sensum communem, tractus autem e converso. Causa autem pulsus est motus spiritus animalis et subtilis sanguinis ab organo imaginationis ad organum sensus communis, qui sanguis distribuitur in alimentum organorum sensuum propriorum. Causa autem tractus proprius motus est ipsarum formarum; formae enim apprehensae a sensu communi secundum suam naturam transeunt ad organum imaginationis." Note that this explanation is different from Avicenna's, who held (*De anima*, III, 8 and IV, 1) that vertigo originates from smoky vapors in the part of the brain devoted to common sense. For a similar explanation of vertigo: Albertus Magnus, De motibus animalium II tr. 1 c. 4, ed. Auguste Borgnet, Paris 1890 (Ed. Paris. IX, 290a).

the spread of animal spirit into members to produce sensation and motion (produced by the rear brain ventricle through the spinal chord): this account is common to both *De homine* and Albert's later works on Aristotle,[224] showing how Albert's interest on pathologies is not medical as such—for he does not teach how to treat diseases—but must be rather thought as the reverse side of the same coin of the *necessitas ex suppositione finis* in natural sciences.[225] But what does that concretely mean?

According to Albert, our sensitive faculties work in a certain way to carry out sense perception, but there may be physical reasons why this might not be always the case. With sensitive faculties, this may well be the case of impairments such as double-sight, opacity of vision, epilepsy and vertigo; the common feature of all these phenomena is their connection to animal spirit. Certainly, *spiritus animalis* is the physiological means through which the soul's sensitive faculties perceives its sensibles and affections or in the organ of sense themselves cause sensible to be misperceived or not perceived at all. If we recall Albert's understanding of spirit as an instrument to the soul or as the vehicle of psychic faculties, we can think of his analysis of sensorial impairments along these lines: in the same manner an instrument serves the artisan to craft an artifact or as a vehicle serves its rider to reach his final end, either may not reach its end due to any of a number of impediments. In the case of spirit, these factors are purely physiological, but still do not essentially affect Albert's teleology of sense perception, which he inherited from Aristotle.

---

224 Albertus Magnus, De homine (Ed. Colon. XXVII/2, 322.47–51): "[…] in epilepsia enim vapor multus, ascendens ad cerebrum implet nervos et suffocat ipsos, ita quod spiritus sensibilis et motivus non potest descendere per ipsos ad organa; quo facto remanent organa exteriora insensibilia et immobilia." Albertus Magnus, De homine (Ed. Colon. XXVII/2, 347.5–11): "Epilepsia enim provenit ex subito motu proveniente ex superflui tate humoris et spiritus infrigidatis in cerebro, quae postea 'descendendo venas tumefaciunt', ut dicit Philosophus in *primo de somno*, quae tumefactae 'coartant porum, ubi respiratio fit', hoc est vocativam arteriam, in tantum quod etiam quandoque suffocantur."

225 Albertus Magnus, De somno et vigilia I tr. 2 c. 8, ed. Auguste Borgnet, Paris 1890 (Ed. Paris. IX, 150b–151a): "[…] frequenter pueri epileptici sunt propter multum humidum quod fertur ad caput eorum: somnus enim simile aliquod est ita causa epileptiae: et ipse somnus quaedam epileptia videtur esse. Propter quod etiam pluribus hominibus principium epileptiae accidit in somno, licet postea forte evigilent: et dormientes quandoque in epileptia consumuntur et moriuntur, suffocato nimis humido: vigilantes enim quoniam non habent humidum nimi ex causa aegritudinis, et ideo minus habent quam illi qui habent ex somno et ex morbo: cum enim multus spiritus evaporationis tam in somno quam in epileptia feratur sursum ad caput, descendens per venam colli et gutturis, facit intumescere venas, et illis intumescentibus coarctatur et praefocatur porus cannae per quem fit respiratio, et sic suffocatur."

4.7 *Teleology of Sense Perception*

In his *De homine*, Albert clearly identifies two different criteria for the hierarchy of sense perception: being (*esse*) and well-being (*bene esse*). Like Aristotle, Albert assesses sense perception according to two aspects: in relation to the animal (*sensitivum*) and in relation to the cognitive content of the sensed object (*sensibile*). In the first case, touch is considered the primary sense, for it is the sense that constitutes the animal and distinguishes it from plants. On the other hand, if we consider the richness of cognitive content, sight becomes the primary sense. Albert corroborates his observation on the hierarchy of senses according to cognitive content relying on the medico-physiological considerations of Costa Ben Luca: the complexity of sight is directly correlated to the purity and brightness of the spirit acting in sight, the highest among all animal spirits that catalyze sense perception.[226]

In the *De anima*, Albert presents a similar view on the hierarchy of senses: sight is the primary sense in relation to cognitive richness (*visus inter omnes sensus maxime est sensus, eo quod habet magis sensibilis cognitionis*),[227] while taste takes first place in relation to the animal's existence (*sensus autem primus, propter quem animal est animal, est tactus*).[228] Taste is not the only sense which Albert considers to exist for the sake of the animal's preservation; he also attributes this purpose to taste (*gustus est ad esse, sine quo non salvatur natura animalis*).[229] In the *De sensu et sensato*, Albert follows Aristotle's remarks concerning the superiority of hearing to sight in learning,[230] claiming that it allows the understanding of things through signs (*res per signa*) and, moreover, the grasping of commonalities and mutual differences between the different habits of these things (*rerum habitudines ad se invicem per convenientiam ipsarum et differentiam, quod non facit visus*). For this reason, hearing is accidentally superior to sight in the process of learning. All in all, Albert's consideration of the teleology of sense perception throughout his corpus remains consistent and is clearly based on an Aristotelian framework with which he was acquainted from early on.

---

226 Albertus Magnus, De homine (Ed. Colon. XXVII/2, 147.68–71): "Item, omnes sensus perficiunt operationes suas per spiritum animalem; sed visus hunc habet purissimum et clarissimum, ut dicitur in *libro de differentia spiritus et animae*; ergo ipse est sensus primus." Costa ben Luca, De differentia spiritus et animae, ed. Wilcox 1985, 159.194–199.

227 Albertus Magnus, De anima III tr. 1 c. 9 (Ed. Colon. VII/1, 176.26–27).

228 Albertus Magnus, De anima II tr. 1 c. 6 (Ed. Colon. VII/1, 73.50–51).

229 Albertus Magnus, De anima II tr. 3 c. 22 (Ed. Colon. VII/1, 130.58–59).

230 Albertus Magnus, De sensu et sensato I tr. 1 c. 2 (Ed. Colon. VII/2a, 23.13–16).

4.8 *Having a 'Better' Sense: A Physiological and Psychological Question*

As we observed in the third chapter, one of the recurring axioms in the thought of Albert the Great is that *homo inquantum homo solus est intellectus* and, because of the immortal and separable quiddity of the human soul, even the lowest faculties (nutritive, sensitive) in humans are superior to those found in all living beings because of their partaking to the rational, immortal soul.[231] If we apply this axiom to his theory of sense perception in his mature works, some tensions appear to emerge, particularly in the case of smell. In chapter 3, we saw how human beings are not only behaviorally, but also essentially distinguished from animals by virtue of their possessing a rational soul. This immortal, separable substance is the final cause for the sake of which body parts are arranged.[232]

If we assume these premises as always true, one might be tempted to conclude that human sense perception must be always superior to animals. This may be the case, for instance, with touch and taste.[233] On the other hand, though, the case of smell (and sight) is somehow at odds with these ideas, though. In the *De homine* and the *De sensu*, Albert proposes a dedicated sets of quaestiones on why, following the lead of Aristotle's *De sensu et sensato*,[234] humans have a less sharp sense of smell (*odorant prave*) than other animals. Based on Aristotle and Avicenna, Albert concludes that humans have a weaker sense of smell than animals, because the species of smell are more easily canceled out than those species stored in the form-bearing faculty. Consistent with this, it is rare that humans oneiric images contain smells too. If they were preserved in a manner that resembled the other sensible species, they would be

231 Albertus Magnus, De animalibus XXI tr. 1 c. 1 n. 4 (ed. Stadler, II, 1323.31–35): "Genus enim proximum est sensibile, et remotum est vivum: et nos ostendimus quod tam sentiens quam vivum participant aliqualiter rationem, licet ipsae potentiae de natura sua sint irrationales. Eodem autem modo etiam vegetativum in sensibili secundum esse alterum et aliud est quam sit ex natura sua in seipso."

232 Albertus Magnus, De homine (Ed. Colon. XXVII/2, 226.28–228.13); De sensu et sensato tr. 2 c. 6 (Ed. Colon. VII/2a, 70.20–75.55).

233 Albertus Magnus, De sensu et sensato tr. 2 c. 6 (Ed. Colon. VII/2a, 70.34–37): "*Tactum autem* habemus *discretissimum* inter omnia *alia animalia*. Cum igitur gustus *quidam tactus sit*, consequitur etiam ut aliis sensibus et aliis animalibus melius et discretius habemus gustum." On the role of complexion in the interspecific (among animals and humans, among different species) differences in sense perception, see Theodor W. Köhler, Homo animal nobilissimum. *Konturen des spezifisch Menschlichen in der naturphilosophischen Aristoteleskommentierung des dreizehnten Jahrhunderts*, 3 voll. (Studien und Texten zur Geistesgeschichte des Mittelalters 114.2/1), Leiden-Boston 2014, vol. 2/1, 240–266.

234 Aristotle, De sensu et sensato, 4 (440b30–441a2).

perceived during dreaming.[235] Therefore, Albert concludes that humans have a less sharp sense of smell than animals (*prave odorant*).

The solution he envisaged in the *De sensu* is slightly different and based on the notion of complexion: sharpness in smell is directly proportional to a drier and hotter complexion, since smelling is a soul power primarily connected with heat and dryness. For this reason, animals with a drier and hotter complexion have a better sense of smell (*prave odorant*).[236] Based on different reasons than in the *De homine*, the conclusion is the same: humans have a weaker olfactory sense than animals, and this should be due to the different complexion of animal spirits, too, which are less receptive than those found in certain animals.

In his *De animalibus*, though, Albert explains how this apparent discrepancy can be reconciled with his claim that humans are essentially superior to all animals because of their intellectual nature (*homo inquantum homo solus est intellectus*) and how this applies to the sensitive faculties of the soul. Albert is conscious, for instance, that lynxes have a better sense of sight and other animals such as spiders have a great sense of smell.[237] On the other hand, he is conscious that humans make a cognitively more complex (hence, better) use of the information gathered via sense perception. Among all species, they are the only one able to abstract universals out of the sensible forms received. For these reasons, Albert distinguishes two senses according to which one can say that an animal or humans has a 'better' sense than other animals (or humans too). In the first case, better refers to the sharpness according to which these sensibles are received. For instance, lynxes (or eagles, too) have a particularly sharp sense of sight rooted in their complexion.[238] The second sense of 'better', though, refers to the cognitive use which is made out of the gathered sense data (*dicitur ideo melius videre nobiliori uti visu sive largioris potentiae super quodlibet obiectum*). From this perspective, no animal can surpass the human cognitive and rational capacities.[239]

In the end, animals can excel over humans in their receiving the sensible species of certain senses. This superiority is due to a better complexional affinity between the complexion better suited for the sense in itself and the complexion of the animal. This is the case with dry and hot animals like spi-

235 Albertus Magnus, De homine (Ed. Colon. XXVII/2, 224.69–227.18).

236 Albertus Magnus, De sensu et sensato tr. 1 c. 15 (Ed. Colon. VII/2a, 53.36–37): "Et animalia quorum complexio magis accidit ad calidum siccum melius olfaciunt quam alia."

237 Albertus Magnus, De animalibus XXI tr. 1 c. 8 n. 46 (ed. Stadler, II, 1345.26–27): "In odoratu autem propter siccitatem capitis et membri quod loco cerebri est, valde excellunt."

238 Albertus Magnus, De animalibus XXI tr. 1 c. 1 n. 4 (ed. Stadler, II, 1323.10–19).

239 Albertus Magnus, De animalibus XXI tr. 1 c. 1 n. 4 (ed. Stadler, II, 1323.19–25).

ders which have a better sense of smell than humans. Human rational abilities, though, put human sensorial capacities above all animals not because of their physiological capacity to receive sensible species, but because of their superior psychological capacity to use these species to elaborate more complex cognitive and rational informations, to devise arts and crafts, produce a meaningful language etc. All these abilities are not shared by animals in the same manner, for the sounds they use and the art-like productions (e.g.: spider webs) they can device proceed from a different set of psychological capacities.

Conclusions

Albert's psychophysiology of sense perception is clearly influenced by the concept of *spiritus animalis*, conveyed by the Galenic medical tradition to the Islamic World (Avicenna, Costa ben Luca etc.) and other relevant sources such as Nemesius' *De natura hominis*. His staunchly Aristotelian, cardiocentric understanding of the relationship of the soul's powers to the soul, read through the lenses of Neoplatonic emanationism, goes hand in hand with a detailed description of the role of brain, cranial nerves and animal spirits in sense perception, common sense and relative impairments. In the externally apprehensive powers, animal spirit (*spiritus animalis*) is charged with the transmission of sensible species to the brain via the pairs of cranial nerves, except in the case of smell. Smell represents a most notable case of sense perception: on the one hand, it highlights the human-specific capacity to perceive odors not only as food-related sensible species but as species of substances lethal or detrimental to the brain's health. On the other hand, smell shows how the partaking of the lower psychic faculties in the separable, immaterial rational soul is nonetheless conditioned by material factors such as complexion. The extreme balance of the latter makes humans less suited than other animals to having a sharp sense of smell. As a consequence, animal spirits in humans carry a cognitively richer sense of smell. By contrast, accuracy in the perception of these odors by humans or certain animal species may be qualitatively inferior to animals that are hot and dry in their complexion, exemplifying once again how material factors affect psychic powers. This fact alone, though, does not change the human superiority in the use of the content gathered from sense perception, determined by the use which can be made out it from the intellectual capacities of the rational soul and by the partaking of the lowest faculties of the soul to the immortal, rational soul.

Albert follows the Galenic imprinting of Avicenna's *De anima* and, most notably, *De animalibus*, in construing his anatomy of sense perception in the

De homine achieved by blending Avicenna with other key sources such as Costa Ben Luca and, to a smaller extent, Nemesius of Emesa. In this work, Albert employed Avicenna's zoological masterpiece to correct the shortcomings of Aristotle's psychophysiology of sense perception. In this manner, Albert succeeded in grafting his theory of animal spirits onto an Aristotelian psychology of sense perception. From a purely historical point of view, the *De homine* represents a case of peculiar interest because it represents one of the first cases—as shown already in the third chapter—in the history of the reception of Avicenna's *Kitāb al-Hayawān* (*De animalibus*) during the 13th century. While close to Avicenna's physiology of sense perception, Albert's *De homine* bases his study of teleology of sense perception—almost absent in Avicennian psychology—on Aristotle's psychology. In the *De homine*, Albert even links the direct proportionality between refinedness of animal spirits and complexity of sensitive operations in sight, and he considers the first sense inasmuch as it provides the richest amount of cognitive data. The link between an apt material composition of animal spirit and the capacity to perceive an immaterial sensible species is a criterion that features in the *De homine* but is also mentioned more extensively in Albert's handling of animal spirit in the *De spiritu et respiratione* and which more generally influences his theory of externally and internally apprehensive powers. Albert's insistence on the relevance of material factors in the explanation of the activities of psychic powers—consistent throughout his works—finds perhaps its best exemplification in the explanation of double vision, long- and short-sightedness, vertigo and epilepsy, on account of the causal role played by animal spirits in the process. In his early works, the study of animal spirits applies to all animals in a comparative manner, though in an understandably less systematic manner than in his mature zoology and psychophysiology.

Albert's Aristotelian paraphrases display a more detailed and extensive treatment of the physiology of sense perception and common sense, this time based on a combination of both Avicenna's *Canon* and *De animalibus*, the latter of which comprises the key epistemic underpinning for Albert's approach to zoology and psychophysiology. While the *spiritus animalis* doubtless plays a key role in his notion of sense perception and common sense, which is close to Avicenna and the Galenic medical tradition, it would be quite inaccurate to reduce Albert's views on the causal role of animal spirits in sense perception and common sense purely to Avicenna's own. Influenced by contemporary debates on the doctrine of so-called "spiritual being" (*esse spirituale*) or "intentional being" (*esse intentionale*), Albert considered animal spirits to be the source of spiritual being in the sensible species of contact senses (touch, taste). Years later he would modify his position, attributing this role to the medium (flesh, tongue) of these two senses. Albert's standpoint on spiritual being in the *De homine* emerges from the critique of other contemporary authors (Rufus of Cornwall, Alexander of Hales, Roberte Grosseteste) who consider light the abstracting agent responsible for the production of spiritual being in all sensible species. Albert rejects this doctrine as countered by empirical evidence (we smell and hear in the dark) and attributes this abstracting action to light only in the case of sight. In his mature production, the issue of spiritual being will come to be of paramount importance in explaining perception in distal senses but will be considered the form *qua* form (or the medium in contact senses) as the agent responsible for the production of spiritual being. While retaining the process of *multiplicatio specierum* as the explanation for the diffusion of the sensible species in the medium, Albert's position on the form *qua* form as the source of sensible species in the medium and the denial of any abstracting role of light and animal spirits bring his doctrine closer to Averroes'.

Albert's indebtedness to Averroean psychology emerges also in his handling of common sense. In the *De homine*, common sense emerges as a root of all five senses at both a psychological and a physiological level, for common sense represents the source of all animal spirits involved in sense perception. At a psychological level, common sense appears to be charged with second-level perceptual awareness, discrimination of sensibles and their association and the perception of common sensibles as its proper object, precisely as in Averroes. In his early works, Albert even points toward a possible role for common sense in structuring the content of sense perception according to a dimension that enables time awareness. Albert's sketchy remarks, though, do not allow the reader to establish whether Albert thought common sense capable of structuring reality in such a way and whether time awareness might be different, as Albert will later state in his *De memoria*, in humans and animals. In his mature works, Albert reverted to a more Aristotelian conception of common sense, maintaining that perception of common sensibles occurs via the different senses.

As we have shown, the relevance of *spiritus animalis* in Albert's theory of the *potentiae apprehensivae deforis* invests not only the efficient dimension of sense perception, i.e. that of which spirit is vehicle (color, sounds, tangibles, flavors, common sensibles), but also the finality of sense perception. In Albert's philosophy of nature, animal spirit are designed by nature in their material, complexional constitution to be the most apt material substrate for a number of activities such as the perception of sensible species typical of the five sense, and a number of operations typical of the common sense. Among them, the association between the common sensibles and the proper sensibles, the judgement, composition or separation of different sensibles between them, and perceptual awareness. Through sight, animal spirit provides the richest amount of cognitive information, useful for practical purposes. Through hearing, animal spirits aid the individual in his learning process, fundamental for his intellectual progress. Through taste, they give the soul information about the suitability of food; through smell, the imprinting on the animal spirits in the brain informs animals and humans on the nature of food and humans alone as to the dangerousness of the source odors come from. By the means of touch, the flow of these spirits sets the human and the animal apart from plants by perceiving excesses in qualities in external objects and avoiding contact with potentially destructive objects. By means of common senses, animals gain awareness of their perceptive activities, of common sensibles, and become able to discriminate between different sensibles. It is not quite clear, by contrast, whether animals can be aware of time in a manner that somehow resembles the human cognition of time as a number of motion. Rather, it seems not, and there is good reason to ask whether this may be due to the human-specific, intellectual capacities. This last question introduces us the next key question, around which will revolve the next chapter, viz. the cognitive divide between humans and animals and its psychophysiological implications.

CHAPTER 7

Animal Spirits and Internally Apprehensive Powers (*potentiae apprehensivae deintus*)—Part 1: Estimation and Memory

During the Middle Ages, the cognitive divide between humans and animals was an object of considerable debate for quite a few decades. The introduction of Aristotelian psychology into the universities sparked debates on the psychology of cognitive powers derived from (imagination, memory) or related to (estimation, cogitation) sense-perception. The peculiar feature of these powers of the soul is that they were able to receive (1) the form of an object or (2) its "meaning" (*intentio*) even with that object not physically acting on the sense organs considered or even being not physically present. For this reason, the powers considered responsible for these activities were called "internally apprehensive powers" (*potentiae apprehensivae deintus*). The psychology of these powers covered two main areas: on the one hand, instinctual reactions, incidental perception and their link to memory; on the other hand, veridical dreams and prophecy. While the first topic represents a natural evolution of problematic questions raised by the Aristotelian text itself, the second proves to be quite a more complex matter.

Resorting to the first topic of discussion, viz. the psychology of instinctual reactions and memory, it is a known fact that Albert, like his contemporaries and followers, had to address one of the most debated questions about human and animal cognition, viz. the upper limit of animal knowledge. Very often, Medieval philosophers discuss whether animals have knowledge of singular substances, or by contrast, their cognition boils down to a merely inchoative sense of individuality (the main view in the Latin Middle Ages). This question was often associated with the interpretation of the difficult Aristotelian theory of so-called 'incidental perception', discussed in *De anima*, III, 1. The debate about the human-animal cognitive divide hinged on a key question of zoology: do animals react instinctively, being aware of the thing to which they are reacting (hence, fleeing or pursuing) as an individual of a certain species? Or is their response purely mechanical? This question can well be exemplified by the case of the sheep that flees the wolf and the dog that runs away from the stick or the mud brick with which it has been previously beaten. In the first case, the fact that a sheep flees the wolf seems to point toward an inborn com-

 | DOI:10.1163/9789004706743_009

ponent of reaction,[1] while a dog that flees a stick because it was previously hit with it seems to denote a reaction stemming from previous experience. The main question is whether this previous experience can make the animal aware that what it is perceiving is an individual different from others of the same species; in other words, whether the dog knows that this stick is different from others by generalizing its features and, thus, grasping a sort of concept of the stick. The possible answers to these thorny questions were explored by the psychologies of Avicenna and Averroes, who tried to give their own interpretations of the Aristotelian cognitive functions of imagination (*phantasìa*) through their famous theories of estimative (Avicenna) and cogitative power (Averroes). The cognitive divide between human and animals in instinctual reactions and memory answers indirectly another key question of Medieval zoology, viz. the degree of flexibility that can be granted to animal behavior, usually chained to a deterministic pattern.

The reflection on the cognitive content responsible for triggering instinctual reactions paved the way to a strictly related topic, viz. memory, its cognitive content and its differences between humans and animals. Intuitively, this difference between humans and animals is most evident in the case of recollection, considered as a human-specific and intellect-related power. On the other hand, though, one needs to ask oneself whether humans also possess a distinct cognitive content/object than animals of paramount importance to explain instinctual reactions and memory. The subsequent question prompts one to wonder whether the presence of such different cognitive content in humans is correlated to a different arrangement in brain structure that sets humans apart from animals. While the psychological implications of this process have already been inquired by Steneck, these studies should be complemented by a few remarks on the teleology of the bodily organs involved, especially the brain ventricles.[2] Moreover, detailed attention to the role of spirit will enable

1 Medieval writers fittingly associate this fact with the etimology of *instinctus* ("placed in") itself: more on this in Stefano Perfetti, "Nelle gabbie del fissismo etologico complessità del desiderio umano e monotonia dell'appetito animale in Tommaso d'Aquino," in *Il Desiderio nel Medioevo*, ed. Alessandro Palazzo, Roma 2014, 111–129.

2 Steneck's excellent work on Albert's psychophysiology explains *how* animal and human brains differ in the localization of psychic activities but does not go to great detail to explain *why*. This is not Steneck's fault, for his contribution is aimed precisely at showing how and that human and animal brains may differ in Albert's system of internal sense and *potentiae apprehensivae deintus*. See: Nicholas H. Steneck, "Albert on the Classification and Localization of Internal Senses," *Isis* 65 (1974/2) 193–211. For a different reading of Albert's theory of internally apprehensive powers, see: Jörg Alejandro Tellkamp, "Albert the Great on the Structure and Function of Inner Senses," in *The Judeo-Christian-Islamic Heritage: Philosophical and*

the reader to better appreciate how the highest cognitive abilities of the soul can be influenced by their material substrate, singling out animals and humans who possess better or worse cognitive capacities according to endogenous (sex, age) or exogenous (climate) causes.

The second topic discussed by Albert, viz. veridical dreams and prophecy, will be touched on in detail in the next chapter and is perhaps even more complex, as it combines Aristotelian psychophysiology with the theological demands of Christian orthodoxy. In this chapter, we shall try to highlight how animal spirits, understood as a vehicle for the faculties of the sensitive soul, can constitute a causal factor in the determination of the cognitive divide among animals and between humans and animals, considering especially the case of Albert. Given that the doctrine of animal spirits (like vital and natural spirits) as vehicle for the faculties of the sensitive soul was broadly accepted in the Latin tradition prior to Albert's works, the main difference will need to be observed at a psychological level, viz. how animal spirits activate different cognitive functions in the brain according to the psychological doctrines of different thinkers. Additionally, we shall question whether the extent of human rational and animal sensitive knowledge is somehow mirrored in the structure of the organ where these faculties are localized, namely the brain. While Albert's physiology of internally apprehensive powers is clearly derived from the medical tradition of the Islamic World, its psychology takes its cue, as with almost any Medieval author, from Aristotle's psychology of imagination and memory.

1 Aristotle on Human and Animal Cognition: Imagination, Memory, and Recollection

In approaching the topic of human and animal cognition, meant as knowledge of singular things and not of human-specific universals, Medieval authors were faced with the tough issues represented by Aristotle's problematic doctrine of *phantasia* and Aristotle's account of incidental perception (*De anima*

---

theological perspectives, ed. Richard Taylor/Irfan A. Omar (Marquette Studies in Philosophy 75), Milwaukee 2012, 305–324. As we shall see, the characterization of Albert's approach to internally apprehensive powers in the *De homine* as "physicalist", opposed to Albert's mature *De anima*, whose subject is explicitly and primarily the soul and its powers (*de ipsa anima et potentiis sive partibus eius*) and not its psychic activities (*Parva Naturalia*) or organs and organic parts (*De animalibus*), does not quite capture Albert's approach to the psychophysiology of internal senses. Rather, a good basis of comparison with the attitude towards bodily processes displayed in his early works could be represented by his *De anima*, *Parva Naturalia* and zoology considered altogether.

III, 1–2). Indeed, Aristotle had been clear in his distinction between the human-specific capacity of the intellectual faculties to produce universals by abstracting them from the "images" or "appearances" (φαντάσματα) of particular things. Animal cognition, by contrast, appeared to be chained to the knowledge of particulars (color, smell etc.) and was therefore unable to generate abstract and utterly immaterial (i.e. stripped of material features) concepts. On the one hand, humans were considered to be the only living beings able to abstract universals from the compound images provided by common sense, viz. *phantasmata*. According to Aristotle, it is not possible to think—viz. have rational activities—without images.[3] On the other hand, though, the images that are passed on to imagination and perceived by it during dreaming activity, and the perception of illusions and fictions, do not represent a distinguished object from the image compounded by common sense. According to some of modern scholars of Aristotle like Thomas Kjeller-Johansen, this is the reason why imagination must not be thought of as a faculty in its own right but rather as an after-function of sense perception, whence the *phantasmata* are derived.[4]

On the one hand, these *phantasmata* can be perceived by chance during non-teleological processes such as dreams or, during the state of waking, within delusions and fictions. On the other hand, these *phantasmata* play crucial roles in human intellection and animal cognition. In human intellectual activity, they represent the substrate of intellectual activity, viz. the object whence universals are abstracted. Moreover, in Aristotelian psychology *phantasmata* represent the cognitive object of the highest animal cognitive capacities. In the *De anima* (III, 7, 431a14–17; III, 11, 434a4–12), Aristotle ascribes to *phantasmata* the capacity to trigger instinctual reactions. These reactions seem to be based on the evaluation of properties (e.g. usefulness or harmfulness) that clearly exceed the scope of the merely sensorial content building up the *phantasmata*. Sharp exegetes of Aristotelian psychology such as Avicenna and Averroes started to wonder whether these non-sensible qualities required the existence of a specific object and, hence, of a specific faculty besides Aristotelian imagination.

1.1 *Instinctual Reactions and Incidental Perception: Aristotle and Non-sensible Qualities (?)*

In the first place, the fact alone that humans and animals pursue or flee certain *phantasmata* according to their good or their bad nature makes Aristotle's

3 Aristotle, De memoria 1 (449b15).

4 Kjeller-Johansen, *The Powers of Aristotle's Soul*, 201: "It is this similarity in content to perception that leads Aristotle to understand phantasia as a function of the perceptual capacity, rather than of a distinct capacity in its own right."

reader question the object of imagination. 'Good' or 'bad' certainly does not fall within the category of sensitive qualities but, apparently, of non-sensitive qualities that are indeed related to a sensible object. While these qualities indeed exceed the nature of a sensorial object, Aristotle never postulates the existence of a faculty whose activity is to perceive these non-sensible qualities. Moreover, it is not quite clear what these non-sensible qualities amount to. On the one hand, these passages may lead the reader into thinking that these non-sensible qualities concern the instance of beneficial or harmful qualities alone. But an attentive reader of Aristotelian psychology may start to question whether these reactions start from a judgment that a thing is harmful or beneficial because it belongs to a certain species. For instance, if a sheep flees a wolf, does it do so because it perceives the wolf (hostility) as hostile or because it identifies it as a wolf and, therefore, as hostile? In short, what does this perception of a non-sensible quality of wolf (hostility) depend on a perception of purely sensorial features (e.g.: grey, barking, furry thing) or something more, like its substantial form? In Aristotle, we find only a short nudge towards the topic. In the *Historia Animalium*, Aristotle tells a very Oedipean story of a Scythian horse that, having been forced by its keeper to have intercourse with its mother (with her face covered to avoid recognition), killed itself after the wrapper had been removed and it recognized its mother as the mare with which it had had intercourse. While this passage seems to point to some combination of moral shame (usually specific to humans and not to animals) and recognition of the individual being based on non-sensible data (in this case, the mare's identity as a mother), Aristotle does not take his explanation any further, either at a cognitive or at a moral level.[5]

In the end, Aristotelian psychology does not seem to answer the question about the nature and kind of perception to which these non-sensible features are subject and in what manner they trigger certain animal behaviors. To be fair, the same question seems to be sparked, although with a different starting point, from a more removed and more problematic topic, viz. incidental perception (*De anima* III, 1–2), which has prompted Aristotelian scholars such as Brentano to speak of 'intentionality' in connection with Aristotle's own doctrines.[6] In the

5 Aristotle, Historia Animalium IX 47 (630b31–631a7). I have noticed this passage thanks to the contribution of: Paloma Rubio Hernandez, "The Role of the *Intentio Individualis* in Albert the Great's Sense Perception Theory," in *The Embodied Soul: Aristotelian Psychology and Physiology in Medieval Europe between 1200 and 1420*, ed. Marek Gensler/Monika Mansfeld/Monika Michalowska (Historical-Analytical Studies on Nature, Mind and Action 11), Cham 2022, 135.

6 On the question of "intentionality" in Aristotle, see at least: Richard Sorabji, "From Aristotle to Brentano: the Development of the Concept of Intentionality," *Oxford Studies in Ancient Phi-*

opening of the third book, Aristotle highlights how the faculties of the sensitive soul—more specifically, common sense—have the capacity to grasp not only the common sensibles (rest and motion, number, size and shape) but also to recognize certain non-sensible qualities in an individual. This is the case, for instance, with the famous example of Cleon's son. When an observer becomes aware that the object of their perception is, for instance, the son of an acquaintance, that person does so via perception of certain sensorial features (look, smell etc.): the child's being Cleon's son, though, is not an essentially sensible property of the subject, so it is a non-sensible quality of the subject that is incidentally related to sense perception which lead us to recognize somebody as Cleon's son.[7]

Incidental perception seems therefore to have a cognitive content which may be comparable to instinctual reactions inasmuch as it involves judgment about a non-sensible quality (the relation of the son to Cleon) related to the content of sense perception. This is precisely the case with the above-mentioned cases of hostility and friendliness, even though it poses a problem in terms of cognitive content of incidental perception: does the knowledge that someone is somebody else's relative (as in our knowledge of Diares' son) entail our knowledge of him as a singular substance, as a man different from other men. If so, then there is an immediate question that concerns the link between this knowledge and intellectual activity, i.e., whether this knowledge of singular substances belongs to humans alone. While this question does not appear in the Aristotelian text, it is far from trivial. In the Middle Ages, the Aristotelian doctrine of incidental perception and animals' logical abilities of predicating the non-sensible qualities of an object will become a battlefield for many erstwhile psychologists in both the Arabic and the Latin Medieval traditions. These authors also had to come to terms with another *aporia* of Aristotelian psychology, viz. the lack of any connection between non-sensible qualities and memory, a topic about which Aristotle remains silent since he lacks the concept of "non-sensible qualities" in the first place.

---

losophy, Supplementary Volume (1991) 227–259. On the cognitive capacities of imagination, our account is loosely based on: Kjeller-Johansen, The Powers of Aristotle's Soul, 199–220; Dorothea Frede, "The Cognitive Role of *Phantasia* in Aristotle," in *Essays on Aristotle's* De anima, ed. Amelie O. Rorty/Martha C. Nussbaum, Oxford 1995, 280–296.

7 Aristotle, De anima III 1 (425a14–b3).

1.2 *Memory and Recollection: Human-Specific or Shared Capacities?*

According to Aristotle, the object of memory coincides with that of imagination, viz. the *phantasmata*.[8] Unlike imagination, though, memory enjoys an essential relation to time since, according to Aristotle, memory is memory of something experienced in the past *qua* past.[9] While the object of memory is distinguished from imagination because of its essential link to the past, *phantasmata* as subjects of the imagination occur to us without being related to a precise temporal dimension but rather are the product of a random, undirected activity. Memory belongs to the affections common to body and soul—like sleep and wake, inspiration and expiration, life and death—and is therefore common to humans and animals. Judging from Aristotle's remarks, though, it is not certain whether the reference to past in humans and animals is the same. Moreover, Aristotle points out that the object of memory can be understood as both an image in its own rights and as an image referring to an extramental object, as in the case of a painting. The main features of the extramental objects that are remembered by us, according to Aristotle, concern their sensible qualities and past-directedness but do not contain a reference to any non-sensible quality of such objects. Moreover, it is not possible to work out from the Aristotelian doctrine of memory whether the cognitive content of memory (*phantasmata*) should be different in humans and animals, viz. whether they can both retain the knowledge of these non-sensible qualities or whether there are some non-sensible qualities, such as individual substances, that can be known and remembered by humans alone.

While the human-specific features of memory are difficult to single out, this seems less to be the case with recollection. In the *De memoria*, recollection emerges an activity clearly distinguished from memory, but it is not clear whether the activity of recollection belongs to the same faculty as memory or whether it constitutes a faculty of its own.[10] Recollection is not explicitly assigned to any specific part of the soul by Aristotle, for recollection is considered a sort of deliberation and deduction. Judging from Aristotle's statements, though, recollection belongs to humans alone because of its syllogistic nature, tied to the activity of the rational soul.[11] All in all, the Aristotelian account of

8 This short analysis of Aristotle's treatment of memory is based on: David Bloch, *Aristotle on Memory and Recollection. Text, Translation, Interpretation, and Reception in Western Scholasticism* (Philosophia antiqua 110), Leiden 2007, 53–135.

9 Aristotle, De memoria 1 (449b15; 449b24–28).

10 Aristotle, De memoria 1 (451a20–21); De memoria 2, (453a4–5).

11 Aristotle, De memoria 2 (453a6–14); Aristoteles Latinus, De memoria et reminiscentia [A.L. XIV.1], ed. Donati (online): "Differt autem ab ipso memorari reminisci non solum

memory retains an aura of opacity as respects its psychological dimension, viz. the relationship between memory and recollection in terms of their mutual relationship, not to mention the fact that, as in the case of imagination, Aristotle does not present a precise physiology of memory. How can memory and recollection be part of the same faculty and are at the same time different in psychological and metaphysical terms? Luciana Repici has clarified that, much as memory appears to be conceptually linked to a good blend that is subject to decay in old age, this idea appears to be too scant in Aristotle's philosophy of nature to allow us to speak of a proper physiology of memory.[12] Moreover, it must not be forgotten that the psychophysiology of the Stagirite pivoted around the key role of the heart while totally ignoring the function of brain and nerves in cognitive processes.

Nonetheless, Philip van der Eijk has highlighted the key role assigned to a highly pure and concocted blood as the physiological correlate to a high degree of intelligence in living beings.[13] This concoction is usually correlated to a good ratio between heat and moisture, found in the most intelligent animals as humans. As pinpointed in a recent work by Cristina Cerami, though, the key role of vital heat in the concoction of blood, viz. its proportional correlation to vital heat, blood concoction and intelligence must not mislead us into thinking that vital heat represents per se a criterion for ranking animal intelligence in the *scala naturae*.[14] Rather, it is the instrumental function of vital heat that matters, for a great amount of vital heat and a finely concocted blood represents a criterion for ranking animal intelligence if and only if it refers to the complexity of cognitive (and rational, in humans) activities which they exert.[15] This discussion on the correlation between blood and cognitive abilities, though, is not systematically related to Aristotle's faculty theory, as it will appear in major authors of the medical tradition like Avicenna.

---

secundum tempus, sed quoniam ipso memorari et aliorum animalium participant multa, sed reminisci nullum, ut dicam, que cognoscuntur animalium, nisi homo. Causa autem est quia reminisci est ut sillogismus quidam: quod aliquid prius vidit aut audivit aut aliquid huiusmodi passus fuit, sillogizatur reminiscens, et est ut questio quedam. Hoc autem quibus et deliberativum meminere accidit."

12 Repici, "Tutto invecchia per opera del tempo," 28–29.

13 Philip van der Eijk, *Medicine and Philosophy in Classical Antiquity: Doctors and Philosophers on Nature, Soul, Health and Disease*, Cambridge 2005, 206–237.

14 Cristina Cerami, "Function and Instrument Toward a New Criterion of the Scale of Being in Aristotle's Generation of Animals," in *Aristotle's Generation of Animals: A Critical Guide*, ed. Andrea Falcon and David Lefebvre (Cambridge Critical Guides), Cambridge 2018, 130–149.

15 On the direct correlation between concoction of the blood and animal intelligence: Freudenthal, *Aristotle's Theory of Material Substance*, 62–65.

All in all, Aristotle's account of memory retains three major problems: the determination of the cognitive content of memory and the non-sensible qualities at play in instinctual reactions and incidental perception; the relationship between memory and recollection; and, last but not least, a discussion of these activities from a detailed, physiological point of view. While the first two defects could be solved by appealing to Avicenna's and Averroes' psychological doctrines Albert found a valuable source for his psychophysiology already in other authors such as Costa Ben Luca and Nemesius, who strongly influenced his localization and physiology of internally apprehensive powers (*potentiae apprehensivae deintus*).

2 Brain Ventricles and Cognitive Faculties in Costa Ben Luca, Nemesius, and Other Sources

As we will try to show, Costa ben Luca's *De differentia spiritus et animae*, Nemesius' *De natura hominis* (and the corresponding parts in John Damascene's *De fide orthodoxa*) and the *Pantegni* constitute in Albert's eyes a valuable complement to Avicennian psychophysiology of brain activities. All of these sources followed the classical, Galenic localization of imagination-related activities (*cellula phantastica*), reason and reason-like activities (*cellula logistica/syllogistica*) and memory (*cellula memorativa*) to the three brain ventricles. This was adopted by Albert already in his earlier works with reference to human anatomy.[16] The only exception to this categorization seems to be represented by Costa Ben Luca's *De differentia spiritus et animae*, which rather speaks of a "common space" (*commune spatium/faḍāʾ muštarak*) between the two ventricles—what we know today as hemispheres—and the posterior ventricle.[17] The main flaw in Costa ben Luca's account, as in the other texts which were heir to or drew on the Galenic medical tradition, is represented by its psychology: it lacks any definitional criteria for the objects of such activities, and thus failed to establish a clear border between rational and cognitive or reason-like activities still belonging to the sensitive soul. While this distinction could

16 Steneck, "Albert on the Classification," 204–205.

17 Costa ben Luca, De differentia spiritus et animae, ed. Wilcox 1985, 153.124–131 [ed. Troupeau 2011, ix.5–9]: "Cerebrum vero dividitur in duas divisiones quarum una est anterior, quae est maior, et altera posterior. Et in illa anteriori duo sunt ventriculi habentes introitum ad commune spatium quod est in medio cerebri. In posteriori vero habetur unus ventriculis (sic) faciens iter ad supradictum spacium quod est commune utrisque ventriculis qui sunt in anteriori cerebro."

be later found in Avicenna's and Averroes' doctrines on the different degrees of abstraction, these sources provided a detailed physiology of brain-located activities, which was often complemented by considerations on the etiopathology of mental diseases in terms of impairments of the different brain areas.

Among Albert's sources, Costa Ben Luca's *De differentia* is indeed a physiological reference text for Albert, as we hope to have shown so far in the case of life and sense perception. On the other hand, the *De differentia* presents the most hypertrophic and sometimes confusing classification of different cognitive, reason-like and intellectual activities located in the middle brain ventricle. On the one hand, Costa ben Luca follows the typical Galenic localization of imagination (*phantasia* [*bantāsiā*/*taḫayyul*]) in the anterior brain ventricle, assigning memory to the posterior one. The remarkable oddity with Costa ben Luca's concept of memory, though, is that it considers this faculty to be human-specific and not shared with animals.[18] This peculiar claim is countered by another and hitherto neglected work by Costa ben Luca, "The Book on the causes of the difference among people in their character, behavior, desires and choices" (*Kitāb min ʿilal iḫtilāf al-nās fī aḫlāqihim wa-siyarihim wa-šahawātihim wa-iḫtiyārātihim*) or, as summed up by Paul Sbath, "The book of characters" (*Le livre des caractères*). In that book, Costa ben Luca gives a list of cognitive activities which he deems human-specific:

> He (sc. God) placed the faculty of intellect (*ʿaql*), cogitation (*fikr*), cognition (*tamyīz*), opinion (*ẓann*), estimation (*wahm*) and vision (*ruʾyā*) in the middle of it (sc. the brain). This is a faculty that is proper to humans [alone], unlike all other animals (*wa-hiyya al-quwwa taḫuṣṣu al-insān dūna sāʾir al-ḥayawān*).[19]

Judging from this passage, memory is not ranked among human-specific faculties (hence, shared with animals), yet the *De differentia* upholds the oppo-

18 Costa ben Luca, De differentia spiritus et animae, ed. Wilcox 1985, 185.511–519 [ed. Troupeau 2011, xxvii.13–xxviii.4]: "Et opera anime sensibilis que dicitur vitalis sunt visus, auditus, gustus, odoratus, et tactus, quinque videlicet sensus corporis, et phantasia ac motus transmutationis de loco in loco voluntarie; et hec opera inveniuntur in omni animato (Ar. *ḥayy*, "living being" more than "animal" [*ḥayawān*]), idest in animalibus et in homine. Opera autem anime rationalis sunt cogitatio (*fikr*) et providentia (*raʾwiyya*), estimatio (*ẓann*) et dubitatio (*šakk*), visus (the Arabic reads instead: *wa-l-ʿazm wa-l-ʿilm*, "deliberation and knowledge," corrupted in the Latin translation) atque memoria. Et hec opera proprie sunt hominis inter cetera animalia."

19 Costa Ben Luca, Kitāb min ʿilal iḫtilāf al-nās fī aḫlāqihim (ed. Sbath 1941, 120.12–13). Translation is mine.

site view, viz. that memory is human specific. In a second passage from the same book, Costa ben Luca opposes the activities of the "animal" or "irrational" (*ḥayawāniyya*) soul to those of the rational (*nāṭiqa*) soul, among which is memory itself. From this passage it is quite difficult to understand whether memory is exercised by humans alone, for it is generically stated that God "placed the faculty of sensation, motion, imagination, memory, cogitation and other activities of the rational soul in the brain" (*wa-ǧaʿala quwwata al-ḥiss wa-l-ḥaraka wa-l-taḫayyul wa-l-ḏikr wa-l tamyīz wa ġayr ḏalika min al-afʿāl min al-nafs al-nātiqa fī l-dimāġ*).[20] The problem seems to lie in the classification of memory and its connection to the "other activities of the rational soul". Based on the evidence found in this book, memory should not be considered human-specific. If we consider the *De differentia*, it is hard to maintain the same view. For this reason, Costa ben Luca's psychology of memory appears to suffer from tension—if not self-contradiction—when it comes to classifying memory. For a Medieval reader like Albert, however, who was acquainted with the *De differentia* alone, Costa ben Luca's position was peculiar but unequivocal: though listing the Aristotelian *De anima* among his sources, the *De differentia* nonetheless classified memory as a faculty of the rational soul.

By contrast, Costa ben Luca's physiology of memory seems to offer several insights upon which Albert will later draw. In the first place, Costa ben Luca considers the purity of animal spirits, derived from a quasi-digestive process, as directly proportional to the cognitive complexity of the operations performed by them in the different ventricles of the brain.[21] Flowing through the different brain ventricles, animal spirits become gradually thinner and thinner and more apt for the different cognitive operations localized in the brain. To allow the passage of animal spirits into the posterior ventricle, Galen explains that the flow of these spirits is regulated by the upward and downward motion of a worm-like particle, which Costa ben Luca himself describes.[22] The peculiar-

20 Costa ben Luca, Kitāb min ʿilal iḫtilāf al-nās fī aḫlāqihim, (ed. Sbath 1941, 124.1–4).

21 Costa ben Luca, De differentia spiritus et animae, ed. Wilcox 1985, 153.134–138 [ed. Troupeau 2011, ix.11–15]: "[…] inde transiens ad alium ventriculum efficitur ibi subtilior et purgatur atque aptatur ad recipiendum virtutem anime, et hoc fit ei quasi digestio et conversio in quemdam spiritum subtiliorem atque clariorem (The Arabic has also *aṣfā*, "purer," omitted in Latin)." See also: Constantinus Africanus, Pantegni [Theorica IV.19], ed. Burnett, 114: "Spiritualis (*sc.* vitalis) vero spiritus cum a corde progreditur, in tel[a]e modum multipliciter diffusus et ibi immorans implicitus tamdiu ibi digeritur quoad depuratus ibi fieret lac, et post hic spiritualis spiritus rete per ambas arterias super rete retortas egreditur, et ad ventriculum prorae cerebri dilabitur, ubi iterum subtiliatus, quod depuratum supererat, eicit per suos meatus, idest palato atque naribus."

22 Costa ben Luca, De differentia spiritus et animae, ed. Wilcox 1985, 153.142–155.156 [ed.

ity of Costa ben Luca's text, though, is that it combines the Galenic function of the cerebellar worm as a tool for regulating the flow of animal spirit from the common space, viz. the Galenic middle ventricle, to the posterior ventricle of the brain, with some curious empirical observations.

According to the *De differentia*, the upward or downward motion of the worm can be aided by certain kinds of posture. For instance, a person who is thinking or who wants to think would keep his head down, fixing the earth at his feet. By contrast, the same person would raise his head when remembering or trying to remember something. According to Costa ben Luca, keeping the head down when thinking causes the worm-like substance to stay lower down and prevents animal spirit from flowing into the rear ventricle, thus triggering the activity of memory. By contrast, we raise our head to facilitate our remembering capacity.[23] In the end, the main problem with Costa ben Luca's psychophysiology of memory is that its psychology is ill-suited to classical Aristotelian faculty theory and, even taken by itself, is not entirely consistent. By contrast, the physiological explanation of the remembering process is detailed and full of curious empirical observations that will not escape Albert's intellectual curiosity.

In like manner, Costa ben Luca's account of reason and reason-like activities presents some major problems in its psychological part, as it does not provide any real distinctive criterion (comparable to the different degrees of abstraction in Avicenna and Averroes) for singling out cognitive and rational faculties alongside with their respective objects. On the other hand, Costa ben Luca's physiology of cognitive activities exemplifies in detail how failures in cognitive faculties are linked to specific brain ventricles, thus creating a palat-

---

Troupeau 2011, x.2–9]. The cerebellar worm is also featured in the account of: Constantinus Africanus, Pantegni [Theorica IV.19], ed. Burnett 1994, 114 [ed. Harbī 2018, 531.1–5].

23 Costa ben Luca, De differentia spiritus et animae, ed. Wilcox 1985, 155.161–166; 157.170–186 [ed. Troupeau 2011, x.14–16, xi.4–12]: "Et ideo accidit ei qui vult recordari alicuius rei ut caput suum valde mergat vel inclinando illud retrovertat et inmotis oculis sursum aspiciat ut haec positio vel figura sit ei quasi auxiliatrix ad aperiendum predictum foramen et ut ipsum corpus vermile possit sursum elevari. [...] Cum ergo homo cogitaverit vel aliquid praeviderit, necesse est ut meatus, idest iter vel foramen quod est inter commune spatium quod iungitur illis duobus ventriculis qui sunt in anteriori parte capitis et ventriculum qui est in posteriori eius, clausus fit ut spiritus videlicet qui est in communi spatio possit facere moras ut confortetur et sit hoc ei quasi augmentum fortitudinis ad cogitandum et intelligendum, idest ut sit fortior in cogitatione et intellectu, providentia ac cogitatione. Et ideo accidit ei qui cogitat mergere caput ad terram et multum aspicere eam et incurvet quasi scriberet aliquam scripturam vel describeret in ea aliquas figuras, ut sit hoc quasi auxilium ad deponendum illud corpus quod diximus esse simile vermiculo super foramen illius decursus per quem transit spiritus ad posteriora capitis."

able source for Medieval scholars of Aristotelian psychophysiology. According to the *De differentia*, brain functions appear to be grouped into three main areas, viz. the brain ventricles. According to the usual Galenic scheme, reason and reason-like activities are placed in the middle ventricle. Unfortunately, the list of cognitive activities provided by Costa ben Luca is quite hypertrophic. Though it is repeated three times, it never features the same list of activities. The first occurrence of reason and reason-like activities appears in the section of the work devoted to the spirit. Here, Costa ben Luca argues that animal spirit causes understanding (*intellectus/fahm*), thinking (*cogitatio/fikr*), providence (*providentia/rawiyya*) and cognition (*cognitio/tamyīz*).[24] The second passage where these activities are mentioned occurs in the recapitulation on the cognitive activities of the middle brain ventricle, which may be damaged by possible impairments of this portion of the brain. In this context, Costa ben Luca argues that thinking (*fikr*), cognition (*tamyīz*) and providence (*rawiyya*) can be damaged if the middle brain ventricle is affected by a pathology.[25] The main difficulties in the Latin versions of Costa ben Luca's work, as is the case with the *Pantegni*, lies in the translation of the terms *fahm* (*De differentia*)[26] and *fikr* (*Pantegni*) with the term "intellect" (*intellectus*).[27] This translation posed a

24 Costa ben Luca, De differentia spiritus et animae, ed. Wilcox 1985, 157.167–170 [ed. Troupeau 2011, xi.2–4]: "Intellectus (Ar. *fahm*) enim et cogitatio (Ar. *fikr*), providentia atque cognitio (Ar. *tamyīz*, "discernment") fit per spiritum qui est in ventriculo qui participatur illis duobus ventriculis qui sunt in anteriori cerebro." Note that the Arabic term *ra'y* "opinion" is left untranslated in the Latin.

25 Costa ben Luca, De differentia spiritus et animae, ed. Wilcox 1985, 165.276–282 [ed. Troupeau 2011, xv.14–xvi.4]: "[…] ille spiritus qui est in anterioribus ventriculis operatur sensus, idest visum, auditum, gustum, odoratum et tactum, et cum his operatur athagil (Ar: *al-taḫayyul*, "phantasy"), quam Greci vocant fantasiam et quod spiritus qui est in ventriculo medio operatur cogitationem et cognitionem atque providentiam; et spiritus qui est in posteriori ventriculo operatur motum et memoriam."

26 Constantinus Africanus, Pantegni [Theorica IV.19], ed. Burnett 1994, 115 [ed. Harbī 2018, 531.7–8]: "Spiritus autem qui ad puppim pertransit motum ibi et memoriam facit; in prora immorans, sensum creat et fantasiam; spiritus medii ventriculi intellectus sive ratio (Ar.: *fikr*, "thinking") fit." Concerning internal senses in the *Pantegni*, see: Ruth E. Harvey, *The Inward Wits: Psychological Theory in the Middle Ages and the Renaissance* (The Warburg Institute Surveys 6), London 1975, 13–21.

27 The possible problem with this passage lies in the term *intellectus* which, while representing a very accurate translation of *fahm*, can become problematic if understood in a strictly philosophical sense, thus reducing the activities of the rational soul to the physical flow of animal spirits. See: Harvey, *The Inward Wits*, 39: "Costa's ideas about the nature of rationality are not clear: he seems to hold that the rational soul is incorporeal, but that reason itself takes place in the central ventricle by the operations of a very pure kind of animal spirit."

challenge to Medieval readers of these texts operating within a framework of an Aristotelian philosophy of nature, especially psychology and noetics. The main problem is represented by the lack of a criterion able to single out activities of the sensitive soul from those of the rational soul. To work out this difference between the quasi-rational and the rational activities of the soul, Medieval authors had to appeal to authorities like Avicenna and Averroes, who combine a detailed anatomy of the brain with clear psychological criteria for singling out cognitive (viz. belonging to the sensitive soul) and rational activities.

The objection that we have applied to the *De differentia* and the *Pantegni* can be similarly applied to other texts such as the *Anatomia Richardi*, the two anatomies attributed to the *Magister Nicolaus Peripateticus* and Nemesius' *De natura hominis*.[28] None of these provides a discussion *ex professo* which allows one to distinguish which activities placed in the middle brain ventricle belong to the sensitive soul and which to the rational soul. The *De motu cordis*, while mentioning the existence of the animal spirit, does not discuss his role in the psychophysiology of cognitive and quasi-rational activities.

In the absence of a criterion suitable to distinguish between the activities of the sensitive and of the rational soul, it is necessary to turn to the psychological part of the work, especially to Costa ben Luca's short chapter on the soul's faculties. According to the *De differentia*, the roster of activities of the rational soul (*anima rationalis/nafs nāṭiqa*) includes thinking (*cogitatio/fikr*), providence (*providentia/rawiyya*), opinion (*aestimatio/ẓann*), doubt (*dubitation/šakk*), and memory (*memoria/ḏikr*).[29] Among these, cogitation (*cognitio/fikr*), providence (*providentia/rawiyya*) and memory (*memoria/ḏikr*) stand out as activities that had been previously listed as activities produced by the action of animal spirits.[30] The overall suggestion that might be gathered from reading this passage is that Costa ben Luca means that all the activities he had

---

28 Nemesius, De natura hominis c. 11 (ed. Verbeke/Moncho 1975, 86.61–64): "[...] excogitativi autem sunt generaliter quidem iudicationes et depositiones et fugae et et impetus actus, specialiter vero intelligentiae intelligibilium et virtutes et disciplinae et atrium rationes et consilium activum et electivum [...]"; Richardus (Anglicus), Anatomia (ed. Tarrasch 1898, 14): "tres sunt cellule, fantastica logistica et memorialis. Fantastica cellula est, que est in anteriore parte capitis, in qua viget ymaginacio; in logistica, que est in medio parte capitis, viget racio; in memoriali, que est in occipitio, viget memoria." Nicolaus Physicus, Anatomia (ed. Redeker 1917, 38): "In anteriori parte constituunt cellulam fantasticam. In medio capitis cellulam logisticam. In posteriori parte capitis constituunt cellulam memorialem. In fantastica cellula dicitur habere sedem immaginatio. In logistica ratio. In memoriali memoria."

29 See above, 356 n. 18.

30 Costa ben Luca, De differentia spiritus et animae, ed. Wilcox 1985, 165.281–282 [ed. Troupeau 2011, xvi.2–4]: "[...] spiritus qui est in ventriculo medio operatur cogitationem (*fikr*)

previously listed as products of the animal spirits in the middle brain ventricle belong to the rational soul. This hypothesis finds strong confirmation in his "Book of characters," which explicitly takes the activities of intellect (*ʿaql*), cogitation (*fikr*), cognition (*tamyīz*), opinion (*ẓann*), estimation (*wahm*) and vision (*ruʾyā*) located in the middle brain ventricle to be human-specific and therefore to belong to the rational soul.[31] Costa ben Luca places the a set of human-specific cogitative power in the middle brain ventricle, like Nemesius and other minor medical sources available to Albert (*Anatomia Richardi* etc.). The activities of the sensitive soul seem to be confined to the anterior part of the brain, where phantasy is localized, while the middle and rear-brain ventricles seem to host activities of the rational soul, among which is memory. The status of the rear ventricle in the *De differentia* is curious, for it seems to host both activities of the rational (memory) and the sensitive (locomotion) soul.

The complexity of the cognitive operations seems to be directly dependent, according to Costa ben Luca, on the purity of the spirit found in the brain ventricles: spirit performing *phantasia* is subtle, but spirit performing the quasi-rational activities located in the middle ventricle is even more so.[32] For this reason, those who possess the most perfect complexion display intellective and cognitive capacities operating at their best.[33] By contrast, the good condition of the brain ventricles determine the operation of these faculties: if the middle brain ventricle is affected, then cognition (*cognitio*/*tamyīz*) and thinking (*cogitatio*/*fikr*) are impaired, but sensation and motion are intact. If the rear ventricle is affected, memory is impaired; if the impairment hits two or three of the ventricles, then the impairment of functions is complete, such as in the case of stroke (*morbus caducus*/ *sakta*) or epilepsy (the Arabic *ṣarʿ* is untranslated).[34] This brief summary of impairments and their relation to brain

et cognitionem (*tamyīz*) atque providentiam (*rawiyya*) et spiritus qui est in posteriori ventriculo operatur motum et memoriam."

31 See above: 350 n. 18.

32 Costa ben Luca, De differentia spiritus et animae, ed. Wilcox 1985, 189.566–573 [ed. Troupeau 2011, xxxi.5–11]: "Postea per spiritum qui est in ventriculis anterioris cerebri fiunt sensus et phantasia propter quod adquisivit de augmentatione subtilitatis supra spiritum qui est in corde, deinde per spiritum qui est in sequentibus ventriculis fit cogitatio et providentia. Et propter id quod acceperit de subtilitate supra alium spiritum, et propter quod necessaria est ei habundancia subtilitatis [...]."

33 Costa ben Luca, De differentia spiritus et animae, ed. Wilcox 1985, 157.188–190 [ed. Troupeau 2011, xi.15–xii.1]: "In quibusdam enim est subtilis et clarus, et hic erit rationalis cogitans dispositor et bone cogitationis (the Latin abbreviates the Arabic: *fa-yakūnu ṣāḥibuhā ʿāqilan mufakkiran sāʾisan mudabbiran mumayyizan*, "his possessor reasons, thinks, commands, rules and discerns" distinguishes)." Translation is mine.

34 Costa ben Luca, De differentia spiritus et animae, ed. Wilcox 1985, 163.261–165.275 [ed.

functions could look, to eyes of a perceptive reader, very similar to Nemesius' study of brain impairments in his section on memory (*De memorativo*).[35] And as we shall see, it is quite close to Averroes' etiopathology of cognitive failures too. All in all, Costa ben Luca's—like Nemesius'—approach to mental activities seems well in keeping with the previously analyzed approach to sense perception, i.e. uses anatomical dissections and the study of impairments of cognitive or sensitive faculties as a confirmation of their localization in certain body parts such as nerves and brain ventricles.

A second and crucial topic in Costa ben Luca's psychophysiology of cognitive and rational activities appears to be the study of endogenous (sex/age) and exogenous (climate) causes that impact the exercise of these activities in humans. Among endogenous factors, Costa ben Luca includes age (children are less sharp in reasoning) and sex (women are cognitively inferior to men). The rationale is as follows: individuals possessing a moister complexion than mature men possess inferior cognitive abilities. This line of reasoning will be quite in line with Avicenna, and with Albert himself. Among exogenous factors, Qustā considers mostly climate. According to him, Ethiopians and Slavs possess weaker cognitive abilities because their surrounding environments are excessively hot or cold. These conditions negatively affect the complexion of animal spirit and its effectiveness and suitableness in people's cognitive activities.[36]

Troupeau 2011, xv.4–13]: "Verbi gratia: Si affuerit impedimentum in parte cerebri media et cetere partes eiusdem cerebri fuerint incolumes, destruitur cogitatio et cognitio tantum et remanent sensus ac motus equales, ut illud quod accidit in melancholica infirmitate, que est commixtio vel turbatio rationis et destructio cognitionis; et si fuerit impedimentum in parte cerebri superiori, destruitur memoria tantum et erunt ceteri actus hominis equales, idest recti. Si vero fuerit impedimentum in duobus ex his ventriculis aut in tribus et occupaverit totum cerebrum, erit impedimentum in cognitione et in sensu atque motu universaliter, ut illud quod accidit caducis et his similibus."

35 Nemesius, De natura hominis c. 12 (ed. Verbeke/Moncho 1975, 88.7–89.1): "Nam anterioribus quidem solum ventriculis secundum aliquem umquam modum laesis, sensus quidem impediuntur, discretivum verum vero adhuc manet servatum; cum autem medium ventriculum solum patitur, discretio quidem fallitur, sensum vero membra manent custodientia secundum naturam sensum; si autem anteriores et medius ventriculus patiantur, discretio simul cum sensibus absciditur; parenkephalide vero patiente, memoria sola deperit, senso et discretio nullo laesis [...] si vero cum anterioribus et posterior patiatur aliquid, et sensus et discretio et memoria simul corrumpetur, insuper et totum periclitatum perire animal. Fit autem hoc manifestum et per alia quidem passiones et multa symptomata, maxime autem ex phrenesi."

36 Costa ben Luca, De differentia spiritus et animae, ed. Wilcox 1985, 189.552–561 [ed. Troupeau 2011, xxx.11–xxxi.1] "Cuius vero complexio membrorum in quibus fuerit spiritus in propria temperatione fuerit minus perfecta, imperfectus erit et spiritus in hoc quod debe-

Consistent with our remarks on sense perception in the previous chapter, Costa ben Luca insists, like many other Medieval authors, on the physiology of mental activities without going into the details of the teleology of these activities. Rather—and this is true of Costa ben Luca's case—these authors start to foresee the different degrees of cognitive abilities within humans based on sex, age and external environment. All in all, the emphasis on climatic influences, the division of brain functions according to the Galenic scheme of ventricles and the insistence on the correlation between complexity of cognitive acts and subtleness of spirit makes Costa ben Luca a valuable source to Albert in a way that blends well with his Avicennian psychophysiology of cognitive activities.[37]

3 Avicenna's Psychophysiology of Inner Senses: Estimation, Memory, and Recollection

The theory of internal senses is arguably one of the most innovative features of Avicenna's influential faculty theory. The Avicennian account of the cognitive faculties (viz. of the sensitive soul) elaborates on aspects of Aristotelian psychology, such as the cognitive role of imagination, and inputs from the Galenic medical learning to construct its own version of the cognitive activities of the sensitive soul. By "internal senses" (*sensus interiores/ḥawāss al-bāṭina*), Avicenna means common sense, retentive imagination or the form-bearing faculty (*formativa, imaginatio/muṣawwira, ḫayāl*), imagination (*imaginativa/mutaḫayyila*), estimation (*aestimatio/wahm*), memory and reminiscence (*memoria, reminiscentia/ḥifz, taḏakkur*). These activities are located by Avicenna within the human body by means of a reworking the tradition Galenic scheme of the three ventricles: common sense and retentive imagination are seated in the front one, imagination and estimation in the middle, memory in the rear. They are defined as internal senses because they do not apprehend their objects directly but only through the activities of the other faculties of the soul.

---

tur ei ex subtilitate, et imperfectiores erunt actus animae hac de causa; et ideo fuerint virutes animae in pueris imperfectae et in mulieribus debiliores, et similiter fierit in gentibus in quarum complexionem vincerit calor vel frigus propter propinquitatem earum a sole vel earum longitudinem ab eo, ut sunt Ethiopes et Sclavi et eorum consimiles."

37 On internal senses in Qusṭā, see also: Judith C. Wilcox, "Qusta ibn Luqa's *De differentia spiritus et animae* in Medieval Considerations of the Internal Senses," in *Corpo e anima, sensi interni e intelletto dai secoli 13.–14. ai post-cartesiani e spinoziani: atti del Convegno internazionale, Firenze, Dipartimento di scienze dell'educazione e dei processi culturali e formativi, 18–20 settembre 2003*, ed. Graziella Federici Vescovini/Valeria Sorge/Carlo Vinti (Textes et études du Moyen âge 30), Turnhout 2005, 55–77.

The main issues with which Avicenna's psychophysiology of internal senses was concerned were at least twofold. On the one hand, Avicenna needed to provide a consistent criterion for singling out the different objects of the different faculties, tackling the problems raised by Aristotle's sketchy remarks on instinctual reactions and incidental perception. For his part, Avicenna sought to find an explanation for the perception of non-sensible qualities (goodness or badness, friendliness or hostility, incidental percepts) of the "appearances" (*imago*/*ḫayāl*) that could blend with the Aristotelian imagination by introducing the faculty of estimation (*aestimatio*/*wahm*), endowed with a proper object, "meaning" (*intentio/maʿnan*),[38] distinct from the object of the form-bearing faculty, the image (*imago*/*ḫayāl*). On the other hand, Avicenna tried to localize these faculties in the brain so as to overcome the flaws of Aristotelian physiology. In the end, the purpose of Avicenna's theory of inner senses was to integrate Galenic medical learning into a framework of Aristotelian philosophy of nature that could account for the mutual relation between cognitive faculties and their objects both psychologically and physiologically.

In our study of Avicennian memory, we shall primarily rely on a recent contribution by Tommaso Alpina, while on estimation we shall take the cue from the studies of Black, Hasse, and, to a lesser extent, Di Martino.[39] In his study, Alpina has highlighted how in his *De anima* (I, I, 5), Avicenna distinguished the faculties of the sensitive soul and the inner senses among them. The first criterion introduced by Avicenna to distinguish these faculties was represented by their object. Common sense and the form-bearing faculty is responsible for the perception (common sense) and storing (form-bearing faculty) of images, while estimation and memory are responsible for the perception (estimation) and storing (memory) of meanings. A second criterion is the distinction between two different kinds of perception (*apprehensio*/*idrāk*) based on

38 We have chosen to stick to the term "meaning" rather than "intention", "notion" or something else because it provides, while not covering the domain of *maʿnā* entirely, one of the most distinctive features of this concept, i.e. the meaning that a certain image has for the perceiver: Rotraud E. Hansberger, "Representation of Which Reality? 'Spiritual Forms' and '*maʿānī*' in the Arabic Adaptation of Aristotle's *Parva Naturalia*," in *The* Parva Naturalia *in Greek, Arabic and Latin Aristotelianism. Supplementing The Science of The Soul*, ed. Borje Bydén/Filip Radovic (Studies in the History of Philosophy of Mind 17), Cham 2020, 103.

39 Tommaso Alpina, "Retaining, Remembering, Recollecting. Avicenna's Account of Memory and Its Sources," in *Memory and Recollection in the Aristotelian Tradition: Essays on the Reception of Aristotle's* De memoria et reminiscentia, ed. Véronique Decaix/Christina Thomsen Thörnqvist (Studia Artistarum 47), Turnhout 2021, 67–92; Di Martino, Ratio particularis, 23–39; Deborah Black, "Imagination and Estimation: Arabic Paradigms and Western Transformations," *Topoi* 19 (2000), 59–75; Dag N. Hasse, *Avicenna's* De anima, 127–153.

psychological and physiological criteria. Common sense and estimation were considered inner senses endowed with "primary perception" because they perceived their object, while form-bearing faculty and memory only received them and therefore possessed "secondary perception".[40] The distinction between primary and secondary perception was also based on the medical associations that saw reception as closely linked to a moist complexion and retention as correlated to a drier one. This is exemplified by empirical evidence, viz. an image which can be seen in water but water is not able to retain an image when it is too rippled.[41]

In this manner, faculties charged with the perception of objects (common sense, estimation) were localized in what were thought to be moister areas of the brain, while faculties responsible for retention (form-bearing, memory) were localized in drier ones. A particular case is represented by the imaginative-cogitative faculty, responsible for the composition and separation of forms and meanings by chance, the only faculty that operated vis-à-vis both objects (forms, meanings) by means of an active and not a passive perception. When commanded by the rational soul, the imaginative power can take on the role of "cogitative" (*cogitativa/mufakkira*) power.

The physiological arrangement provided by Avicenna for the localization of these faculties in the brain is the following: common sense and form-bearing faculty in the front ventricle, imaginative-cogitative and estimation in the middle, memory in the rear-most. The criterion that Avicenna uses to single out the objects of these faculties is his theory of abstraction. According to this theory, the objects of the faculties of the sensitive soul are distinguished according to their degree of detachedness from material appendages.

The first apparent degree of abstraction is represented by the sensible forms as perceived by common sense. Yet, according to Avicenna, this is not a kind of true abstraction (*non abstrahit formam a materia abstractione vera/lā yanziʿu al-ṣūra ʿan al-madda nazʿan muḥkaman*) because common sense requires matter *qua* object of sensation.[42] According to Avicenna, what counts as abstraction strictly speaking requires not only the abstraction of form from matter, but the existence of this form in the human mind independent of its existence

40 Avicenna, Liber de anima I c. 5, ed. Van Riet 1972, 85.88–92 [ed. Rahman 1959, 43.1–5]: "Sed virium ab intus apprehendentium, quaedam apprehendunt formas sensibiles, quaedam vero apprehendunt intentiones sensibilium. Apprehendentium autem quaedam sunt quae apprehendunt et operantur simul, quaedam vero apprehendunt et non operantur, quaedam vero apprehendunt principaliter et quaedam secundario."

41 Avicenna, Liber de anima I c. 5, ed. Van Riet 1972, 88.25–29 [ed. Rahman 1959, 44.9–11].

42 Avicenna, Liber de anima II c. 2, ed. Van Riet 1972, 116.83–117.87 [ed. Rahman 1959, 59.12–15].

in the physical world. This is what happens in the case of the form-bearing (or retentive imagination) and imaginative faculties, the first real degree of Avicennian abstraction.[43] In the case of the imaginative-cogitative faculty, its object does not necessarily have a truth value, for it does not necessarily perceive objects that we have sensed in the extramental world but can be the result of the random process of composition and division of forms and images so typical of the imaginative faculty when it is not reined in and directed by the rational soul. The sensed objects and the products of random imaginative activity (e.g. dreams) are then stored into the posterior part of the front brain ventricle. While retentive imagination is lodged in the posterior part of the front brain ventricle, the imaginative-cogitative power is hosted in the anterior part of the middle brain ventricle. Finally, the third degree of Avicennian abstraction is represented by the object of "meaning,"[44] while the fourth and highest degree of abstraction is represented by the universal, abstracted from the image provided by imagination with the cooperating aid of the Agent Intellect.[45]

The flow of animal spirit from cranial nerves to the front brain ventricle and then through the brain ventricles themselves represents the physiological correlate of psychological abstraction. In Avicennian psychophysiology, the attainment of every degree of abstraction is causally correlated to a correct flow of the animal spirit through the brain ventricles. As appears to be clear, the motion of animal spirit in the brain activates the faculties of the soul capable of producing an increasingly more abstracted knowledge of the extramental thing, moving from the image to the universal. To be clear, this process can be conditional upon impediments such as aging and illnesses which, as shown in chapter 4, do impact the exercise of cognitive and intellectual activities over the years. Like Costa ben Luca, Avicenna follows Galen in placing a worm-like particle at the entrance of the rear brain ventricle to regulate the flow of animal spirits to and from this ventricle, thus triggering the faculty of memory.[46]

---

43 Avicenna, Liber de anima II c. 2, ed. Van Riet 1972, 117.88–93 [ed. Rahman 1959, 59.14–60.1]: "Sed imaginatio (Ar. *ḫayāl*) sive imaginari (Ar. *taḫayyul*) est cum abstrahitur forma a materia maiore abstractione: ipsa etenim capit formam sine materia, ita quod, ad hoc ut haec subsistat in illa, non est necesse existere materiam, quia materia sive sit absens sive destructa, esse tamen formae stabile erit in imaginatione, et tunc apprehensio ipsius erit abstrahens eam a colligatione quam habet cum materia, abstractione perfecta."

44 Avicenna, Liber de anima II c. 2, ed. Van Riet 1968, 119.18–25 [ed. Rahman 1959, 60.19–61.5].

45 Avicenna, Liber de anima II c. 2, ed. Van Riet 1968, 120.26–41 [ed. Rahman 1959, 61.5–17]. This first section on the Avicennian division of internal senses is based on: Alpina, "Retaining, Remembering, Recollecting," 70–77.

46 Avicenna, Liber de anima III c. 8, ed. Van Riet 1972, 270.73–271.88 [ed. Rahman 1959, 153.5–

Avicenna's account of inner sense had the indisputable merit of being the first to try and build an account which blended medical anatomy and philosophical psychology. Through its introduction of the notion of meaning (*maʿnā*/*intentio*), he sparked a long-lasting debate about the human-animal cognitive divide, with a special focus on the cognitive content of meanings themselves.

3.1 *Instinct and Incidental Perception: Animal Spirits and the Human-Animal Cognitive Divide*

In his psychology, Avicenna introduced the concept of meaning to account mainly for the perception of non-sensible qualities such as connotational attributes (e.g. goodness or badness, friendliness or hostility) and for the perception of non-sensible qualities in the object of internal senses. "Meaning" as a non-sensible quality of the object applies not only to the case of a sheep that flees from a wolf or a dog that fears the stick it has been beaten with in the past, but also to some cases of incidental perception.[47] Unlike Aristotle, though, Avicenna does not discuss the case of our knowledge of Cleon's son as such and does not seem to be hinting at any animal capacity to perceive an individual of a certain species. As highlighted by Deborah Black, though, if we consider that the status of "meaning" according to Avicenna is relational, viz. that a meaning is such—a non-sensible, connotational attribute of the object—only in relation to the subject,[48] the class of "meaning" as a synonym for "non-sensible quality" encompasses many categories and may implicitly cover individual substances too. This problematic implication of Avicenna's theory of estimative power will be discussed at length by many Medieval authors and strongly criticized by Averroes.

On the one hand, Avicenna conceives an *intentio* as a sensible species that we do not sense but which we associate with our sense perception. This is the

154.3]. On the role of the cerebellar worm, see also: Avicenna, Canon III fen 1 c. 1, ed. Venezia 1507, fols. 166^{ra-rb} [ed. New Delhi 1981–1996, vol. 3/1, 5.15–6.10].

47 On these examples, see: Black, "Imagination and Estimation," 60–61.

48 Deborah Black, "Estimation (*Wahm*) in Avicenna. Estimation (*Wahm*) in Avicenna. The Logical and Psychological Dimensions," *Dialogue. Canadian Philosophical Association* 32 (1993/2), 226: "For Avicenna, then, the status of an object as a form or an intention is not absolute, but determined by the relations that obtain in a given act of perception between the subject and the object perceived. Some percepts, like friendliness and hostility, will always be intentions, because they can never affect any of the five external senses as such; incidental percepts, however, become intentions whenever their perception is not conveyed by their own sensible form itself, e.g., by the sweet taste of the honey to the tongue, but instead, by some other form which they accompany, such as the honey's yellow colour." A similar position in: Di Martino, Ratio particularis, 115.

case, for instance, of yellow and sweet in the case of honey: we judge the yellow thing we are seeing even though, strictly speaking, in that moment its being sweet is a non-sensible quality for us.[49] Naturally, the judgemental capacity of estimation can be prone to error for we may, for instance, infer that that yellow thing we are seeing is bile and not honey.[50] Whether meanings *qua* non-sensible qualities of objects in relation to subjects might encompass individual substances *qua* non-sensible qualities of extramental images is difficult to tell, at least based on the textual evidence Avicenna gives. The only evidence that may guide us here comes not from Avicenna's own views but from his reporting of an episode from Aristotle's *Historia Animalium* (IX, 47, 630b31–631a7). In his short chapter devoted to animal habits and behaviors (*De animalibus* VIII), Avicenna argues that camels and horses may have intercourse with their mothers if induced by humans to do so. If they become aware of such an act, though, they may try to kill the humans who induced such an act or commit suicide themselves.[51] This could happen, one may infer, if and only if an animal recognized its mother as a camel (or a horse) and then as its mother, i.e. distinct from other camels or horses, thus triggering a reaction of moral shame, something completely foreign to animal psychology, at least according to Aristotle. Nonetheless, Avicenna does not draw any explicit conclusion from this example concerning the nature and content of the estimative faculty.

A second element of controversy sparked by Avicenna's discussion of the faculty of estimation concerns the kind of reactions triggered by the activity of the estimative faculty, viz. their concrete exemplification. In the Latin Middle Ages, the staple example for instinctual reactions was the sheep-wolf episode. The fact that the sheep flees from the wolf (actually, all wolves with no possible exception) emphasized an inborn component of animal reaction to perception, yet one that does not represent the full spectrum of animal behavior. In his psychology, Avicenna explains that there may be some examples of animal behavior, such as the dog fearing sticks due to past trauma, that allow for some leeway in the experiential component of estimation. To be clear, Avicenna does not go so far as to associate this kind of behavior with an identification of the object (e. g. stick) as potentially harmful or dangerous *qua* belonging to a certain species. This issue remains latent in Avicennian psychology and represents the major object of Averroes' criticism; but, as we shall

49 Black, "Estimation (*Wahm*) in Avicenna," 226.

50 Avicenna, Liber de anima IV c. 3, ed. Van Riet 1968, 35.97–100 [ed. Rahman 1959, 182.14–183.1].

51 Avicenna, Abbreviatio de animalibus VIII c. 7, ed. Venezia 1508, fol. 40ra [ed. Cairo 1970, 138.17–139.3].

see, the criticism of the Corduban thinker is primarily based on the Arabic adaptation of the *Parva Naturalia*. A further element of difference between Avicenna's and Aristotle's respective psychologies is represented, as we showed in the last chapter, by the fact that Avicenna assigns to estimation the ruling function above all faculties of sense perception and, among other functions, that of second-level perceptual awareness, assigned by Aristotle, Averroes and Albert to common sense.[52] This ruling function is explicated in many ways: estimation works as a means of the rational soul to command the activity of the imaginative faculty during the production of dreaming, thus aiding in the production, perception and interpretation of truthful or truth-embedded dreams (e.g. divinatory dreams). When it is the estimative power alone that commands the imaginative-cogitative power, these dreams become non-veridical phenomena, because of the non-involvement of the intellect.[53] Dag Hasse has shown how sometimes estimation accounts for reflex-like reactions (not really occurrences of instinctual reaction), such as the eye that closes before dust enters it,[54] or for other reactions such as that of the child who instinctively clutches his mother's breast.[55]

Animal spirit emerge as a crucial psychophysiological concept in this context, for their flow through the brain ventricles facilitates a highly complex set of cognitive activities shared by both humans and animals and their relative cognitive content. The fact that these abilities may be shared, however, does not mean that their operation is completely identical between any two animals, between humans and animals or between different human beings. There are differences among humans that depend on climate and sex, as stated in Costa ben Luca, for instance. These remarks are found mostly in Avicenna's *Canon* and are therefore human-specific.

For instance, humans who inhabit hotter regions suffer an evaporation of their spirits because of the excessive heat upon their bodies, which makes them thin-skinned and makes them age early.[56] The second connection between

---

52 Di Martino, Ratio particularis, 27: "l'estimation est le chef, *al-šayḫ*, de l'âme sensible et la dirige en servant de toutes le autres faculté sensibles […] L'estimation est, enfin, la seule facultè qui puisse, le cas échéant, percevoir les actions des autres facultés."

53 Black, "Estimation (*Wahm*) in Avicenna," 227 n. 40.

54 Avicenna, Liber de anima IV c. 3, ed. Van Riet 1968, 37.19–38.27 [ed. Rahman 1959, 183.16–184.3]. On this passage, Hasse, *Avicenna's* De Anima, 136.

55 Hasse, *Avicenna's* De Anima, 134.

56 On the relationship between climate and cognitive activities, see: Avicenna, Canon I fen 2 doctr. 2 c. 11, ed. Venezia 1507, fol. 32[ra] [ed. New Delhi 1981–1996, vol. 1, 156.1–4]: "Loca habitabilia calida nigrificant capillos intantum et crispant quod eos cum colliguntur piperi similes efficiuntur. digestiuam debilitantes. et cum in eis maxima fuerit resolutio et minuetur

spirit and exogenous factors such as climate is articulated within Avicenna's take on seasons. In this part of the *Canon*, he argues that the spring season is more congruous to the nature of spirit and blood,[57] since the complexion of spring is not hot and wet, but temperate.[58] According to Avicenna, the elemental qualities of season (winter is dry and cold, spring is temperate, summer is hot and dry, fall dry and wet) can influence the material composition of animal spirits, even though this assertion is spelled out only in the two above-mentioned passages. This discussion will be expanded by Albert the Great. The influence of inborn complexion on behavioral responses, more specifically the theme of passions, can be also found in "On cardiac remedies" (*De medicinis cordialibus*/*'Adwiya qalbiyya*), whose subject goes well beyond the scope of this book. Suffices it here to say that in this work Avicenna links the complexion of blood and spirit (*qua* byproduct of breathed-in air, and blood) to certain behavioral and passional dispositions.[59] Given the relevance attributed to the dichotomies between thinness and thickness, moistness and dryness in the formation of animal spirits, these remarks correlate to a differentiation of cognitive capacities among individuals according to complexion in blood and spirit.

As we observed in Chapter 5, the cognitive capacities of humans are not equal at all time according to Avicenna (and Albert too). Children have lesser cognitive abilities than youngsters and adults because of an excessive amount of moisture which negatively influences the complexion of animal spirits. By the same token, incidental moisture caused by digestive failure during old age negatively impacts one's cognitive and rational abilities by blocking the flow of animal spirits into the brain. In a comparable manner, women were thought to be moister than males, giving them lesser cognitive capacities than men.

---

humiditas cito adueniet senium sicut in terra nigrorum. Illis namque qui ibi morantur aduenit senium annis triginta et eorum corda sunt timida propterea quod spiritus multum resoluitur." On Avicenna's theory of places, see the above-mentioned chapter (I, 2, 11). On complexion in general, see at least: Avicenna, Abbreviatio de animalibus, XII c. 1, ed. Venezia 1508, fols. 44vb–45rb [ed. Cairo 1970, 192–197]; Avicenna, Canon I fen 1 c. 1, ed. Venezia 1507, fols. 2ra–3rb [ed. New Delhi 1981–1996, vol. 1, 38–43].

57 Avicenna, Canon I fen 2 doctr. 2 c. 6, ed. Venezia 1507, fol. 29va [ed. New Delhi 1981–1996, vol. 1, 147.1]: "Ver autem cum est secundum suam complexionem est melius tempus proportionale complexioni spiritus et sanguinis [...]"

58 Avicenna, Canon I fen 2 doctr. 2 c. 3, ed. Venezia 1507, fol. 28rb [ed. New Delhi 1981–1996, vol. 1, 142.4–5]: "Et dicemus quod veris complexio est temperata: neque est sicut quidam existimant calida et humida. certificare vero istud multa inuestigatione est in parte naturalis philosophie. Credere autem debemus quod ver est temperatum."

59 See the section on spirits about "On cardiac remedies" in: Jacquart, "La notion philosophico-médicale," 25–31.

This view, which lent itself to misogynistic readings, would endure through the Latin Middle Ages in some part because of its close connection to generation, where the material role of the female was causally correlated to her colder and moister complexion, as opposed to the hotter and drier complexion of the man as formal and efficient cause.

The picture that emerges from Avicenna's account of human and animal cognition/instinctual reactions is one in which animals may modify their behavior based on past experience but do not possess any plasticity in their behavior. Unlike humans, they do not possess the capacity to willfully restraining their behavior and modify it rationally. This is because humans may possess a variety of dispositions towards a certain behavioral response rooted in their physiological constitution (humors, spirits etc.) due to endogenous (sex) or exogenous (climate) factors. Thanks to their rational capacities, though, humans are not deterministically chained to their complexion, while animals are. While behavioral response can differ between humans and animals given the same external stimulus (because of human rational abilities, education etc.), the cognitive content of meanings is shared by both humans and animals. Therefore, the kind of information transmitted to the brain area dedicated to estimation is similar, at least from the point of view of the perceived content.

3.2 *Memory, Meanings, and Animal Spirits*

As we have tried to explain in the previous paragraphs, Avicennian psychology of memory relates the object of memory (meanings) to estimation in the same way the object of retentive imagination (images) relates to common sense. Like the form-bearing faculty that attains its objects (images) through the "first perception" actualized by common sense, in a similar guise memory is able to receive meanings from the activity of first perception actualized by estimation. According to Avicenna, the symmetry between retentive imagination and memory continues with their functions: just like retentive imagination stores images compounded of sensible species derived from common sense, so memory stores meanings in the posterior brain ventricle.[60] From this point of view, both the physiological and the psychological mechanism of perception and storing forms and meanings are symmetrical, thus rendering two storehouses

60 Avicenna, Liber de anima I c. 5, ed. Van Riet 1972, 89.53–90.60 [ed. Rahman 1959, 45.11–15]: "Deinde est vis memorialis et reminiscibilis; quae est vis ordinata in posteriori concavitate cerebri, retinens quod apprehendit vis aestimationis de intentionibus non sensatis singulorum sensibilium. Comparatio autem virtutis memorialis ad virtutem aestimationis talis est qualis comparatio virtutis quae vocatur imaginatio ad sensum, et comparatio huius virtutis ad intentiones est qualis comparatio illius virtutis ad formas sensibiles."

of cognitive content in the brain: images in the front brain ventricles, meanings in the rear one. The Avicennian account of memory clearly departs from Aristotle's psychology in its introduction of the innovation of meanings, of the estimative faculty and of the lack of time-directedness of memory. Memory in the strict sense is not memory of something past *qua* past but is the storing of meanings and their relation to a sensible form.[61] There is an element of Aristotle's account of memory retained in Avicenna's psychology, though. As shown by Alpina, when Avicenna distinguishes between memory and recollection, he considers the latter as a human-specific activity.

The recollective process seems to be based on a different idea than memory, i.e. the inference of the prior existence of something—be it a sensible form or its related meaning—that has vanished. According to Aristotelian psychology, it is the very act of inference (and not of judgement) that sets recollection apart from mere memory. Therefore, the nature of the process of recollection leads from something known (the remembered form or meaning) to something unknown, allowing the latter to be known. Since the act of recollection can only be deliberate and deliberation belongs to the rational soul alone, recollection, according to Avicenna, is the sole provenance of humans.[62] To actualize their capacities of recollection, humans need the assistance of both theoretical and practical intellect: while the latter is responsible for deliberation, the former triggers the recollecting process through the aforementioned inference, so that we may remember a particular thing.[63]

There are two ways by which recollection can occur, depending on the status of the first thing known, i.e. form or meaning. If we consider recollection, the retrieving process might start from the inference of something lost to our mind by two paths. The first occurs by our reviewing the simple or complex images stored in retentive imagination and then turning to memory *stricto sensu* in order to find the related *intentio/maʿnan*.[64] This process, though, can also happen the other way round, i.e. from the meaning to the sensible form(s).[65] In

---

61 Avicenna, Liber de anima IV c. 3, ed. Van Riet 1968, 39.43–45 [ed. Rahman 1959, 184.16–17]. This study of the Avicennian account of memory is based on: Tommaso Alpina, "Retaining, Remembering, Recollecting," 77–81.

62 Avicenna, Liber de anima IV c. 3, ed. Van Riet 1968, 40.61–41.68 [ed. Rahman 1959, 185.9–15]. Alpina, "Retaining, Remembering, Recollecting," 81–84.

63 Avicenna, Liber de anima V c. 1, ed. Van Riet 1968, 78.32–34 [ed. Rahman 1959, 207.12–13]: "[…] haec virtus (*sc.* practical intellect) transsumit ex virtute iudicante de universalibus: inde enim transsumit maximas propositiones ad id quod cogitat et concludit de particularibus."

64 Avicenna, Liber de anima IV c. 1, ed. Van Riet 1968, 9.14–10.20 [ed. Rahman 1959, 167.12–17].

65 Avicenna, Liber de anima, IV c. 1, ed. Van Riet 1968, 10.20–26 [ed. Rahman 1959, 168.2–4].

both cases, the key role in the recollective process is played by the estimative faculty, commanded by the rational soul. This is what happens in recollection from a psychological point of view. A whole another story, however, is represented by the physiology of recollection and memory.

Avicenna's physiology of memory and recollection appears both in his psychology and, very briefly, in his medicine.[66] In the *De anima* we find some useful remarks in his discussion of the psychology of recollection, whose activity is compared to learning in order to differentiate between the two. Avicenna holds that while recollection is a translation from something known to something unknown in order for the latter to be known, learning is, by contrast, similar in terms of mechanism but different in method and in the nature of the thing attained. On his account, recollection attains in the future the likeness of what memory had attained in the past. Learning, on the other hand, consists in attaining something new in the future, independent of past experience. In terms of method, the transfer of learning proceeds from the unknown to the known necessarily: given certain premises, something necessarily follows, while memory proceeds by way of indications that not everyone shares. To clarify the problem, Avicenna gives examples of mnemonic techniques, such as the case of remembering a specific book by remembering the master who read it to him. While this inferential process might work for a specific person, this does not ensure the universal and necessary nature of this process, for somebody else might remember the same book in relation to a different experience he had of it.[67]

In this framework, the link to animal spirits is provided by a series of considerations on the natural disposition toward a good memory. According to Avicenna, the effectiveness of the act of remembering seems to depend not on the rapidity by which the vermiform substance is lifted (as in Costa ben Luca),[68] but rather on complexional criteria. Alpina has shown that, according to Avicenna, there are five groups which can be considered when speaking of memory, recollection and learning: (i) those for whom learning is easier than recollection, (ii) those for whom, by contrast, recollection is easier than learning, (iii) those who are strong in retention but weak in recollection, (iv) those who are strong in recollection and weak in retention, and (v) those who are

66 Avicenna's medical explanation of estimation (*Canon*, lib. I, fen 1, doctr. 6, c. 5) is strictly in keeping with his psychological approach: for a brief comparative study, see: Pormann, Avicenna on medical practice, 106–107.

67 Avicenna, Liber de anima IV c. 3, ed. Van Riet 1968, 41.81–84 [ed. Rahman 1959, 186.8–11].

68 Costa ben Luca, De differentia spiritus et animae, ed. Wilcox 1985, 155.153–161 [ed. Troupeau 2011, x.7–14].

strong in understanding but weak in retention.[69] Avicenna grounds his analysis on the dichotomy between moist (reception)-dry (retention), thus highlighting how individuals of a drier complexion are better suited to retain (hence, memorize) while others are better suited to receive some information. On the one hand, there seem to be individuals which are able to gather a lot of information from the non-sensible qualities they sense in estimation but, given their moist complexion (e.g. women and children), are more forgetful than others. By contrast, people who are less forgetful and of a drier complexion may be prone to gather less information about the extramental world.[70] This line of reasoning may well apply to the complexion of animal spirits and the careful distinction of the cognitive abilities based on their complexion. Human complexion and its influence on the complexion of animal spirits, though, does not exhaust the sufficient criteria for the inclusion of an individual in one group rather than another (forgetful, etc.). Inclusion in one group rather than another must take into account a plurality of factors, such as motions and concerns of the soul, passions and desires and age.[71]

In the end, Avicenna's psychophysiology of memory is rooted in his theory of complexion, which is supposed to explain the rationale behind the localization of inner senses and the different degrees of cognitive abilities. Consistent with his approach, he emphasizes the physiological correlation of complexional balance to the process of memory and recollection: individuals with a better blend in terms of dryness retain better, those with a moister one perceive (receive) better. In the case of memory, the storing of meaning or, vice versa, the recalling of meaning from the rear brain ventricle, associated with forms once memory has occurred, has a physiological account of its own, based on the key role of the Galenic cerebellar worm. In either case, the faculty which supposedly rules the opening and lowering of this fleshy particle and thus commands the flow of animal spirits to and from the rear brain ventricle is the chief faculty of the sensitive soul, viz. estimation.

Up to now we have seen how for Avicenna the cognitive divide between humans and animals does not hinge, as in Aristotle, on the divide between the perception of images and the abstraction of universals. While maintain-

69 Avicenna, Liber de anima IV c. 3, ed. Van Riet 1968, 42.86–43.12 [ed. Rahman 1959, 186.13–187.12].

70 Alpina, "Retaining, Remembering, Recollecting," 84.

71 The crucial relevance of the ethical and emotional component in the explanation of mental activities parallels that which distinguishes, according to Avicenna, the action of a sorcerer and that of a real prophet in his account of "motive" prophecy (*De anima*, IV, 4), see: Hasse, *Avicenna's* De Anima, 160–164.

ing the classic, Aristotelian human-specific capacity of abstracting universals, Avicenna attributes to animals a further object of cognitive activity which they shared with humans, viz. meanings. In Avicenna as in Aristotle, the cognitive divide between humans and animals seems to hinge on psychological factors alone, as in the case of recollection *qua* activity guided by the intellect through the estimative faculty. To be sure, the fact that these psychological activities are shared does not mean that they are the same between any animals and humans or within the same human at any stage of life. To be clear, however, we do not find in Avicenna's works an arrangement of body parts, especially in the structure of the brain, meant for hosting the human-specific operations of the rational soul and which sets human brain apart from animal brain.

All in all, Avicenna's psychophysiology of internal senses gives Albert two main advantages. In the first place, it presents a clearcut distinction between the objects of internal faculties classified according to the different degrees of abstraction that account for both sensible and non-sensible objects of mental activity, a latent and problematic aspect in Aristotelian psychology. In this manner, Albert is able to expand on Aristotle's psychology without blurring the line of demarcation between human and animal cognition. Secondarily, it offers him a detailed physiological survey of the nature of animal spirits within cognitive activity—primarily in humans—as a means of understanding the different degrees of human and animal intelligences in the *scala naturae*. In Avicenna's account, the complexion and thinness of spirits blends, though with a different perspective, with Aristotle's insistence on the correlation between the thinness and concoction of blood and intelligence. While Aristotle had emphasized the correlation between the thinness and the degree of concoction of blood as a physiological criterion correlated to animal intelligence, Avicenna insisted on the typically Galenic notion of balance (*aequalitas*/*iʿtidāl*) in complexion as a means of doing so. Given the origin of spirit from blood, Avicenna's emphasis on the correlation between the thinness of animal spirit and increasing degree of cognitive abilities in individuals falls in line with both the Aristotelian and the Galenic traditions.[72] While his physiology of cognitive activities was destined to meet with a large consensus in both East and West, mostly thanks to the diffusion of his handbook of medicine, the *Canon*, Avicenna's psychology of cognitive activities would meet the fierce criticism in Averroes, often represented as his philosophical arch-rival.

72 A similar but more summarized account of the internal senses can be found in: Algazel, Metaphysica pars 2 tr. 4 c. 4, ed. St. Clair 2005, 69.205–71.266 [ed. Cairo 1936, 46.1–48.5].

4 Averroes' Critique of Avicennian Estimation: A Turning Point in Medieval Psychology

Averroes' critique of Avicenna's theory of estimation represented a relevant source for the Medieval Latin understanding of instinctual reactions, incidental perception and knowledge of singular substances. As the recent studies of Rotraud Hansberger have shown, Averroes' criticism of Avicenna was strongly based on the *Vorlage* for Averroes' *Epitome of Parva Naturalia*, namely the Arabic adaptation of Aristotle's *Parva Naturalia*. Likely produced in the Kindī circle, this key work presented only a part of Aristotle's *Parva Naturalia*: the *De sensu et sensato, De memoria, De somno, De insomniis, De divinatione per somnum, De longitudine et brevitate vitae*. As shown by Hansberger, Averroes' critical remarks about animal cognitive capacities were combined with a physiology strongly based on Galenic anatomy.[73] Comparing Averroes and his source, Hansberger has convincingly demonstrated how the criticism which Averroes directs at Avicenna's faculty of estimation is based on two assumptions. The first, further confirmed in his "The Incoherence of incoherence" (*Taḥāfut al-taḥāfut*), rests on the perceptive capacity of imagination, which is able to trigger instinctual reactions. Averroes does away with the faculty of estimation because, on the one hand, he regards imagination as a faculty with sufficient cognitive power to trigger instinctual reactions (as implicitly in Aristotle).[74] The second reason for his rejection of Avicenna's theory of estimative power rests on a different concept of "meaning" (*intentio/maʿnan*) than Avicenna's.[75] This concept appears to be rooted in the Arabic adaptation of the *Parva Naturalia*, which consider the cognitive content of meanings to be very close to the individual *qua* individual of a certain species.[76] Avicenna's meanings as connotational attributes (Hasse) or non-sensible qualities (Black) are much broader in scope and, most importantly, do not require the assistance

73 For a more general survey, see: Hansberger, "*Kitāb al-Hiss wa-l-mahsūs*," 143–162.

74 On Averroes' doctrine of imagination, see also the recent contribution by: Deborah Black, "Averroes on Imagination as A Cognitive Power," in *Contextualizing Premodern Philosophy Explorations of the Greek, Hebrew, Arabic, and Latin Traditions*, ed. Katja Krause/Luis Xavier López-Farjeat/Nicholas A. Oschman, New York 2022, 341–360.

75 Hansberger, "Averroes on the Internal Senses," 138.

76 Averroes, De memoria et reminiscentia, ed. Shields/Blumberg 1949, 80.65–69 [ed. Blumberg 1972, 55.6–11]: "Intentio enim, que comprehenditur per cogitationem, est spiritualis. Et hoc non accidit alicui animalium nisi homini, quia non habent intellectum, et tantum comprehendunt descriptiones et cortices rerum." See: Hansberger, "Averroes on the Internal Senses," 151. For a comparison between Averroes' passage and the Arabic adaptation of the *Parva Naturalia*, see: Hansberger, "Averroes on the Internal Senses," 148–149.

of the human-specific intellect to be perceived. Again, Hansberger has shown that the concept of meaning displayed by Averroes results very likely from the reworking from the at times inconsistent doctrine of intentions and 'spirituality' conveyed by his *Vorlage*, the Arabic adaptation of the *Parva Naturalia*.[77]

Now, Averroes' account is a bit problematic when it comes to two different aspects of the psychology of individual intentions, viz. their abstraction (or discrimination from the content of sense perception) and their perception. On the one hand, Averroes understands the abstraction of individual intentions (*intentiones*/*maʿānī*) from the cogitative power to require the assistance of the human-specific intellect.[78] On the other hand, Averroes' account on the faculty that is charged with the full perception of meanings bristles with doctrinal tensions because of its peculiar characterization of memory. In her contribution "Memory, Individuals and the Past in Averroes' Psychology," Deborah Black has clearly shown that the main feature of the faculty of memory in Averroes is its perceptual capacity, as opposed to the role of passive receptacle of intentions which it plays in Avicenna.[79] Memory is therefore not only able to store intentions, but it also turns out to be capable of presenting and perceiving individual intentions that are already embedded within the content of sense perception. These intentions are gradually "rinsed" or "cleansed" until they are separated (or abstracted) by cogitative power from the content of sense perception as the core from its outer wrap.[80] In her recent contribution on Averroes' internal senses, Hansberger has rightfully stressed that Averroes employs for this activity of the cogitative power, viz. the abstraction of the intention and its discrimination from images of sense perception, the term "perception" (*appre-*

77 Hansberger, "Representation of Which Reality?," 99–121.

78 Hansberger, "Averroes on the Internal Senses," 153–154: "Thus, Avicenna's estimative faculty is here rejected not as usurping the function of the cogitative faculty as such, but as an addition that is unnecessary because it is supposed to represent the function of the cogitative faculty in animals, where, however, it is not really present at all." Averroes, De memoria et reminiscentia, ed. Shields/Blumberg 1949, 54.12–16 [ed. Blumberg 1972, 39.13–40.2].

79 Deborah Black, "Memory, Individuals, and The Past in Averroes' Psychology," *Medieval Philosophy and Theology* 5 (1996), 173: "To remember something is not primarily to recognize it as a past object of perception but to comprehend it as this particular thing. I cannot remember something unless I recognize it, not as a random collection of sensible qualities, but as a determinate individual distinct from all other such individuals, or in Aristotle's terms, as an *eikon*, as well as a *phantasma*."

80 Black, "Memory, Individuals," 183: "[...] an abstractive internal sense faculty does have some awareness of the intentions and images that it is combining, even if it is only through the faculty of memory itself that the individual intention is fully grasped as a distinct object from the image."

hensio/*idrāk*) as well. While it is indeed true that Averroes' understanding of memory peculiarly stresses its perceptive capacity, it must not be ignored that the perception of a meaning begins when the cogitative power, aided by the human intellect, abstracts the individual intention (*intentio*/ *maʿnan*), discriminating it from the simple image of the thing perceived. In other words, much as this individual intention is fully perceived and grasped in memory, its process of being perceived already begins with the activity of the human-specific cogitative faculty.[81] Further, the fact that cogitative power has a less sharp perception of the individual intention than memory strikes a chord with the (not unproblematic) Averroean doctrine of "spirituality". Averroes applied this ranking to single out the objects of the different faculties of sense perception and cognitive powers. These degrees are represented by: (1) the form perceived from the external senses, (2) the form in the common sense, (3) the form presented in imagination, (4) the intention abstracted and partially perceived by the cogitative power and (5) the intention stored in memory.[82] Hansberger has convincingly shown how this fivefold ranking of "spirituality" is based on a reworking of the Arabic adaptation of the *Parva Naturalia*.[83]

Besides its notable difference from Avicenna in how it presents the functions of memory—i.e. the rejection of the power of estimation—there is a further feature of Averroes' account of memory that separates it from Avicenna's (and Aristotle's too). According to Averroes, if the cognitive content of memory in humans is meanings and meanings are abstracted from the human-specific cogitative power, then it follows that animals cannot share the same object of memory with humans.[84] Aristotle had spoken of images with reference to the past qua past, Avicenna of intentions with no reference to the past. Both cases, though, foresaw a shared cognitive content of memory, at least in terms of object. In Averroes, this seems not to be the case. Given that Averroes seemingly attributes memory to animals, the best way to get out of this hermeneutical problem would be to assume that the cognitive content of memory differs

81 Hansberger, "Averroes on the Internal Senses," 150 n. 47.

82 Averroes, De memoria et reminiscentia, ed. Shields/Blumberg 1949, 58.6–59.2 [ed. Blumberg 1972, 42.11–43.4].

83 Hansberger, "Averroes on the Internal Senses," 148–149.

84 Averroes, De memoria et reminiscentia, ed. Shields/Blumberg 1949, 52.62–64 [ed. Blumberg 1949, 39.7–9]: "Et in animalibus rememorativis est simile intellectui: ista enim virtus (sc. cogitativa) est in homine per cognitionem, et ideo investigat per rememorationem." It seems at least that some animals have recollection-like capacities, intimately linked to the existence of a mnemonic power. Therefore, it would seem that at least some animals have memory, even though they neither perceive nor store "meanings" (*intentiones*/*maʿānī*) in memory like humans. But this is still a problematic point in Averroes' doctrine.

between animals (images) and humans (meanings) because of their different psychological capacities. Still, this is a very charitable interpretation of Averroes' position.

All in all, the picture of animal behavior in Averroean psychology is one in which animals are deterministically chained to their non-rational nature (as in Avicenna, to be fair). The degree of plasticity attributed to their cognitive capacities is less than in Avicenna, though, for animals are capable only of fixed behavioral responses and have no room at all to modify their behavior according to experience. Moreover, humans appear to possess a human-specific seat for the capacity to perceive meanings with the help of intellectual faculties, i.e. the cogitative power in the middle brain ventricle. These meanings are fully grasped by the faculty of memory, which appears in humans alone to be primarily linked to the perception of an individual as such as the base for any remembering process. This is confirmed by cases of possible impairments in the brain ventricles, especially in the middle brain ventricle, where cogitative power is hosted.[85] In the end, the perceptive nature of memory, the primary association of an intention to a thing *qua* individual of a certain species, and the localization of a human-specific cogitative power in the middle brain ventricles will all be aspects of Averroes' theory that, as we will try to show in the upcoming section, will be relevant to Albert's own psychology of internally apprehensive powers.

5 Animal Spirits and *potentiae apprehensivae deintus* in Albert the Great: Estimation and Memory

Albert's psychology of the so-called "internally apprehensive powers" (*potentiae apprehensivae deintus*) owes much to his Arabic sources, especially Avicenna and Averroes. In this respect, he was a child of his time. This is because during the first quarter of the 13th century, Avicenna's *De anima* theory of inner senses and doctrine of estimation strongly permeated debates on human and animal instinctual behavior. Dag Hasse has clearly shown how this doctrine permeated the 13th-century psychological production, and how Albert was no exception to this general trend. Later on, from the late 1230s and early 1240s, the entry of Averroes' psychology into the Latin Middle Ages played the role of a crucial reagent—almost as in a chemical reaction—to Avicenna's theory.

85 Averroes, De memoria et reminiscentia, ed. Shields/Blumberg 1949, 57.47–58.55 [ed. Blumberg 1972, 42.3–10].

The interaction between these two sources strongly shaped 13th-century exegesis of Aristotelian psychology. Following the lead of Avicenna and Averroes, medieval authors would debate the cognitive content of *intentiones* and of the faculty of estimation, whose existence was largely accepted, though not by all authors.[86]

The first scholar to betray a particularly fine-grained understanding of Avicenna's theory of estimation was John Blund in his *Tractatus de anima*.[87] A second author who did not limit himself to simply reporting Avicenna's theory of estimation with the sole sheep-wolf example was John of La Rochelle. In the short chapter on estimation in his *Summa De anima*, John faithfully reported almost all the examples (the sheep, the dog and the stick, the child made standing) of estimation from the Avicennian text, thus emphasizing both the inborn and the experience-related dimensions of this faculty.[88] Dag Hasse has shown how the adoption of Avicennian estimation was very common at this time in both Paris and England, witnessed by authors like Hugh of Saint Cher, Robert Grosseteste, Roland of Cremona, the *Summa Halensis*, William of Auvergne,[89] Philip the Chancellor (*Summa de bono*) and Peter of Spain (*Quaestiones super de animalibus*).[90] These authors tended to emphasize, through the classic sheep-wolf example, the inborn and deterministic component of certain animal reactions. By placing the emphasis on the sheep-wolf example, medieval scholars emphasized the latter, as convincingly shown by Stefano Perfetti.[91] This very common doctrine concerning animals' innate and experience-based reactions to non-sensitive qualities came to Albert's attention just as it did to many of his contemporaries. The main issues of this theory were represented by three key philosophical problems: (1) the psychological mechanism underlying the perception of non-sensible qualities in instinctual reactions and incidental perception, (2) their link to estimation and memory and (3) the cognitive content of memory and estimation themselves.

---

86 Among those who denied the existence of the estimative faculty we find Pseudo-Peter of Spain, Adam of Buckfield (who identified estimation with Aristotelian opinion), and the anonymous *Quaestiones in tres libros De anima* by Vennebusch, which conflated it with the Aristotelian faculty of imagination. Quite different is the position of the anonymous author of a commentary on the *De anima* edited by Ferdinand van Steenberghen, who attributes to both humans and animals a cogitative power: on these authors: Hasse, *Avicenna's* De anima, 152–153.

87 Hasse, *Avicenna's* De anima, 145–146.

88 Hasse, *Avicenna's* De anima, 147.

89 Hasse, *Avicenna's* De anima, 147.

90 Philip the Chancellor, Summa de bono, De bono gratie in homine, pars 2 q. 3 (ed. Wicki 1985, vol. 1, 588.141–146).

91 Perfetti, "Nelle gabbie del fissismo etologico," 116–117.

To come to grips with these questions, contextualize Albert's take on the topic and gauge the influence of his Arabic sources, it is necessary to understand his concept of *potentiae apprehensivae deintus*. Steneck has noted how Albert defines as internally apprehensive those powers which perceive their object not only without matter attached to it but without the subject being present to the perceiver at all.[92] Secondarily, it is crucial to understand how Albert defined the cognitive content of the *intentio*, a major issue in the Avicennian-Averroean debate on human and animal instinctual reactions and incidental perception. In the 13th century the most common view is perhaps that represented by Aquinas, according to whom estimative power is the animal counterpart to human cogitative power, which he also defines as *ratio particularis*. In Aquinas' view, estimative power does not equal the function of human cogitative power because, as in Averroes, humans alone are capable of perceiving individuals *qua* individuals of a certain species; in other words, only humans can gain knowledge of singular substances through cogitative power.[93] On the other hand, Aquinas considered animals capable of perceiving meanings through estimative power, contrary to Averroes' opinion, but not of perceiving individuals as individuals belonging to a certain species (this time, Averroes' genuine doctrine) and for pragmatic purposes only. Very often in Late Medieval animal psychology, the reactions produced by animal estimative power are stereotypically reduced to the sheep-wolf example. By contrast, other texts close to the Oxfordian university environment such as those John Peckham and Geoffrey of Aspall would follow Avicenna's psychology more closely, considering estimation a faculty shared between humans and animals. For instance, in the *Tractatus de anima* John presents the example of animals which find the right cure after several attempts thanks to their estimative power, and other authors like Peter of Spain link animal behavior to the factors listed by Avicenna (*cautela naturalis, experientia, similitudo*).[94] In a different manner, Geoffrey argues that animal behavior works either because of natural instinct, as in the sheep-wolf example, or by habit (*ex assuetudine/ex consuetudine*), as in the case of dogs.[95]

92 Steneck, "Albert on the Classification," 199–200: "An internally apprehensive power is a power that is able to receive the form of an object-Albert's definition of apprehension—even when that object is not physically before it. This is distinguished from an externally apprehensive power, which can only receive the form of an object when that object is present."

93 On this feature of cogitative power in Aquinas: Black, "Imagination and Estimation," 68; Di Martino, Ratio particularis, 90–101.

94 Hasse, *Avicenna's* De anima, 153 (on Peckham); Köhler, Homo animal nobilissimum, pp. 202–203 (on Peter).

95 Galfridus de Aspall, *Questions on Aristotle's Physics I 20 n. 14, ed. Silvia Donati/Cecilia Tri-*

This second model of animal psychology leaves more leeway for the role of experience in modifying animal behavior. By contrast, they consider cogitative power not as a power in its own rights, but according to the Avicennian understanding of it; that is, they consider cogitative power to be the mode of activity of the imaginative power when subject to the command of the rational soul. To be clear, both explanations of animal ethology displayed in these texts are deterministic. The differences are a matter of nuance, of the concessions made to the (relative) plasticity of animal mind in the light of experience. According to the first understanding of estimation, influenced by Aquinas' doctrine, animal instinctual reactions are almost reduced to an inborn, God-given set of prearranged reactions exemplified by the sheep-wolf example. The second case, though, gives animals more leeway to change their behavior, although neither in a rational nor in a free manner. This position is closer to Avicenna's and, also, Albert's mature doctrine. In both cases, animals are never said to acquire a knowledge of individual substances like humans and never act freely or rationally.

While these are the historical philosophical coordinates of Albert's writing on cognitive activities and memory, it is necessary to remember that his *De homine* already puts meaningful effort into building a physiology of cognitive activity linked to the causal role of animal spirits. In his later works, the complexion of animal spirits will represent one of the most crucial notions for determining the existence or non-existence of certain psychic capacities in animals. Albert's *De homine* was written at the beginning of the 1240s, a period during which Averroean psychology was gradually increasing. This context is detectable in Albert's theory of abstraction, which represents the cornerstone for distinguishing the different objects of internally apprehensive powers. As we will see, this theory rests on a fusion between Avicenna's and Averroes' psychology with the flow of animal spirits through the brain as its physiological correlate.

---

fogli. English Translation by E. Jennifer Ashworth/Cecilia Trifogli (Auctores Britannici Medii Aevi 26/1), Oxford 2017, 78.22–24: "Ad secundam dicendum quod aut hoc faciunt ex naturali instinctu, sicut agnus in fugiendo lupum, aut ex assuetudine, et sic est de cane. Ex consuetudine enim applaudit domino suo et huiusmodi similia facit." These are just a few examples to highlight two possible trends (valuing animal experience or downplaying animal behavior to a set of inborn reactions) detectable in much broader corpus of text awaiting further investigation: for a first general survey on 13th-century Aristotelian commentaries, see: Köhler, Homo animal nobilissimum, 320–325 (on cogitation), 357–374 (on estimation), 375–391 (on memory).

5.1 *Abstraction, Animal Spirits, and Brain Anatomy in the Early Albert*

Albert's psychophysiology of the internally apprehensive powers can be best understood in relation to his psychology, namely his take on the cognitive difference between humans and animals and their psychological structures. In the case of psychology, Deborah Black has shown how Albert's understanding of abstraction changes slightly over time because of the different roles assigned to imagination or, as he calls it, *phantasia*.[96] Moreover, Nicholas Steneck has demonstrated how Albert's localization of internally apprehensive powers differs between humans and animals throughout his works. Steneck's study, blended with Black's (along with Di Martino's and Decaix's) studies,[97] will represent a crucial guide for elucidating the causal and teleological role of animal spirits in Albert's account of instinctual reactions, incidental perception, cognition of singulars and memory. Let us start with Albert's views on abstraction.

Black has clearly demonstrated that Albert does accept the existence of the Avicennian faculty of estimation, although he partially blends its psychological account with Averroes' understanding of "meaning". According to Avicenna, meanings *qua* non-sensible qualities are perceived as related to the subject of the sensible species and not to represent a cognitively more refined version of sensible species, as in Averroes. Meanings, in Averroes' view, are "rinsed" or "cleansed" by the human-specific cogitative power. Albert's standpoint seemingly represents a fusion between his two main Arabic sources. On the one hand, he maintains that humans and animals are both capable of perceiving meanings through estimative power, like Avicenna. On the other hand, he emphasizes the nature of meaning as an object that is extracted—the verb employed is *elicitare*—from the sensible species and then perceived.[98] This seems a concession to Averroes, who sees meaning as embedded within the content of sense perception, of which meanings represent a more spiritual and purer version. As noted by Black, the peculiarity in Albert's account of estimation lies in the faculty charged with the "extraction" of meanings: in the *De homine*, this faculty is the compositive imagination (*phantasia*), equivalent to the Avicennian imaginative-cogitative (*imaginativa*/*mutaḫayyila*). In his later *De anima*, though, Albert will turn to a more Avicennian stance, attributing the elicitation of meanings from images (*elicitare ex imaginibus*) to the estimative power. The understanding of abstraction in the early Albert differs from his

96 Black, "Imagination and Estimation," 63–66.

97 Steneck, "Albert on the Classification and Localization," 193–211; Di Martino, Ratio particularis, 132–136.

98 Black, "Imagination and Estimation," 65; Hasse, *Avicenna's* De anima, 149–150.

later thought in terms of psychological faculties and, as we shall see, in terms of brain localization of the different degrees of abstraction.

According to Hasse, this function of *phantasia* stems from a forcible reading of al-Ġāzālī's *Intentions of the Philosophers*, in which *phantasia* translates the Arabic *mufakkirah*, "cogitative" power in the Avicennian—and not Averroean—sense. Albert forcibly attributes to this *phantasia* the capacity of extracting *intentiones* from the images resulting from the compounds of sensible species and common sensibles produced by common sense and stored in retentive imagination (*imaginatio*).[99] According to Gianfranco Fioravanti, there may be a lexical and doctrinal reason for the distinctive use of *phantasia* and *imaginatio*. The first and doctrinal reason is dependent on issues of linguistic disambiguation: by avoiding the interchangeability and possible confusion of the terms *imaginatio* and *imaginativa*, Albert wants to avoid confusion between retentive imagination (*imaginatio*) and the faculty responsible to produce dreaming images, mental fictions and visions (*phantasia*). Fioravanti has shown how this attempt at linguistic disambiguation in terms related to imagination and *phantasia* is shared by other 13th-century authors. Second, the lexical distinction between *imaginatio* and *phantasia* could have been rooted, as Fioravanti argues, in the Graeco-Latin version of Aristotle's *De anima* itself. In James of Venice's Greek-Latin translation, the Greek *phantasia* is translated either as *ymaginativa* or *phantasia*.[100] Fioravanti has convincingly shown how this distinction plays well into the hand of many authors (like Albert) who wanted to avoid issues of clarity in the exegesis already complex Aristotelian doctrine of *phantasia* by distinguish its retentive capacity (*imaginatio*) from its productive/poietic one (*phantasia*) through clearly different terms.

Relying on a different rendering of the Graeco-Latin translation of the same Greek term, Albert succeeded in harmonizing the Avicennian function of combining forms and meanings attributed to the imaginative-cogitative faculty (*imaginativa*/*mutaḫayyilah*) with the originally Aristotelian idea that *phantasmata* may be true or false. Dag Hasse has shown how Albert takes true and false judgements to be parts of estimation as well as universal forms, thus modifying the original Avicennian framework of estimation.[101] Yet, as we shall see,

---

99 Hasse, *Avicenna's* De anima, 148 n. 370.

100 Gianfranco Fioravanti, "*Phantasia*: tra Aristotele e i Peripatetici," in *Immaginario e immaginazione nel Medioevo. Atti del convegno della Società italiana per lo studio del pensiero medievale* (*S.I.S.P.M.*), *Milano, 25–27 settembre 2008*, ed. Maria Bettetini/Francesco Paparella/Roberto Furlan (Textes et Études du Moyen Âge 51), Louvain-La-Neuve 2009, 165–170.

101 Albertus Magnus, De homine (Ed. Colon. XXVII/2, 295.18–28). Hasse, *Avicenna's* De anima, 149: "For Albert, the term *intentiones* can refer to anything derived from sense data, i.e. not

this is true only of his mature works. However, the lexical distinction between *imaginatio* and *phantasia* will be maintained throughout Albert's later works as well. Still, in the *De anima*, we witness a different understanding of *phantasia*, now charged with the production of oneiric images, delusions, but stripped of any role in the extraction or perception of *intentiones*, typical of the early *De homine*.[102]

Albert's changing views on the faculty specifically tasked with the extraction and perception of meanings indirectly modifies his theory of abstraction and its physiological correlate in the flow of animal spirits through the brain. In his study of Albert's classification and localization of internally apprehensive powers, Steneck has clearly highlighted how Albert presents a view on the degrees of abstraction that differs from Avicenna inasmuch as it emphasizes how each degree of abstraction is essentially embedded within the sensible image (*in imaginibus apprehensis*). According to Albert, there is a very specific criterion according to which *potentiae apprehensivae deintus* are separated from *potentiae apprehensivae deforis* (five senses and common sense): those psychic powers have essential knowledge of forms, which are abstracted from matter and from the presence of matter itself (*forma abstracta a praesentia materiae*).[103] As we have briefly highlighted and as we shall see in more detail, the reference to forms and, more generally, the derivative nature of any cognitive content (meanings included) from sensible forms represents the hallmark of Albert's theory of degrees of abstraction, likely inspired by Averroes' doctrine of spiritual being.

Like Avicenna, for Albert the first degree of "true abstraction" (even though in his scheme this is the second degree, right after the forms perceived by the senses and common sense) equates with the absence of the material appendages (*appendices materiae*) of the object perceived. True abstraction beings with the work of the faculties of retentive imagination and *phantasia*; the third degree of abstraction is represented by meanings and the fourth by universals. This view is found in both the *De homine* and the *De anima*.[104] The difference in the psychological processes of extracting and perceiving meanings will be mirrored on a fluctuating localization of the physiological localization of the third degree of abstraction (meanings) in the brain ventricles. According to Steneck,

only to the harmful and useful, but also to the true and wrong (*phantasia*), and—perhaps following Averroes—to the universal form (intellect)."

102 Hasse, *Avicenna's* De anima, 150.

103 Steneck, "Albert on the Classification," 200.

104 On abstraction in the *De homine* and the *De anima*, see: Steneck, "Albert on the Classification," 200–203.

Albert presents three views on the localization of internally apprehensive powers that are consistent throughout his works and not mutually contradictory.

The first localization provided by Albert is based on the medical tradition of authors like Nemesius, Costa ben Luca, Constantine the African and, to an underestimated extent, even Averroes. Following these authors, he describes the front brain ventricle as *cellula phantastica*, the middle as *logistica/syllogistica/cogitativa* and the rear as *memorativa*.[105] This localization, broadly speaking, applies to both animals and humans. The second and third localizations of internally apprehensive powers derive from a rearrangement of various sources and distinguish a localization of faculties in animals (the second localization) and humans (third localization). In animals (*bruti*), internally apprehensive powers are localized following the Avicennian model: common sense and retentive imagination (*imaginatio*) in the front brain ventricle, *phantasia* and estimation in the middle brain ventricle, memory in the posterior ventricle.[106] By contrast, humans present a localization of internally apprehensive powers that seemingly holds all intellect-related activities in the middle brain ventricle, thus separating them, as shown by Steneck, from other activities of the sensitive soul. In this manner, common sense comes to be seated in the front of the anterior ventricle, retentive imagination in the middle of it, while its posterior part is designed to host the powers of *phantasia* and estimation.[107]

From a teleological point of view, the middle brain ventricle seems to acquire two different purposes in animals and humans: in animals, it enables the exercise of the highest cognitive activities of the sensitive soul. In humans, the middle brain ventricle appears specifically designed to host the intellect-related, human-specific activities such as abstraction of the universal and the starting point of the process of recollection. This is witnessed at least is a couple of passages in the *De homine*. In the first, Albert discusses Costa Ben Luca's (in the following paragraph, also Johannitius') that "intellect" (*intellectus*, Ar. *fahm* "understanding") and cogitation are produced by the animal spirit in the middle brain ventricle.[108] To this he replies that intellect is itself accidentally seated in the middle brain ventricle, because it is from the images (*phantasmata*) carried by the animal spirit flowing into that brain ventricle that it carries out its

105 Steneck, "Albert on the Classification," 204–207.

106 Albertus Magnus, De homine (Ed. Colon. XXVII/2, 305.25–40).

107 Steneck, "Albert on the Classification," 204.

108 Albertus Magnus, De homine (Ed. Colon. XXVII/2, 463.17–20): "Hoc etiam videtur Constabulus dicere in libro *de differentia spiritus et animae* sic: 'Intellectus et cogitatio atque providentia et cognitio fiunt per spiritum, qui est in ventriculo priori et in medio'."

abstractive function.[109] This point is further confirmed by Albert's take on the corruption of *phantasia*: if imagination is corrupted, so are intellectual activity (*ratio*) and memory.[110] While intellectual activity is the act of no bodily part, as in Aristotle, rational activities are influenced by a healthy disposition of the brain, as emphasized by most of the medical tradition, especially Avicenna. For this reason, if the physiological disposition of the brain is an unhealthy one, intellectual activities of the soul become impaired in their operations.[111] In like manner, Albert emphasizes the fact that recollection *qua* human-specific activity is principled by the interaction of intellect and cogitative powers in the middle brain ventricle.[112]

Note that Albert's understanding of cogitative power is not exempt from tensions with and misunderstandings of the doctrines of Averroes. For instance, in his *De memoria* and his *De animalibus* he argues that Averroes improperly called "cogitative power" the faculty according to which animals flee harmful things (*fugiunt nociva*) or pursue beneficial ones (*persequuntur convenientia*).[113] This was certainly not Averroes' genuine position, who attributed this role to the perceptive capacity of imagination. On the other hand, we can now see how Averroes' human-specific localization of cogitative power in the middle brain ventricle does play a role in Albert's psychophysiology of cognitive activities.[114]

Having elucidated the problematic relationship between the four degrees of abstraction and the localization of the different faculties in the brain, we are

---

109 Albertus Magnus, De homine (Ed. Colon. XXVII/2, 463.64–68): "Ad duo sequentia dicendum quod intellectus dicitur esse in media cellula et in spiritu mediae cellulae, non quod sit actus illius partis corporis, sed quia species abstrahit a phantasmatibus, quae ut in vehiculo sunt in spiritu illius cellulae."

110 Albertus Magnus, De homine (Ed. Colon. XXVII/2, 292.22–25): "Spiritus enim animalis transit ab anteriori parte ad mediam et postremam; ergo si corruptus est in anteriori, corruptus transit ad mediam et ad postremam; et ita corrupta phantasia corrupta erit ratio et memoria."

111 Albertus Magnus, De homine (Ed. Colon. XXVII/2, 292.70–293.3): "Ad primum ergo dicendum quod spiritus in anteriori parte corrupta non corrumpitur corruptione propria, et ideo quando transit a priori parte ad mediam et postremam, non remanet corruptus, quia corruptionem non habuit per se, sed per accidens."

112 Albertus Magnus, De homine (Ed. Colon. XXVII/2, 307.31–32; 309.69–72): "Isaac in libro *de diffinitionibus*: 'Recordatio est inquisitio adventus rei iam oblitae a virtute cogitativa'. […] De quinta et ultima dicendum quod ipsa considerat ordinem ipsius reminiscentiae ad rationem, dicitur ibi virtus cogitativa eo quod cogitat principia reminiscentiae, ut ex ipsis accipiatur ultimo quaesitum."

113 Steneck, "Albert on the Classification," 205.

114 On the possible criteria of harmonization among the first, second, and third localization of internal senses in Albert, see: Steneck, "Albert on the Classification," 209.

now in a better position to understand how the physiology of internally apprehensive powers (viz. the flow of animal spirits in the brain) correlates to the psychological differences between humans and animal in the cases of instinctual reactions, incidental perception and memory based on the key faculties of *phantasia* (in Albert's early works) and estimation (in the Aristotelian paraphrases).

5.2 *Practical Animal Intellect? Estimation and Animal Spirits in the Early Albert*

In the *De homine* Albert ascribes to *phantasia* the task of the "extraction" (*elicitare*) of *intentiones* from images. Deborah Black has noted how the aim of Albert's peculiar position on estimation, often seen as analogous to practical intellect, is to find a midway between objections that reject the purely cognitive value of estimation and its crucial role in triggering instinctual (also, motive) responses. According to Albert, this midway can be found in attributing to estimation a role analogous to practical intellect in humans. As practical intellect derives its object from theoretical intellect and triggers motion in humans, in the same manner *phantasia* extracts meanings from sensorial content and transmits it to estimation which, in turn, triggers instinctual reactions (e.g.: flight).[115] The result is that estimation is made dependent upon *phantasia*, thus completely adulterating the gist of the Avicennian theory.[116] In the *De homine*, both instinctual reactions and incidental perception are therefore the result of a perceptive process which, unlike in Avicenna, is performed primarily by *phantasia* and only at a second moment transferred to estimation as a hybrid cognitive and motive power analogous to human practical intellect.

All in all, from the few pages devoted to estimation in the *De homine*, we can gather that the content of estimation comprises judgments about the sensible object, essentially derived from the sensible image of it (as in Averroes) but perceived also by animals (as in Avicenna). The object concerns both connotational attributes, non-sensible qualities such as goodness and badness, and, as in Aristotle, truth or falsehood.[117] The activity of estimation seems to

---

115 Black, "Imagination and Estimation," 63–64.

116 Albertus Magnus, De homine (Ed. Colon. XXVII/2, 294.8–15); Black, "Imagination and Estimation," 64.

117 Albertus Magnus, De homine (Ed. Colon. XXVII/2, 295.22–28): "Alio modo accipiuntur per intentiones numquam separatas a sensibilibus, quae sunt hic et nunc, et sic accipiuntur duobus modis, scilicet prout sunt principium veri et falsi in particularibus tantum, et sic sunt phantasiae; vel prout determinant nocivum vel conveniens in appetibilibus, et sic sunt aestimativae."

produce, according to Albert, a judgement that should have, in Medieval language, *sententia*, a sort of (anachronistically speaking) propositional content. According to Albert, this judgement that predicates something of the sensed object is issued when estimation receives the *intentio* from *phantasia*: but what examples does Albert proffer as illustration? Unfortunately, the *De homine* only presents us with the classic example of the sheep who flees from the wolf; according to a quote in John Damascene's *De fide orthodoxa*, animals are said to act according to their nature.[118] Albert follows the classic sheep-wolf example, though inserting his notable ideas on meaning as encompassing both truth and falsehood, which will resurface in the case of memory and recollection. Like many of his contemporaries, Albert endorses the view that animals react to certain stimuli univocally (*uno modo*), being thus incapable, as stressed by Anselm Oelze, of changing their behavior.[119] In the Aristotelian paraphrases the situation will be different, with more substantial concessions given to animals' cognitive capacities.

From a physiological point of view, the upshot of Albert's shift from Avicenna are manifold: from the point of view of faculty localization, estimation enjoys an equivocal localization, viz. the posterior part of the middle brain ventricle in animals, the posterior part of the anterior one in humans. A similar discourse concerns the localization of *phantasia*, placed in the front of the middle ventricle in animals and in the posterior part of the anterior ventricle in humans. These different localizations must be taken into account when considering the action of animal spirits: they activate similar cognitive faculties (e.g. estimation, *phantasia*) that are equivalently localized between humans and animals. This is an aspect of Albert's psychophysiology of internally apprehensive powers, which is well worth emphasizing, as this view is not shared by most of his contemporaries at the time of the *De homine*'s writing. In the *De anima*, Albert's position on the relationship between *phantasia* and estimation has changed.

118 Albertus Magnus, De homine (Ed. Colon. XXVII/2, 295.41–51): "Ad aliud dicendum quod aestimativa magis imitatur naturam in movendo quam scientiam et artem intellectus practici. Unde licet uno modo ovis accipiat inimicum in omni lupo, tamen non utitur inquisitione et consilio in modo fugiendi quemadmodum homo; sed omnis ovis fugit uno modo omnem lupum. Et hoc est quod dicit Damascenus quod 'bruta potius aguntur a natura quam agant'. Et idem dicit Avicenna quod instinctu naturae moventur bruta et non ad rationem apprehensi, et ideo rationem universalis non considerant."

119 Anselm Oelze, *Animal Rationality. Later Medieval Theories 1250–1350* (Investigating Medieval Philosophy 12), Leiden-Boston 2018, 70–71.

5.3 *Going Back to Avicenna? Albert's Mature Psychophysiology of Estimation*

In Albert's *De anima*, estimation proper becomes the faculty responsible for the extraction of meanings from sensible images.[120] Although maintaining his old analogy between estimation and practical intellect, Albert now substitutes the role played by *phantasia* with retentive imagination. Now the analogy is once again based on the fact that estimation derives its object from retentive imagination, as it did with *phantasia*, although with a slightly different emphasis. On the one hand, we learn that estimation possesses a hybrid motive and cognitive nature (*non penitus apprehensiva sed motiva*), as opposed to the purely cognitive nature of retentive imagination.[121] Moreover, Albert argues that estimation is certainly an active power when it comes to extracting meanings from sensible forms but, in a certain respect, it is also passive. This is because, as Albert had maintained throughout his works, meanings are embedded within the content of sense perception (they are extracted from it). For this reason, retentive imagination transmits, at least to some extent, meanings to the estimative faculty, which does not act by itself but only through the meaning it can extract from the image passed on by the imaginative power (*non agit per se, sed per intentionem, quam elicit a forma acquisita*).[122] This is a much more passive view on the psychological capacity of estimation if compared to Avicenna's own views.

As noted by Black, the true peculiarity of *phantasia* in Albert's *De anima* is that it becomes the highest faculty among the *potentiae apprehensivae deintus* in the places of estimation. This faculty is possessed only by the most perfect animals and,[123] moreover, charged with the capacity to rule over and control

---

120 Albertus Magnus, De anima III tr. 1 c. 2 (Ed. Colon. VII/1, 157.47–54).

121 Albertus Magnus, De anima III tr. 1 c. 2 (Ed. Colon. VII/1, 167.58–63): "Oportet igitur dici, quod sicut intellectus practicus se habet ad speculativum, ita se habet aestimativa ad imaginationem; et ideo haec virtus non penitus apprehensiva, sed et motiva est per hoc quod determinat, ad quid movere debet animal et a quo fugere." Black, "Imagination and Estimation," 64–65.

122 Albertus Magnus, De anima III tr. 1 c. 2 (Ed. Colon. VII/1, 167.82–86, 167.90–94): "Est autem aestimativa magis activa quam imaginativa, quoniam elicere intentiones est aliquid agere et magis perfectum est quam speculari imagines solas, sicut si speculum animatum imagines sibi impressas speculari diceremus. [...] Licet autem plus actionis habeat aestimativa quam imaginativa, tamen proprietatem passivae potentiae habet in hoc quod non agit per se, sed per intentionem, quam elicit a forma acquisita." Black, "Imagination and Estimation," 65.

123 Albertus Magnus, De anima III tr. 1 c. 3 (Ed. Colon. VII/1, 168.53–57).

the other internally apprehensive powers of the soul.[124] In other words, while in Avicenna the highest among the animal faculties (estimation) is such because of its object (*intentiones*), which are the most abstract among the objects perceived by the animal's sensitive soul, in Albert's *De anima* the criterion according to which sensitive faculties are ranked does not relate to degrees of abstraction but rather to the degrees of activity and passivity of a sensitive faculty in itself. If we consider Albert's and Avicenna's accounts of imagination and estimation, there are two remarks which need be made: the first concerning the elemental and complexional characterization of the brain and the second concerning the cognitive content of *intentiones*.

Nicholas Steneck has shown how Albert's anatomical localization of psychic faculties in the brain is physiologically correlated to a difference in terms of elemental and complexional qualities. In particular, the anterior part of the anterior brain ventricle is characterized as more pliable (*mollior*), while imagination requires a dry location (because of the medical correlate: dryness-retention), while phantasy and estimation, because of their perceptive capacities (to receive), require a moist complexion.[125] This criterion remains consistent throughout Albert's works. By contrast, he seems to introduce into the Aristotelian paraphrases a new physiological criterion for the physiology of internally apprehensive powers to differentiate the complexion of the portion of the brain designated for *phantasia*. According to him, there are two kinds of perception, that of sensible forms and that of meanings, and their preservation is based on the classical medical assumption that drier complexion is better suited for retention. In addition to the perception of sensible species and meanings, Albert individuates a further activity of the internally apprehensive powers of the soul, which is the composition and division of sensible species and meanings, assigned to the faculty of *phantasia*, whose complexion

124 Albertus Magnus, De anima III tr. 1 c. 3 (Ed. Colon. VII/1, 168.62–76). Note that Albert here is suggesting that all internally apprehensive powers of the soul are rooted in *phantasia*: this is because, according to him, *phantasia* is best understood in terms of *motus a sensu secundum actu factus*, see: Albertus Magnus, De homine (ed. Colon. XXVII/2, 38.67–72); De anima II tr. 1 c. 7 (Ed. Colon. VII/1, 74.68–71), i.e. that it derives its cognitive content from sense data. Hasse (*Avicenna's* De anima, 150) has shown that while this may be a way to suggest that his teaching is in agreement with Aristotle, historically this is wrong. On Albert's attribution of the key role in the animal soul to *phantasia*, see: Black, "Imagination and Estimation," 64: "Whereas Avicenna repeatedly declares that estimation, as the principal judge in the animal soul, uses all the other internal senses as its instruments and can to this extent claim even their activities as its own, Albert substitutes phantasy for estimation as the root power within the sensible soul."

125 Steneck, "Albert the Great on the Classification," 209.

is a dominated by a "spiritual heat". While the connection between heat and activity may be more intuitive, the details of what Albert designates as 'spiritual' heat remain relatively obscure.[126]

A second, and relevant, development in Albert's psychophysiology of cognitive activities concerns the scope of animal knowledge. In the *De anima*, the cue to analyze Albert's position on estimation and the cognitive content of *intentiones* can be taken from his account of incidental perception, as in the famous Aristotelian example of Diares' son. Albert's take on the matter blends together two different examples: the first concerns instinctual reactions such as the sheep fleeing the wolf, and the second is our perception of somebody being Diares's son. According to Albert, both cases are explained by prior knowledge of a substantial form. This is a considerable difference from his prior account in the *De homine*.[127] In his Aristotelian paraphrases, Albert shows to have developed his ideas about the cognitive content of animal knowledge. As Deborah Black notes, according to Albert instinctive reactions and incidental perception would seem to be rooted in previous knowledge of the individual substance (whether this substance amounts to the actual substantial form is unclear and yet to be established).[128] Since "no wolf would ever have pity over its offspring unless it had knowledge both of this individual and of the fact that this individual is its offspring."[129] Albert argues:

> [...] the intention is not a part of the thing like the form, but rather it is the species of the whole knowledge of the thing. Thus, because the intention is abstracted from the whole and is the signification of the whole, it can be predicated of the thing; for the intention of the colored thing, which

126 Benedict Ashley, "Albert the Great on the Cogitative Power," in *A Companion to Albert the Great: Theology, Philosophy and the Sciences*, ed. Irven Resnick (Brill's Companions to Christian Tradition 38), Leiden-Boston 2013, 311.

127 Albertus Magnus, De anima II tr. 3 c. 4 (Ed. Colon. VII/1, 102.4–8): "Et tale est, quod accipimus hunc esse filium Deonis et esse agnum vel hominem, aliud autem esse lupum vel leonem, secundum quod substantiales formae mediantibus sensibilibus et non separatae ab ipsis apprehenduntur. Et iste gradus propinquus est cognitioni et numquam est sine aestimatione et collatione." Oelze, *Animal Rationality*, 71–72.

128 Black, "Imagination and Estimation," 64: "This suggests that Albert believes that there is some unifying thread that underlies the perception of affective qualities and incidental perception and that necessitates interpreting both as instances of the perception of intentions, and that unifying thread seems to lie in their shared presupposition of some ability to grasp an object as an individual [...]." For a recent contribution, see: Paloma Rubio Hernandez, "The Role of the *Intentio Individualis*," 133–161.

129 Albertus Magnus, De anima III tr. 1 c. 2 (Ed. Colon. VII/1, 167.44–47): "[...] nec umquam lupus miseretur nato suo, nisi habeat cognitionem et huius individui et quod hoc individuum est natus eius."

> is in the eye, makes known the whole thing, just as the intention which is in the imagination makes known the particular which is not present.[130]

In this passage, though, the *intentio* must not be taken in its cognitive sense of subject of estimation but, rather, from an ontological point of view, as the cognitive, immaterial object related to a thing (*intentio rei*) as opposed to the form of the hylomorphic compound (*forma rei*). According to Albert, the intentional being of cognitive objects such as sensible species (color, smell. etc.), makes it possible to know the entire thing, thus making the entire sensitive and cognitive apparatus concerned with the perception of the individual in its diverse features (sensible, non-sensible). The judgment passed by estimation concerns the thing *qua* whole and, following this judgement about the thing *qua* whole, determines the appropriate kind of behavior. Paloma Hernandez has convincingly shown that, unlike Averroes, Albert already roots the knowledge of individual substances in the common sensibles (e.g. size [*magnitudo*]) perceived by common sense.[131] This perception of common sensibles leads the estimative power to produce a judgement of the extramental object, as maintained by Deborah Black, for it is (also) through these common sensibles that a child judges someone to be his or her father or a lamb recognizes its mother.[132]

Prima facie, Albert's understanding of the human-animal cognitive divide tends to stress the commonalities between humans and animals. Moreover, it assigns a highly complex structure to animal cognition, capable of grasping individual substances. In this way, Albert's position is notably different from most of his contemporaries, who do not assign to animals the capacity to grasp individual substances. The most optimistic among his contemporaries—e.g.

130 Albertus Magnus, De anima II tr. 3 c. 5 (Ed. Colon. VII/1, 102.36–42): "Et ideo intentio non est pars rei sicut forma, sed potius est species totius notitiae rei; et ideo intentio, quia abstrahitur de toto et est significatio totius, de re praedicatur; intentio enim colorati, quae est in oculo, totam rem notificat, sicut et intentio, quae est in imaginatione particulari non praesente." This translation by Black in: Black, "Imagination and Estimation," 65.

131 Albertus Magnus, Physica I tr. 1 c. 6 (Ed. Colon. IV/1, 11.65–71): "Acceptio autem rationis permixtae sensibus aut aestimationis in brutis est super rei naturam, in qua sunt accidentia, quae sunt sensata propriorum sensuum, et cuius est magnitudo subiecta sensibilibus propriis, quae accipitur per sensum communem. Et per illam (*sc.* magnitudo) accipitur puer, quod vir homo est pater et non asinus, et agnus accipit, quod ovis est mater et non lupus." Rubio Hernandez, "The Role of the *Intentio Individualis*," 157.

132 Black, "Imagination and Estimation," 64–65: "The picture that emerges is one in which the estimative faculty is imbued with the power to recognize any sensible species as a representation of a complete individual, a power which it has in virtue of its ability to perceive the sensible species as possessing intentional or cognitive, rather than real, being. And once it has apprehended that individual as an integral whole, it is able in turn to judge whether it is an enemy or a friend, and to adopt the appropriate stance towards it."

John Blund, Oxford masters such as Peckham and other anonymous authors of English manuscripts—maintained that indeed humans and animals could share the perception of *intentiones*, that this perception was carried out by a shared faculty (estimation), but they never went so far as to state that animals grasp individual substances.[133]

Still, Albert's psychology of animal cognitive capacities did not see, like Aquinas or all other Medieval authors, any freedom in animal behavior. Quoting a passage from John Damascene's *De fide orthodoxa* that he had already employed in the *De homine* to explain animal behavior, Albert maintains in his zoology that animals act in accord with their nature.[134] While leaving an open door to the role of experience in some changes in animal behavior, Albert's psychology of animal instinctual reactions remains, as Aristotle's, Avicenna's and Averroes', strictly deterministic. These considerations are key to understanding Albert's account of memory throughout his works.

5.4 *Memory and Recollection in the* De homine

There have been plenty of studies that have assessed Albert's psychology of memory in depth.[135] Fewer have taken into account Albert's mature psychophysiology of theory of internally apprehensive powers.[136] In our case, the unescapable point of reference for the study of Albert's psychophysiology is once again Nicholas Steneck's study on the localization of internally apprehensive powers. To be clear, Albert's doctrine on memory in his

133 Hasse, *Avicenna's* De anima, , 193. 153. I plan, however, to focus more precisely on this issue in future research contributions.

134 Albertus Magnus, De animalibus VIII tr. 1 c. 1 n. 1 (ed. Stadler, I, 571.9–15): "Mores [scil. animalium] (…) non habent principium nisi per modum naturae. Propter quod etiam Ioannes Damascenus animalia magis agi dicit a natura quam agenti aliquid ex libertate animae aut ymaginatione aut cogitatu, cuius signum est quod ea quae sunt unius speciei, aut similes habent mores aut parum valde differentes."

135 Veronique Décaix, "What Is Memory Of?," 153–167; Jörn Müller, *Albertus Magnus über Gedächtnis, Erinnern und Wiederinnerung. Eine Philosophische Lekture von* De Memoria et Reminiscentia *mit Übersetzung* (Lectio Albertina 17), Münster 2017; Di Martino, Ratio particularis, 132–136; Henryk Anzulewicz, "*Memoria* und *reminiscentia* bei Albertus Magnus," in *La mémoire du temps au Môyen Age*, ed. Agostino Paravicini Bagliani (Micrologus Library 12), Firenze 2005, 163–200.

136 For instance, Tellkamp, "Albert the Great on the Structure and Function of Inner Senses," 305–324, argues that Albert's approach to internal senses in the *De anima* is less "physicalist" than in the *De homine*. As we shall see, Albert's Aristotelian paraphrases (zoology, *Parva Naturalia*) display an extensive handling of the psychophysiology of internally apprehensive powers. The *De anima*, because of its purpose, to handle the soul and its powers (not its activities in the body), is less detailed from a physiological point of view. This assumption, though, stands already as true for Aristotle's *De anima* compared to his *Parva Naturalia* and zoology.

De homine clears up any possible theological misunderstanding by distinguishing a more appropriate (*magis propria*) meaning, which he equates with the Avicennian and philosophical understanding, completely different from memory in a theological, Augustinian sense.[137] This is because the scope of Augustinian memory encompasses past, present and future. For this reason, its study (at least, according to Augustinian doctrine) belongs to the activities of the rational and not the sensitive soul.[138] According to Albert, the faculty of memory also includes recollection, part of the same faculty, though differing, as Averroes holds, in the modality of its activity (*reminiscentia est virtus sensibilis animae, et est in subiecto eodem cum memoria, sed differunt ratione*)[139] *qua* dependent on the rational activity of the soul.

While Albert agrees that Avicenna seems to provide the most fitting definition of memory in terms of object (*intentiones*) and localization (posterior brain ventricle) common to animals and humans, he exhibits a number of difference with the Avicennian account of memory that help elucidate Albert's own doctrine of memory. In the first place, Albert conceives of *intentiones* as non-sensible qualities extracted from images in a manner that is rather different from Avicenna and closer to Averroes. Secondarily, while according to Avicenna the storing part of memory (*conservativa*/*ḥāfiẓa*) derives its objects from estimation,[140] the early Albert derives meanings primarily from the abstracting action of *phantasia*, which then passes meanings on to estimation and memory. The third and perhaps most striking difference between Albert's and Avicenna's (also, Averroes') accounts of memory constitutes the cognitive content of memory itself. As noted by Carla Di Martino and Veronique Decaix,[141] the objects involved in the remembering (not recollecting) process, characterized by the act of comparison (*collatio*) between forms and meanings, are three: the reference to past time (*tempus praeteritum*), the sensible image (*imago*) that guides the soul to the attainment of the sought-after object (i.e. the thing perceived in the past) and the *intentio* by which memory grasps the *imago*.[142]

As Albert maintains in the *De homine* and in the *De bono*, memory stores forms (endowed with past-directedness, unlike forms of retentive imagination)

---

137 Albertus Magnus, De homine (Ed. Colon. XXVII/2, 300.41–51); Di Martino, Ratio Particularis, 78.

138 Albertus Magnus, De homine (Ed. Colon. XXVII/2, 301.40–47).

139 Albertus Magnus, De homine (Ed. Colon. XXVII/2, 310.54–56).

140 Albertus Magnus, De homine (Ed. Colon. XXVII/2, 301.65–69): "Ad aliud dicendum quod memoria per se recipit intentiones non sensatas, ut dicit Avicenna, et per illas devenit in imagines sensatas et per imagines in rem. Et ideo ab Algazele diffinitur a conservatione intentionum, quia intentiones proximiores sunt memoriae."

141 Veronique Décaix, "What is Memory Of?," 153–167; Di Martino, Ratio particularis, 132–136.

142 Albertus Magnus, De homine (Ed. Colon. XXVII/2, 304.61–73).

and meanings altogether.[143] In this manner, Albert singles out the content of memory from retentive imagination not only in terms of past-directedness—as in Aristotle—but also in terms of the coexistence of forms and meanings, a feature alien both to Avicenna's and to Averroes' psychology. In this manner, one can see how the richness of the cognitive content vehiculated by animal spirits flowing through the brain ventricles notably differentiates Albert's account of memory in both humans and animals from his medico-philosophical sources. On the other hand, the cognitive divide between humans and animals, if we consider the object of memory, appears to be similar to Avicenna's. As we have shown, Albert's early account of estimation seems to be limited to the sheep-wolf case study. Thus, the cognitive difference between humans and animals amounts not only to the human-specific capacity of abstracting and perceiving, but tacitly also the perception of individual substances already known via the common sense.

Returning to Albert's physiology of cognitive activities, we find traces of influences of the medical tradition in the postulation of the conventional link between good memory and a dry complexion. This fact explains why children—as in Aristotle—remember less than adults or, by contrast, why elderly people tend to suffer memory failure. Both cases are explained by the overabundance in moist complexion (essential in children, produced accidentally in elderly people).[144] Albert, like Avicenna and Costa ben Luca, places much emphasis on the pivotal role of complexion in his discussion of the motion of the cerebellar worm. In the medical tradition, the key psychological role of this organ consists in making the content of memory (in this case, *intentiones* and forms) accessible to other cognitive faculties able to produce a judgment about them (estimation) by way of comparison (*collatio*) with present experiences. Referring to Costa Ben Luca's physiological explanation of memory, Albert takes the slowness in the motion of vermiform substance to be a material cause that conditions the process of reminiscence (also, memory), together with the ignorance of the subject (*ignorantia ex parte reminiscentis*) and the difficulty of the sought-after object (*pauca reminiscibilia*). Unfortunately, he does not identify what actual faculty is responsible for the motion

143 Albertus Magnus, De bono tr. 4 q. 2 a. 2 (Ed. Colon. XXVIII, 247.62–68): "Praeterea, habitum est in tractatu *De anima* quaestione *De memoria*, quod memoria etiam sensibilis animae non tantum conservat imagines, sed etiam intentiones acceptas ex imaginibus illis. Ergo videtur, quod imagines loci corporalis non valeant sufficienter, sed deberet etiam tradere artem, qualiter deveniremus in intentiones abstractas ab illis." On this passage, see: Oelze, *Animal Rationality*, 174.

144 Albertus Magnus, De homine (Ed. Colon. XXVII/2, 306.14–27); Anzulewicz, "Memoria *und* reminiscentia," 190–192.

of this tiny yet crucial part of the brain.[145] A way out of this problem could be constituted by relying on Albert's consideration of estimation as the faculty responsible for judgement passed about forms and meanings: we assume that the faculty able to grasp meanings at the present moment is also the same that needs to gain access to the cognitive content of past experiences stored in memory, therefore estimation is reasonably the faculty responsible for opening the cerebellar worm separating the middle and posterior brain ventricles.[146]

If we go on to assess Albert's psychophysiology of recollection, we may find that Albert's ideas on recollection in the *De homine* are much richer than his sources. This is because Albert considers recollection a human-specific activity that involves not only forms and meaning but also universals, (*propter hoc habet et universalia et imagines et intentiones particularium*). To carry out recollection, Albert argues that the intellect needs to review and order individually perceived objects to reach the correct conclusion. This operation he ascribes to the cogitative power, not defined as an internally apprehensive power on its own.[147] Rather, cogitation must be viewed, as stated above, as a set of psychic

145 Albertus Magnus, De homine (Ed. Colon. XXVII/2, 311.57–68): "Quorum unus (*sc.* impedimentorum) est malitia complexionis, quae impedit velocitatem motus ab uno in aliud; secundus est constrictio viarum spiritus animalis a prima cellula in mediam et a media in postremam. Hae enim cellulae distinguuntur quadam caruncula, quae facta est ad modum vermis, quae sita est in ostiis cellularum, ut dicit Constabulus; et haec quandoque impedit discursum spiritus per cellulas, et ideo nitentes reminisci et meditari movent caput quandoque elevando, quandoque deprimendo, quia tunc natura nititur ad removendum carunculam illam, ut libere transeat spiritus." Costa ben Luca, De differentia spiritus et animae, ed. Wilcox 1985, 155.161–166; 157.170–186: "Et ideo accidit ei qui vult recordari alicuius rei ut caput suum valde mergat vel inclinando illud retrovertat et immotis oculis sursum aspiciat ut haec positio vel figura sit ei quasi auxiliatrix ad aperiendum praedictum foramen et ut ipsum corpus vermile possit sursum elevari. [...] Cum ergo homo cogitaverit vel aliquid previderit, necesse est ut meatus, idest iter vel foramen quod est inter commune spatium quod iungitur illis duobus ventriculis qui sunt in anteriori parte capitis et ventriculum qui est in posteriori eius, clausus fit ut spiritus videlicet qui est in communi spatio possit facere moras ut confortetur et sit hoc ei quasi augmentum fortitudinis ad cogitandum et intelligendum, idest ut sit fortior in cogitatione et intellectu, providentia ac cogitatione. Et ideo accidit ei qui cogitat mergere caput ad terram et multum aspicere eam et incurvet quasi scriberet aliquam scripturam vel describeret in ea aliquas figuras, ut sit hoc quasi auxilium ad deponendum illud corpus quod diximus esse simile vermiculo super foramen illius decursus per quem transit spiritus ad posteriora capitis".

146 Albertus Magnus, De homine (Ed. Colon. XXVII/2, 294.10–15): "Si autem stricte sumatur, tunc aestimativa est virtus sequens phantasiam et diversa ab ipsa, et est determinans imitationem vel fugam in intentionibus apprehensis; quae, inquam, intentiones coniunctae sunt compositioni et divisioni phantasmatum, non tamen sunt acceptae a sensibus."

147 Albertus Magnus, De homine (Ed. Colon. XXVII/2, 305.33–40): "In homine autem media

activities which require the cooperation between the rational soul and the faculties of the sensitive soul such as memory (in the case of recollection) and imagination (in the case of intellectual abstraction). This is a concept which is very close to the Avicennian concept of cogitative power because of its essential link with the intellective faculties.

To sum up, Albert's early psychophysiology of memory exhibits a development from his sources (Aristotle, Avicenna, Averroes) in psychological terms (cognitive content of memory and recollection) while maintaining a staunchly Galenic localization of internally apprehensive powers. The typical feature of Albert's account, as remarked by Steneck and which cannot be stressed enough, is that the arrangement of the faculties of the sensitive soul in the brain ventricles is at variance in animals and humans. In the latter, the middle brain ventricle hosts intellect-related and human-specific activities. In this way, the activation of different psychological powers thanks to the flow of animal spirits through the brain in Albert's thinking acquires different, richer (and at times not unproblematic) nuances than in his sources. This kind of approach can be seen also in his mature works.

5.5 *Memory, Animal Spirits, and the Aristotelian Paraphrases*

As shown by Steneck, Albert maintains in his *De spiritu* and in his *De animalibus* a localization of the different psychic powers in the brain, which is both in keeping with the *De homine* and which differentiates the structure of human and animal brains.[148] In the *De memoria*, Albert follows Averroes' localization

---

cellula capitis rationalis sive logistica vocatur, eo quod in illa operetur vis quaedam, quae a Gregorio Nixeno et Damasceno dicitur excogitativa, et similiter ab Avicenna et Algazele dicitur cogitativa. Et est vis inquirens et disponens et ordinans singluaria, et secundum hoc phantasia et aestimativa erunt in posteriori parte primae partis cerebri." Albertus Magnus, De homine (Ed. Colon. XXVII/2, 307.31–32; 309.69–72): "Isaac in libro *de diffinitionibus*: 'Recordatio est inquisitio adventus rei iam oblitae a virtute cogitativa'. [...] De quinta et ultima dicendum quod ipsa considerat ordinem ipsius reminiscentiae ad rationem, dicitur ibi virtus cogitativa eo quod cogitat principia reminiscentiae, ut ex ipsis accipiatur ultimo quaesitum."

148 Albertus Magnus, De animalibus I tr. 3 c. 1 n. 528 (ed. Stadler, I, 188.29–34): "Iste autem ventriculus vocatur a quibusdam coniunctio duorum ventriculorum, eo quod defert spiritus recipientes formas a priore in posteriorem, in quo sunt virtutes servantes, sicut diximus: et ideo iste ventriculus optime attribuitur cogitativae et ymaginativae: quia illae fiunt in transitu et motu spirituum, sicut scitur ex hiis quae nos de differentia spiritus et animae diximus." See also: Albertus Magnus, De animalibus XII tr. 2 c. 4 n. 130–132 (ed. Stadler, I, 849.5–36); Albertus Magnus, De spiritu et respiratione I tr. 2 c. 4, ed. Auguste Borgnet, Paris 1894 (Ed. Paris. IX, 236b). As Steneck notes, this latter description is based on Avicenna, Canon, III 1 (Steneck, "Albert on the Classification," 207 n. 58) but does not locate compositive imagination (*phantasia*) and estimation in the middle brain ventricle of humans.

of internal senses, which causes some tension within his account of cogitative powers.[149] Paraphrasing Averroes' famous theory of the different degrees of spirituality, Albert places common sense (second degree of 'spirituality') in the front brain ventricle with imagination (which he considers here, like Averroes, to be the same as *phantasia*), the *virtus distinctiva* or *cogitativa* in animals (*Quartus locus est, in organo virtutis distinctivae, quam vocat cogitativam brutorum, quae distinguit intentiones a figuris rerum*) in the middle ventricle and memory in the posterior ventricle.

The first problem seems to arise if we take *cogitatio* as applied to animals and humans at face value. For Albert, this faculty is the Averroean equivalent of the Avicennian estimative. This is either a misreading of Averroes, who attributed cogitation to humans alone, or a tacit attempt at harmonizing the positions of the two philosophers by forcibly equating the human *cogitativa* and the animal *virtus distinctiva*, viz. estimation, which allows the extraction of meanings from the content of sense perception.[150] The latter option could be equally palatable if we consider Albert's localization of internal senses: on the one hand, cogitation in humans is localized in the same place as animal estimation (middle brain ventricle) and, moreover, cogitation is equated with estimation in a passage from the *De animalibus*. In the end, we seem to find two meanings of cogitative power according to Albert: strictly speaking, it is represented by the set of relational activities which involve intellect and sensitive soul (intellection, recollection). In a looser sense, usually connected to the exegesis of Averroes' text, it is equated with animal estimation and reason-like activities (thus, *cogitativa*) and localized in the middle brain ventricles in both humans and animals. In the end, this doctrine would harmonize the patently conflicting positions of both authors.[151] Albert does not draw a difference between the cognitive content of estimation (hence, of memory) in humans versus animals, but he does posit different perceptions of time.[152] His initial idea that memory contains forms, past-directedness and intentions remains consistent throughout his writings.

---

149 Albertus Magnus, De memoria et reminiscentia tr. 1 c. 1 (Ed. Colon. VII/2a, 114.11–34).

150 Albertus Magnus, De animalibus VII tr. 1 c. 1 n. 2 (ed. Stadler, I, 496.23–28): "In quibusdam autem eorum invenitur etiam quaedam vis, quae est sicut cogitativa et compositiva sive collativa in homine: et haec vocatur a quibusdam cogitativa sensibilis, a quibusdam autem fantasia componens et dividens. Ab aliis autem vocatur aestimativa, et hoc nomen est ei magis conveniens, sicut in hiis, quae *de Anima* disputata sunt, dictum est."

151 Note that a similar strategy, viz. blending together Averroes' position with Avicenna's without dealing with the criticism of the latter by the former, has already been observed in the case of the origin of veins (chapter 3.1).

152 Albertus Magnus, De memoria et reminiscentia tr. 1 c. 2 (Ed. Colon. VII/2a, 116.61–117.8); Veronique Decaix, "What Is Memory Of?," 165.

Apart from these terminological issues, Albert's account of memory in the *De memoria* struggles with further and doctrinal tensions. In this work, Albert advocates for a conception of memory that is strongly based on his Arabic sources, Avicenna and Averroes, against the unclarity of Aristotle's words and the hermeneutical mistakes of his Latin predecessors.[153] According to him, the remembering process according to Avicenna and Averroes mainly consists of four different stages: the attainment of the sensible form in the common sense (*sensus communis*); (2) the preservation of this form in the soul; (3) the separation of intentions from the form; (4) the act of memory by recombining form and intentions.[154] A latent and major difficulty in this presentation of Avicenna's and Averroes' views is that they are—once again—not critically compared. Moreover, Albert does not see an apparent problem with the attribution to Averroes of a position according to which meanings and forms are stored in memory, as opposed to meanings only. As we may recall, this was precisely Albert's position in the *De bono* and the *De homine*, which he now seeks to put under the aegis of Averroes (and Aristotle), finding agreement between two of the main authorities (the third is Avicenna) on memory.[155] As highlighted by Carla di Martino, one of the most relevant changes in Albert's mature account of memory is that the faculty responsible for the combination and separation of forms and meanings is not estimation (Avicenna's position), nor is cogitation (Averroes') the faculty responsible for combining meanings and forms, but rather memory itself (*componendo ista duo refertur ad res extra per ipsa*).[156] This position stems from Albert's separation of two usually conjoined functions: (1) the composition/division of forms and meanings with (2) judgmental activity. While these functions had been usually associated with the faculty of estimation (in Avicenna) or cogitation (Averroes), in Albert the function of estimation is reduced to its judgmental function concerning meanings. If we take a step back and observe the overall framework of Albert's *potentiae apprehensivae deintus*, we see how estimation is no longer, compared to Avi-

153 Albertus Magnus, De memoria et reminiscentia tr. 1 c. 1 (Ed. Colon. VII/2a, 113.15–20): "Quia autem, ut mihi videtur, omnes fere aberraverunt Latini in cognitione harum virtutum quas memoriam et reminiscentiam appellamus, ut aestimo propter verborum Aristotelis obscuritatem, ideo primo volumus ponere planam de memoria sententiam Peripateticorum, antequam Aristotelis sententiam prosequamur."

154 Di Martino, Ratio particularis, 134.

155 Albertus Magnus, De memoria et reminiscentia tr. 1 c. 1 (Ed. Colon. VII/2a, 114.48–53): "Ex hiis igitur patet quod conservativa secundum Averroem conservativa conservat tam imagines quam intentiones, sed memoria componendo ista duo refertur ad res extra per ipsa. Hoc est igitur quod Averroes et alii dicunt esse sententiam Aristotelis."

156 Di Martino, Ratio Particularis, 135.

cenna, the main cognitive faculty: as shown before, *phantasia* has taken up its role as the highest cognitive faculty in animals and humans, while Albert's estimation extracts *intentiones* from the content of sense perception and judges the external objects through the cognitive content of these meanings. Like Avicenna, Albert takes the estimative faculty to be shared by humans and animals, but does not assign to it the role of perceptive awareness (proper of common sense) and does not consider it the highest power of the sensitive soul,[157] a role assigned to *phantasia* inasmuch as it is the only active power of the sensitive soul. Moreover, estimation is no longer a power charged with "first perception" as in Avicenna. Rather, it depends on the sensible forms which are passed on to it by the other faculties, making it, to some extent, a passive faculty (Avicenna would have spoken of "second perception") inasmuch as it derives its content from retentive imagination by way of extracting (*elicitare*) meanings from images.

In Albert's mature *De anima*, *phantasia* plays a key role in the recollective process (*recordatio*) too. Recollection begins with an act of reason, by inferring the existence of a now forgotten but previously existing form or meaning. This act, according to Albert, proceeds under cogitative power. According to Albert, the combining-and-dividing activity of *phantasia* facilitates recollection, since it leads one form into another and then into another. The ultimate goal of this process is to attain the sought-after object.[158] Albert's account of recollection in the *De memoria* seems to follow a similar path to the *De homine* and the *De anima*, arguing for the essential dependence of recollection on an act of reason that starts from something existing in us (*iam pridem existit in nobis*) and proceeds to the attainment of the object through an inferential process (*ad inferendum quaesitum*). This quasi-syllogistic, inferential process proceeds differently through different associations. For this reason, the capacity of *phantasia* to present the rational faculties with a great number of forms and meanings to be reviewed plays into the hands of the recollective process. Picking up from Aristotle's account of memory, Albert adds that recollection can be conditioned by further factors such as passions and complexion: children, for instance, recollect more but remember less.[159] As shown in the excellent book *Animal Rationality* by Anselm Oelze, Albert discusses the particular behavior of vultures, which deliberately return to a place they have observed: to the site of

157 Di Martino, Ratio particularis, 138–142.

158 Albertus Magnus, De anima II tr. 2 c. 7 (Ed. Colon. VII/1, 157.74–98).

159 Albertus Magnus, De memoria et reminiscentia tr. 2 c. 3 (Ed. Colon. VII/2a, 123.20–129.66). On Albert's explanation of recollection: Anzulewicz, "Memoria *und* reminiscentia," 192–193; Oelze, *Animal Rationality*, 169.

a carcass, for instance. Until now, we have seen how, according to Albert, recollection is due to an "act of reason" (*actus rationis*) requiring the cooperation of the cogitative power which, in the strict sense, is proper of humans alone. This position, typical of the *De homine*, *De anima*, *De memoria* and in line with the rest of the Peripatetic (Aristotle, Avicenna, Averroes) tradition,[160] is met by an apparent aporia in the *De animalibus*; namely, when considering the behavior of vultures which return to a place they have already been. The question is, do they do this rationally or irrationally?

As for any Medieval exegete of Aristotelian psychology, Albert's answer is negative. On the other hand, Albert does not minimize animal cognitive capacities by refraining from giving a psychological explanation of this quasi-recollective process in vultures.[161] Quite the opposite: Albert denies animal recollection by distinguishing the different starting points of human and animal behaviors. According to Albert, while humans can recall any object at will, animals, by contrast, start to search in the thesaurus of forms and meanings whenever their appetites push them to.[162] As long as they are sated, they feel no need to look for more food: once their appetite prompts them, they start looking into their mind to find a representation of a suitable thing to satisfy their hunger. For this reason, the act displayed by vultures seems to be fixed, not determined by free choice, as Albert himself had argued in the *De animalibus*, following Nemesius' *De natura hominis*.

For these reasons, one can see how the mechanism of animal memory in Albert's Aristotelian paraphrases is based on two different causal explanations: the first depends on a cognitive activity, i.e. the association of a meaning perceived in the past with an image (or vice-versa), as in his early *De homine*. Additionally, Albert argues that whenever an animal connects forms and mean-

---

160 Albertus Magnus, De anima II tr. 2 c. 7 (Ed. Colon. VII/1, 201.36–40): "Et inde est, quod cogitativa indiget intellectus, quae dicitur sic, quia cogitat particularia ad eliciendum ea quae sunt in ipsa, et illa non est sine ratione, sed ratio in ea egreditur et iuvatur sensibilibus potentiis." Albertus Magnus, De anima III tr. 2 c. 19 (Ed. Colon. VII/1, 206.25–28): "[…] reminiscentia fit per conversionemad particularia, quae sunt in phantasmatibus, et fit mediante cogitativa, quae est actus rationis conferentis de particularibus." Albertus Magnus, De memoria et reminiscentia tr. 2 c. 1 (Ed. Colon. VII/2a, 124.22–28): "Tertium autem est quod Arabes vocant virtutem distinctivam, quae est componens omnia haec et attribuens rei omnibus modis quibus attribui possunt per simile et contrarium et per locum et tempus et aliis modis, ut eliciatur secundum intentum quod cecidit in oblivione, sicut inferius monstrabimus. Et quoad hoc solum reminiscentia solis rationabilibus convenit."

161 On vultures, see: Albertus Magnus, De animalibus VIII tr. 2 c. 6 (ed. Stadler, I, 612–614); De animalibus XXI tr. 1 c. 2 (ed. Stadler, II, 1326).

162 Oelze, *Animal Rationality*, 173–174.

ings to the past, it regains awareness of an object's cognitive value, e.g. a certain pleasant food has been previously eaten is remembered as palatable.[163] The second and additional causal explanation of animal memory and memory-like processes shows how animal behavior, much as it is chained to their nature, can be triggered by their appetitive faculty once animals cease to be sated. In the case of humans, memory or thinking can be aided by the motion of the head. Paraphrasing a passage from Costa ben Luca's *De differentia*, Albert argues that the downward motion of the head favors thinking by closing off rearward conduction into the posterior brain ventricle. Vice-versa, raising the head helps to lift the worm-like particle connecting the middle and posterior brain ventricles.[164] While there is no mention of which faculty should be responsible for the motion of these particles, there appears to be a way out of the question, as in the *De homine*. If we consider the nature of the act of memory itself, we see that the faculty that has access to the thesaurus of information stored in the posterior brain ventricle to produce an estimative judgment is estimation. It is therefore estimation, as in Avicenna, that can gain access to the rear brain ventricle and therefore command the cerebellar worm. There are two notable differences between Albert and Avicenna here: the content of memory encompasses forms, past-directedness and intentions in the same organ. Moreover, the faculty that composes forms and meanings is memory and not estimation, unlike in Avicenna, but unlike in Albert's early *De homine* too. Like Avicenna, Albert devotes an explicit section to the physiological conditions of recollection, largely in keeping with the results of the *De homine*, emphasizing the relevance of moisture in recollection and dryness in memory, as highlighted by a recent study of Marilena Panarelli. This study has the merit, much in line with the present book, of underlying the legacy of Costa Ben Luca and Avicenna in Albert's discussion of the biological foundations of cognitive capacities and

163 Oelze, *Animal Rationality*, 171: "it is not the case that they (sc. animals) do not perceive the past at all. However, they notice the past only by recognising something. For instance, they realise that the stable to which they are currently returning is the stable in which they lived "in the past" (*in praeterito*). Their perception of the past is thus linked to a particular object and its function is to recognise this object."

164 Albertus Magnus, De spiritu et respiratione I tr. 2 c. 4, ed. Auguste Borgnet, Paris 1894 (Ed. Paris. IX, 237a): "Sic igitur spiritus discurrit quidem a corde ad cerebrum, et ibi opera animalia et animalium virtutem operatur transeundo de cellula in cellulam colatus et digestus et quasi sublimatus, sicut dictum est. Hujus autem signum est, quod cum homo intente de aliqua re cogitat et meditatur, tunc elevat et deprimit caput, quasi ipso motu natura ostendat se laborare et adjuvare ad hoc ut facilius elevata caruncula vermiculari spiritus fluat de cellula in cellulam cum formis intellectus, imaginationis, et recordationis." Costa ben Luca, De differentia spiritus et animae, ed. Wilcox 1985, 155.161–166; 157.170–187.

their differences among humans.[165] As can be reasonably understood, this distinction between people better suited to remembering or recollecting based on the dry-moist dichotomy can be traced back to the influence of Avicenna's psychophysiology of memory.

To gain a complete and broader picture of the role of animal spirits in instinctual reactions, incidental perception, knowledge of individual substance, memory and recollection, we need to connect Albert's mature psychology (Aristotelian paraphrases) to his doctrine of animal spirits. A better study of their relevance can be conducted against the background of Albert's understanding of the *scala naturae*, laid out in book XXI of his *De animalibus*. This book, a unique addition to Aristotle's zoology, offers a remarkable correlation between the complexion of animal spirits and the complexity of animals' cognitive capacities, especially those of estimation and memory. Further insights will also come from Albert's *De natura loci* will help us understand intraspecies differences among humans as caused by exogenous factors (climate) alongside endogenous ones (sex and age).

5.6 *Ranking Cognitive Capacities: Animal Spirit and the* scala naturae

In both Avicenna and Costa ben Luca, as in many authors indebted to the Galenic medical tradition, the ranking of cognitive activities in animals depends on two factors: the psychology of animals (i.e. the faculties possessed by them) and the complexion of the spirit. The latter factor can be influenced by endogenous or exogenous factors: animal spirits, a refined and concocted version of vital spirits produced out of air and blood, are themselves composed of humors. Spirit can also impact the psychological differentiation of cognitive activities in animals (interspecific difference) and among humans themselves (intraspecies differences). For this reason, we need to turn to Albert's ranking of the cognitive activities presented in book XXI of *De animalibus*, from which we can understand the impact of complexion of spirit on interspecies differences within animal cognition.

In book XXI of his *De animalibus*, Albert presents his version of the *scala naturae*, ranking animals according to their intelligence as following: humans, pygmies, apes, quadrupeds, land birds, water birds, fish, reptiles, worms and *larvae*. As demonstrated by the recent studies of Gabriella Zuccolin, Evelina

165 Albertus Magnus, De memoria et reminiscentia tr. 2 c. 7 (Ed. Colon. VII/2a, 135.20–137.30); Marilena Panarelli, "Bodily Prerequisites of the Mind: The Spirit as the Highest Product of Digestion," in *The Embodied Soul: Aristotelian Psychology and Physiology in Medieval Europe between 1200 and 1420*, ed. Marek Gensler/Monika Mansfeld/Monika Michalowska (Historical-Analytical Studies on Nature, Mind and Action 11), Cham 2022, 115–132.

Miteva, and Anselm Oelze, the case of pygmies is particularly remarkable.[166] According to Albert, pygmies are a peculiar class of animals with an estimative faculty apparently superior to all other animals and seem therefore to possess a "shadow of reason" (*umbra rationis*).[167] In the *De animalibus*, Albert compares the cognitive abilities of pygmies to humans with damaged brain ventricles (*moriones*).[168] The difference between humans and pygmies, though, is essential: while the former can actually go back to the exercise of their rational abilities if they recover from their illness, the latter have no possibility whatsoever of achieving rationality, rooted as it is in the essence of the human being (*homo inquantum homo solus est intellectus*). To understand their cognitive abilities, Albert draws a parallel between humans, pygmies and apes hinging upon three progressive degrees of cognitive ability: (1) attentively comparing (*diligenter attendere et conferre*) the cognitive content (images, meanings) related to the extramental world, (2) contrasting and comparing this information to gain experience (*ex collatis memoria retentis experta sumere*) from it and (3) abstracting the universal out of this information (*ex illis universale depuratum accipere*). While this third and highest degree of cognitive activity depends on the rational soul and is human-specific, the first two appear to be possessed by animals too. According to Albert, the second degree of cognitive ability belongs to pygmies, who are better than other animals, especially apes, in relying on previous experience. Not that other animals are incapable

166 On pygmies, see: Gabriella Zuccolin, "Monkeys, Pygmies, and Human Beings. *Sensus disciplinales* and the Hierarchy of Living Beings in Albert the Great," in *Aristotle's* De sensu *in the Latin Tradition, 1250–1650*, ed. Mattia Mantovani/Roberto Zambiasi/Gabriella. Zuccolin (Micrologus Library, Special Issue 31), Firenze 2023, 85–104; Evelina Miteva, "Pygmies, Twins, Monsters: Human Nature on Its Borderline in Albert the Great," in *The Embodied Soul: Aristotelian Psychology and Physiology in Medieval Europe between 1200 and 1420*, ed. Marek Gensler/Monika Mansfeld/Monika Michalowska (Historical-Analytical Studies on Nature, Mind and Action 11), Cham 2022, 73–90; Oelze, *Animal Rationality*, 154–155.

167 Albertus Magnus, De animalibus XXI tr. 1 c. 2 n. 13 (ed. Stadler, II, 1329.5–11): "Virtus autem illa animae quam umbram rationis quamdam vocavimus superius, innominata quidem est a philosophis, sed circumloquendo cognoscimus quod haec vis aliquid potentiae addit super aestimativam. Cum enim aestimativa brutorum iudicet de intentionibus quae cum sensibilibus accipiuntur, ista plus facit quoniam res fert huiusmodi intentionis ad memoriam et elicit expertum et utitur illo postea ad quae confert."

168 Albertus Magnus, De animalibus XXI tr. 1 c. 2 n. 12 (ed. Stadler, II, 1328.26–32): "Et ideo pigmeus nichil omnino percipit de rerum quiditatibus nec umquam percipit habitudines argumentorum: et sua locutio et [corr. est; Anselm Oelze] sicut locutio morionum qui naturaliter stulti sunt eo quod non perceptibiles sunt rationum. Sed in hoc est differentia quod pigmeus habet privationem rationis ex natura, morio autem habet per accidens ex melancolia vel alio accidente non privationem rationis, sed potius privationem usus rationis."

of drawing on past experience, though. For instance, weasels that have been poisoned by snakes learn that endive leaves can help them fight off the poison.[169] The difference is that, unlike pygmies, weasels, apes and other animals are progressively less able to rely on past experience.[170]

In more than one passage, Albert emphasizes how the different degrees of animal cognitive capacities depend on the sharpness in the faculty of estimation, which needs to be pure and endowed with healthy organs (*pura est et boni organi*). In both cases animal spirits are involved. In the first case, purity of estimation depends on the animal spirit's balanced (*temperatum*) complexion. In the second, if the organs of the animal are of a melancholic (viz. dry and cold) nature, they are not able to receive the spirit in all its clarity. Vice-versa, a healthy complexion of organs, closer to the most perfect balance—found in humans alone—makes for the highest cognitive abilities in animals, found in pygmies, then in apes and, in gradually descending order, in all other animals.[171]

As shown by Oelze, the complexion of spirit allows Albert to single out animals that have memory and those that are deprived of it on purely physiological grounds. For instance, the organs of flies are too cold and watery to store impressions, and this excess in coldness and moisture "immobilises the animal spirit" (*immobilizat spiritum animalem*), thus impeding the existence of the act of memory itself.[172] The practical consequence is that flies do not remember blows previously dealt by humans, and therefore will fly near humans regardless of possible harm to themselves. This passage shows how Albert, based on animal behavior, infers the quality of the complexion of animal spirit from the lack of behavior that would be in accord with any animal endowed with memory. If flies keep coming around us even though they have been beaten by us, this means that they have no memory or organs (thus, no material instruments) suitable to sustain memory. In the *De animalibus*, Albert reiterates the link between complexity of cognition and balance in the complexion of animal

169 Albertus Magnus, De animalibus XXI tr. 1 c. 2 n. 11 (ed. Stadler, II, 1327.32–34).

170 Albertus Magnus, De animalibus XXI tr. 1 c. 3 n. 15 (ed. Stadler, II, 1330.8–12).

171 Albertus Magnus, De animalibus XXI tr. 1 c. 4 n. 22 (ed. Stadler, II, 1333.35–1334.1): "In aestimatione autem animalium est praecipue quia aestimatio componitur in ipsis ex apprehensione fantasiae et appetitu desiderii, et habet se in sensibili anima sicut prohaeresis in rationalibus, et haec bene pura est et boni organi, tunc efficitur boni iudicii de intentionibus rerum. Dico autem puram quando spiritus eius est purus et clarus et calidum eius est temperatum neque pigrum ex frigido neque commiscens operationes ex nimis calido. Organum autem bonum est non melancolicum quod bene suscipit spiritus in eadem claritate qua deferuntur ad ipsum in spiritu."

172 Albertus Magnus Metaphysica I tr. 1 c. 6 (Ed. Colon. XVI/1, 8.14–28). On flies, see: Oelze, *Animal Rationality*, 170.

spirits to explain that sea birds, for instance, have some deficiencies in their estimative faculty because of their melancholic complexion.[173] Moreover, he directly links the brightness (*luciditas*) found in the animal spirits of air birds and reptiles to the physiological cause of their sharp cognitive abilities.[174]

As can be inferred from these paragraphs, complexion of animal spirits is a key aspect to understanding interspecies differences in a framework that is clearly influenced by the Galenic theory of complexion and humors. This doctrine, primarily transmitted by Avicenna, was used also by Albert to detect intraspecies differences among humans according to exogenous factors (climate) and endogenous ones (sex, age). As research conducted by Henryk Anzulewicz has highlighted,[175] this causal link between the nature of spirit, exogenous factors (theory of climate) and cognitive abilities is best investigated in Albert's *De natura loci*. In this work, Albert differentiates cognitive and rational abilities in humans according to the places they live, highlighting also the seasons best suited to the exercise of their cognitive capacities.

In the first place, Albert divides the earth into seven climate zones. Since each is defined on the basis of average climatic conditions, they are divided into two extremes and a middle zone. The result is that, according to Albert, the inhabitable regions of earth number twenty-one. Albert posits the relation of a

173 Albertus Magnus, De animalibus XXI tr. 1 c. 5 n. 27 (ed. Stadler, II, 1336.22–25).

174 Albertus Magnus, De animalibus XXI tr. 1 c. 5 n. 26 (ed. Stadler, II, 1336.1–9 [on birds of prey]); De animalibus XXI tr. 1 c. 5 n. 27 (ed. Stadler, II, 1336.22–29 [on marine birds]); De animalibus XXI tr. 1 c. 7 n. 43 (ed. Stadler, II, 1343.35–36 [on the correlation between shiny spirits and shrewdness]).

175 Henryk Anzulewicz (forthcoming, 2023), "Psychophysiology, Geographical Regions, and Climates: Albert the Great on the Natural Prerequisites of Scientific Aptitude in Human Beings," in *Medieval Science between Emergence and Inheritance: Albert the Great and His Arabic Sources* (Philosophy in the Abrahamitic Traditions, forthcoming, 2023) and bibliography therein. I would like to the author for sharing his draft version of his paper in advance with me. Other contributions on the biological roots of cognitive and behavioral differentiation in Albert, see: Alessandro Palazzo, "*Hi sunt natura servi.* Natural Inequality and Biology in Albert the Great's *Politica*," in: Homo—natura—mundus. *Human Beings and their Relationships. Proceedings of the XIV International Congress of the Société Internationale pour l'Étude de la Philosophie Médiévale, July 24–28, 2017, Porto Alegre, Brazil*, ed. Roberto Hofmeister Pich/Alfredo Carlos Storck/Alfredo Santiago Culleton (Rencontres de Philosophie Médiévale 22), Turnhout 2020, 613–631; Carlos Steel, "A Philological Diet for Philosophers. Aristippus' Translation of Book IV of Aristoteles' *Meteorology* and Albert the Great," in *L'Antichità classica nel pensiero medievale. Atti del convegno della Società Italiana per lo Studio del Pensiero Medievale (SISPM). Trento, 27–29 settembre 2010*, ed. Alessandro Palazzo (Textes et études du Moyen âge 61), Porto 2011, 79–106; Gianfranco Fioravanti, "*Politiae Orientalium et Aegyptiorum.* Alberto Magno e la Politica aristotelica," *Annali della Scuola Normale Superiore di Pisa*, 9 (1979 [series 3]), 195–246.

place (with its climate) to a place's content to the form-matter relation.[176] Each of the seven climates is differentiated in a threefold manner, in accordance with the mean and the two border zones at each extreme. Each of these climatic segments influences in its own unique way a human being's conception, prenatal development, and growth, composition of bodily physiology, and disposition of psychic faculties. For these reasons, Anzulwicz has rightfully called the reader's attention to the places at which Albert directly links the material composition of spirit to the places where those spirits dwell.

Inhabitants of very hot regions have a very frail bodily constitution, tend to age early and suffer from physiological and mental weakness. Ethiopians (inhabitants of the second tropical climate zone) are black-skinned, light-weighted body, and suffer from "simple-mindedness" because of the extreme heat and the evaporation of their spirit.[177] By contrast, Albert attests that the inhabitants of the first tropical climate zone, along the equator, possess a strength of sense perception and a skill in invention caused by the agitating warmth and consequent sharpness of their spirits. In Albert's view, outstanding philosophers, mathematicians, and magicians owe their special abilities to the geographical, regional, and climatic conditions of their residence. This is because, in Albert's view, their skills are generated by strong radiation emanating at a perpendicular angle from the heavenly bodies.[178] By contrast, inhabitants of colder regions, such as Daciens, Goths, Slavs and Parthians, are distinguished by their mental dullness and loutishness. This mental inferiority is due to physiology, since the coldness of these regions results in a condensation of liquids in the organs that slows down the physiology of sense perception and obstructs at times the reception of sense impressions.[179] The climatic conditions in the southern regions (4–5 climate) benefit their inhabitants more than

176 Anzulewicz, "Psychophysiology, Geographical Regions, and Climates."

177 Albertus Magnus, De natura loci tr. 2 c. 3 (Ed. Colon. v/2, 27.2–4): "Aethiopes nigerrimi leves quidem sunt corpore et fatui mente propter defectum et evaporationem spiritus animalis."

178 Albertus Magnus, De natura loci tr. 2 c. 3 (Ed. Colon. v/2, 26.85–92): "In operationibus autem animalibus, qui sub aequinoctiali sunt, vigent propter subtilitatem spirituum et plus in inveniendo propter calidum movens et acumen spirituum eorum. Cuius signum est, quia praecipui in philosophia in India fuerunt et praecipue in mathematicis et magicis propter fortitudinem stellarum super climata illa super quae perpendiculares radios proiciunt planetae."

179 Albertus Magnus, De natura loci tr. 2 c. 3 (Ed. Colon. v/2, 27.52–57): "Operationes autem eorum animales non vigent propter spissitudinem, quippe umor eorum est piger et spissus nec oboedit motui et receptioni formarum animalium. Sunt igitur tales hebetes et stolidi, nisi hoc sit ex studii exercitio. Sed quando moventur, diu durant et efficiuntur multo meliores post exercitium."

anywhere else, as they benefit the physiological processes and cognitive capacities of the human beings residing in them, because of the balanced ratio of heat and moisture in these places. For this reason, inhabitants of these regions have a healthy constitution, are well-versed in sciences and arts, exhibit ethically noble behavior and are long-lived.[180]

Having established the impact of each climate zone on intraspecific differences in cognitive and rational activity among humans, Albert takes also into account the role of seasons in a short passage of his *De sex principiis*. In this work, he argues that the melancholic individual can exercise his cognitive and rational abilities better in spring (hot and wet) than in other seasons. This is because the complexion induced by the spring season counterbalances the dry and cold qualities typical of a melancholic body. The spirit receiving the sensory forms exhibits the highest possible level of agility on account of the season's warmth and humidity. By the same token, winter is the best season for choleric individuals, summer for phlegmatic individuals, autumn for sanguine people. Conversely, when measured by the cognitive performance of a particular complexion type, summer is the least favorable time for the choleric individual, winter for the phlegmatic, spring for the sanguine, and autumn for the melancholic.[181]

As for endogenous factors, Stefano Perfetti has recently shown how Albert understands the cognitive and rational abilities of children in comparison to those of animals. Children's not yet fully-formed rational capacities are used as a tool for understanding the animal mind. Albert lays out a model of child development according to three progressive stages: first potentiality, second potentiality and full actuality.[182] The first degree corresponds to disposition towards accomplishing a certain act, the second degree of potentiality to the knowledge of the tool necessary to carry out a certain activity, while the successful exercise of the activity in itself equates to full actuality. The main point in Albert's comparison is to emphasize how, contrary to humans, who achieve a full degree of activity in their rational capacities, animals retain a more crystallized limit to their cognitive capacities, which can only sporadically evolve by referring to the experience of particular things, as in the weasel-endive example. As emphasized by Perfetti, this can be a weak comparison,[183] but it is the point that Albert wants to make in terms of cognitive differences between humans and animals.

180 Albertus Magnus, De natura loci tr. 2 c. 3 (Ed. Colon. v/2, 27.57–70).

181 Albertus Magnus, De sex principiis tr. 4 c. 6 (ed. Colon. I/2, 43.39–44.4).

182 Perfetti, *Nature imperfette*, 30–35.

183 Perfetti, *Nature imperfette*, 32.

According to the Dominican master, differentiation of cognitive activity applies to women as well. At the beginning of the eight book of the *De animalibus*, pivoting around animal behavior, Albert argues that the moist nature of women makes them quicker in learning and more solicitous towards their children. Albert retains typical misogynistic descriptions such as women's being envious, their being deceitful (an idea borrowed from Avicenna) and their being less apt to motion because of the minor ratio between heat and moisture within them, which is less abundant in men.[184]

As we can observe, the complexion of animal spirits represents a cornerstone of Albert the Great's psychophysiology of estimation and memory. The vehicle of the internally apprehensive faculties, animals spirits carry a cognitive content in animals broader than was supposed by Albert's Arabic predecessors. In his mature works, the cognitive content vehiculated by animal spirits encompass both non-sensible qualities (meanings) as instances of connotational attribute, truth and falsehood. By contrast, a noteworthy difference is represented by the fact that individual substances are presented to estimation via the previous action of common sense, as shown by Rubio Hernandez. Heir to the Galenic doctrines of complexion, Albert argues that the prevailence of certain primary qualities in the complexion of animal spirits explains certain interspecific differences in cognitive powers among animals, as in the case of flies lacking memory because of their excessively cold animal spirits. Moreover, Albert puts in direct correlation the degree of purity and brightness of animal spirits with a higher apprehensive power of the estimative faculty, better

184 Albertus Magnus, De animalibus VIII tr. 1 c. 1, nn. 5–6 (ed. Stadler, I, 573.14–39): "Generaliter autem loquendo feminae omnium animalium sunt leviores ad instruendum et astutiores quoad hoc quod citius capiunt: et hoc ideo est, quia sunt molliores, et sunt magis avarae et sollicitae circa nutrimentum filiorum. Mas autem e contra est maioris irae et magis silvestris et minoris astutiae. Passiones autem istae sexuum magis manifestantur in homine quam in alio animali: homo enim inter omnia animalia magis attingit naturae perfectum: et ideo etiam magis manifestatur, quod in aliquo deflectitur a perfecto secundum naturam. In homine igitur mulier facilius inclinatur ad pietatem sive benevolentiam quam vir, et citius lacrimatur, et magis est invida, et magis contentiosa et litigiosa et rixosa, et diligit castigare sive criminari, et nequitia animae maior in femina quam in viro ⟨est⟩. Est etiam debilis spei sive fiduciae et mendax magis quam vir et inverecunda, et facile decipitur, eo quod est mali iudicii, et habet ex frigiditate congelante aliquid fortis memoriae in malis et ex frigido comprimente: et quia corpus eius nec consumit neque repellit a se superfluum, est pigra et segnis. Hiis autem addit Avicenna, quod in rebus honestis et bonis et regimine sunt magis stultae et magis recolentes res viles, et sunt fraudulentae, et fraudes recipientes ab aliis et magis paratae ad malas consuetudines: propter quod etiam proprium subiectum non habet in eis virtus et honestas: femina enim universaliter est gravis motus, et mas econtra mobilior, et magis confidens et audax et maioris virilitatis."

able not only to perceive and apprehend *intentiones* but also to gather experience from them. In the end, the balance in the humoral complexion of animal spirits, influenced both by endogenous (sex, age) and exogenous (climate) factors, appears to be among the fundamental physiological factors, together with humoral complexion, explaining the different possibilities for the exercise of cognitive and rational activities in human beings and in the *scala naturae* of animals' cognitive capacities.

Conclusions

Animal spirits constitute the cornerstone upon which Albert founds his theory of "internally apprehensive powers" (*potentiae apprehensivae deintus*), which undergo some doctrinal developments through the years. Faced with intrinsic tensions within the Aristotelian account of *phantasia* and memory, Albert develops a psychophysiology of *phantasia*, estimation and memory deeply influenced by his Arabic medico-philosophical sources. In the first place, he individuates within the theory of abstraction the criterion for singling out the objects of the internally apprehensive powers. While the first (common sense), second (imagination) and fourth (intellection) degrees of abstraction remain constant throughout his works, the borders of the third (meanings) degree of abstraction vary from the *De homine* to the mature *De anima*. In the former, meanings are extracted from images by *phantasia*, while in the Aristotelian paraphrases this task is assigned to estimation. The shift in the faculty responsible for the extraction of meanings also affects the physical localization of the cognitive objects (images, meanings) in the different brain areas. The physiological correlate of the internally apprehensive powers, which is to say the flow of animal spirit through the brain ventricles, acquires a different function in terms of activation of certain psychic capacities in well-defined brain areas.

The difference in the faculty responsible for the extraction of meanings does not change the shared nature of meanings themselves, though. Quite the contrary, Albert's position among his contemporaries seems to place him among those who pushed animal cognitive, quasi-rational capacities to their highest level, conceding that animals could attain the knowledge of singular substances and could modify their behavior according to past experience, although neither freely nor rationally. This position represents a doctrinal development compared to his early *De homine*, which adopts a view on animal instinctual responses closer to the theological treatises of that time (except for John of La Rochelle's *Summa de anima*), which emphasized the inborn component of estimative power, exemplified by the typical sheep-wolf illustration.

Albert's psychophysiology of estimation highlights an enrichment in the cognitive content vehiculated into/through the appropriate areas of the brain by animal spirits. Given the shared nature of the cognitive content of meanings, the divide between humans and animals is to be situated around the perception and apprehension of meanings (linked to particulars) by animals and of universals by humans. Even more than Avicenna, Albert makes more concessions for animal cognitive powers. In his account of memory, for instance, Albert maintains that its object contained three main aspects: images, meanings and past-directedness of images. Apart from the cognitive content of memory, a relevant difference between Albert's and Avicenna's accounts of estimation and memory resides (1) in the more passive role attributed to the activity of estimation, dependent on phantasy (*De homine*) or retentive imagination (*De anima*) and (2) in the different localizations of these powers in humans and animals; the two authors give different teleological values to the middle brain ventricle. While in animals it hosts estimative power and *phantasia*, in humans it appears to be designed to host those intellect-related activities (intellection, recollection) identified by Albert by the term *cogitativa*. Albert inherits too from the medical tradition the role of the Galenic cerebellar worm, whose motion (commanded by estimation) is key to understanding the psychology of memory and estimation. Moreover, Albert borrows from Costa ben Luca some curious empirical observations on the causal correlation between head motion, lifting or lowering the cerebellar worm and the activities of intellect and memory. His complexion of animal spirits allows him to better shape his *scala naturae*, thus determining interspecies differences in animal cognition and highlighting different the different dispositions among humans towards scientific and intellectual activities.

For instance, in his Aristotelian paraphrases Albert takes the case of human suffering, an impairment of the functions placed in the middle brain ventricle (viz. cogitative power, intellection etc.), as a comparative tool for understanding the nature of pygmies as non-rational animals endowed with a "shadow of reason" (*umbra rationis*). Moreover, Albert identifies within the cold complexion of animal spirits in flies the physiological causes that incapacitate their memory-related activities. The brightness of spirits explains why birds of prey and snakes have a particularly sharp estimativefunction, as opposed to water birds, whose phlegmatic complexions makes for a less confuse activity of estimation. Likewise, intraspecies difference among humans are explained, in a manner clearly inspired by Avicenna's *Canon*, on the grounds of their different habitations, sexes and ages.

In the end, the elemental and complexional characterization of spirits, so crucial to ranking the cognitive capacities of animals and humans according to endogenous (sex, age) and exogenous (climate) factors, plays a pivotal role in the second part of Albert's key issue pertaining to internally apprehensive powers. This is true for estimation and memory as much as it is true, as we shall endeavor to show in this last chapter of our book, in the case of the psychophysiology of sleep, dreaming and visions.

CHAPTER 8

Animal Spirits and Internally Apprehensive Powers (*potentiae apprehensivae deintus*)—Part 2: Sleep, Dreams, and Prophecy

In his introduction to his article on the Latin reception of Aristotle's *De somno et vigilia*, Cristophe Grellard reports the conflicting views of two characters from Geoffrey Chaucer's *The Nun's Tale*, Chantecler and lady Pertelote, on the true significance of dreams. The latter mocks Chantecler's belief in the truth value of dreams, advocating instead an explanation that sees them as mere symptoms of bodily states. This physiological interpretation is rooted by lady Pertelote in the Aristotelian psychology and physiology of dreaming.[1] John Kruger's book, *Dreaming in the Middle Ages*, clearly shows that during the 13th century Aristotle's psychology and physiology of dreaming influenced debates on dreaming and God-given dreams[2] in combination with Late Antique and classic sources such as Macrobius and Augustine.[3] Aristotle's skepticism on the matter of God-given dreams, though, constituted a major stumbling block for any adherents to the Christian faith. In the Middle Ages, the 13th century displayed a flourishing literature around the theme of dreams, especially God-given ones, in theological treatises on the human soul (e.g.: Albert's *De homine*) and in the commentaries on Aristotle's three treatises (*De somno, De insomniis, De divinatione per somnum*) known under the name of *De somno et vigilia*.[4] The

1 Cristophe Grellard, "La récéption médiévale du *De somno et vigilia*. Approche anthropologique et épistémologique du rêve, d'Albert le grand à Jean Buridan," in *Les* Parva Naturalia *d'Aristote. Fortune antique et médiévale*, ed. Christophe Grellard/Pierre-Marie Morel, Paris 2010, 220–221.

2 Note that in this chapter we shall use the term "God-given" to refer to dreams and prophecies that are derived from direct, divine revelation and which concern mostly and primarily acts derived from human free will (especially in Albert). By contrast, we shall use interchangeably the adjectives "divinatory" or "veridical" to indicate the kind of dreams that contain foreknowledge of future events which, though originating with God as their first cause, are ruled by the rules of nature and, primarily, of hypothetical (or suppositional) necessity. In the case of Aristotle, the adjective "divinatory" covers both realms.

3 Steven Kruger, *Dreaming in the Middle Ages* (Cambridge Studies in Medieval Literature 14), Cambridge 1992, 83–87.

4 For a list of commentaries and *quaestiones* on Aristotle's *De somno*, see: Sten Ebbesen, Véronique Décaix, Christina Thomsen-Thörnqvist, "Questions on *De sensu et sensato, De memoria and De somno et vigilia*," *Bulletin de Philosophie Médiévale* 57 (2015). On the influ-

 | DOI:10.1163/9789004706743_010

challenge posed to Medieval readers of Aristotelian psychology was, in the first place, to find a way to harmonize Aristotle's psychology and Christian orthodoxy. To do so, sources from the Islamic World such as Avicenna, Averroes, Maimonides etc. would have acquired a crucial relevance. This is witnessed by the well-known Medieval adage, based on Avicenna's *De anima*, that morning dreams are the most truthful ones.

Yet the prognostic value that could be attributed to dreams was anything but uncontroversial in the Late Middle Ages. In her book on prophecy in the 13th century, Anna Rodolfi has shown that divinatory practices were condemned, as in Aquinas' case, as sin per se,[5] as they found unquestionable rejection and condemnation within the Bible itself (Deuteronomy 18, 9–14). The real issue at stake here was the prognostic value of dreams: assuming that a human could foretell any future event on the base of divinatory dreams would put his cognitive abilities on equal footing with God's own foreknowledge of future things, a patently unorthodox claim. This is why handbooks of divinatory dreams and divinatory practices were so strictly forbidden and attacked in the Middle Ages. In like manner, Medieval authors who were theologians themselves (like Albert) yet who did not reject the existence of divinatory dreams foretelling future natural events (not based on human free will, then) had to adopt a strategy based on the careful delimitation of the epistemic boundaries of such dreams to ward off possible charges of unorthodoxy.

Moreover, the complex problems posed by the exegesis of Aristotelian psychophysiology of dreaming and God-given dreams were matched and combined with another sensitive issue, viz. the rational explanation of God-inspired prophecy. In her book on the epistemological status of prophecy in the 13th century, Anna Rodolfi has convincingly stressed how Albert's production is situated at the beginning of a new way to understand prophecy from a gnoseological point of view, strongly influenced by the role of the Aristotelian faculty of imagination and its revisitation operated by Avicenna, Averroes and Maimonides. Until Albert's work, the main issues in Latin prophetology were constituted by the reception, mode of understanding and communication of the object of prophetic foreknowledge through the classical doctrine of the *speculum aeternitatis*.[6] In the Dominican master's theory of prophecy, notes

ence of Aristotle on contemporary theological production, see Kruger, *Dreaming in the Middle Ages*, 89–99 (on Bartholomaeus Anglicus and other authors), 116–122 (on John of La Rochelle, Raoul de Longchamps and William of Varouillon).

5 Anna Rodolfi, Cognitio obumbrata: *lo statuto epistemologico della profezia nel secolo 13* (Micrologus Library 74), Firenze 2016, 43.

6 Rodolfi, Cognitio obumbrata, 105–128.

Rodolfi, we find far more relevance imputed to the Aristotelian claim about the key role of images than in previous authors, who often referred to the traditional notion of *speculum aeternitatis*. Often equated with God's essence or God's wisdom, this entity, at times understood as endowed with proper will, explains the communication of the content of prophetic revelation. The concept of *speculum aeternitatis* was itself considered capable of communicating the object of prophecy to the prophet. The role of images in this process is therefore merely concomitant, since these served the prophet as a mere tool for articulating the prophecy. On the other hand, the origin of the foretold event was thought to come from the *speculum aeternitatis*, an intelligent being endowed with will. Albert's attitude toward the role of images in prophetic revelations is, as we shall see, a bit different. Paraphrasing Aristotle, prophecy cannot exist without images.[7] Yet Albert has to deal with a set of issues that concern the compatibility between the images originated from sense perception and stored in imagination (the so-called "*imago a quo*") and their epistemic role in the intellect of the subject (*in quo*), whose activity is a necessary requirement for prophetic revelation, and never occurs without the aid of the intellect. In the first place, he had to explain and distinguish between prophecies related to natural events—called *prophetia naturalis* in Aquinas—and God-inspired prophecy, just like he had to distinguish between God-given dreams and divinatory or prognostic dreams. The different solutions envisaged by Albert through his different conceptions of image will be quite revealing as regards Albert's understanding of the relationship between philosophy and theology throughout his works.

The central role played by images in Albert's psychophysiology of sleeping and dreaming will emerge in his considering the physiological processes (viz. flow of animal spirits) underlying the psychological activities of dreaming and prophecy and, on the other hand, of sleep. For this reason, the flow of animal spirits (*qua* vehicles of the images) will come to constitute the key physiological correlate, explaining sleep as both an affection of the sensitive part of the soul and the activity of dreaming as a causal product of the activity of imagination. Moreover, spirits (not just the animal ones) constitute the crucial physiological notion for understanding both the mechanics and the teleology of sleep as an incapacitation of sensitive activities and as a restoration of the amount of heat needed to perform sensation and locomotion (in moving animals) during the waking state. The interaction between the psychophysiology of sleep, dreaming and prophecy presents us with a case study straddling medicine, philosophy

7 Rodolfi, Cognitio obumbrata, 26.

and theology. Throughout this chapter, we will observe the different argumentative strategies and methodological approaches employed by Albert to fashion his psychophysiology of sleep in accordance with the Christian tradition, either *via* the harmonization of Aristotle with Christian authorities or by means of a careful distinction in the object of divinatory dreams and prophecy as true according to the laws of nature or as true according to a theological explanation. Building on the previous studies of Rodolfi, Palazzo, Donati, and Söder (the latter two, Donati and Söder, are key for their study of the role of spirits),[8] we shall seek to construe a more detailed psychophysiology of dreams, divinatory/veridical dreams connecting Albert's own doctrine of sleeping, dreaming and divinatory dreaming to the notion of spirit in a manner that is more systematic than the studies of Söder and Donati. To begin our study, we shall focus on Aristotle's own doctrine of sleeping and dreaming.

1 Aristotle's Psychophysiology of Sleeping and Dreaming

Unlike his teacher Plato, Aristotle was notoriously skeptical about God-given dreams. According to the Stagirite, their existence is puzzling from a teleological point of view in the first place. if the divinity were to gift the foreknowledge of future things to randomly selected humans, regardless of their moral and intellectual qualities, this thing alone would prove destructive in teleological terms, let alone if God-given dreams were communicated to animals, as dreaming pertains to the sensitive soul.[9] In her introduction to the Aristotelian text we shall follow in this section, Luciana Repici has shown that, if the exercise of

---

8 Silvia Donati, "Albert the Great as a Commentator of Aristotle's *De somno et vigilia*: The Influence of the Arabic Tradition," in Parva Naturalia *in Greek, Arabic and Latin Aristotelianism. Supplementing the Science of the Soul*, ed. Börje Bydén/Filip Radovanovic (Studies in the History of Philosophy of Mind 17), Cham, 2018, 169–209; Silvia Donati, "Dreams and Divinatory Dreams in Albert the Great's *Liber De somno et vigilia*," in *Contemplation and Philosophy: Scholastic and Mystical Modes of Medieval Philosophical Thought. A Tribute to Kent Emery, Jr.*, ed. Roberto Hoffmeister Pich/Andreas Speer (Studien und Texte zur Geistesgeschichte des Mittelalters 125), Leiden-Boston, 2018, 178–215; Alessandro Palazzo, "*Philosophi aliter loquuntur de prophetia quam sancti*. Alberto il Grande e la profezia naturale," in *Immaginario e immaginazione nel Medioevo. Atti del convegno della Società italiana per lo studio del pensiero medievale* (*S.I.S.P.M.*), *Milano, 25–27 settembre 2008*, ed. Maria Bettetini/Francesca Paparella/Roberto Furlan (Textes et Études du Moyen Âge 51), Louvain-La-Neuve 2009, 179–201; Joachim Söder, "Albert der Große über Sinne und Träume. Beobachtungen am Traumtraktat vom *De homine*," in *I cinque sensi*. (Micrologus Library 10), Firenze 2002, 239–250.

9 Aristotle, De divinatione per somnum 2 (463b12–22); Aristoteles Latinus, De divinatione per somnum (ed. Drossaart Lulofs 1947, 38.17–40.8); Repici, "Introduzione," in Aristotele, Il sonno

intellectual capacities were not to grant the communication of divine revelations to the individuals who exercised them, God-given dreams would just be a matter of luck and chance.[10] Given the shared nature of dreams as an affection of the sensitive soul, Aristotle opts for another, naturalistic explanation of oneiric images. According to him, dreams are causally linked to external reality as (1) causes or (2) signs of events or, on the other hand, (3) coincidences or (4) a combination of all three categories.[11] We shall come back to this classification later; for we now need to briefly elucidate the role of images and the faculty of imagination in Aristotle's theory of dreaming.

In Aristotle's view, the poietic function of the faculty of imagination is responsible for the chancy appearance of images during sleep or wakefulness. These images have, strictly speaking, an origin in previously perceived images, but do not have a truth value, for they often represent fictional beings or unreal situations. During the state of wakefulness, such images can be judged as false when they appear via our opinion (δόξα). Since sleep is an incapacitation of sensitive and rational activities of the soul, the latter of which comprehends opinion, then our perception of such images is misled into thinking that dreaming images are true images.[12] According to Aristotle, oneiric images are, physiologically speaking, the byproduct of residual motions produced by perceptual activity left in the blood and in the sense organs. These residues (καταλείμματα), left in the sense organs and imprinted in the blood, flow back to the heart during the process of sleep.[13] Contrary to the bigger motions produced by sense perception during wakefulness, their residues remain small and fractionated in the blood and in the sense organs. By flowing back to the heart during sleep, the perception of these small motions is actualized as if they were actual and stronger perceptions. This is the reason, argues Aristotle, we dream of thunders and thunderbolts if we have perceived a small echo in our ears. Since we lack the capacity of consideration and judgment during sleep, these images (thunders, thunderbolts) are perceived in an altered manner.[14]

---

e i sogni. Il sonno e la veglia, I sogni, La divinazione durante il sonno, ed. Luciana Repici, Venezia 2003, 44–45; Aristotle, *On sleep and dreams. Text and translation with introduction and notes by D. Gallop*, Warminster 1996, n. 49, 162–163.

10 Aristotle, De divinatione per somnum 2 (462b20–26); Luciana Repici, "Introduzione," 40–41; Aristotle, *On sleep and dreams*, 156 n. 44.

11 Aristotle, De divinatione per somnum 1 (462b26–463b11).

12 Kjeller-Johansen, *The Powers of Aristotle's Soul*, 208–210.

13 Aristotle, De insomniis 3 (462a8–31).

14 Aristotle, De divinatione per somnum 1 (463a10–16); Aristoteles Latinus, De divinatione per somnum (ed. Drossaart Lulofs 1947, 34.19–36.6).

According to Aristotle, the perceptual content of a dream is always similar to the sensation which gave rise to it and, moreover, the activity of dreaming always implies a motion in the faculty of imagination, whose outcome is the production of the oneiric image.[15] In this sense, dreaming is an affection of the sensitive soul *qua* imaginative, since it is imagination that produces the oneiric images out of these small residues left in the blood.[16]

Let us now take up again Aristotle's theory of dreaming and its classification of dreams, which will be particularly relevant in the early production of Albert the Great (*De homine*). The first instance of a causal link between dreams and reality is that of 'cause'. According to Aristotle, a thing is said to be the cause of another in the same manner that the moon is related to the sun as the cause of its eclipse, or toil is linked to fever as its cause. To explain how is it possible to understand dreams as causes of events, Aristotle evokes the existence of the so-called "straight dreams" (εὐθυονειρία).[17] This particular category of dreams consists in the transposition in dreaming images of activities that (1) we have or we have not yet completed, (2) we are on the verge of completing or (3) we are still concerned with. For this reason, the images of sense perception tied to these activities somehow cause a successive related dreaming image. In this sense, "straight dreams" can also act as reminders of the commencement or conclusion of a planned but hitherto unprincipled action. According to Aristotle, the causal nature of straight dreaming is explainable in causal terms given the linear connection between sleep and waking. Since sleep begins when wakefulness ceases, likewise imaginative motions produced during sleep can prepare the ground for actions performed during the day, or the other way round.[18]

The second category that Aristotle employs to analyze the relation between dreams and reality is that of signs. For instance, having a dried tongue is a sign of fever, or the moon's entering the sun's shadow is a sign of the imminent eclipse. If dreams are to be correlated to events as signs, doctors should heed the description of dreamt images: for instance, the dream of walking in the

15 Aristotle, De anima III 3 (428b10–17); Aristoteles Latinus, De anima [A.L. XII.1], ed. Decorte–Brams (online): "Sed quoniam est, moto hoc, moveri aliquid ab hoc, fantasia autem motus quidam videtur esse et non sine sensu fieri, sed in his que sentiunt et quorum sensus est, est autem fieri motum ab actu sensus, et hunc similem necesse est esse sensui, erit utique ipse motus neque sine sensu contingens neque non in sensibilibus esse, sed multa est secundum ipsam facere et pati habens, et esse et veram aut et falsam." Kjeller-Johansen, *The Powers of Aristotle's Soul*, 210.

16 Aristotle, De divinatione per somnum 1 (462a15–16).

17 Aristotle, De divinatione per somnum 1 (463a21–25).

18 Repici, "Introduzione," 43.

fire would suggest that one's body temperature has risen.[19] Since the smallest bodily alterations can be perceived during sleep, physicians should consider the study of dreams a diagnostic tool for incipient yet hitherto unnoticed diseases.[20]

The third and last category applied to the link between oneiric images and reality is that of accidental correlation. According to Aristotle, dreaming of a future event is just as much a coincidence as the appearance of an eclipse during a stroll, since the former is neither the cause nor a sign of the latter, and vice-versa. The relation between some dreams and reality is, in other words, just a matter of pure chance, as in the case of the eclipse.[21]

As Luciana Repici has convincingly shown, dreams and dream divination, dreams appear as product of chance rather than of divine design.[22] In the best-case scenario, they can be connected to reality as their sign or cause, but are highly unlikely to mirror future events or vehiculate divine revelation. Aristotle's skeptical stance on the prognostic value of dreams is exemplified by his analysis of the physiological conditions correlated to the occurrence of such dreaming. According to him, the occurrence of veridical dreaming is directly proportional to the frequent occurrence of dreams, as in the case of melancholics. Like a player who chances to score higher when rolling dice more often than others, in the same manner melancholics end up having a higher likelihood of dreaming veridical dreams. The actual existence of veridical dreams is, according to Aristotle, a matter of pure luck and chance.[23]

Philip van der Eijk has noted how Aristotle's standpoint on the chance nature of divinatory dreams has been developed as a correction of the Democritean doctrine on dreams with which Aristotle was, to some extent, sympathetic. According to Democritus, divinatory dreams are the result of idols (*ydola* [Lat.]/ εἴδωλα [Gr.]) and emanations (*defluxiones*/ἀπόρροιαι) that originate from external objects and end up affecting the dreaming subject.[24] This

19 Aristotle, De divinatione per somnum 1 (463a15–16).

20 Aristotle, De divinatione per somnum 1 (463a5–21).

21 Aristole, De divinatione per somnum 1 (462b26–463a3); Aristoteles Latinus, De divinatione per somnum (ed. Drossaart Lulofs 1947, 34.5–10): "[…] necesse igitur sompnia vel causas esse vel signa eorum que fiunt vel accidentia, vel omnia vel aliqua horum vel unum solum. Dico autem causam quidem velud lunam defectus solis et laborem febris, signum vero defectus stellam subintrare, asperitatem vero lingue febriendi, accidens vero aliquo ambulante deficere solem: neque enim signum deficiendi hoc est neque causa, neque defectus ambulandi. Ideo accidentium nullum neque semper fit, neque ut multum."

22 Repici, "Introduzione," 70–72.

23 Aristotle, De divinatione per somnum 2 (463b12–463b22).

24 Philip van der Eijk, *Medicine and Philosophy in Classical Antiquity: Doctors and Philoso-*

motion is principled, according to Democritus, by these idols and emanations travelling in the atmosphere, largely unnoticed during the daytime. Given the absence of external stimuli during sleep, the motions triggered by these idols are then perceived by the sentient subject during sleep.[25] This explanation does ring a bell with Aristotle's explanation of dreaming images as the result of small motions in the blood which, during sleep, are subject to an altered perception, triggering dreams. In the end, Aristotle's explanation of dreaming provides a naturalistic explanation that depicts dreaming as an accidental and non-teleological process, an explanation apparently ill-suited to any Medieval monotheist.

By contrast, the process of sleep has a key and teleological role in the quest for animal survival. Defined as an incapacitation of sense perception, sleep is distinguished from loss of consciousness or fainting because of two main factors: its final cause and its efficient cause. From a teleological point of view, sleep is required to restore the heat lost during the state of waking. Since a continuous activity of perception and locomotion is impossible, sleep exists as an affection of the sensitive soul for the sake of partially restoring and bringing back to the heart the heat used up in the process of sense perception, locomotion and, more generally, during the state of wakefulness.[26] The efficient cause that differentiates, according to Aristotle, sleep from any other kind of incapacitation or privation of sensorial activities resides in the vapors produced by the stomach during digestion.

In the wake of the digestive process, exhalations from the nutritive substances rise up to the brain, which, counterbalances the heat of these vapors through its cold and moist complexion. These vapors, cooled down by the brain's cool mixture, descend down to the heart, pressing the vital heat from the upper and outside part of the body to the lower, inner part and also affecting the blood vessels. Given the strict causal correlation between the presence of heat and motion, the downward and inward motion of heat during sleep is empirically confirmed, according to Aristotle, by people moving their feet when they sleep.[27] According to Aristotle, sleep causes all sense perception (as well the awareness of sense perception) to cease. For this reason, Aristotle argues that

---

phers on Nature, Soul, Health and Disease, Cambridge 2005, 201; Aristotle, De divinatione per somnum 2 (464a1–20); Aristoteles Latinus, De divinatione per somnum (ed. Drossaart Lulofs 1947, 42.1–21).

25 Van der Eijk, *Medicine and Philosophy*, 201–202.

26 Aristotle, De somno 2 (455b13–34).

27 For Aristotle's description of the causal relation between nutrition and sleep: Aristotle, De somno 3 (456a30–458a10).

sleep is the affection of the "first perceiver" (πρῶτον αἰσθητικόν), which could be identified with common sense, who is also charged with a second-level perceptual awareness.[28]

This passage from the second chapter of the *De somno*, though, sparked serious controversies for two reasons. In the first place, it became controversial to understand sleep, based on Aristotle's remarks, as a deactivation of the first perceiver, with which the deactivation of touch is somehow correlated. Whether the deactivation of the sensitive powers as a whole, viz. of common sense, is merely concomitant with the deactivation of touch or the "first perceiver" had to be considered primarily dependent on the sense of touch, thus making sleep an affection of touch and, concomitantly, of common sense. In the *De somno*, Aristotle suggests that the sense of touch can exist in animal beings without necessitating other external senses. By contrast, the reverse is not the case. In the end, it seems that common sense can be ascribed to an animal as soon as it is endowed with the sense of touch. For this reason, Aristotle argues:

> For there exists a single sense-faculty, and the master sense-organ is single, though its being differs for the perception of each kind of thing, e.g. of sound or colour. And this accrues to an animal at the same time as the sense of touch, in particular. For this can exist separately from the other sense-organs, whereas the others are inseparable from it. These points have been discussed in our studies on the soul. Given these things, it must plainly follow that waking and sleep are an affection of this. That is why they belong to all animals. For it is only touch that belongs to all of them.[29]

Now, there may be two interpretations of this passage, as we shall see. The first and more deflationary reading points to a concomitance between the deactivation of common sense and touch (when the former is no longer active, so is the latter). The second and stronger one posits a dependence of common sense on touch. The upshot of giving preference to the first over the second interpretation would be reflected on the faculty that is primarily connected to

---

28 Aristotle, De somno 2 (455a4–b2); Repici, "Introduzione," 14–16; 56–58.

29 Aristotle, De somno 2 (455a15–25). This translation by David Gallop in: Aristotle, *On sleep and dreams*, 69. See: Aristoteles Latinus, De somno c. 2 (ed. Drossaart Lulofs 1943, 4.11–27): "Est autem et quedam communis potentia que sequitur omnes, que quod videt et audit et percipit […], hoc autem simul aptikon, id est ei que per tactum exercetur, maxime inest (hoc enim separatur ab aliis organis sentiendi, alia vero ab hoc inseparabilia. Dictum est autem de hiis in hiis que sunt *De anima* speculationibus): manifestum igitur quoniam huius est passio vigilia sompnusque. Ideoque omnibus inest animalibus: nam et tactus solus omnibus."

second-level perceptual awareness. If sleep is an affection of the common sense inasmuch as it depends on the deactivation of the sense of touch, then the faculty of common sense would seem to have perceptual awareness depending on the activity at least of the sense of touch. On the other hand, a milder reading of this passage might postulate that sleep is an affection of common sense and, concomitantly, of the sense of touch as a proximate cause. Moreover, it remains difficult to understand according to which of its functions (if any) common sense can play the role of perceptive awareness during dreaming.[30]

In the end, Aristotle's psychophysiology of dreaming presented Medieval readers with a solid naturalistic and teleological account of sleep, while downplaying if not downright rejecting the existence of God-given dreams. In this context, the role of primary importance in the physiological transmission of dreaming images is assigned to blood, without any connection to the concept of *pneuma*. The central role played by images in Aristotle's theory of dreaming, though, would largely survive in the Medieval tradition, which had to come to grips with its problematic consideration of God-given dreams. It was the philosophy of the Islamic World that was the first to face such a problem.

2 The Psychophysiology of Sleep, Dreaming, and Prophecy in Avicenna and the Arabic Philosophical Tradition

Until the discovery of the anonymous adaptation of Aristotle's *Parva Naturalia* in Arabic, little was known about the transmission of the *Parva Naturalia* in the Arabic tradition. To be sure, we knew through Averroes the works that made up this corpus: a first section on sense and sensation, followed by one on memory, then a third section on sleep and wakefulness (*bāb al-nawm wa-l-yaqaẓah*) and a last and fourth one on length and brevity of life. Given the likely context of its production (the Kindī circle), it is intuitive to imagine that the section on sleep, dreaming and divinatory dreams underwent notable doctrinal changes to be harmonized with Islamic orthodoxy.

30 On this passage, see the different interpretations of Gregoric and Kahn. Kahn takes common sense to be a sensitive power inasmuch as it is one in being and is multiplied in all the special senses through which it exits (Kahn, *Sensation and Consciousness in Aristotle's Psychology*, 16–22). By contrast, Gregoric distinguishes two notions of common sense: the first one is, *strictu sensu*, closer to Kahn's interpretation. According to Gregoric, though, there is also a broader notion of common sense as a sensorial power not only belonging to the five senses but to be connected to the imagination as well. See: Gregoric, *Aristotle on the Common Sense*, 52–61 (on the two understandings of common sense), 163–173 (on common sense in sleep).

In her article "How Aristotle came to believe in God-given dreams", Rotraud Hansberger has traced the influence of this section of the Arabic adaptation of the *Parva Naturalia* on crucial authors like al-Kindī and al-Farābī and on the disputed "Epistle on dreams" (*Risāla al-manāmiyya*), usually ascribed to Avicenna.[31] While the paternity of this latter work may be questionable, it is doubtless that Avicenna's handling displays a similar rationalistic approach, seeking to harmonize Aristotelian psychology and Islamic orthodoxy, as in the case of al-Kindī, al-Fārābī and, later on, Averroes.[32] The peculiarity of the Islamic tradition lies in its emphasis on both the cognitive (prognostic) and the socio-political function of the prophet, viz. to communicate the divine revelation and to found upon it a political community with its laws. These laws are crucial to disciplining the relationship between humans: as argued in a recent and key contribution by Olga Lizzini, "without prophecy and its fundamental socio-political function, the human species could not be permanent."[33]

While it is well-known that the Arabic philosophical tradition in general (plus Maimonides)[34] exerted a key influence on Albert's psychology of divinatory dreams and prophecy, in this section we shall focus primarily on Avicenna, given the strong medical and psychophysiological background of his approach to dreaming, very influential in Albert. From a physiological point of view, it was Avicenna's doctrine (combined with Ghazalian insights) that exerted the strongest influence on Albert given its strong emphasis on the physiological component (animal spirits) crucial to our study. Nonetheless, we shall touch on relevant comparisons between Avicenna and other authors such as al-Kindī and Averroes whenever needed. The aim will be to highlight the differences within Albert's sources from the Arabic philosophical tradition when their doctrines play a relevant role in Albert's psychophysiology.

31 Rotraud E. Hansberger, "How Aristotle Came to Believe in God-given Dreams: The Arabic Version of *De divinatione per somnum*," in *Dreaming Across Boundaries. The Interpretation of Dreams in Islamic Lands*, ed. Louise Marlow Cambridge, Mass./London 2008, 50–77.

32 Olga L. Lizzini, "Representation and Reality: On the Definition of Imaginative Prophecy in Avicenna," in *The* Parva Naturalia *in Greek, Arabic and Latin Aristotelianism. Supplementing the Science of the Soul*, ed. Borje Bydén/Filip Radovic (Studies in the History of Philosophy of Mind 17), Cham 2020, 150.

33 Lizzini, "Representation and Reality," 136.

34 Concerning the impact of Maimonides on Albert, see: Caterina Rigo, "Zur Rezeption des Moses Maimonides im Werk des Albertus Magnus," in *Albertus Magnus. Zum Gedenken nach 800 Jahren*, ed. Walter Senner/Henryk Anzulewicz (Quellen und Forschungen zur Geschichte des Dominikanerordens, NS 10), Berlin 2001, 29–66. We have chosen not to provide an in-depth study of Maimonides' position insofar as it does not present any doctrine on the connection between spirit, sleep and dreaming and prophecy.

In her contribution, Olga Lizzini has shown that the central element of Avicenna's doctrine on God-given dreams and prophecy resides in the role played by the image, which he inherits and reworks from Aristotle into a template of Aristotelian psychology consistent with Islamic orthodoxy. Moreover, a distinctive element of the Arabic philosophical tradition that reaches its climax with Avicenna is the causal correlation between the medico-philosophical, psychological explanation of prophecy and its connection to the theological and metaphysical structure of the cosmos.[35] Avicenna foresees three different categories of prophecy: the intellectual, the imaginative and the motive one. In this chapter we shall confine ourselves to the imaginative one, linking its psychological and metaphysical doctrinal requirements to its material substrate, viz. spirits and humors.

In his psychology, Avicenna seeks to blend the medico-philosophical requirements for a perfect imaginative faculty with a philosophical psychology of prophecy based on the Aristotelian gnoseology, hinged on the role of the image. Unlike Aristotle, who considered imagination a single faculty, Avicenna individuated a retentive or form-bearing (*imaginatio/muṣawwira*) faculty, responsible for storing images derived from sense perception in the front brain ventricle,[36] and an imaginative-cogitative faculty, responsible for the combination and separation of forms and meanings and seated in the front of the middle brain ventricle.[37] According to Avicenna, the images stored by retentive imagination can be of different nature, though. Carla Di Martino has shown that they can be complex and result from the unifying activity action of the common sense on sensible forms (e.g. we store yellow and sweet when we perceive honey).[38] Especially in humans, retentive imagination can also store forms produced by the action of the imaginative-cogitative faculty (*imaginativa/mutaḫayyila*). On the other, hand, objects of retentive imagination can also encompass images of a simple nature, related to a single sensation.[39] On its

35 Lizzini, "Representation and Reality," 135–138. On Kindī, Farābī and Avicenna, see also: David Bennett, "Avicenna's Dreaming in Context," in *Forms of Representation in the Aristotelian Tradition. Volume Two: Dreaming*, ed. Christina Thomsen Thörnqvist/Juhana Toivanen (Philosophia Antiqua 162), Leiden-Boston 2022, 97–104 (on al-Kindī, al-Farabī and the *Risāla Manāmiyyah*), 91–97, 104–108 (on Avicenna).

36 Avicenna, Canon I fen 1 doctr. 6 c. 4, ed. Venezia 1507, fol. 24vb [ed. New Delhi 1981–1996, vol. 1, 128.20–21]. For a more detailed description of the brain ventricles, Avicenna, Canon III fen 1 c. 1, ed. Venezia 1507, fol. 165va–166rb [ed. New Delhi 1981–1996, vol. 1, 4.3–6.10].

37 Avicenna, Liber de anima I c. 5, ed. Van Riet 1972, 89.44–48 [ed. Rahman 1959, 45.2–6].

38 Di Martino, Ratio particularis, 31.

39 Di Martino, Ratio Particularis, 28.

part, the imaginative-cogitative faculty enjoys a particular state within Avicennian faculty theory: it is the only active faculty, but on the other hand it seems to be also the only one without a clear teleological purpose.

Yet, following in the footsteps of illustrious predecessors such as al-Kindī and al-Farābī, Avicenna maintains the crucial role of this faculty in his rationalistic explanation of divinatory dreams and prophecy. The role of imagination, although with due differences, was later followed by Averroes too.[40] Unlike other authors of the Arabic philosophical tradition, Avicenna's rationalistic approach to the different kinds of prophecy (imaginative, intellectual and motive) is grounded in an ontological proof provided in his *Metaphysics*.[41] According to Avicenna, imaginative-cogitative power (viz. cogitative when it is commanded by the rational soul) plays a preparatory role in the process of intellection, which can command its activity and lead it to refuse to perceive deceitful images. The productive activity of the imaginative-cogitative power can also be reined in by the subject's focus on its activity of sense perception. Its focus on images and sensible species (also, meanings) existing in the extramental world leads him to pay attention to those incidentally produced by the imaginative faculty.[42]

40 As we shall see, these differences concern the modality of communication between the Universal (in Averroes) or Active (in Avicenna) Intellect and the dreamer. On Averroes's position, we shall rely mainly on Rotraud E. Hansberger, "Averroes on Divinatory Dreaming," in *Forms of Representation in the Aristotelian Tradition. Volume Two: Dreaming*, ed. Christina Thomsen Thörnqvist/Juhana Tovainen (Philosophia Antiqua 162) Leiden 2022, 110–149. Moreover, it must not be forgotten that Averroes did not distinguish between a retentive imagination and an imaginative-cogitative faculty like Avicenna did. On Averroes, see also: Richard Taylor, "Averroes and the Philosophical Account of Prophecy," *Studia Graeco-Arabica* 8 (2018), 287–304.

41 Amos Bertolacci, "The Metaphysical Proof of Prophecy in Avicenna," in *Prophecy and Prophets in the Middle Ages*, ed. Alessandro Palazzo/Anna Rodolfi (Micrologus Library 103), Firenze 2020, 39–75. In this section, we will treat mainly the "imaginative" prophecy (*De anima*, IV, 2) and, more briefly, the intellectual prophecy (V, 6) by comparing it to the imaginative prophecy. A further and last kind of "motive" (IV, 4) prophecy is presented in Avicenna's *De anima*: on this classification within Avicenna's prophecy theory, see: Lizzini, "Representation and Reality," 134 n. 3; Hasse, *Avicenna's* De anima, 154–155.

42 Avicenna, Liber de anima IV c. 2, ed. Van Riet 1972, 16.7–15 [ed. Rahman 1959, 171.17–172.3]: "Deinde virtus imaginativa est virtus quam aliquando retrahit anima a sua actione propria duobus modis: uno, cum occupatur anima sensibus exterioribus et convertit virtutem formalem ad operandum in sensibus exterioribus et movet eam per id quod reddit ei de illis, ita quod non permittit imaginativam cogitare, sed ut imaginativa impediatur a sua propria actione, et formalis etiam impediatur ne possit coniungi imaginativae, et id quo utraque eget de sensu communi est pertinax ad impediendum sensus exteriores; et hic est unus modus."

This second case, viz. the rational soul's refusal to allow false images, represents a key to understanding Avicenna's rationalistic explanation of God-given dreams and prophecy in general. If the judging activity of the rational soul is absent when dreaming images are perceived, it is possible that we can perceive false images without judging them as such. This is the case, for instance of a flying man or of an emerald mountain, as Avicenna argues in the *Canon*.[43] The deceitful appearance of similar non-truthful images can be due to a plurality of factors such as drunkenness, mental disorders or peculiar celestial configurations.[44] In both cases, it is clear that the absence of the intellect leads the sensitive soul to judge them as true images without their being necessarily so. Because of this, the presence of the rational soul is necessary to grasp the true meaning of a dream image or an image that appears to us during the state of waking. In *De anima* IV, 2 Avicenna provides an account of veridical, God-given dreams and prophecy based on the key role of the faculty of imagination, recently studied by Olga Lizzini. Her contribution, together with Jacquart's study on Avicenna's concept of spirit, will be the starting point for our study on animal spirits in Avicenna's psychophysiology of sleep, dreams and prophecy.

According to Avicenna, as with many philosophers (al-Kindī, al-Farābī, Averroes), prophethood requires an extraordinarily balanced complexion of the imaginative faculty as a preliminary requirement. The immediate upshot of this idea is the direct involvement of animal spirits *qua* vehicle of forms and meanings. To be of the most perfect quality among humans, the imaginative faculty of the prophet needs to be adequately sustained by a most perfect complexion of animal spirits, the thinnest among bodily substances and therefore the most apt to be receptive to the soul's highest and most complex activities. In prophets, the imaginative faculty is so strong and powerful, as we shall see, that it allows not only the grasping of perfect representations of particular events, but also allows the prophet to overcome sensorial activities and have

---

43 Avicenna, Canon I fen 1 doctr. 6 c. 5, ed. Venezia 1507, fol. 25ra [ed. New Delhi 1981–1996, vol. 1, 128.28–29]: "et [sc. the imaginative faculty] singulas percipit formas quemadmodum ad eam perveniunt a sensu; et alias formas ab eis diversas sicut hominem qui in monte volat smaragdino (Ar.: "a flying man, and an emerald mountain" [*ka-insan yaṭīru, wa-ǧabal min zumurrud*])." The Arabic refers to two distinct objects, while the Latin blends them together into a single image.

44 Avicenna, Liber de anima IV c. 2, ed. Van Riet 1968, 13.70–75 [ed. Rahman 1959, 170.5–11]: "Cum autem contigerit ex aliqua causa, aut ex imaginatione aut ex cogitatione, aut ex aliqua figurarum caelestium, ut appareat in formali aliqua forma, et intellectus fuerit absens aut cessans ab inspiciendo, possibile est tunc describi in ipso sensu communi quem ostendimus, ut audiat et videat colores et sonos qui non habent esse extra nec aliquid de illis est extra." Lizzini, "Representation and Reality," 141.

visions during the state of waking in the same manner that people perceive dream images during sleep. In his psychology (*De anima* IV, 2), Avicenna is not very explicit about the content of these visions, arguing that they can be both visual and audible.[45] By contrast, as recently highlighted by Amos Bertolacci, more details are provided in Avicenna's metaphysical account of prophecy. In *Metaphysics*, X, 1, Avicenna specifies that the content of veridical dreams can be either be God's voice, the sight and voice of an angel, the anticipation of future events or the disclosure of hidden meanings of present and past events.[46]

In a recent study, Olga Lizzini has linked Avicenna's prophecy theory to his theory of "preparedness of matter" (*praeparatio*/*istiʿdād*): in the same manner as a substantial form can be received from the "Giver of Forms" when matter achieves a degree of suitable organization, the imaginative faculty can receive the forms flowing from the celestial world when its material substrate (complexion) is suitably prepared. Preparedness of matter, though, is only one of the two conditions that need to be fulfilled in order for imaginative prophecy to occur. The second is the conjunction between the soul of the prophet and the soul of the celestial intelligences or "operative angels",[47] which presupposes specific physiological and psychological requirements.

In physiological terms, the conjunction requires perfection in the physiological complexion of the prophet. In his study on the role of imagination in the Avicennian account of prophecy, Dimitri Gutas has convincingly shown that the communication of future events can occur through the mediating action of angels, identified with the celestial intelligences (often called "operative angels"). This communication of future events can be justified on the psychological ground of what Avicenna considers a "congenic similarity" (*muǧānasa*) between human souls and angelic souls. This congenic similarity is rooted in their shared possession of the practical intellect, which represents the side of

45 Avicenna, Liber de anima IV c. 2, ed. Van Riet 1968, 19.59–61 [ed. Rahman 1959, 173.18–20]: "[…] multotiens apparet similitudo et videtur eis quod id quod apprehendunt sit locutio illius imaginis veluti verba audita quae tenent et legunt."

46 Avicenna, Philosophia prima X c. 1, ed. Van Riet 1980, 523.19–30 [ed. Marmura 2005, 435.15–436.1]. A more detailed comparison between Avicenna's views on prophecy in psychology and metaphysics in: Bertolacci, "The Metaphysical Proof," 57–61.

47 Lizzini, "Representation and Reality," 142: "The ultimate cause of visions—both in the case of an ordinary human being and of a prophet—is the reception of flux and the conjunction with the celestial world, a conjunction which is not sought as such and which can involve many elements—intelligibles (*maʿqūlāt*), warnings (*inḏārāt*), poetical inspirations (*šiʿr*), etc.—and which is always also explained in terms of the material preparation or disposition (*al-istiʿdād*) of the receiving substance, or its attitude (*al-ʿāda*) and constitution (*al-ḫulq*)."

their soul which is oriented towards matter.[48] The possibility of the conjunction between the prophet and the angel during sleep is phrased by Avicenna in negative terms: if there is no obstacle to the contact of the human and the angelic soul during sleep, they may come into contact.[49] In the Avicennian psychological account of prophecy, the congenic similarity between human and angelic soul is further aided by the theoretical intellect, which acts as a mediator between the two.[50] The final result is that the prophet has access to knowledge of future events, which are shared by the celestial souls.[51] On the other hand, Avicenna considers visions occurring during the waking state to be due to an extraordinary strength of the soul. This strength makes the prophet's soul able to incline towards the intellectual world without losing contact with the physical one.[52]

In his recent contribution on Ghazalian prophecy, Marco Signori has clearly shown how this key thinker and theologian of the Islamic World was strongly indebted to the Avicennian conception of philosophy. Avicennian influence emerges clearly and unmistakably in al-Ġazālī's "Intentions of the Philosophers" (*Maqāṣid al-falāsifa*). Here al-Ġazālī welds an Avicennian theory of prophecy to classical themes of Islamic theology such as the profession of faith and some passages from the Qurʿān.[53] In particular, the "Intentions of the Philosophers" adheres to the main points of Avicennian doctrine about the

48 Lizzini, "Representation and Reality," 138; Dimitri Gutas, "Imagination and Transcendental Knowledge in Avicenna," in *Arabic Theology, Arabic Philosophy: From the many to the One: Essays in Celebration of Richard M. Frank*, ed. James E. Montgomery, Leuven 2006, 337–354.

49 Lizzini, "Representation and Reality," 138.

50 Lizzini, "Representation and Reality," 146.

51 Lizzini, "Representation and Reality," 137–138: "By virtue of imaginative prophecy, the prophet has access to the knowledge and therefore the recognition and the representation of particular events—of the past, the present and the future—that belong to the celestial souls (the terms *al-ġayb* and *al-muġayyabāt*—'invisible' or 'unseen' things—seem in fact to indicate hidden events in particular). Such access is possible because of a sort of conjunction between the prophet's soul and the celestial spheres, with which the prophet (or more generally the human being) shares a sort of common structure […]."

52 Avicenna, Liber de anima IV c. 2, ed. Van Riet 1968, 18.46–50 [ed. Rahman 1959, 173.9–12]: "Contingit autem aliquibus hominibus quod haec virtus imaginativa sit creata in illis fortissima et praevalens, ita ut non dominentur ei sensus nec formalis resistat ei, et quod anima eorum sit fortissima, ita quod propter hoc quod contemplatur intellectum et id quod est supra intellectum, non destruatur eius condescensio ad sensus." Lizzini, "Representation and Reality," 141; Hasse, *Avicenna's* De Anima, 159.

53 Marco Signori, "Prophecy and the Authority of the Prophet in al-Ġazālī's *Maqāṣid al-falāsifa*," in *Prophecy and Prophets in the Middle Ages*, ed. Alessandro Palazzo/Anna Rodolfi (Micrologus Library 103) Florence 2020, 90–95.

conjunction between angelic souls of the prophet and of angelic intelligences as the cause of prophetic revelation.[54] As highlighted by the recent study of Hansberger, Averroes' doctrine is a bit different inasmuch as it is based on the Arabic adaptation of the *Parva Naturalia*. This adaptation, likely produced by the Kindī circle, stresses the role of the "Universal Intellect" (viz. the Avicennian Active Intellect) in the production of the forms fundamental for divinatory dreaming.[55] This text usually associates dreaming with a nobler perception of reality than that of actual sense perception, thus paving the way to some of its assertions on the truthfulness of God-given dreams through the mediation of the Universal intellect. It is not a coincidence that the author of this re-adaptation reverses the Aristotelian definition that sleep is a privation of waking (also, of sense perception) into the more radical definition that being awake is the privation of sleep.[56] This claim is reversed by Averroes, who goes back to the genuine Aristotelian definition.[57]

In her paper, Hansberger demonstrates how Averroes frames his understanding of God-given and veridical dreams into a framework characterized by the Neoplatonic emanative scheme of forms from the Universal Intellect to the sublunar realm. In this structure of the reality, the flow of forms from the Universal Intellect affects the imagination of those who sleep.[58] The reception of these forms occurs, as clearly highlighted by Hansberger, in a manner that is congruous with the nature of the recipient. The Universal Intellect does not convey the singular form or meaning related to the future event but a universal one, since cosmic intelligences are only capable of thinking in universals. Rather, the flowing of the form from the Universal Intellect triggers a reaction in the imaginative faculty of the soul, which receives this form as a particular in a way analogous to the reception of forms in matter.[59] Averroes, though, restricts the number of people which can foretell future events. In the first place, foreknowledge of future events requires not only a good reception of the universal form—hence, a physiologically perfect imaginative faculty and a perfect humoral balance[60]—but an additional epistemic factor: preparatory knowledge. As Hansberger stresses, Averroes emphasizes the role of prepara-

54 Signori, "Prophecy and the Authority," 92, 96–98.
55 Hansberger, "Averroes on Divinatory Dreaming," 113–114.
56 Hansberger, "Averroes on Divinatory Dreaming," 122.
57 Hansberger, "Averroes on Divinatory Dreaming," 124–125.
58 From this point of view, Averroes detaches himself from the Arabic adaptation of the *Parva Naturalia*, which holds that all internal faculties (imagination, cogitation and memory) are affected at the same time: Hansberger, "Averroes on Divinatory Dreaming," 129.
59 Hansberger, "Averroes on Divinatory Dreaming," 136.
60 Hansberger, "Averroes on Divinatory Dreaming," 139.

tory knowledge in the foretelling of future events to liken the process of knowledge acquisition from dreams to that rooted in empirical experience.[61] Given a certain set of premises applied to a particular case, the prophet can reach the conclusion about a future event, just like a physician can foretell future diseases based on a universal principle applied to a singular case.[62] Moreover, Averroes restricts this capacity for foretelling future events to premises that concern events related to our interests, lands we know etc. The range of our predictions is therefore limited to those we have experienced in the past and thus cannot concern the whole of reality, as the Arabic adaption of the *Parva Naturalia* seems, by contrast, to suggest.[63] In Averroes' view, the entities that seem to be primarily involved in the communication of the universal forms are angels, which may identified in his production with cosmic intelligences.[64] In spite of this little commonality, Averroes' model of communication of divine revelation notably differs from Avicenna's and Ġazālī's conjunction between the soul of the prophet and that of angelic intelligences. As shown by Hansberger, this difference may well be influenced by the *Vorlage* of Averroes' commentary, viz. the Arabic adaption of the *Parva Naturalia*.

In all cases of God-given dreams and visions, for philosophers of the Islamic World the crucial role of the complexion of the imaginative faculty seems to involve (though tacitly) the role of animal spirits. The purer, the clearer and the more perfect in complexion (as the rest of the imaginative faculty and its organs), the easier the conjunction of the prophet's soul with those of angelic ones. In the case of dreaming images, Avicenna proceeds also to explain which are the most truthful ones. Avicenna's answer will be long-lasting: morning dreams are the most truthful dreams.[65] Avicenna proceeds to give a physiological reason for why this is so. Sleep, as in Aristotle, is primarily caused by the activity of digestion and, in physiological terms, presupposes the backflow of spirits towards the inner organs. The backflow of spirit is meant to strengthen and restore the heat that has been dispersed during wakefulness and through the activities of sense perception and locomotion. This vital heat

61 Hansberger, "Averroes on Divinatory Dreaming," 137: "As Averroes apparently envisages the process, the Agent Intellect provides a universal intelligible that, in combination with knowledge of certain sensible premises, brings about perfect knowledge of the causal chain leading up to a certain event—manifesting itself in the dream representing the outcome."

62 Hansberger, "Averroes on Divinatory Dreaming," 137.

63 Hansberger, "Averroes on Divinatory Dreaming," 137–139.

64 Hansberger, "Averroes on Divinatory Dreaming," 145.

65 Avicenna, Liber de anima IV c. 2, ed. Van Riet 1968, 32.39–40 [ed. Rahman 1959, 180.13–14]: "Praeter hoc etiam non sunt vera somnia plerumque nisi quae videntur in mane."

is now required for digestive activity, and digestion is primarily perfected in the heart, whence spirits flow (during wake) and whither they flow back (during sleep). For this reason, sleep is characterized at a physiological level as a retreat of spirits into the inner parts of the bodies,[66] an idea well explicated in al-Ġazālī's definition of sleep as "withdrawal (*retractio*/*inḥiǧās*) of the spirit towards the internal parts [of the body]".[67]

If we resort to Avicenna's account of sleep, we can observe how the digestive process causes the individual to fall asleep because of the vapors arising from the stomach. Once the digestive activity nears its end, and vapors begin to rarefy, the images produced by the activity of the imaginative faculty and perceived by common sense are less subject to misperception. The fact that excesses in moisture make dreams less truthful or less perceptive chimes well with Avicenna's statement that children, of notoriously moister complexion than adults, cannot dream properly until they are 4 years old.[68] In Avicenna, as in most of the medical tradition, an excessively moist complexion of spirits prevents the subject from seeing its object clearly, as if it were placed on rippled water. This is the case, for instance, of sensorial impairments generated by the excessive moist exhalations found in baths or, in the case of sleep, of the cold and moist digestive vapors that cause animals to fall asleep. These ideas can also be found in Averroes' *De somno*, from his explanatory paraphrase of his *Parva Naturalia*, where melancholics are considered the class of humans most prone to veridical dreaming because of their complexion. By contrast, too much moisture blocks the flow of animal spirits and impedes the perception of dreaming images.[69]

---

66 Avicenna, Liber de anima IV c. 2, ed. Van Riet 1968, 34.74–80 [ed. Rahman 1959, 181.18–182.2]: "Qui autem est ex curis, fit propter hoc quod cibaria et humores cum coniunguntur intus, indigent ut adveniat eis spiritus cum toto calore naturali ad hoc ut possit calor efficere digestionem perfectam et relinquat exteriora. Sed qui est ex inoboedientia instrumentorum, fit cum nervi imbibuntur et oppilantur ex vaporibus et cibis qui diffunduntur in eis ita ut digerantur, et spiritus refugit movere propter multam humiditatem."

67 Algazel, Metaphysica pars 2 tr. 5 c. 2, ed. Muckle 1933/1967, 188.17–19 [ed. Cairo 1936, 66.5–6]; Avicenna, Liber de anima IV c. 2, ed. Van Riet 1968, 33.66–69 [ed. Rahman 1959, 181.12–14].

68 Avicenna, Abbreviatio de animalibus IV c. 3, ed. Venezia 1508, fol. 33[va] [IV, 2, ed. Cairo 1970, 64.6–7]: "[…] et pueri non somniant recte ante 4 annum; et multi homines non somniant antequam sint magni et quidam numquam." Michael Scot's translation splits the Arabic chapter of Hayawān, IV, 2 into two distinct parts: IV, 2 (Hayawān, IV c. 2, 61.1–64.12), and IV, 3 (Hayawān, IV c. 2, 64.13–66.7).

69 Averroes, De somno et vigilia, ed. Shields/Blumberg 1949, 118.48–119.54 [ed. Blumberg 1972, 86.9–12]: "Et quidam homines sunt verioris sompnii quam quidam propter diversitatem eorum in hac virtute ymaginativa; et isti sunt habentes complexiones melancholicas frigi-

While excesses in moisture, be they the product of endogenous (as in children) or exogenous (as in sense impairments or climatic conditions) causes, are an obstacle to correct perception, a moist complexion makes people better suited to perceive dream images. This fact alone, though, does not necessarily render people who dream often able to retain oneiric images, as a moist complexion is not associated with a good memory. By contrast, good memory occurs more frequently in people with a drier complexion.[70] For this reason, a perfectly balanced complexion in terms of both moisture and dryness is needed in the case of extraordinary individuals such as prophets, who both dream and retain the dream image, which they comprehend or interpret and then communicate. According to Avicenna, there are certain dreams that retain a medical meaning, since they indicate an alteration of bodily state, as in Aristotle. For instance, dreaming of fire indicates elevated body temperature, while the opposite is true if one dreams of diving into cold water.[71] Avicenna, though, does not go into greater detail to analyze the prognostic value of dreams from a medical point of view, viz. systematically connecting medical doctrines to the explanation of God-given dreams and prophecy, and neither do Averroes, al-Kindī or al-Ġazālī.

According to Avicenna, awareness of the perception of dream images is granted by the role of common sense.[72] In authors like al-Kindī, al-Farābī, Avicenna and Averroes, the deactivation of the faculty of common sense during sleep does not concern only its relationship with the external senses. In Farābī, Avicenna and Averroes, the faculty of common sense is the faculty charged with the perception of dream images as if they were objects of actual sense percep-

das et siccas: humiditas enim cooperit virtutes et oppilat vias spirituum et facit sompnum similem morti, ita quod fere nichil ymaginatur dormiens."

70 Avicenna, Liber de anima IV c. 2, ed. Van Riet 1968, 32.47–51 [ed. Rahman 1959, 180.19–181.1].

71 Avicenna, Liber de anima IV c. 2, ed. Van Riet 1968, 30.6–16 [ed. Rahman 1959, 179.9–17]: "Naturalia autem sunt quae veniunt ex temperantia virtutum humorum cum spiritu gerente virtutem formalem et imaginativam: primum etenim non repraesentat nec occupatur nisi per hoc, et aliquando etiam repraesentat id quod est in corpore et quod est illi accidens, sicut cum virtus expulsiva spermatis movetur ad expellendum, virtus vero imaginativa repraesentat tunc formam quam anima solet diligere et commisceri cum ea; habenti vero famem, repraesentat fercula, et cui fuerit necesse expellere superfluitatem, repraesentat locum ipsius, et cui acciderit quod aliquod membrum eius calescat aut infrigidetur calore aut frigiditate, repraesentat ei quod illud membrum mittitur in ignem aut in aquam frigidam."

72 Avicenna, Liber de anima IV c. 2, ed. Van Riet 1968, 18.34–45 [ed. Rahman 1959, 172.19–173.8].

tion.[73] In general, these authors share the assumption that the essence of sleep can be better captured—in positive terms—by the redirection and strengthening of internal senses rather than in negative terms of privation or deactivation of sense perception. This is the case with al-Kindī's claim that sleep reinvigorates the faculty of imagination, or with Averroes claim that sleep allows the better exercise of cogitation, a claim corroborated by the case of intense thinking, which leads to insensibility to external stimulation. Moreover, Averroes characterizes dreaming as a sort of reverse perception from the point of view of motion: while sense perception starts from the senses and ends in memory, here the process would seem similar. Yet, adds Averroes, it must not be forgotten that the two processes are not exactly symmetrical, for dreaming ensues from the motion of imagination (and not memory) towards common sense.[74]

Avicenna and al-Ġāzālī hold that sleeping does not completely deactivate common sense, but only insofar as it represents the root of all sense perception. According to Avicenna, the perceptive capacity of the soul remains active, if not strengthened, during sleep. This emphasis on the stronger and clear perception of images produced by the imaginative faculty once the soul is not distracted anymore from sense perception—which, tacitly, seems to be regarded as inferior—is close to Averroes' own ideas and, more generally, to the view of the internal senses introduced by the Arabic adaptation of the *Parva Naturalia*.[75]

The emphasis on the physiological component of dreaming, to the extent that it can influence the truthfulness of the image perceived during sleep, is typical of Avicenna's take on the mechanisms of differentiation among cognitive abilities presented in the *Canon*. The fact that morning dreams are the most truthful ones will become a commonplace in the Middle Ages. His emphasis on the causal correlation between humoral complexion, perfection of the imaginative faculty and quality of the prophetical, God-given images can be retrieved—if only to a minor extent—in Averroes' *Parva Naturalia* and, more generally, in most of the Arabic philosophical tradition and in other authors

73 Hansberger, "Averroes on Divinatory Dreaming," 129: "Curiously, all three thinkers (viz. Farābī, Avicenna and Averroes) share the adaptation's assumption that the common sense also continues to be active, in its function as 'screen' on which the dream images are played out—a striking development considering that Aristotle argues, in *De somno et vigilia*, that sleep must affect the primary perceptive part."

74 Averroes, De Somno et vigilia, ed. Blumberg/Shields 1949, 91.57–92.65 [ed. Blumberg 1972, 63.10–64.6]. Al-Kindī, Liber de somno et visione, ed. Nagy 1897, 14.5–16.20 [ed. Abū Ridā 1950–1953/1999, 295.6–296.12].

75 Avicenna, Liber de anima IV c. 2, ed. Van Riet 1968, 17.26–18.45 [ed. Rahman 1959, 172.11–173.8].

such as Maimonides.[76] The particularity of Avicenna's approach, in the case of prophecy, lies both on the very strong emphasis on the medico-physiological component of prophecy and on its methodology. In psychological and metaphysical terms, there are clear doctrinal differences between the Avicennian-Ghazalian doctrine on the conjunction between human and angelic souls and Averroes' doctrine of divination and prophecy as the reception of a divine flux of forms turned into particular images. The uniqueness of Avicenna's doctrine rests on his justification of the phenomenon of prophecy on metaphysical and ethical grounds: without prophecy and the prophet Muhammad, the existence of humankind in itself would be endangered.

More generally, the Arabic philosophical tradition confronted Albert with a strong emphasis on the material factors (complexion, spirits) in the determination of oneiric images. The doctrines of Avicenna, Averroes (to a lesser extent, al-Ġazālī) succeeded in bring into agreement Muslim faith, philosophical psychology and Galenic medicine. These rationalizations of divine revelation through dreams and prophecies and the emphasis on their medical and biological component represented perhaps one of the most important products of the philosophical psychology of the Islamic World, one that would strongly influence Medieval Latin theology and philosophical psychology.

3 God-Given Dreams, Dreaming, and Sleep in Albert the Great

Pretty much in keeping with previous authors of the 13th century, Albert's approach to the theme of God-given dreams and prophecy in the *De homine* has to find a compromise between the teachings of Christian faith and philosophical doctrines. As we pointed out in the introduction, Albert's psychophysiology of God-given dreams, prophetic visions and sleep is situated at the intersection between three main issues: the prognostic value of dreaming images, the rational explanation of prophecy and the psychophysiology of sleep as an affection

76 The perfection of the complexional component of the animal spirit is also crucial in Avicenna's explanation of how prophets use imagination to communicate universal truths grasped by their intellects through rhetorical speeches, poetry and Holy Scriptures. Further, in intellectual prophecy (*De anima* V, 6), philosophers endowed with a particular disposition of intellect (*intellectus sanctus*/*ʿaql qudsī*) that enables them to attain a great number of intelligibles in a short period of time by conjoining the Active Intellect also need a strong imaginative faculty to translate into symbols the intelligible truths they are able to comprehend: Lizzini, "Representation and Reality," 135–137 (on intellectual prophecy), 143–148 (on the prophetic use of symbols); Hasse, *Avicenna's* De Anima, 163–165 (on intellectual prophecy).

of common sense/touch and dreaming as an affection of imagination (also, common sense). Unlike in his mature works, such as the theological *Quaestio de prophetia* and the philosophical *De somno et vigilia*, where Albert operated a sharp distinction between the theological and the philosophical discourses on prophecy (*philosophi de prophetia aliter loquuntur quam sancti*), the *De homine* opts for a different solution. At the beginning of his theological career, Albert sought to harmonize Aristotelian psychology and Cristian doctrines, relying on a peculiar explanation of Aristotelian theory of dreams. His main element of innovation, as pointed out in the introduction following Rodolfi's lead, is to link dreams and prophecy to the Aristotelian concept of image.

3.1 *Harmonizing Aristotle and Christian Orthodoxy: God-Given Dreams and Prophecy in the* De homine *and* De quattuor coaequaevis

In the 13th century, the introduction of Aristotelian theories of sleeping and dreaming was not unproblematic. Among the most famous pages of philosophical literature on the topic is Roger Bacon's claim that the third book of Aristotle's *De somno et vigilia*, viz. the *De divinatione per somnum*, provided one of the reasons for the excommunication of Aristotle's natural philosophy from the universities.[77] Other colleagues of Albert were even more radical than Bacon in their negative judgment about philosophical and naturalistic explanations of dreams. In his *Commentary on the Sentences*, Bonaventure claimed that it was necessary to stick to Christian beliefs and the Holy Scriptures, regardless of any possible reconstruction of what philosophers could have possibly said about dreams.[78] In his book *Dreaming in the Middle Ages*, though, Steven Kruger has shown how the penetration of Aristotelian psychology of dreaming into 13th-century thought is well-documented in authors prior to Albert such as Bartholomeus Anglicus and illustrious colleagues of Albert like John of La Rochelle, master of theology for the Franciscan Order at Paris in the 1230s.[79] While Aristotelian doctrine certainly caused some trouble for its exegetes, among which Albert himself stood, the three treatises on sleep, dreams and dream divination were still being used at Paris in the works of some major theologians (e.g. John of La Rochelle) when Albert wrote his *De homine*.

---

77 Kruger, *Dreaming in the Middle Ages*, 87.

78 Bonaventura, In quatuor libros Sententiarum, II, d. 7 pars 2 a. 1 q. 3 ad 3, ed. PP. Collegii a S. Bonaventura, Quaracchi 1882–1889 (Opera omnia I–IV), vol. II, 195b: "Ad illud quod obiicitur de somniis, in quibus praevidentur futura; dicendum, quod quidquid de hoc senserint philosophi, quorum positiones longum esset hic retexere, quod de eis dicit sacra Scriptura indubitanter est tenendum."

79 See above: 405 n. 4.

Albert's discussion of dreams, prophetic dreams and prophecy in the *De homine* takes a cue both from philosophical (Aristotle, Averroes) and theological (Augustine, Gregory the Great) authorities, arguing that dreams are always the product of images.[80] In this manner, he begins to detach himself from some of his contemporaries (Philip the Chancellor, Hugh of St.-Cher, etc.) who strongly depended on the notion of *speculum aeternitatis* as a means to transmit God-given dreaming images.[81] In her volume on the epistemic status of prophecy in the 13th century, Anna Rodolfi has convincingly shown that Albert's intellectual production represents a turning point inasmuch as it strongly relies on Aristotelian gnoseology revisited by the authors of the Arabic philosophical tradition, and on the key role played by images. As we will show following her accurate account, images will play a crucial role in Albert's study of dreams, God-given dreams and veridical/divinatory dreams. Moreover, this approach will help us elucidate the role of spirit in the psychophysiology of dreams, visions and prophecy because of its bing the vehicle of the images found in imagination. The strategy adopted by Albert has been clearly analyzed by Kruger in his *Dreaming in the Middle Ages* and more recently by Anna Rodolfi. Both scholars have shown how Albert grounds his rationalistic explanation of God-given dreams and prophecy on a peculiar interpretation of Aristotle's categorization of dreams. In the decisive *quaestio* on the topic, Albert endorses the Aristotelian view according to which there are dreams that convey no meaning and dreams that can be related to reality (*dicendum cum Philosopho quod somnia dupliciter fiunt, quandoque habent significationem et quandoque non*) as a sign, cause of future events or merely incidentally.

Albert begins his analysis with the category of dreams as causes of the knowledge of future events or, at least, as causally related to extramental reality. On the one hand, dreams appear to be related to reality in terms of causality, viz. they are either caused by endogenous factors or, on the other hand, can be the cause or principle of some other activity. This twofold understanding of causal relationships between dreams and reality, viz. active (dreams are the principle of waking activities) or passive (dreams are effects caused by the activities we think of, we set out to do or we are on the verge of completing during the state of waking) goes back to the Aristotelian doctrine of "linear dreams". In keeping with Aristotle's approach, Albert argues that dreams can come either from the body (*ex parte corporis*) or from the soul (*ex parte ani-*

80 Anna Rodolfi, "Sogno e profezia in Alberto Magno," in Scientia, Fides, Theologia. *Studi di filosofia medievale in onore di Gianfranco Fioravanti*, ed. Stefano Perfetti, Pisa 2011, 201.

81 On these authors, see at least: Rodolfi, Cognitio obumbrata, 49–103, and the bibliography therein.

mae). In the case of dreams from a part of the body, Albert welds the Aristotelian doctrine according to which dreams of being on fire or in snow are caused by an alteration of body temperature with medical theories.[82] This is a well-established connection in 13th- (even late 12th-) century philosophical production.[83] According to physicians, dreaming images are caused by motions in the internal humors and dreaming images depend therefore on the humoral complexion of each individual. Melancholics dream horrible dreams because of the frequent internal motions to which they are subject.[84] The emphasis on the frequency of dreaming in melancholics is typically Aristotelian. In the same manner as melancholics dream according to their complexion, choleric people (hot and dry by complexion) dream that they are burning (*cholerici frequenter somniant se ardere*), while phlegmatic people (moist and cold) dream that they are swimming or submerged in water (*flegmatici natare vel submergi in aquis*).[85]

Having elucidated the manner according to which dreams are causally related to the body, viz. as effects of bodily states and humoral complexion, Albert goes on to explain how dreams are causes from the point of view of the soul (*ex parte animae*). In this case, dreams are causes as respects possible objects of knowledge (*scibilia*) or of future actions (*operabilia*). In the first case, dreams may be prefigurations of the final results of an intellectual activity (*inquisitio*) with which we are occupied while awake and which we have not yet completed (*in eis manifestantur futura scibilia*). If we consider future actions (*operabilia*), though, dreams may arise from bodily dispositions that can lead us to dream of future actions which we may be willing to undertake.[86] Again, this idea pretty much chimes with Aristotelian ideas on linear dreaming.

Albert proceeds to divide the categorization of dreams as signs into three parts: in the first case they can be signs of future events according to astrological configurations at the time they occur. Albert argues that some dreams may take on different meanings if they are perceived with the new moon (*in novilunio*) or alongside other astrological configurations. In the second case, dreams may come from alterations in the elemental mixture of the human body and be the signs of incipient meteorological phenomena: dreaming of fish, for instance, may indicate an upcoming rain. In the last case analyzed, dreams can repre-

82 Albertus Magnus, De homine (Ed. Colon. XXVII/2, 384.26–34).

83 Marta Fattori, "Sogni e temperamenti," in *I sogni nel Medioevo: Seminario Internazionale, Roma 2–4 ottobre 1983*, ed. Tullio Gregory, Roma 1983, 87–109.

84 Albertus Magnus, De homine (Ed. Colon. XXVII/2, 373.66–78).

85 Albertus Magnus, De homine (Ed. Colon. XXVII/2, 384.34–39).

86 Albertus Magnus, De homine (Ed. Colon. XXVII/2, 384.51–58).

sent a sign from the point of view of the dreamer (*ex parte somniantis*). This can take many forms. For instance, dreaming of fire might metaphorically signify future anger, and so on. These three kinds of dreams, combined with the dreams coming from the soul (*ex parte animae*), provide limited foreknowledge of future events.[87]

In his book, Kruger has shown how Albert harmonizes the Aristotelian classification of dreams with Christian orthodoxy. To do so, he forces biblical revelations into the third Aristotelian category of dreams, viz. coincidences. According to Albert, God-given dreams can be considered coincidences not because their relation to reality is incidental, but because the foreknowledge of the future events signified by them overcomes any possible (semiotic, causal) interpretive capacity of the subject and, moreover, the causal origin of these dreams cannot be traced to any natural phenomenon, as held by Kruger.[88] Among these dreams feature also angelic revelations, as in the Arabic philosophical tradition (Avicenna, al-Ġazālī, Averroes)[89] and episodes of biblical dreams.[90] In her article on Albert's doctrine on dreaming and prophecy, Rodolfi has shown how the downside of Albert's position is that it lends itself to an immediate and strong critique: is the realization of God's will in any way incidental, if biblical dreams fit into the category of dreams as coincidences? She showed clearly how, to avoid this major theological inconvenience—to put it very mildly—Albert drew on a typical theological concept, viz. the prophecy

87 Albertus Magnus, De homine (Ed. Colon. xxvii/2, 384.59–385.9). On the above listed passages, see: Kruger, *Dreaming in the Middle Ages*, 107.

88 Albertus Magnus, De homine (Ed. Colon. xxvii/2, 385.22–28): "Accidens autem sive incursus est 'de his quae procul sunt' tempore vel loco et quorum notitia 'supra nostrum videtur esse prudentiam', ut dicit Philosophus, sicut est de navali bello futuro, vel de regnis subvertendis, vel de toto statu mundi futuro, sicut in somnio Nabuchodonosor. Et notitia huiusmodi somniorum non fit sine revelatione, ut infra determinabitur." Kruger *Dreaming in the Middle Ages*, 112: "Such an assertion misrepresents Aristotle, but the pressure of Christian dream theory is here too strong to allow Aristotle an undistorted voice." Rodolfi, "Sogno e profezia," 203.

89 Albertus Magnus, De homine (Ed. Colon. xxvii/2, 385.66–70): "Ad ultimum dicendum quod omnia somnia illa quae non manifestantur in signis corporalibus vel causis, supra nostram sunt prudentiam, ut dicit Philosophus, et sunt revelationes ab intelligentiis factae humanis animabus, ut infra determinabitur."

90 Albertus Magnus, De homine (Ed. Colon. xxvii/2, 357.1–9): "Ad aliud dicendum quod revelatio somnialis quae fit per angelos, habet aliquid de esse prophetiae et aliquid de esse somnii; et secundum quod est habens esse somnii, non sistit in intellectu, sed in imaginibus praeparatis a phantasia, sicut quod Daniel vidit bestias ascendentes de mari, et Pharao spicas et boves; sed secundum quod habet esse prophetiae, sic principaliter est in intellectu, et sic differt a somnio, sicut determinari habet in *quaestione de prophetia*." This last reference is perhaps to his later *Quaestio de prophetia*.

of menace. By prophecy of menace Albert meant the kind of prophecy that is not in itself immutable (it is a conditional threat), but rather depends on the free will of the individual. A classic example is constituted by Jonah's prophecy: if the citizens of Nineveh citizens had continued to pursue their despicable behavior regardless of God's threat to destroy their city, God's punishment would have necessarily occurred within 40 days. Since the citizens of Nineveh opted for a change in their behavior, they avoided this otherwise-necessary eventuality.[91] Among the examples of God-given dreams proposed by Albert we find the dreams of the Egyptian pharaoh interpreted by Joseph as a warning about seven coming years of prosperity followed by an equal number of years of famine, the latter of which would still have occurred if the Pharaoh had acted differently.[92]

In like manner, Albert held that accidental dreams could be realized if the prior conditions, tied essentially to the free will of those involved, did not change, while in case of a voluntary change operated by the subject, the realization of prophetic dreams was itself not necessary anymore.[93] Albert closes the section on the prognostic value of dreams by mentioning the skeptical Aristotelian arguments against God-given dreams (the case of melancholics, dreams in animals). He then tries to reach a final agreement between the two perspectives by bending Aristotle's position to the necessities of Christian doctrine, arguing that Aristotle's criticism relates only to the dreams which can be explained according to the laws of nature and leaves some leeway for the occurrence of prophetic dreams, though Aristotle's position is much more skeptical than Albert makes it to be.[94] In Albert's eyes, the domain covered by dreams, prophetic dreams and prophecies in the *De homine* is shared by Aristotle and Christianity from an epistemological point of view, this resulting from adaptations and forced readings of Aristotle's positions for the sake of harmonizing them with the teachings of Christian faith. In his early theology, Albert strives to find an agreement between the *sancti* and *philosophi* on the issue of veridical dreams. Purely mental revelations are classed by Albert under the

91 Rodolfi, "Sogno e profezia," 204–205.

92 Genesis 41, 1–7.

93 Albertus Magnus, De homine (Ed. Colon. XXVII/2, 386.66–387.6): On this passage, see: Rodolfi, "Sogno e profezia," 204–205.

94 Albertus Magnus, De homine, (Ed. Colon. XXVII/2, 392.62–68): "Ad id autem quod primo obicitur per auctoritatem Aristotelis, dicendum quod Aristoteles tangit causam frequentis somnii, et probat hoc esse a natura infirmitatis et melancholiae et non a merito gratiae vel virtutis. Dicit tamen quod non semper hoc est falsum quod somnia sint a deo, et per hoc innuit quandoque somnia esse per revelationem divinam." Rodolfi, "Sogno e profezia," 206.

title *prophetia*, as opposed to dreams that have an essentially image-related content.[95] The latter, according to Albert, are not necessarily true, as Aristotle, after all, had actually maintained.[96] Albert goes on by arguing that revelations through dreams are upheld by both theological and philosophical authorities such as Aristotle, Socrates[97] and the most prominent Arab (Avicenna, al-Ġazālī, Averroes) and Jewish (Maimonides, Isaac Israeli) philosophers. In this manner, Albert can conclude that dreams, especially morning dreams, can have a prognostic value, even though this value need not always be accurate.[98] Still, it remains unclear whether the truth content of these dreams and prophecies can pertain to the acts of free will object of divine knowledge alone, or only to dreams explainable in terms of natural causation. In the early Albert, these two areas are not methodologically separated, as they become in his mature works.

As we observed in the beginning, Albert's *De homine* intersects three major issues in dreaming and prophecy: the truth value of dreams, a rationalistic account of prophecy, the definition of sleep and dreaming. Until now, we have observed how Albert integrates the account of God-given dreams into the Aristotelian canvas of dreams as signs, causes or coincides. Until now, little has been said as respects prophecy. In the *De homine*, Albert places much of his

95 Albertus Magnus, De homine (Ed. Colon. XXVII/2, 386.53–65): "Ad primum autem dicendum quod revelationes fiunt duobus modis, scilicet secundum veram intelligentiam voluntatis divinae, et secundum permixtionem phantasmatum. Et primo modo non est somnium, ut supra habitum est, sed prophetia, de qua infra disputabitur. Secundo vero modo est revelatio non significans rem futuram simpliciter, sed secundum condicionem aliquarum causarum naturalium vel voluntariarum, quae causae cum non sint necessariae, non erit necessarium futurum, quod significatum est. Hoc tamen semper intelligendum est quod somnium sit in proprietatibus phantasmatum, quia sine phantasmatibus non est aliquod somnium."

96 Albertus Magnus, De homine (Ed. Colon. XXVII/2, 386.50–52): "Solutio: Consentientes Philosopho dicimus quod somnia non eveniunt de necessitate, sive sint causae sive signa sive incursus accidentales futurorum."

97 On Socrates as a philosophical authority who endorses the existence of divinatory dreams and prophecy (and the sources on Socrates' character), see: Thomas Ricklin, "Albert le Grand, commentateur: L'exemple du *De somno et vigilia III, 1*," *Freiburger Zeitschrift für Philosophie und Theologie* 45 (1998), 31–55.

98 Albertus Magnus, De homine (Ed. Colon. XXVII/2, 382.19–31): "Solutio: Dicimus cum sanctis et philosophis quod somnia frequenter aliquid significant in futurum, et maxime, sicut dicit Avicenna, quae sunt in fine dormitionis, sicut illa 'quae videntur in mane, omnes enim cogitationes hac hora quiescunt, et motus humorum sunt finiti', et ideo cum imaginativa non sit impedita per corpus nec impediatur ab aliis actibus animae, tunc obsequium eius, quod praestat animae, melius est quam potest, et ideo in talibus imaginibus frequenter praevidentur futura. Somnium tamen licet principaliter sit futurorum, tamen est etiam praeteritorum et praesentium, quae latent cognitionem nostram, ut dicit Alfarabius (*sc.* Averroes)."

emphasis on dreams and dream images, linking it explicitly to prophecy in the *De quattuor coaequaevis*. In the *De homine*, Albert had endorsed the view that God-given prophecies come from angelic revelations, cross-referencing his handling of the topic in the *De quattuor coaequaevis*. There, he based his discussion on the modality of angelic revelation to humans on the classical Augustinian scheme of three visions: *corporalis*, *spiritualis* and *intellectualis*. To explain the modality of prophetic revelation, though, it is necessary to step back and understand angelic prophetic foreknowledge according to Albert and then to link angelic knowledge to human prophetic capacity.

In Albert's *De quattuor coaequaevis*, angels are identified with cosmic intelligences, a view he will later reject. These angels have foreknowledge of future events because they know directly the causes of future events that are embedded within nature (*indita in natura*) and which are considered, following Augustine, the *rationes seminales* of future events. These *rationes* are neither particular nor universal, for they are considered to exist in relation to the causes of future events as their paradigm (*in intelligentiis sunt rationes et paradigmata causarum illarum*) and not as forms derived from intellectual abstraction. This is the mode of causation proper to the principle of arts and crafts, which guides the production of artifacts.[99] Angels know that certain events may occur and, if they do not occur, they again know the reasons why. The example employed by Albert is particularly meaningful, viz. generation. If the generation of some substance occurs, angels know, quite literally, the reasons (*rationes*) why. If it does not, they know what kind of impediment might have caused the failure in the process.[100] Angels are endowed with the knowledge of natural events (*habent scientiam naturalem de rebus*). By contrast, the foreknowledge of future events depending on human free will can be attained by angels if and only if it is communicated to them by God.[101] Therefore, angelic prophetic foreknowledge encompasses both the prognostication of future events according to the laws of nature and those events that occur according human action. While the former can be known by angels, knowledge of the latter and its communication to humans requires a prior communication by God, who alone knows future contingents subject to free will.

99 Albertus Magnus, De quattuor coaequaevis tr. IV q. 24 a. 2, ed. Auguste Borgnet, Paris 1895 (Ed. Paris. XXXIV, 480a).

100 Albertus Magnus, De quattuor coaequaevis tr. IV q. 24 a. 2, ed. Auguste Borgnet, Paris 1895 (Ed. Paris. XXXIV, 480b).

101 Albertus Magnus, De quattuor coaequaevis tr. IV q. 24 a. 2, ed. Auguste Borgnet, Paris 1895 (Ed. Paris. XXXIV, 481a): "[…] ideo talia non cognoscunt Angeli per naturam, sed si cognoscunt, hoc erit per revelationem […]".

Now that we have elucidated the mode of prophetic foreknowledge of angelic or cosmic intelligences, let us turn to their capacities to communicate future events to humans. According to Albert, this communication goes back to the threefold Augustinian kinds of vision: corporeal, spiritual and intellectual.[102] Of the three kinds, the most perfect is the intellectual, which occurs because of the abstraction of the species by the angelic intelligence in cooperation with the human soul.[103] In opposition to the corporeal vision, which is created *ex nihilo* in the soul of the prophet, intellectual vision takes place through the reordering of previously existing images in the human soul, viz. in its retentive imagination.[104] In the case of intellectual vision, the angelic soul cooperates with the human intellect to reveal hidden intelligible realities (*dare autem illuminationem est dare revelationem de aliquo intelligibili occulto*), a formulation that echoes Gregory the Great. Given that in Augustine, the intellectual vision does not require the sustaining of any image, so Albert does not—and rightfully so—explain according to what mechanism purely intellectual content can become an image. The problem, though, concerns mostly the 'spiritual' and 'corporeal' kinds of vision.

To begin with, one is left to wonder whether the *rationes seminales* mentioned by Albert as objects of angelic knowledge can be equated with the species of intellectual vision. In the *De homine*, Albert addresses how the human imaginative faculty can produce images out of cognitive content provided by angelic revelation. In this passage, Albert refers to his handling of angelic knowledge and argues that these "forms of composition" (*formae compositionis*), as opposed to forms of abstraction, are particular and not universal. This is because the forms of composition, which can be reasonably equated with what Albert had called the exemplary or paradigmatic causes (principles of arts and craftsmanship) in the *De quattuor coaequaevis*, belong to the activ-

102 Albertus Magnus, De quattuor coaequaevis tr. IV q. 34 a. 2 ed. Auguste Borgnet, Paris 1895 (Ed. Paris. XXXIV, 524b).

103 Albertus Magnus, De quattuor coaequaevis tr. IV q. 34 a. 2, ed. Auguste Borgnet, Paris 1895 (Ed. Paris. XXXIV, 524b–525a): "Unde sicut intellectus agens species abstrahit a phantasmatibus et unit eas intellectui possibili secundum ordinem ad scientiam et opus, sic intelligentia angelica lumine suo abstrahit species a phantasmatibus, et componit cum intellectu animae secundum ordinem ad revelationem occultorum."

104 Albertus Magnus, De quattuor coaequaevis tr. IV q. 34 a. 2 ed. Auguste Borgnet, Paris 1895 (Ed. Paris. XXXIV, 525a): "Si autem quaeratur, Ubi sint phantasmata a quibus abstrahit? Dico, quod sunt in anima, et possunt fieri in anima utroque duorum modorum: potest enim Angelus facere in ea de novo per visionem sensibilem vel imaginativam. Potest etiam in ea facta naturaliter ordinare et componere et dividere secundum exigentiam revelationis secundum quam intendit illuminare. Et hos modos dicit Augustinus miros modos, quibus Angeli visa sua imprimunt in animas."

ity of the practical intellect, which is concerned with particulars.[105] According to Albert, it is the nature and transmission of these *formae compositionis* (or *rationes*, in Augustinian language) that grants the foreknowledge of future events to human prophets thanks to angelic mediation. This mediation is explicated via cases of intellectual, spiritual and corporeal vision, all derived—though differently—from the communication of these forms to human beings. This clearly evince a strong Augustinian influence, though combined with a peculiarly Aristotelian emphasis on the need for the image as a substrate of prophetic intellectual visions, as clearly detailed in Rodolfi's studies.

On a completely different note, while prophecy possesses a psychology of its own, it does not possess a physiology comparable to the detailed study of dreaming psychophysiology provided in the *De homine*. The physiological mechanisms that undergird prophecy can only be inferred from Albert's psychophysiology of dreaming, which shares the same rationale, viz. the appearance of prophetic images as a case of 'reverse perception' involving a motion of animal spirits from imagination (and its organ) to common sense (and its organ). The analogy, though, stops here, because animal spirit also flows back to the organ of imagination for the storing of dream images.

3.2 *The Physiology of Dreaming in the* De homine

In the *De homine*, Albert began to carve out his physiology of dreaming, influential in his mature works such as the *De somno et vigilia*. Following Aristotelian psychology, dreaming is both qualified as an "affection of the common sense and *phantasia*" (*passio sensus communis et phantasiae*).[106] From the perspective of a deep connoisseur of Aristotelian psychology like Albert, it was uncontroversial to relate to the capacity of the imagination of randomly producing dream images. Rather, there are two main problems, viz. how to combine Aris-

105 Albertus Magnus, De homine (Ed. Colon. XXVII/2, 390.68–391.14): "Posset etiam dici quod intelligentia infundit intellectum simplicem, et huic intellectui simplici imaginatio format imagines. Tertio modo potest dici, ut supra determinatum est in quaestione de scientia angelorum, quod scilicet sunt formae compositionis et formae abstractionis; et formae compositionis deserviunt intelligentiae activae, formae autem abstractionis deserviunt intelligentiae contemplativae. Cum autem omnis activus agat circa particularia, necesse est formas activi intellectus esse particulares, licet sint simplices. Qualiter autem forma simplex possit esse particularis, infra determinabitur in quaestione de intellectu practico. Et tales sunt formae intelligentiarum, quas imprimunt in animas et ab anima ipsa propter naturam imaginativae efficiuntur corporales."

106 Albertus Magnus, De homine (Ed. Colon. XXVII/2, 361.20–23): "Solutio: Dicendum quod somnium est passio phantasiae et sensus communis secundum relationem, quam habet ad sensus proprios; et motus somnii incipit a phantasia et terminatur ad sensum [...]"

totle's statements on the localization of dreaming images in the blood filling the sense-organs and how to reconcile the definition of sleep as a deactivation of the sensorial powers and dreaming as an affection of the sensitive part of the soul *qua* imaginative part. But let us not jump the gun; we should begin with Albert's physiology of dreaming, briefly analyzed in a contribution by Joachim Söder, in relation to the productive capacity of imagination and its link to the localization of dreaming images in the *De homine*.[107] For the rest of our contribution, we shall rely on the studies of Silvia Donati in the case of sleep, enlarging her study of the role of spirit in sleep as a deactivation of the sensitive powers to the role of spirit in dreaming and dream formation.[108]

The first step consists in understanding Albert's line of reasoning in the physiology of dreaming images: where do they come from? Following Averroes, Albert argues that dreaming is a motion that is principled in imagination and ends in common sense, for it involves perceptual awareness and perception of sensibles that are derived from the imaginative part of the soul.[109] Yet we must remember that, unlike Albert and Avicenna, Averroes did not distinguish between a retentive and an imaginative-cogitative power of the sensitive soul. To solve this apparent contradiction, it is necessary to refer to Albert's own wording of the *quaestio*, which asks whether dreaming is an affection of

107 Joachim Söder, "Albert der Große über Sinne und Träume," 239–250.

108 Donati, "Dreams and Divinatory Dreams," 178–215. Other contributions on Albert's doctrine on dreams and prophecy are: Sebastià Giralt, "*Aristoteles imperfectus*: Natural Divination, Dream and Prophecy in the Latin Middle Ages (1210–1310)," in *Die mantischen Künste und die Epistemologie prognostischer Wissenschaften*, ed. Alexander Fidora, with the collaboration of Katrin Bauer (Beihefte zum Archiv für Kulturgeschichte 74), Köln-Weimar-Wien, 2013, 23–59; Marianne Schlosser, Lucerna in caliginoso loco. *Aspekte des Prophetie-Begriffes in der scholastischen Theologie* (Veröffentlichungen des Grabmann-Institutes zur Erforschung der mittelalterlichen Theologie und Philosophie. Neue Folge 43), Paderborn 2000, 92–97 (on Albert); Bert Roest, "Divination, Visions and Prophecy according to Albert the Great," in Media Latinitas. *A Collection of Essays to Mark the Occasion of the Retirement of L.J. Engels*, ed. R.I.A. Nip/ H. van Dijk/E.M.C. van Houts/C.H.J.M. Kneepkens/G.A.A. Kortekaas (Instrumenta Patristica et Mediaevalia 28), Turnhout 1996, 323–328; Tullio Gregory, "I sogni e gli astri," in *I sogni nel Medioevo, Seminario Internazionale. Roma, 2–4 ottobre 1983*, ed. Tullio Gregory, Roma 1985, 111–148.

109 Albertus Magnus, De homine (Ed. Colon. XXVII/2, 359.4–10): "Solutio: Dicendum quod sicut dicit Alfarabius, in vigilia motus sensibilium est a sensu in imaginatione, ita quod principium est a sensu et finis in imaginatione. In somnio autem motus est sensibilium praeacceptorum ab imaginatione ad sensum, ita quod principium est ab imaginatione et finis ad sensum, et ideo somnium est imaginationis ut a quo est principium motus eius." Averroes, De somno et vigilia, ed. Shields/Blumberg 1949, 98.65–99.9. See also: Albertus Magnus, De homine (Ed. Colon. XXVII/2, 363.43–54). On the interpretation of these passages: Söder, "Albert der Große über Sinne und Träume," 242–243.

imagination/phantasy (*passio imaginationis sive phantasiae*). In this manner, it appears clear that Albert is referring to the productive function of Aristotelian imagination, and so was Averroes. In this manner, dreaming appears to be an affection of the imaginative faculty with a physiology of its own. This physiology, then, seems to detach itself from Aristotle's own remarks that dreaming images are found in sense-organs. Rather, these images come clearly from that ventricle of the brain where retentive imagination is localized. Christina Thomsen has shown how, to reconcile Aristotle with the Arabic philosophical tradition and the Galenic localization of imagination in the front brain ventricle, Albert either interpolates *De Insomniis* 2, 460b2–3 or follows an interpolated manuscript of the *translatio vetus*. The final outcome is that desired by Albert: Avicenna's brain-centered localization of images stored in imagination is put under the aegis of Aristotle.[110]

From a physiological point of view, the production of oneiric images derives from the motion of spirits in the brain from the organ of *phantasia* to the organ of common sense and back. Concretely, this means that dreams occur when animal spirits flow from the organ of *phantasia* to the organ of common sense, seated in the front brain ventricle.[111] This makes the mechanism of dreaming somehow the reversed analogous to vertigos, produced by the motion of spirits from the common sense to retentive imagination and back. Both processes consist in two opposite motions (*tractus and pulsus*) between retentive imagination to common sense which occur when we are awake (vertigo) and when we are asleep (dreaming). In the case of dreaming, the backflow of spirits is caused by the natural attraction of sensible forms to imagination where they are usually stored.[112]

110 Albertus Magnus, De homine (Ed. Colon. XXVII/2, 367.63–66): "Praeterea dicit Aristoteles quod accepta sensibilia manent in cella; cella autem est concavitas cerebri, ut dicit Avicenna; ergo non manent in organis sensuum proprium." On this passage, see: Christina Thomsen Thörnqvist, "How Dreams Are Made: Some Latin Medieval Commentators on Dream Formation in Aristotle's *De insomniis*," in *Forms of Representation in the Aristotelian Tradition. Volume Two: Dreaming*, ed. Christina Thomsen Thörnqvist/Juhana Tovainen (Philosophia Antiqua 162), Leiden-Boston 2022, 158.

111 Albertus Magnus, De homine (Ed. Colon. XXVII/2, 371.55–64): "Dicendum quod somnium est phantasma 'a motu simulacrorum', ut dicit Philosophus, qui motus descendit a loco phantastico et tangit sensum communem et revertitur ad phantasiam. Cum enim fiat evaporatio a loco digestionis ad cerebrum, subtilis sanguis elevatus descendit ad anterius capitis, cum quo progreditur spiritus animalis deferens simulacra a loco phantasiae ad organum sensus communis, quae cum movent sensum communem, fit somnium, et videtur sentiri forma delata a phantasia ad sensum communem."

112 Albertus Magnus, De homine (Ed. Colon. XXVII/2, 372.26–36). Thomsen Thörnqvist, "How Dreams Are Made," 161.

Even if Albert is silent about this, there is a remark worth making at this point. Given the different localizations of *phantasia* in humans and animals shown in the previous chapter, the areas of the brain involved in human and animal dreaming are therefore different. While in humans dreaming can be localized in the front brain ventricle alone (as in Averroes, for instance), dreaming activity in animals covers both the anterior ventricle and a portion (the anterior part) of the middle one, as in Avicenna. This difference is the result of Albert's equivocal arrangement of internally apprehensive powers in the brains of humans and animals.

Once they are perceived by common sense, these images go back to the organ of *phantasia*, meanwhile being stored in the retentive imagination.[113] The remembering of these images is primarily connected to the complexional constitution of individuals: those of a drier nature remember dreams better, those with a moister complexion receive dreaming images but tend to be forgetful. In the *De homine*, dreaming images are causally correlated to the humoral balance of fluids such as spirits. If spirits have an excessively moist complexion, then the result is the dreaming of horrible, monstrous images, as with children, because of the smoky vapors which arise from an excessive amount of moisture.[114] As we have seen above, Albert considers melancholics—like Aristotle—to be the humans that dream more often because their constitution predisposes them to have frequent and strong internal motions. These frequent motions often result in horrible dreams or nightmares.[115] Just as melancholics dream horrible and obscure dreams because of their complexion, so choleric individuals often dream about burning (*cholerici frequenter somniant se ardere*) and phlegmatic individuals dream about swimming or drowning in water (*flegmatici natare vel submergi in aquis*).[116] As we pointed out above, Albert here endorses the medical view that dreams result from the complexion and come therefore from the body and the disposition of the dreamer.

During the activity of dreaming, the illusion that the object of perception is perceived as if it were real is effected by the role of common sense, reached by the sensible forms carried by animal spirits. According to Albert, common

---

113 Albertus Magnus, De homine (Ed. Colon. XXVII/2, 372.38–42): "[…] motus ille corporis est, quod est subiectum et vehiculum illarum formarum, et hoc est spiritus animalis qui descendit ab imaginativa ad sensum communem ea sensu communi revertitur ad ipsam cum formis sensibilibus."

114 Albertus Magnus, De homine (Ed. Colon. XXVII/2, 373.52–60).

115 Albertus Magnus, De homine (Ed. Colon. XXVII/2, 373.66–78).

116 Albertus Magnus, De homine (Ed. Colon. XXVII/2, 384.34–39).

sense is not completely deactivated during sleep. Rather, it is deactivated in the part of its power that concerns its connection to the external senses. Its 'turning' to the interior sense remains always active during sleep and ensures, as in Avicenna, Averroes and al-Ġazālī, a rudimental awareness of perceptive activity taking place.[117] This means that humans—or animals that have common sense—perceive certain dream images and are aware—at least in a rudimental manner—when they are perceiving an image as if it was real. Without the assenting role of the intellect, they would not be able to judge whether these images are true or false.

In sleep and dreaming, the faculty of common sense plays a crucial role both because of its apperception (second-level perception) exercised during dreaming and its role in the deactivation of the senses which coincides with the act of falling asleep. As Silvia Donati has shown, both the *De homine* and his mature *De somno et vigilia*, Albert ascribes to the faculty of common sense a key, double-faced role: on the one hand, awareness of perceiving dream images as if objects of sensation through its conversion to the interior (*conversio ad interius*). On the other, common sense in relation to outer sense perception appears to be primarily connected to sleep as a deactivation of the sensitive powers. The main question that Albert needs to answer, especially in the light of Aristotle's text, is: how, precisely? Moreover, if the deactivation of common sense seems to be concomitant with that of touch, what is the relationship between the two faculties in the definition of sleep as an affection of common sense.

3.3 *Spirits, Sleep, and Common Sense: An Aristotelian Conundrum*

In the case of sleep, Albert follows Aristotle closely by endorsing the view according to which sleeping is materially produced by the vapors produced during digestion.[118] In keeping with Avicenna, he confirms that morning

117 Albertus Magnus, De homine (Ed. Colon. XXVII/2, 361.56–66): "Sed hoc non placet, sed potius secundum Alfarabium et Avicennam et Algazelem dicendum est quod motus somnii sistit ad organum sensus communis, et ille in somnis immobilitatur per accidens secundum conversionem ad actum exteriorem; non enim immobilitatur per se, sed non influit virtutem sensus et spiritum sensibilem organis sensuum propriorum propter hoc quod ipsa organa propriorum sensuum immobilitata sunt ex frigiditate somni. Unde adhuc sensus communis per se manet in actu secundum conversionem ad interius."

118 Albertus Magnus, De homine (Ed. Colon. XXVII/2, 345.26–31): "Efficiens autem est evaporatio nutrimenti in loco cerebri infrigidata; materialis autem, ut in qua est somnus, sunt membra exteriora sensibilia et motiva; formalis vero est defectus spiritus sensibilis et motivi et caloris in exterioribus; finalis autem est quies organorum et recuperatio virtutis sensibilis et motivae." Söder, "Albert der Große über Sinne und Träume," 243.

dreams are the most truthful ones since they are almost not affected at all by the smoky vapors arising from digestion. From a teleological point of view, sleep is conceived as a process meant for recovering the heat lost during the activities (sense perception, motion) exercised during the waking state. This restoration of lost heat occurs through a backflow of spirit (*retractio spiritus*) from the peripheral regions of the body to the key organs, especially the heart.[119] Albert borrows the idea of sleep as a backflow of spirits from Avicenna, Averroes and al-Ġazālī and recasts it into his psychophysiology of sleeping, strongly influenced by Aristotle. He maintains that this backflow of spirit explains the motion of peripheral, inferior limbs such as feet during sleep,[120] and that this backflow correlates to the deactivation of the five senses and common sense *qua* root of all perceptive powers. Now Albert has to solve a question deriving from the Aristotelian text, a question he would not find in Avicenna: namely, what is the correlation between common sense and touch in the process of falling asleep. In the *De somno*, Aristotle had maintained that the deactivation of all sense during sleep was somehow correlated to the deactivation of common sense.

Silvia Donati has shown that Albert takes a cue from a passage in Aristotle's *De somno* (2, 455a15–25) where the sense of touch is said to exist in animals without necessarily entailing the existence of other external senses, while the reverse is not the case. It would seem that, according to Aristotle, common sense can be ascribed to an animal as soon as it is endowed with the sense of touch.[121] These passages present a major hermeneutical dilemma, i.e. whether Aristotle is here pointing to a concomitance between the deactivation of the sense of touch and the deactivation of common sense when sleep ensues. Put more strongly, Aristotle is postulating a sort of dependence of common sense

119 Albertus Magnus, De homine (Ed. Colon. XXVII/2, 319.1–2): "Algazel autem sic: 'Somnus nihil est nisi retractio spiritus ab exterioribus ad interiora'." Albertus Magnus, De homine (Ed. Colon. XXVII/2, 342.53–58): "De hoc etiam dicit Avicenna quod somnus 'est ex lassitudine propter hoc, quia id quod vocatur spiritus, dissolutus et debilitatus est, et refugit ad interiora, et sequuntur eum virtutes animales'. Et ex hoc iterum habetur quod labor est causa somni." See: Algazel, Metaphysica pars 2 tr. 5 c. 2, ed. Muckle 1933/1967, 288.17–19; Avicenna, Liber de anima IV c. 2, ed. Van Riet 1968, 33.66–69.

120 Albertus Magnus, De homine (Ed. Colon. XXVII/2, 320.17–24): "Contra tertiam diffinitionem, quae est Algazelis, sic obicitur: (1) Spiritus conducit calorem, quia ubi est multiplicatio spiritus, ibi est multiplicatio caloris; ergo et e converso ubi multiplicatur calor, ibi multiplicatur spiritus; sed in multis dormientibus multiplicatur calor in extremitatibus manuum et pedum vehementer; ergo et spiritus; ergo non est generale quod somnus sit retractio spiritus ad interiora."

121 Aristotle, De somno 2 (455a15–25); Aristoteles Latinus, De somno c. 2 (ed. Drossaart Lulofs 1943, 4.11–27).

from touch. As shown by Donati, Albert's interpretation of sleep as an affection of common sense in the *De homine* opts for the second and stronger reading:

> However, detailed inspection will lead one to say that sleep is not an affection of the common sense except incidentally (namely in so far as it makes the sense-spirit flow into the proper senses), and not such that the proper organ of the common sense is immobilized by sleep.[122]

According to Albert, sleep is an affection of common sense only accidentally. This is because common sense is certainly the ultimate root of the sensitive power, but this power is such if and only if touch is active. If touch is not active, then no other sense is, and thus neither common sense. The strong reading given by Albert makes touch the mediating factor between the activation of common sense and the activation of the other senses: without touch being active, no other sense is.[123] As we have stated above, sleep is therefore not an affection of common sense in itself (*somnus non est passio eius*) but primarily of touch, and then only accidentally of common sense, for its deactivation is concomitant to the deactivation of touch. From a physiological point of view, this can be observed by the fact that animal spirits flow first from the organ of common sense then to the sense organs. This fact makes the reverse mechanism (viz. sleep) primarily dependent on the deactivation of sensory powers, touch, and then, only in turn and only insofar as it relates to the external senses, on the deactivation of common sense.[124] This argument seems to depend, as

122 Albertus Magnus, De homine (Ed. Colon. XXVII/2, 333.46–50): "Si quis autem subtiliter vellet intueri, diceret quod somnus non est passio sensus communis nisi per accidens, scilicet inquantum influit spiritum sensibilem sensibus propriis, et non sic, ut immobilitetur proprium organum sensus communis per somnum." This translation by Donati in: Donati, "Dreams and Divinatory Dreams," 188.

123 Donati, "Dreams and Divinatory Dreams," 188: "Albert's distinctive view on the sense of touch as a central factor in the phenomena of sleep and waking rests on his notion of touch as a mediating factor between the common sense and the other external senses: for Albert, as the common sense is the ultimate root of the perceptive power, the sense of touch, likewise, is the basic foundation of all the external senses. This entails that the activity of the other external senses is grounded in the activity of the sense of touch, in the sense that the activation and the deactivation of the sense of touch also cause the activation and the deactivation of the other senses, respectively."

124 Albertus Magnus, De homine (Ed. Colon. XXVII/2, 333.13–50). On the interpretation of this passage, see: Silvia Donati, "Albert the Great as a Commentator of Aristotle's *De somno et vigilia*: The Influence of the Arabic Tradition," in: Parva Naturalia *in Greek, Arabic and Latin Aristotelianism. Supplementing the Science of the Soul*, ed. Börje Bydén/Filip Radovanovic (Studies in the History of Philosophy of Mind 17), Cham 2018, 204–205.

shown by Silvia Donati, on Albert's ideas about the role of the nervous tissues with which the sensitive nerves are imbued. According to him, tactile powers belong primarily to the tissues of these nerves since they are particularly sensitive to cooling down and therefore are the first to be incapacitated by the cold digestive vapors that cause sleep.

Touch is therefore a power that maintains a mediating role enabling the perception of other senses once the tissues of these nerves are active. Its power seems to be more basic than all other specific powers of the external senses. This causes the physiological causes of sleep to become clearer, even though not in the detail as in his mature *De somno et vigilia*. The cold vapors primarily affect these tissues of the sensitive nerves which, strictly speaking, belong to touch. These tissues' loss of heat and the consequent deactivation of their tactile power leads to an incapacitation of both the exercise of all sensitive powers and of the diffusion of animal spirits into the body through the nerves. This picture of Albert's peculiar understanding of touch will appear, though with some doctrinal changes, in his mature *De anima* and *Parva Naturalia*.[125]

In the broader picture of his early treatment of sleep and dreams, Albert assigns a crucial role to animal spirit that, through its motion in the brain, explains in physiological terms the production of dream images and the deactivation of common sense and its dependence on the sense of touch. In his early theology, though, dream images encompass both cases of God-given dreams, forcibly read into the Aristotelian category of accidental dreams—with due theological precaution, i.e. reference to the prophecy of menace—and dreams that can be explained according to the laws and constraints of natural philosophy. While the *De homine* could be designed, in the words of Henryk Anzulewicz, as a dialogue between doctors of the Christian faith and philosophers (Ort des Dialogs zwischen *sancti* und *philosophi*),[126] this is a place where their doctrines are harmonized. By contrast, in the mature Albert the theological and naturalistic explanations of dreams and prophecy become sharply separated, since philosophers speak of prophecy in a manner that is different from saints (*philosophi aliter loquuntur de prophetia quam sancti*).[127]

125 More on Albert's position on touch and common sense in: Donati, "Albert the Great as a Commentator," 188–193.

126 Henryk Anzulewicz, "Anthropologie des Albertus Magnus als Ort des Dialogs zwischen den *sancti* und *philosophi*," in *Dialog. Auf dem Weg zur Wahrheit und zum Glauben*, ed. Frano Prcela, Zagreb-Mainz 1996, 47–53.

127 Alessandro Palazzo, "*Philosophi aliter loquuntur de prophetia*," 179–201.

3.4 *God-Given Dreams, Prophecy, and Divinatory Dreams in Albert the Great's Mature Works*

The recent investigations of Anna Rodolfi have clearly shown the sharp epistemological distinction posited by Albert between God-given dreams or God-inspired prophecies and dreams or prophecies related to natural events, hence the famous definition of *prophetia naturalis* by Aquinas. As we stressed in the introduction, Albert did so to prevent divinatory dreams from acquiring a truth value very close to God's absolute foreknowledge, even of events subject to human free will. Albert's distinction between God-given dreams or dreams with prognostic value but limited in scope to events predictable according to the laws of natural philosophy is based on a distinction in terms of object. As shown by Rodolfi, Albert individuates as objects of divinely inspired dreams and prophecies those events, viz. future contingents, that depend on human free will.[128] Given that their realization depends on human free will, they can be known by God alone. By contrast, dreams and prophecies that can be explained according to the laws of natural philosophy concern those events whose realization concerns the realm of nature and, therefore, of hypothetical necessity. We may call these dreams and prophecies 'veridical' or 'divinatory' to distinguish them from God-given dreams. In his third book of the *De somno et vigilia*, Albert individuates a ranking of divinatory dreams and visions that succeeds in unifying the discourse on veridical dreams and prophecy, not handled in the same place in his early summa (*De homine, De quattuor coaequaevis*). Perhaps influenced by the list of eleven degrees of dreams and prophecy given by Maimonides, this list reads:

1. Dreams arise in part from human process ("*a proprio corde*" [from one's own heart]) and involve the "*forma coelestis*" [celestial form] only weakly; such dreams do not elucidate the future except "*implicite et confuse*" [entangledly and confusedly] and they are frequently misleading ("*frequenter fallens*") (p. 191);
2. Only an indistinct and confused light reaches the soul, and the images it presents are inappropriate ("*inconvenientia*") to the message or truth that the dream should convey (p. 191);
3. The dream's import is shown in appropriate metaphors that can be reliably interpreted (p. 191);
4. The dream is not metaphorical, but rather expresses its meaning in images of what will actually occur—"*sicut res evenit*" (p. 191);
5. Dream images are presented and explained by someone who appears in

128 Rodolfi, Cognitio obumbrata, 40–48.

the dream; the explanations, however, are not explicit but rather made "*in verbis et figuris*" [in words and figures] (p. 191);

6 The dream images are explicitly interpreted by someone within the dream, "*qui docet recte futuram visionis significationem*" [who teaches correctly the future signification of the vision] (p. 191);

7 Without images, intellective truths are presented to one who is asleep (p. 192);

8 The action of the "*forma coelestis*" is so strong that it moves someone who is awake to see a vision; the vision, however, like the second kind of dream, is expressed "*in simulacris obscuris*" [in obscure images] (p. 192);

9 A waking vision, parallel to the third kind of dream, reveals future things "*per formas expressas et de facili adaptabiles*" [through forms clear and easily fitted (to their meaning)] (p. 192);

10 A waking vision occurs in which the images seen are also explained (p. 192);

11 A waking, imaginative vision occurs in which future and occult things are clearly understood (pp. 192–193);

12 A waking vision directly presents images of future events (p. 193);

13 "*Bona occulta*" [concealed good things] are revealed, which are correctly understood by the waking seer (p. 193).[129]

This list is strikingly detailed and reflects a clear differentiation in object from Albert's work explicitly devoted to prophecy from a theological point of view, viz. the *Quaestio de prophetia*. In that work, Albert had maintained the crucial role played by images, though in a different manner. In the case of the naturalistic explanation of divinatory or prognostic dreams and prophecy, the image involved in the dream is retrieved from past experience. This image can be rudimentary and just an association of basic features (red and hot, yellow and sweet etc.) without the need always to be complex. Sometimes it can be, but not always. Still, the image produced by the composing-dividing activity of imagination (Albert would say *phantasia*) may be composed or divided in its features but reflects the structure of past perception. For instance, we may dream of a red sweet thing though we have not perceived these features together but separately (e.g. the red was attached to a hot thing, the sweet to a yellow one), but they are not changed.

In the *Quaestio de prophetia*, Albert presents a different view of the role of images in his theory of God-given dreams and prophecies, arguing that the

129 Albertus Magnus, De somno et vigilia III tr. 1 c. 10, ed. Donati, 118–128 (ed. Auguste Borgnet, Paris 1890 [Ed. Paris. IX, 190b–193a]). This translation by Kruger in: Kruger, *Dreaming in the Middle Ages*, 120.

species stored in imagination receive a new order.[130] This new order is achieved through the action of divine illumination which, in collaboration with the human (rational) soul produce dreams which sometimes need a proper interpretation, as in the case of the Pharaoh and Nabuchodonosor.[131] Rodolfi has shown that this new paradigm, strongly based on Aristotelian gnoseology, subverts one a key notion of Aristotle, viz. that our knowledge is derived from an image, but this image is rooted on our experience. Technically, the image can be seen in two respects, as an image related to the object of previous knowledge (*a quo*) and as an object related to the subject of this knowledge, viz. the human with its intellective capacities (*in quo*). As Rodolfi has convincingly shown, the function of an image *in quo*, viz. as a foundation of the intellectual activity and any rational inference, is retained. By contrast, the function of the image *a quo* vanishes precisely because of this rearrangement of images produced by the joint action of divine illumination and rational soul, which alone is capable of doing so.[132] In Albert's *De somno et vigilia*, though, both the function *a quo* and the function *in quo* of images are preserved, even though these images may be metaphorically representative of future events subject to the natural laws of hypothetical necessity.

3.5 *Divinatory Dreams and Natural Prophecy: Between Psychology and Astrology*

In the opening lines of the third book of the *De somno et vigilia*, Albert presents an unusually harsh criticism of the Stagirite. In beginning his paraphrase and reworking of Aristotle's *De divinatione per somnum*, the Dominican master targets Aristotle's explanation of divinatory dreams as lacking in philosophical argumentation (*carens probatione*). It is doctrinally scant (*habet parum philosophiae*) inasmuch as it does not provide a full-blown explanation of the different species of dreams he mentions (signs, causes, coincidences) and,

130 Albertus Magnus, Quaestio de prophetia: *visione, immaginazione e dono profetico*, 2.4.1.1., ed. Anna Rodolfi (Per Verba 24) Firenze, 2009, 76: "Dicendum, quod probabilius est, quod species in imaginatione ordinentur et componantur, quam novae imprimantur. Unde omnes species imaginabiles a sensu receptae sunt, sicut vult Augustinum; et Philosophus dicit in III *De anima*, quod 'phantasia est motus a sensu secundum actum factus'."

131 Albertus Magnus, Quaestio de prophetia, 2.4.1.1 (ed. Rodolfi 2009, 78): "[…] nec lux abstrahens nec anima per se componit et ordinat, sed anima illustrata divina luce; et non est inconveniens, quod talis anima notitiam habeat de futuris, vel perfecte sicut prophetae, vel imperfecte sicut Nabuchodonosor et Pharao, qui cognoverunt, quod aliquid signaretur e futuris, sed nescierunt quid."

132 Rodolfi, Cognitio obumbrata, 114–118, especially 118; Rodolfi, "Introduzione" in: Albertus Magnus, *Quaestio de prophetia*, XXVII-XXIX.

moreover, it ignores the astrological component of the process. Despite this harsh criticism of Aristotle's account, Albert still sees his explanation of divinatory dreams as the closest to the truth of any earlier philosopher (*plus appropinquat veritati quam aliquid quod ante vel post aliquis scripserit philosophorum cuius scripta ad nos devenerunt*).[133] This seems an almost ironic statement given Aristotle's well-known skepticism about God-given dreams. Any perceptive reader of Aristotle might wonder about the real meaning of Albert's words, more particularly, about the passages that enable Albert to put his theory of divinatory or veridical dreams and 'natural' prophecy under the aegis of Aristotle. In the first place, Albert proceeds to divide dreams that originate with God and the stars as direct causes (and thus have a prognostic value) and dreams originating from internal motions within the dreamers themselves.[134]

In a recent study, Silvia Donati has shown how Albert's complex psychological and astrological framework of dreaming forcibly places his theory of divinatory dreaming into Aristotle's mouth. To begin with, tracing the historiographical roots of such bold a claim—that Aristotle provides, among all philosophers, the closest to a truthful account of veridical dreams—leads us to Albert's 'creative' reading of a passage in Aristotle's *De divinatione per somnum* where the Stagirite refers to the Democritean theory of dreaming, with which Aristotle was somehow sympathetic. According to Democritus, "idols" and "emanations" may be the cause of dreams. According to Albert, this passage represents a great opportunity to harmonize Aristotle with his emanative cosmology, based on Albert's reworking of Kindī's theory of rays (*De radiis*). As shown by Donati, the underlying assumption according to which Albert uses

133 Albertus Magnus, De somno et vigilia III tr. 1 c. 1, ed. Donati 2020, 68 (ed. Auguste Borgnet, Paris 1890 [Ed. Paris. IX, 178a]): "Et hoc ipsum quod de divinatione dicit Aristoteles breve quidem est et imperfectum et habet in se plurimas dubitationes. Dico autem breve quia carens probatione, sed simplex et parum philosophiae habens videtur esse narratio, nec species somniorum neque probationem somnia aliquid significare in se continens. Imperfectum autem est quoniam, licet sine magicis et astronomicis non possit ars interpretandi somnia ab aliquo adipisci, tamen ex solis physicis sufficienter scitur ex qualibus et quibus simulacris consistit somnium de quo debet esse divinatio, et hoc nec ab Aristotele nec a quoquam philosophorum determinatum invenitur. Plurimas autem habet dubitationes quia et in incerto relinquit causam talium somniorum nec etiam eius quod dicit, quod similitudines inspiciens est iudex somniorum, determinavit modum et causam. Tamen hoc quod dicit Aristoteles plus appropinquat veritati quam aliquid quod ante vel post aliquis scripserit philosophorum cuius scripta ad nos devenerunt."

134 Albertus Magnus, De somno et vigilia III tr. 1 c. 4, ed. Donati 2020, 84 (ed. Auguste Borgnet, Paris 1890 [Ed. Paris. IX, 181b]): "Potest enim esse a deo vel a stellis vel a causa quae est in nobis; ex quacumque autem causa sit, per experta dicimus in somniis esse divinationem ex phantasmatum similitudinibus acceptam et coniecturatam."

this passage to put his explanation of veridical dreams and prophecy under the aegis of Aristotle rests on his belief that the external factor triggering a dream must also be the cause of the future event prefigured by the dream itself.[135] For him, dreams may have a prognostic value because they are caused by the very same agent that will produce a certain event or effect in the sublunar realm. This assertion is, to some extent, similar to Aristotle's genuine doctrine.[136]

The first factor that plays right into Albert's hands is of a lexical nature. According to Democritus, the key role in the production of dreaming images is played by "emanations" (Gr. ἀπορροίαι, *Lat. defluxiones*) of atomic particles affecting the subject during sleep. Given the fundamental value that Albert attributes to the metaphysic of *fluxus* as a paramount structure of reality, it is not hard to see how Albert could have seized this opportunity to bring Aristotle into agreement with his astrologized interpretation of prognostic dreams (true according to the laws of natural philosophy, to be clear). In her article, Donati has clearly shown how Albert reads this passage on the role of *defluxiones* in Democritus together with another passage in *De anima* II, 7. In that passage, Aristotle voices his criticism of those who considered light to be an effluence from a body and thus a corporeal entity. While modern exegetes of Aristotle take this passage to be a critique directed against Empedocles' views on light, Albert connects this idea—wrongly or, rather, in a consciously forceful manner—to the aforementioned passage of the *De divinatione per somnum*. Because of the lexical similarity between the terms *defluxus* (*De divinatione*) and *defluxiones* (*De anima*), Albert reads into Aristotle's position the idea that Aristotle is actually hinting at the role of astral influences in the production of oneiric images.[137] Clearly, this is a forced harmonization of the Aristotelian text with Albert's cosmological model and his theory of rays as carriers of celestial forms based on al-Kindī's theory of rays.

---

135 Donati, "Dreams and Divinatory Dreams," 187.

136 Donati, "Dreams and Divinatory Dreams," 187: "In Albert's scheme, the logical structure underlying this kind of divinatory dreams seems to be basically the same as that which we have detected in Aristotle's theory: the 'precognitive' value of such dreams is based on the fact that the external factor triggering a dream is also the cause of the future event prefigured by the dream."

137 Donati, "Dreams and Divinatory Dreams," 191–192: "For Albert, this terminological similarity is an indication that, in the two passages, Aristotle is talking about the same thing, namely the light rays coming to the sublunary world from the stars. Given that in his doctrine the light rays are the physical medium carrying the astral influence from the stars to the sublunary world, for him, Aristotle's reference to emanations as the causes of divinatory dreams is an indirect allusion to the astral origin of divinatory dreams. In conclusion, according to Albert, the doctrine ascribed to Democritus in the passage of the *De divinatione* is a theory which suggests that divinatory dreams are triggered by the stars exerting

Albert argues that these influences go unnoticed by humans for most of the day, viz. when sense perception is active. When sense perception is deactivated, though, celestial influences can be perceived more strongly, during the state of sleep. In this manner, Albert is able to put, through a 'creative interpretation' of this Aristotelian passage, his ideas on divinatory dreams, based on complex astrological theory, under the aegis of Aristotle.[138] The main difference between Albert's and Aristotle's ideas is that the "effluences" (which Aristotle preferred to call "motions") that were hinted at in his *De divinatione per somnum* were limited to the sublunar realm. This happens, for instance, when we dream of thunder because our ears have detected slight vibration. This fact, though, has nothing to do with any astral influence whatsoever.[139] For Albert, by contrast, this fact is an opportunity to seize and read into Aristotle's words his theory of astral influences expounded more extensively in the *De fato* and in the *Physics*, drawing heavily on the Hermetic concept of fate.[140]

Albert's cosmological account of veridical dreams cannot be thought of as isolated from his psychology of dreaming, though, for his theory of astral influences strictly intersects the psychological dimension of the phenomenon. In this, Albert betrays his strongest commonalities and differences with the philosophers of the Islamic world with whom he was so well-acquainted, especially Avicenna, al-Ġazālī and Averroes. In terms of doctrinal similarities, Albert shares with these Islamic authors (and Maimonides) an emphasis on the perfection of the faculty of imagination from a physiological (complexion) point of view. On the other hand, Alessandro Palazzo has clearly shown how Albert decidedly rejected both the Avicennian-Ghazalian conception of prophecy as well as that passed on by Averroes, Isaac Israeli and Maimonides, because they postulate direct action on the part of the heavenly intelligences—and not of the heavenly bodies—on the human soul. According to Albert, heavenly intelligences are per se incapable of operating physically. To act on

---

their influence on the sublunary world through the light rays that reach the earth from the celestial spheres. This view is basically shared by Aristotle, the only point of disagreement being Democritus' corpuscular theory of light."

138 Donati, "Dreams and Divinatory Dreams," 193.

139 Aristotle, De divinatione per somnum, 1, 463a10–16; Aristoteles Latinus, De divinatione per somnum (ed. Drossaart Lulofs 1947, 34.19–36.6).

140 The connection between fate and divinatory dreams is tackled by Palazzo in: Alessando Palazzo, "The Scientific Significance of Fate and Celestial Influences in Some Mature Works by Albert the Great: *De fato, De somno et vigilia, De intellectu et intelligibili, Mineralia*," in Per perscrutationem philosophicam. *Neue Perspektiven der mittelalterlichen Forschung*, ed. Alessandra Beccarisi/Ruedi Imbach/Pasquale Porro (Corpus Philosophorum Teutonicorum Medii Aevi, Beiheft 4), Hamburg 2008, 56–60.

the sublunar world through their power, then, heavenly intelligences require a medium. In Albert's eyes, this medium is identified with the heavenly bodies and the rays which act as a vehicle of their power and of the forms they pour onto the sublunar world.[141] In the wake of these criticisms of his sources, Albert proceeds to develop a unified naturalistic explanation of both divinatory dreams and chance occurrences of dreaming.

To do this, Albert resorts to his theory of emanation and combines it with the psychology of the imaginative faculty. According to him, veridical dreams, visions and prophecy are caused by the reception of the celestial form (*forma coelestis*) from the human imaginative faculty. To reach the human body, this *forma coelestis* had been transmitted by the rays issuing from the heavenly bodies and the stars. By contrast, dreams as signs or causes or chance occurrences are principled by the activity of the imaginative faculty alone. In the case of veridical or divinatory dreams, though, the subject receives the celestial form according to its capacity (*secundum potestatem suscipientis*). Having been affected by the flow of the celestial form, imagination looks for the representations that are as close as possible to the celestial form. This form uses the imaginative power as a tool to manifest itself to the subject and facilitate the knowledge of future events, even though to different (possible) degrees.[142] If we consider the previous ranking of images by Albert, we can see how the clarity of the image and its non-metaphorical character stand in direct causal correlation to a correct reception of the celestial form. Since this reception is due to the most perfect faculty of imagination, it follows that the clarity of images does depend on the physiological constitution, in combination with other factors such an exemplary ethical behavior and exercise of the intellect. According to Albert, divinatory dreams, which lay at the lowest level of this division of dreams, visions and prophecy, are metaphorical because they essentially rely on an image (*phantasma*) that is other (*alienus*) in relation to the future event. The image that appears to man during sleep is not the celestial form of the future event itself, but rather what from the thesaurus

141 Alessandro Palazzo, "*Philosophi aliter loquuntur*," 192. Albertus Magnus, De somno et vigilia III tr. 1 c. 8, ed. Donati 2020, 108–110 (ed. Auguste Borgnet, Paris 1890 [Ed. Paris. IX, 188b]); Silvia Donati, Dreams and Divinatory Dreams, 189: "Albert's basic objection against this theory positing a sort of non-physical action of the intelligences on human imagination is that it violates the principles of natural philosophy, according to which the celestial movers act on the sublunary world only through the mediation of the celestial bodies […]."

142 Albertus Magnus, De somno et vigilia, III tr. 1 c. 9, ed. Donati 2020, 116 (ed. Auguste Borgnet, Paris 1890 [Ed. Paris. IX, 190a–b]). For the interpretation of this passage, see: Donati, "Dreams and Divinatory Dreams," 200–201.

of stored images comes closest in signification to the celestial influence that will cause the future event.[143]

The more this celestial form is correctly perceived and the more the representations fetched from retentive imagination match the content of the celestial form, the more closely the dreaming images and visions resemble events that are bound to happen, at least in terms of natural causation. This is the case, for instance, of a child who has been crushed and killed by a mill wheel in the same manner that Albert—so he says—had dreamt the previous night.[144] The fact that a man is able to see clearly events that may occur in the future depends, as in the Islamic philosophers, on a perfectly balanced state of the imaginative faculty and its material components (organs, animal spirits etc.). This time, though, the link between the complexion of spirits and the kinds of dreaming or prophetic images is very specific and detailed in its extent.[145] According to Albert's own theory, this allows a suitable reception of the celestial form, which in turn fetches the most suitable image from the imaginative faculty and causes divinatory dreams or visions, which are higher in ranking than dreams because they are perceived during the state of waking. Clearly, this perfect reception of the celestial form needs to be matched by intellectual

143 Albertus Magnus, De somno et vigilia III tr. 1 c. 4, ed. Donati 2020, 84–85 (ed. Auguste Borgnet, Paris 1890 [Ed. Paris. IX, 182a]); Donati, "Dreams and Divinatory Dreams," 204: "At the time of the dream, this individual event cannot be grasped by means of its own representation, since it does not exist yet. Neither can it be grasped by means of the representation of its convertible cause, since the external factor triggering a divinatory dream—the celestial form that affects the imaginative power by affecting the body—is only a remote, universal cause that needs to be further specified by a variety of additional factors in order to actually produce the individual event represented by the dream. Given that divinatory dreams cannot supply appropriate representations of the future events they signify, it follows that they represent their objects by means of images that properly belong to something else ('*phantasmata aliena*') and that are related to those objects by a more or less close similarity."

144 Albertus Magnus, De somno et vigilia III tr. 1 c. 10, ed. Donati 2020, 120 (ed. Auguste Borgnet, Paris 1890 [Ed. Paris. IX, 191b]): "Ego etiam somniavi me stare super ripam cuiusdam fluminis, in quo suspensum super aquam fabricatum fuerat molendinum, videbaturque mihi quod puer in aquam caderet et veniens ad rotam molendini deprimeretur et necaretur ibidem. Evigilans autem, cum hoc sociis referrem, antequam finita esset recitatio, nuntiatur adesse mulier magno eiulatu nuntians hoc idem per omnia quod videram in somnis, cum tamen et mater et puer omnino fuissent mihi ante hoc incogniti nec umquam noveram vel videram easdem personas."

145 Note that we are here only dealing with the correlation between the temperament of spirits and dreaming or prophetic images. A full-fledged analysis of the psychophysiology of dreaming should encompass not only spirits but also humors and the more general temperament of the body in the formation of dream images.

powers, which are crucial for a correct interpretation to the received dreaming images.[146] Given these philosophical premises, the interpretation of divinatory dreams acquires a scientific dignity of its own. Dream interpretation is therefore a conjectural science (*scientia coniecturalis*), as stated by Ptolemy in his *Quadripartitus*.[147]

According to Albert, the highest degrees of knowledge of future events, viz. prophecy, is gifted only to a very restricted group of individuals who were born for this. Moreover, their condition is not a mystical one, but "a scientific ascetism consisting in the cultivation of the intellectual virtue and the sciences; moral improvement is the complement of a life dedicated to science."[148] For these reasons, divinatory or veridical dreams often have a metaphorical value only and need a science of their own to be interpreted accordingly.[149]

In the second treatise of the third book of the *De somno et vigilia*, Albert deals with those dreams that may have a prognostic value but do not originate from the action of the celestial form conveyed by rays. This analysis follows the division of dreams into those caused by God and the stars and those caused by endogenous factors within dreamers themselves.[150] As in the *De homine*, Albert had maintained in his paraphrase of the Aristotelian *De insomniis* that the production of oneiric images is physiologically produced by the internal motions of spirit from the organ of *phantasia* to that of common sense.[151]

To be correctly understood and interpreted, though, these dream images need to be further divided into signs and causes. In the second case, viz. dreams as causes, Albert follows the usual medical teachings according to which certain dream images are causally correlated to the humoral complexion, bodily affections or passions and good or bad condition of bodily members (*complexio et passio et accidens animi et corporis bonum et malum frequentissime demonstratur in somnio*). This is the case, for instance, of choleric people who dream

146 Albert, like Avicenna, presents a doctrine of intellectual prophecy in his *De intellectu et intelligibili*: Palazzo, "The Scientific Significance," 67–72.

147 Albertus Magnus, De somno et vigilia III tr. 2 c. 5, ed. Donati 2020, 162 (ed. Auguste Borgnet, Paris 1890 [Ed. Paris. IX, 202b–203a]).

148 Palazzo, "The Scientific Significance," 65.

149 On Albert's metaphorical conception of divinatory dreams and theory of divinatory practice, see: Donati, "Dreams and Divinatory Dreams," 202–209.

150 Albertus Magnus, De somno et vigilia III tr. 1 c. 4, ed. Donati 2020, 84 (ed. Auguste Borgnet, Paris 1890 [Ed. Paris. IX, 181b]).

151 Albertus Magnus, De somno et vigilia, II tr. 1 c. 1, ed. Auguste Borgnet, Paris 1890 (Ed. Paris. IX, 159a); II tr. 1 c. 4, ed. Auguste Borgnet, Paris 1890 (Ed. Paris. IX, 162b–163a).

of having pitch poured on their stomach or being burned in fire because they are filled with burned-out humors.[152]

Other dreams, as in the case of Aristotelian linear dreams, are either principled by our activities during waking hours or prompt the beginning or completion of activities while awake. This is the case of a sick patient who did not know how to act regarding his spleen ache. Having dreamt of undergoing phlebotomy between two fingers, he finally resorted to practicing phlebotomy and eventually healed.[153] This is an example which Albert likely inherits from the Galenic medical tradition, even though it is still unclear how. In the second of the Aristotelian *De divinatione per somnum*, Albert maintains, as he had done in the *De homine*, the key role played by the complexion of spirits in the determination of the oneiric image. To be clear, this role is derivative inasmuch as it depends on the humoral complexion of the individual. Still, spirit is worth mentioning as a concomitant causal factor at play in the process. In his explanation of why melancholics are particularly prone to dream veridical dreams, or just to dream in general, Albert links their humoral balance to the action of spirits.

The humoral complexion of the living being determines, in short, the complexion of spirit and is directly correlated to the kind of perceived dream images. We find an exemplification of this concept in Albert's paraphrase of the *De insomniis*, where he correlates an excessively moist complexion with horrible, monstrous images.[154] The example of nightmares in children had been explained in the *De homine* in the exact same way, viz. by appealing to the moist complexion of spirit. The moister the complexion, the greater the amount of vapors produced by the action of heat on moisture and, in the end, the more horrendous the images which appear during sleep. In his paraphrase of the *De insomniis*, Albert also relates the case of dreaming images in ill and drunk people to the complexion of spirits. According to Albert, in ill patients the matter of fever, viz. humors, spirits or radical moisture,[155] produces vapors (*fumo sursum*

---

152 Albertus Magnus, De somno et vigilia III tr. 2 c. 1, ed. Donati 2020, 146 (ed. Auguste Borgnet, Paris 1890 [Ed. Paris. IX, 198b]): "*Et* somniamus nos *ambulare per ignem et calefieri vehementer parvo calore facto circa partes* aliquas corporis, et praecipue si in interioribus corporis fit, sicut somniavit quidam sibi fundi in ventrem picem ardentem et se aestuare in igne picis, eo quod choleram adustam nigram incensam habuit in ventre et hanc emisit cum surrexit a somno."

153 Albertus Magnus, De somno et vigilia III tr. 2 c. 2, ed. Donati 2020, 148 (ed. Auguste Borgnet, Paris 1890 [Ed. Paris. IX, 199b]).

154 Albertus Magnus, De somno et vigilia II tr. 2 c. 1, ed. Auguste Borgnet, Paris 1890 (Ed. Paris. IX, 170a–b).

155 Usually, fevers were divided into putrid (affecting humors), ephemeral (affecting spirits) and hectic (affecting radical moisture). More on this in chapter 4.

evaporat) and mixes the forms stored in imagination with each other because of its heat (*caliditate sua formas commiscet*). Albert then proceeds to add a particular it had been missing in the previous account of the *De homine* when speaking of nightmares, viz. the motion of spirits during the dreaming activity. The thickness of spirits correlated to these smoky vapors. impedes them to flow to the central brain ventricle, viz. the *cellula logistica*, and therefore prevents reason to judge these images as delusions (*detinet rationem contradicentem*). The same explanation is true in drunk people. In their case, the heat of wine upheaves a great amount of smoky moisture from the moist provided by wine itself, thus blocking the flow of animal spirits from the *cellula logistica* and generating horrendous dreaming images. Because of the heat which moves these images, they see them as trembling and rotating images. In melancholics, the occurrence of nightmares and horrendous dreaming images is caused by blackish evaporations produced by the (scarce) amount of heat over moisture. Because of the thick complexion of spirits in the brain, correlated to the dry and moist complexion of melancholics, these blackish exhalations impede the flow of animal spirits to the middle brain ventricle and cause nightmares.[156] Given that melancholics are cold and wet, they do not produce a great amount of vapor, the primary obstacle to a correct perception of dream images (*frigidum et siccum non vaporat multum, et ideo non multum videtur movere phantasmata*). In turn, this complexion affects spirits, which derive from blood (itself a bodily humor). Because of a particularly dry complexion, dream images, often occurring in melancholics, are easily retained.[157]

All in all, we can observe how Albert's psychophysiology of dreaming has changed little over time. Certainly, it has clearly demarcated the domain within which dream images are the product of the motion of animal spirits in the brain, making sure that this detailed medico-philosophical description of dreaming does not clash with Cristian orthodoxy, the existence of God-given dreams and prophecies concerning acts of human free will.

3.6 *Common Sense, Touch, and Sleep. The Role of Spirits*

In his psychophysiology of sleep, Albert recurrently defines sleep, from a physiological point of view, as a *retractio spiritus ad interiora*. This definition is not

156 Albertus Magnus, De somno et vigilia II tr. 2 c. 1, ed. Auguste Borgnet, Paris 1890 (Ed. Paris. IX, 170b–171a).

157 Albertus Magnus, De somno et vigilia III tr. 2 c. 4, ed. Donati, 156 (ed. Auguste Borgnet, Paris 1890 [Ed. Paris. IX, 201b]): "Calidum igitur paucum quod est in melancholicis elevat vaporem et spiritum debilem, quod, si multum esset, consumeret debiles spiritus. Et quia calor non commiscet huiusmodi spiritum, ideo imagines bene tenent se in ipso."

particularly original: as a matter of fact, Silvia Donati has noted that Adam of Buckfield's commentary on the *De somno et vigilia* conveys a very similar idea.[158] The gist of Albert's doctrinal exegesis of the Aristotelian text concerns two main and peculiar points, though: the relationship between common sense and touch in the definition of sleep and the definition of sleep as an affection of the common sense and as a deactivation of the five senses. In both cases, Silvia Donati has highlighted how the medico-philosophical concept of spirit plays a key role in these processes.[159]

As we have shown in the previous section, following Silvia Donati, Albert's interpretation of Aristotle's *De somno*, 2 takes sleep to be primarily an affection of touch and then, incidentally, of common sense. In his *De anima* and especially the *De somno*, his position has become more moderate. On the one hand, Albert maintains that sleep depends physiologically on an incapacitation of common sense to send the physiological component of perceptive power (viz. animal spirits) out into the peripheral limbs. This incapacitation is due to the cold vapors produced by digestion. In this manner, the tissues of these sensitive nerves (belonging to touch) are incapacitated and so is external sense perception.[160] At the end of the section devoted to common sense and the sense of touch, Albert concludes that sleep (and waking) are affections of common sense insofar as it is related to the nerves of the external sense, which, in turn, are informed by tactile power.[161] Finally, Albert concludes his analysis by arguing that sleep is an incapacitation of common sense (not accidentally, as he had held in the *De homine*) and of the organ of touch insofar as the organ of touch is the foundation of all peripheral senses. In this manner, sleep primarily (*primo*) belongs to the faculty of common sense, while it immediately (*proxime*) belongs to the sense of touch *qua* foundation of all peripheral senses.[162]

---

158 Donati, "Albert the Great as Commentator," 197.

159 Donati, "Albert the Great as A Commentator," 194–201.

160 Albertus Magnus, De somno et vigilia I tr. 2 c. 1 ed. Auguste Borgnet, Paris 1890 (Ed. Paris. IX, 138b–139a).

161 Albertus Magnus, De somno et vigilia I tr. 2 c. 1 ed. Auguste Borgnet, Paris 1890 (Ed. Paris. IX, 138b–139a); Donati, Albert the Great as a Commentator, 193: "Albert maintains that sleep and waking are affections of the common sense insofar as it is related to the nerves of the external senses, and those are informed by the tactile power."

162 Albertus Magnus, De somno et vigilia I tr. 2 c. 1, ed. Auguste Borgnet, Paris 1890 (Ed. Paris. IX, 139a): "Sic igitur patet quod somnus est primo passio sensus communis influentis virtutes particularibus: et per hoc ligantur ne hauriant formam sentiendi, et proxime est tactus secundum quod est fundamentum organorum animatorum, et per consequens est sensuum propriorum omnium [...]" Donati, "Albert the Great as a Commentator," 187.

While these are the theoretical grounds according to which Albert explains the relationship between common sense and sleep, this is only half of the story. For him, as for many authors, common sense is involved both in sleep, defined as an affection of common sense, and in dreaming too. This is because during dreaming we see images and hear sounds as if we were actually seeing and hearing them, thus taking them at face value and inferring their truth even when they are delusions. On the one hand, Aristotle had considered sleep a deactivation of common sense, which Albert explained primarily on the grounds of the coldness affecting the tactile power seated in the nervous tissues. On the other hand, the perception of dreaming images testified to some rudimental degree of perceptive awareness, though without the ability to identify them as true or false.

Silvia Donati has higlighted how, to solve the contradiction between a common sense that should be simultaneously deactivated and active during sleep and dreaming, Albert appealed to his Arabic sources, as he had done in the *De homine*. This time, he draws on Averroes and on al-Kindī's *De somno et visione* (*Risāla fī māhiyya al-nawm wa-l ruʾyā*), rather than Avicenna or al-Ġazālī. To explain the role of common sense during the perception of dream images, Albert argues that sleep is indeed an incapacitation of senses and common sense inasmuch as it relates to the external senses. There is a further aspect of common sense, though, that is connected to the awareness of perceiving illusions and images that come from within. From this perspective, defining sleep as a "withdrawal" (*introitus*/*ġuwūr*) of common sense maintains, on the one hand, the deactivation of common sense as a power related to the external senses but, on the other, it does not entail the total deactivation of common sense inasmuch as it is related to the internal senses.[163]

This understanding of sleep as a reinforcement of internally apprehensive powers is typical of the Arabic philosophical tradition and is closely linked by Albert to the motion of animal spirits: it is true that the functions of spirits are deactivated insofar as they are related to the perception of the five senses and common sense as perceptual awareness and perception of common sensibles. On the other hand, the retreat of spirits from the organs of sense perception does not presuppose a complete ceasing of their activity. Rather, their activity is now turned towards possible activities of internally apprehensive powers such as imagination (or the imaginative faculty, in Avicenna) responsible for the production of dreaming images. In Averroes and Kindī, common sense has no exclusive dependence on the external senses and their organs; rather, when

163 Donati, "Albert the Great as a Commentator," 202–205.

those sense powers are deactivated, their activity is invigorated during sleep.[164] In this sense, it seems that Albert prefers to understand sleep in a negative sense, as in Aristotle, as the redirection and consequent strengthening of common sense by the concentration of the animal spirits flowing back to the brain. Therefore, when he needs to tackle the apparent contradiction between the two definitions of sleep, he argues that, while Aristotle defines sleep as an affection of the common sense, he is referring to the relation of common sense to outer senses. By contrast, Averroes and Kindī refer to the relation of common sense to the internal senses.[165]

Thus, the two definitions are not really at odds with each other, for they capture two reverse sides of the same coin. This is a conclusion that the early Albert had already shared in his *De homine*, though without stressing that common sense is reinvigorated by sleep and without really problematizing the compatibility of the doctrines of Aristotle and Averroes. Moreover, while maintaining that the production of oneiric images is triggered by the flow of animal spirits from the organ of *phantasia* to the seat of common sense, this explanation is not built upon the astrologizing interpretation of divinatory dreams and creative reading of Aristotle's text showcased in Albert's mature *De somno et vigilia*, meant to provide a unified physiological account of divinatory, veridical dreams, dreams as signs or causes and chance dreams.

Among medieval readers of Aristotle's *De somno et vigilia*, Albert's position was not shared by everyone. The studies of Luca Bianchi have shown how it was influential, for sure, given its division of dreams between those originating in God and the heavenly spheres and those produced by the human body. Moreover, Bianchi has shown how authors like Boethius of Dacia narrowed down the scope of the prognostic value of dreams to the sole medical field, thus detach-

164 More on this, in: Donati, "Albert the Great as a Commentator," 202.

165 Albertus Magnus, De somno et vigilia I tr. 1 c. 9, ed. Auguste Borgnet, Paris 1890 (Ed. Paris. IX, 135b–136a): "Alfarabius (viz. Averroes) autem diffiniens somnum dicit, quod est introitus sensus ad interiora, et immobilitas sensus omnimoda. Vigilia autem est exitus sensus ad exteriora: et iste Philosophus vult, quod tam somnus quam vigilia sint motus, et neutrum quies et immobilitas: quod contrarium his quae dicta sunt, esse videtur. Sed ad hujus solutionem hoc oportet supponere quod inferius ostendemus, quod scilicet somnus sit in animali secundum primum sensum, quod est sensus communis, et hoc est, quod comparat sensata particularia et propria, sicut prius diximus in libro *de Anima*. Cum autem spiritus sit formarum sensibilium vehiculum, et recurrit spiritus ad interiora, virtus vecta ad interius adhuc comparat ea sicut primum: et tunc magis quando ab exterioribus abstrahitur et abducitur [...]". On the role of Kindī and Averroes, see: Donati, "Albert the Great as a Commentator," 202–205. On Albert's psychophysiology of dreams, see also: Adriana Caparello, *Senso e interiorità in Alberto Magno*, Roma 1993, 153–196; Steneck, "Albert on the Psychology of Sense Perception," 284–285.

ing himself from Albert's approach.[166] His pupil Aquinas, for instance, accepted the existence of what he called a "natural prophecy" (*prophetia naturalis*). The meaning of "natural" is pretty much in keeping with Albert's ideas of the suppositional necessity to which natural prophecy and veridical dreams—never referred to acts of free will—can be subject.[167] This line was followed by Siger of Brabant too.[168] As demonstrated by Kruger and Grellard, Medieval authors had different and nuanced positions on the topic. On the one hand, we find authors like Jean of Jandun, who endorses a causal framework in which dream images result from the representation (by the imaginative faculty) of dream images mirroring bodily alterations caused by astral influences. In this context, medical teaching is subordinated to a correct understanding of astrology in deriving correct medical prognostications from dreaming.[169] By contrast, authors like Boethius of Dacia and John Buridan proved more skeptical about the astrological framework and restricted the prognostic value of dreams to medical science alone. Among these authors, Adam of Buckfield represents a notable exception, since he seems to endorse Aristotle's skepticism on the matter. As noted by Kruger, Adam's position proves to be the exception rather than the rule.[170] Albert's position is certainly relevant, for he would be often read in the universities as an authority (*pro auctore ipso allegatur*), much to the disdain of his contemporary Roger Bacon. On the other hand, though, his influence on many authors (Aquinas, Boethius of Dacia) produced different and nuanced outcomes. While the role of stars and the heavens became commonplace in the Middle Ages, the manner according to which they could influence the body of the dreamer and the prognostic value of dream images resulting from these became the object of a vivid debate across European universities.

Conclusions

Albert's psychophysiology of sleeping, dreaming and prophecy represents a turning point in the Medieval Latin tradition. Based on Aristotelian gnoseology and heavily influenced by his Arabic sources and Maimonides too, Albert's

166 On the impact of Albert's theories on authors like Boethius of Dacia, Simon of Faversham and James of Douai, see: Luca Bianchi, "Introduzione," in: Boezio di Dacia, *Sull'eternità del mondo. Sui sogni. Sul Sommo bene*, ed. Luca Bianchi (La coda di paglia 41), Milano 2017, 137–168.

167 Rodolfi, Cognitio obumbrata, 129–148.

168 Grellard, "La récéption médiévale du *De somno et vigilia*," 237.

169 On Jean de Jandun, see: Grellard, "La récéption médiévale du *De somno et vigilia*," 232–237.

170 On Adam, see: Kruger, *Dreaming in the Middle Ages*, 86–87.

account of God-given dreams, divinatory dreams and prophecy (either Biblical revelation or natural prophecy) emphasizes the role of images in the processes of dreaming and visions. In his early *De homine*, Albert shaped his psychophysiology of dreams and prophecy on a harmonization of Aristotle and Christian orthodoxy. This mediation was achieved at the price of a forcible identification of the Aristotelian categorization of dreams as equating to God-given dreams and prophetic revelations. In his mature works, Albert posited a sharp distinction between future contingents (future events based on human free will) and events that can be prognosticated according to the laws of natural philosophy and hypothetical necessity. While the former pertain to the category of God-given dreams and Biblical revelations, the latter are welded into a complex astrological and medico-philosophical model. The theological handling of God-given dreams and prophecies found in the mature *Quaestio de prophetia* represents a clear step away from traditional notions such as the *speculum aeternitatis* in favor of a more Aristotelian understanding of these phenomena.

Albert's emphasis on the key role of images, though in a much more complex context, will be expounded in the *De somno et vigilia* according to the doctrines of natural philosophy. Silvia Donati has shown how Albert's account of dreams is given here through a series of 'creative' readings of the Aristotelian text that aims at a consistent account of dream causation either from God and the stars or from the dreamer (*ex parte corporis* or *ex parte animae*). Building on the studies of Donati and Söder, we have highlighted how animal spirits come to play a key role insofar as they are (by definition) vehicle of the faculties of imagination and *phantasia*, and of the images perceived during dreaming. This is because dreaming, in Albert's eyes, is the product of animal spirits flowing from the organ of *phantasia* to common sense (and back) into the organ of retentive imagination. This psychophysiological doctrine, strongly influenced by Albert's Arabic sources, was carved out in the *De homine* and was reiterated consistently in his mature paraphrase of Aristotle's *De somno et vigilia*. While in the former it was quite unclear whether the psychophysiology of dreaming as a flow of animal spirits could apply to theological cases too, in Albert's Aristotelian paraphrases this doctrinal tension is solved. In the *De somno*, Albert fashions a complex astrological and medico-philosophical model of sleep and (veridical or divinatory) dreaming that applies strictly to natural philosophy and its laws of suppositional necessity. In this manner, the psychophysiological role of spirits as carrier of images becomes more clearly linked to the process of dreaming and prophetic images and their truth value.

In Albert's *Quaestio de prophetia*, the images required to foretell the future result from a reordering of those images possessed by the human soul in collaboration with divine action. In this manner, these images are able to signify future events subject to human free will. In this case, the psychophysiological model of the *De homine* does not apply. By contrast, in the *De somno et vigilia*, the role of images and animal spirits is more restricted. In this case, the images carried by the flow of animal spirits into the brain can only signify future events according to the laws of natural causation. These images represent the closest likeness to the celestial form (*forma coelestis*) carried by rays and affecting the prophet during sleep or waking (in case of visions or prophecy). This celestial form, in Albert's view, is the agent responsible for future events. The univocity of the agent causing both dream and future events grants a prognostic value to such dreams, visions or prophetic revelations. Often, the signification conveyed by animal spirits and the faculty of imagination is metaphorical and requires proper interpretation. According to Albert, the correct reception and interpretation of these dreams, visions or prophetic revelation depends on three key conditions: (1) organic, biological perfection of the imaginative (*phantasia*) faculty and perfect exercise of (2) intellectual and (3) ethical virtues. The highest ladder of this scale is represented by the prophet, at least according to the naturalistic understanding of the term.

Not all dreams are God-given or veridical, though. As Albert maintains, there are certain dreams that can be related to causes or signs of future events. In particular, there may be a causal connection between humoral complexion and certain kinds of dreams as their physiological effects, or as the effects of altered body states. As we hope to have brought to the fore, there is a hitherto neglected character of animal spirits that fits precisely into this space of Albert's psychophysiology of dreaming. In both his early and his mature works, there is a clear and remarkable causal correlation between the humoral complexion of the spirit and dream images, especially monstrous visions and nightmares. According to Albert, an excessive amount of moisture in the complexion of animal spirits can either prevent dreaming (as in children) or produce horrendous and monstrous visions in ill or drunk people. Because of the many vapors generated from the action of heat over moisture in ill or drunk people, these images are horrendous or, as in children, cannot be perceived. Moreover, Albert goes back to Aristotle and the medical tradition to show that the scarcely moist complexion of melancholics causes them to dream frequently and remember these dream images. In general, these assertions about the causal correlation between the amount of moisture and the clarity (if not the pleasantness) of an image perceived during sleep chime well with general ideas about the correlation between smaller amounts of moisture and vapors and the clarity of images in dreams (and visions). This is the case with the fortunate Avicennian adage that morning dreams are the most truthful ones.

Albert agrees with Aristotle that the fading away of digestive vapors is temporally proximate to our awaking, since falling asleep is caused by digestive vapors. Sleeping is a teleological process meant to restore the heat in the inner parts of the body through the backflow of spirits and which is caused by these cold digestive vapors that rise up into the brain. These vapors affect primarily those tissues of the sensitive nerves which possess tactile power. Given their sensitivity to cold, the deactivation of the tactile power seated in these tissues triggers a deactivation of all other senses. The peculiarity of Albert's doctrine is that he regards this tactile power to be the foundation for the activity of the sensitive powers for all of the other four senses. In the early *De homine*, Albert becomes more radical in his reading of the relationship between touch and common sense in the definition of sleep as a deactivation of senses. Since these tissues are imbued primarily with tactile power, sleep is primarily considered an affection of touch, whereas common sense is affected only by way of concurrence, viz. insofar as it is related to the outer senses. In the mature *De somno et vigilia*, Albert only maintains that sleep is an affection that pertains primarily to common sense *qua* root of all sensitive powers but, proximately, the sense of touch *qua* possessing that tactile power which, once deactivated, deactivates sense perception in all other senses.

The role of animal spirits turns out to be relevant in another issue concerning Albert's psychophysiology of dreaming. More specifically, Albert has to explain the compatibility of sleep defined as a deactivation of the Aristotelian "first sensitive", viz. common sense, and the rudimentary awareness active during dreaming. This possible tension within Aristotelian psychology is solved by Albert's appeal to the doctrines of his Arabic sources (Avicenna, Averroes, al-Kindī), according to which the deactivation of the faculty of common sense during sleep concerns only its relationship to the external senses and not the internal ones. Rather, the redirection of the sensory powers and their vehicle, viz. animal spirits, to the interior strengthens the activity of the internally apprehensive faculties and their physical substrate. By introducing this interpretation of sleep, Albert is able to reconcile Aristotle and the Arabic philosophical tradition, maintaining a unity in his account of the functions and activities of sleeping and dreaming.

Albert's handling of dreaming, divinatory dreams and prophecy gives perhaps the clearest example of the complex interweaving of disciplines and doctrines that permeate his work and his changing epistemological attitude. In the early theological works, the tendency to reconcile philosophy and Christian faith in agreement is palpable, and so is the opposite tendency of the mature Albert, who tends to separate theological and philosophical understandings of dreaming and prophecy. In his early works, the medico-philosophical explanation of dreams and visions is applied in a context where Christian orthodoxy is forcibly harmonized with Aristotelian dream theory. In his mature works, Albert operates a clear-cut distinction between God-given dreams and prophecies and those dreams and visions that can reveal future events dictated only by the course of nature. Far from upholding a theory of double truth, Albert carefully distinguishes the theological and philosophical realms, integrating into the latter much of the medical learning he had found so extensively in his Arabic sources.

Final Remarks

In this book we have demonstrated how Albert the Great represents an outstanding example of fertile interaction between medical doctrines and an Aristotelian philosophy of nature. His pervasive use of medical doctrines was made possible by the existence of a clear epistemic model borrowed from Avicenna (more precisely, Avicenna's *De animalibus*) and complemented by the doctrines presented by other influential sources such as Costa Ben Luca's *De differentia*, Alfred's *De motu cordis*, Nemesius' *De natura hominis*, Averroes' *De corde* and so on. Albert's psychophysiology was indeed deeply influenced by medico-philosophical doctrines from the Islamic World such as spirit, radical moisture, formative power and complexion. This influence is already clearly visible at an early stage of his intellectual production in the *De homine*. On the other hand, we have highlighted how Albert did not restrain himself to passively inheriting certain medical or zoological notions. Rather, he actively reshaped and reworked these concepts to build a unified account of the material substances at play in the physiological processes of the sublunar world. This approach emerges from his mature Aristotelian paraphrases and mirrors what Joan Cadden has rightfully called Albert's "attempt at a universal physiology". In Albert's mature psychophysiology, this is also the case with the concept of spirit. Previously associated with breathing animals, spirit becomes in the Aristotelian paraphrases the vaporous byproduct of the action of heat over moisture and plays the role of vehicle of life and of the soul's powers and instrument of the soul in all ensouled beings, plants included.

Introduced into the Medieval Latin world by the wave of Graeco-Latin and Arabo-Latin translations of medico-philosophical texts, the concept of spirit had begun to influence the doctrine of so-called "mediums" for the union and the action of the human soul in the body prior to Albert's composition of the *De homine*. The studies of Magdalena Bieniak have clearly shown how this is the case with Philip the Chancellor's *Summa de bono*, the *Summa Halensis* and other authors like John of La Rochelle, Roland of Cremona and, we may add, William of Auvergne. According to its usual medical tripartition, spirits were divided into natural, vital and animal and thought to exist mainly in humans as the physical instrument (together with vital heat) best suited to carry out the different activities of the immaterial soul in the body of living beings. Avicenna, following and reinterpreting Aristotle's different concept of *pneuma*, thought that these spirits existed in both humans and animals. In his early theology, Albert seized on Avicenna's zoological understanding of spirit and blended it with his emanative structure of psychology, strongly based on Avicenna's psychology.

 | DOI:10.1163/9789004706743_011

While sharing the consideration of spirit as a medium for the operation of the soul in the body-soul relationship, Albert diverged from most contemporary theology on the issue of the relation between the soul and its faculties, viz. whether the soul could be identified with or was ontologically distinguished from them. The Dominican master committed his psychology to a strongly Avicennian and emanative structure of the soul, which distinguished the soul from its powers. The relationship between soul essence and soul powers was qualified by Albert through the Neoplatonic term of *fluxus*, which distinguished the soul essence (source of this emanation) from the soul powers derived from it. This psychological structure of living beings would have remained consistent throughout Albert's entire career. By contrast, many contemporary authors (John of La Rochelle, *Summa Halensis*) went in a different direction, for example by endorsing the doctrine of the 'plurality of forms'. Apparently, this position was quite common (at least among Franciscans) in the milieu of Paris' theological faculty, where Albert studied and likely finished writing his *De homine*. The result was a different understanding of the causal role of spirit in the psychological structure of the living being.

In the Aristotelian paraphrases, Albert developed a concept of spirit that notably moved away from his sources, even while retaining some elements of continuity. By defining spirit as the vaporous effect of the action of heat over moisture found in all living beings, Albert successfully merged the medico-philosophical concept of spirit (a zoological concept) with doctrines found in the pseudo-Aristotelian *De plantis*, emphasizing the vaporizing action of heat over moisture in plants. In his botany (*De vegetabilibus*) and in his psychophysiology (*Parva Naturalia*), Albert pushed the similarities between plants and animals as far as possible in terms of physiological structures and material substances involved in life-related phenomena. Plants have an organ analogous to the heart placed in their roots which allows the spirit to spread into their bodies through vein-like channels, very often situated in the plant's marrow. At least, this is true of their most complex species. Like animals and humans, plants depend for their longevity (also, growth) on radical moisture, possess proper ages defined by the opposition of primary qualities, are generated through formative power placed in the spirit contained in the seed and have a humoral complexion of their own. Still, this complexion is different and inferior to those of humans and plants since plants possess an inferior degree of physiological complexity (*organizatio*). Unlike humans, plants do not have pulsatile blood vessels like arteries (*venae pulsatiles*), designed to carry vital spirits, a fact which makes it impossible to single out vital and natural spirits in plants as one can in breathing animals and humans. Like bloodless animals endowed with connate pneuma (*spiritus complantatus*), plants lack a physiological structure comparable to veins and arteries in the most complex animals. Plant physiology reveals Albert's outlook on psychophysiology. On the one hand, Albert strongly supports a unifying perspective on the material substances at play in plant physiology, which makes it closer to human and animal and human psychophysiology. This is the case with spirit, radical moisture and formative power. On the other hand, any explanation of plant psychophysiology must come to terms with the different physiological structure of plants, which makes them more similar to the less complex animals, viz. bloodless animals, than are perfect animals, or humans. Beyond a reasonable doubt, the extension of spirit to plants represents the true marker of Albert's mature psychophysiology and sets him apart from contemporary commentary on Aristotelian psychology (*De anima*), psychophysiology (*Parva Naturalia*) and zoology. Still, Albert's general understanding of the concept of spirit proves to be much indebted to his Arabic sources, a debt which goes back to Albert's early theology.

This book has covered two areas where this influence is substantial: in Albert's views on life-related and perception-related activities. Nonetheless, there are still doctrinal fields in which the role of spirit has not yet been fully investigated. This is the case with growth, locomotion, emotions (*passiones*), and the relation between spirit and humoral complexion, all topics which deserve further scholarly attention. While these fields await further research, it is time for us to try and gather the most important results of this long historic-philosophical journey. To do so, we must turn our attention from plants to humans and animals. In the introduction, we argued that this detailed study of Albert's psychophysiology would help us draw some relevant methodology-related and content-related conclusions. Now it is time do so. On the one hand, we shall consider Albert's use of medical doctrines in the light of his understanding of the relationship between natural philosophy and medicine and the historical circumstances that prompted him to make use of medical and medico-philosophical doctrines. As respects the concept of spirit more specifically, we shall posit some final considerations on its role in the (1) physiology of life-related and perception-related processes, (2) the body-soul relation and the psychosomatic integrity of the living being and (3) the place of spirit within a cosmos ruled by divine providence (*opus naturae est opus intelligentiae*).

1 Methodology: Albert the Great and the Role of Medicine

The doctrine of spirit represents a good case study for understanding Albert's attitude toward medical doctrines and medico-philosophical concepts. Inherited mostly from his Arabic sources, the concept of spirit is carefully rethought and reshaped in his Aristotelian paraphrases. This first reception and successive reshaping takes place in keeping with Albert's attitude towards medicine, viz. an open yet epistemically careful one. On the one hand, Albert welcomes much of Galen's anatomy and physiology as early as his *De homine*. Much of this anatomical material was provided by Avicenna's *De animalibus*, of which the *De homine* represents one of the earliest examples of Latin reception, by Costa Ben Luca's *De differentia spiritus et animae* and, to a lesser extent, by Nemesius' *De natura hominis* and Alfred's *De motu cordis*. Following Avicenna's lead, Albert rejects Galen's disputed thesis on the active agency of female semen, which would imply the coexistence and co-agency of two formal causes in a single mixing process. This position will be consistently present in the mature *De animalibus*, following again the lead of Avicennian zoology. Like Avicenna, though, Albert suffered from major epistemic problems of compatibility between his accounts of generation and of length and shortness of life in his Aristotelian paraphrases. By adopting the Avicennian theory of radical moisture and accepting (like Avicenna) that part of the male semen entered the material composition of the embryo, Albert ended up breaking the classically hylomorphic scheme of generation set up by Aristotle. This major problem appears clearly in his *Quaestiones super de animalibus*, where Albert divides masculine semen into two parts. The first contains spirit and formative power and does not enter the material composition of the embryo while informing embryonic matter. The second part of the masculine semen coincides with seminal moisture, which according to Albert equates to radical moisture and becomes therefore part of embryonic matter and the physiological basis for its growth. Though breaking the rules of Aristotelian hylomorphism in generation, Albert's equation between seminal and radical moisture allowed the Dominican master to construe a universal physiology of the cycle of birth-growth-aging-decay-death so crucial on many issues of his psychophysiology. Perhaps this was the price Albert was willing to pay in exchange for the infringement of Aristotle's hylomorphic scheme of generation. This is problem, caused by the role of radical moisture, is shared by Peter of Spain at the time.

The second field of typical medico-philosophical controversy is represented by the theme of the origin of veins. Albert's strongly cardiocentric understanding of the soul and its faculties led him to rejects Galen's doctrine according to which veins originate from the liver and not from the heart. This rejection, only implicit in his early works, became explicit in his mature *De animalibus* and *Quaestiones super de animalibus*. While in the *De animalibus* Albert's rejection was based on Averroes' *De corde*, in his *Quaestiones super de animalibus* much of his *solutio* seems to have been beholden to Avicenna's *De animalibus*. Both Averroes and Avicenna reject Galen's position, but based on different premises: Averroes based his rejection on the cardiocentric nature of nutrition. By contrast, Avicenna insisted on the cardiocentric understanding of embryonic development as a grounds for rejecting Galenic doctrine. All in all, the solutions proposed by the Arabic authors and endorsed by Albert represent two sides of the same coin vis-à-vis Aristotelian cardiocentrism. Besides these medico-philosophical controversies, Albert valued his medico-philosophical sources from the Islamic World as a tool for correcting and complementing Aristotelian works (e.g. *De longitudine et brevitate*) and supplementing the unavailable Aristotelian *De iuventute et senectute* and *De respiratione* by way of paraphrastic reworkings. This is the case with the *De iuventute et senectute*, a paraphrastic reworking of Avicenna's chapters on human age in the *Canon* and *De animalibus*, and of the *De spiritu et respiritu et respiratione*, inspired by Costa Ben Luca's *De differentia spiritus et animae*.

Despite these clear debts as regards methodological issues and Albert's substantial use of them, his understanding of natural science differs notably from Avicenna, his main Arabic source, Averroes and Aristotle himself. This is because Aristotle had forbidden in his *Parts of animals* to deal with parts and functions in animals rather than producing a redundant list of species with their proper organic and psychological features. On the other hand, Albert consciously included books on plants, animals and minerals, categorized not only according to their parts and functions but also according to their species, in his works on natural sciences (botany, zoology and mineralogy). Aristotle, Avicenna (and Averroes too) would have all forbidden this. To be fair, Albert was conscious that this approach was not so much in keeping with Aristotelian methodology (*licet enim hic modus non omnino sit philosophicus*), but he nonetheless decided to consider these books an integral part of his extensive exegesis of the Aristotelian corpus. A further element that marks the distinction between Albert and his Arabic sources consists in the addition to Aristotle's *De animalibus* of an entire book devoted to the study the animal body (*De animalibus*, book XXII) as a whole, or a book dedicated to ranking animals according to their cognitive capacities (*De animalibus*, book XXI). All in all, the main difference that constitutes the irreducible divide between Albert and his Arabic sources (Avicenna, Averroes) is represented by his understanding of the scope of zoology, botany and mineralogy as a combination of the study of living beings and minerals according to parts and organs but also species.

The inquiry undertaken in this book has shown how the reasons for Albert's inclusion of medical learning in his natural philosophy rest on his abstract considerations of the process of healing as the primary product of the agency of nature (and not of art). On the other hand, other chapters have made clear how Albert's use of medico-philosophical sources was common within the context of the last period of his early intellectual formation (Paris) and, on the other hand, how his use depended on the modalities of his paraphrastic work on the Aristotelian corpus. Albert's use of these medico-philosophical texts was driven by both historical and doctrinal factors but presented an original outcome. Like many previous authors at Paris, beginning with Philip the Chancellor, Albert introduced the concept of spirit into his discussion on the activities of the soul's powers in the body. Previous authors of theological treatises or *summae*, though, were nowhere as detailed as Albert in his psychophysiological account.

On the other hand, the remarkable aspect of Albert's mature production is his detailed approach to the genuinely Aristotelian understanding of the principles (not the scope, though, as seen above) of natural sciences. According to Albert, medical texts could fill in gaps in the Aristotelian corpus, complement and reinterpret Aristotelian teachings in a manner that was suitable to natural science (*secundum quantum confert scientiae naturali*) and to facilitate a correct understanding of what he considered the "Aristotelian corpus" (*ut Aristotelem competenter intelligere possint*). Clearly, this inclusion of medical doctrines was not without problems, as in the case of the compatibility between his account of generation and his theory of radical moisture. Nonetheless, Albert represented an authority for philosophers and physicians for decades to come, a child of his time but with strong peculiarities of his own. Albert's extensive, detailed and self-conscious use of medico-philosophical notions to shape his 'universal physiology' represents a peculiar case study within 13th-century philosophy of nature. His inclusion of medical teachings in works of natural philosophy rests clearly on distinct theoretical motivations (*De animalibus*), partially found in Albert's earlier works (*Commentary on the Sentences*) or correlated to statements as to why medicine meets natural scientific inquiry halfway (*Parva Naturalia*). As stressed above, though, his use of medical learning was also driven by concrete, historical circumstances in both his early works and his Aristotelian paraphrases and was influenced by his key Avicennian model.

A third and last factor that conditioned Albert's use of medico-philosophical sources was represented by his approach to sensitive topics straddling philosophy and theology such as God-given dreams, divinatory dreams and prophecy. In his early works, some of which (e.g. the *De homine*) even precede his *inceptio* as *magister theologiae* in Paris in 1245, Albert sought a harmonization between Aristotle's psychophysiology of dreaming and the teachings of the Christian faith. This harmonization was achieved at the expense of forcibly including God-given dreams and prophetic revelations into the Aristotelian category of dreams as coincidences. In this manner, Albert's detailed psychophysiology of dreaming as a motion of animal spirits in the brain straddled the realm of God-given dreams, veridical dreams related to future natural events, dreams as signs or causes of future events, and random occurrences. The outcome of this harmonizing approach is that the *De homine* did not provide any explicit remark as to the context in which this physiological explanation of dreaming could have been applied. By contrast, Albert's mature production was characterized by a careful distinction between objects of God-given dreams and divinatory dreams related to future natural events and based on a sharp distinction between theological and philosophical inquiries into dreams and prophecy (*philosophi aliter loquuntur de prophetia quam sancti*) that he developed in following works, viz. the *De somno et vigilia* and the *Quaestio de prophetia*.

In the *Quaestio de prophetia*, God-given dreams and prophecies concerned future contingents subject to free will and they could not be reduced to laws of causation ruling natural events. Rather, they were explained by the peculiar reordering and rearrangement of previously stored images in the human soul, now given new meaning thanks to the cooperation between God and the human intellect. By contrast, the complex psychophysiological model of dreaming which Albert had developed in the *De homine* was to be largely retrieved in Albert's mature *De somno et vigilia*. To be clear, Albert's mature *De somno et vigilia* does display some doctrinal changes, but those changes primarily concern the role of common sense and touch in the definition of sleep, not the physiological explanation of dreaming as a motion of animal spirits from the organ of *phantasia* to the organ of common sense and back. In his *De somno et vigilia*, Albert presented an influential distinction between dreams originating with God and the stars and those that originate in the dreamer, either from the part of the body or of the soul. In this context, Albert structured the classification of divinatory dreams, visions and prophecy derived from God or the stars according to an ascending ladder of 13 different degrees. This classification was meant to provide a link of causal correlation between dreams, visions and prophecy and future reality. According to Albert, the clarity of the images that revealed future events—according to the laws of natural philosophy—was directly proportional to the capacity of the imaginative faculty to receive the image itself (*secundum potestatem suscipientis*). For this reason, the first requirement of a good prophet was thought to be a perfect complexion of the imaginative faculty, which involves, among other things, the concept of animal spirit *qua* vehicle of the imaginative faculty. To give a correct interpretation of the image produced in the body by the reaction of the imaginative faculty to the reception of the "celestial form" (*forma coelestis*), a good prophet needed also to possess a perfectly actualized intellectual capacity. In this manner, he would have been able to give a correct interpretation of his dreams. If not (or if the images was an insufficiently clear one), he would himself have needed a good interpreter of his dreams or visions.

According to Albert's *De somno et vigilia*, the prognostic value of divinatory or veridical (not God-given) dreams, visions and prophecies triggered by God and the stars is limited to future occurrences of natural events subject to the laws of hypothetical necessity. Thus, all images that predict events based on free will are excluded from any naturalistic explanation. The prognostic value of dreams, visions and natural prophecy is therefore rooted in their primary agential causes: given the identity between the agential cause of a dream and that of the future events, those dreams may have a prognostic value. By contrast, other dreams, viz. those stemming from the dreamer (either body or soul), do not foretell any future event but are rather indicative of the humoral complexion of the individual (dreams from the body) or cause future events by prompting our enacting of a hitherto unaccomplished or expected action (dreams from the soul). In these latter cases, especially in dreams from the part of the body, Albert emphasizes how the imbalance in certain elemental qualities of the spirits, especially moisture, produces horrible images and nightmares in ill or drunk people which have no prognostic value at all. By contrast, people of melancholic complexion dream often and tend to retain these dreams more than others because of their dry complexion. The gradual decrease in the amount of digestion-related vapors, the efficient cause of sleeping, explains why morning dreams, like in Avicenna, are the most truthful dreams (in naturalistic terms). They are so because dreaming images which appear at morning are deprived of those excesses in smoky vapors that make perception and sense perception unreliable, causing images to mix up in the brain as in other cases such as epilepsy and other examples of misperception or sense impairments as those occurring in baths, caused again by an excess in vapors.

All in all, Albert's use of medico-philosophical sources and of the concept of spirit in his doctrine of sleeping and dreaming develops throughout his career. In his early works, the distinction between the psychophysiology of God-given dreams and divinatory or veridical dreams is not so clear, including the physiological correlations of these dreams. By contrast, in his mature works Albert draws a clear line between the realms of theology and natural philosophy based on the notion of future contingents so as to keep carefully at a bay any reductionistic, naturalistic and purely physiological explanation of God-given dreams. Rather, the physiology of dream images based on the complexion of spirits reveals, certain bodily states or psychic intentions, but does not concern divine revelation but only dreams with limited prognostic validity or just chance occurrences of dreaming. This careful distinction in the object of prophecy according to theology and natural philosophy, and the consequent applicability of the psychophysiological model developed from his Arabic sources, sets Albert apart from his Arabic forerunners themselves. While

Avicenna, al-Ġazālī and Averroes had woven the psychophysiology of dreaming into the fabric of a rationalizing explanation of God-given—and not just divinatory dreams and prophecy, Albert developed a careful distinction between God-given dreams and prophecy and a purely natural kind of prophecy, viz. related to future natural events. Moreover, all of these authors had mostly relied on the concept of complexion to explain the nature of dreaming images, as most 12th-century in the Latin West began to do. Albert pushes his medico-philosophical explanation of dreaming a step further, linking the complexion not only of the body but of the (animal) spirits to the process of production of the dreaming (or prophetical) image and the dreaming image itself.

2 The Role of Spirits

Spirit can be legitimately considered one of the best candidates for the concept that best clarifies Albert's outlook on psychophysiology. Its role as the proper vehicle of life and of the soul's powers, and as instrument of God and of the soul makes it a pivotal notion in Albert's psychophysiology. This key role of spirit helps us bring to the fore three main aspects of Albert's psychophysiology which deserve attention: (1) the strict interdependence of all physiological processes in all ensouled beings, (2) the relationship between the psychosomatic unity of the living being and the causal explanation of the psychophysiological process and (3) the inseparable relation among any material substance and the providential structure of the cosmos (*opus naturae est opus intelligentiae*).

2.1 *Physiology of Life and Perception*

The strict interdependence of physiological processes in all living beings emerges clearly from the Aristotelian metaphor used by Albert to read the relationship between the three kinds of spirits (vital, natural, animal). According to Albert, vital spirit is related to natural and animal ones in the same manner the different souls are related to each other, viz. as a triangle is contained in a square. This analogy emphasizes the non-separability, gradual complexity and interrelations among the different psychophysiological sets of activities in a living being. In concrete terms, we have observed how the actions of the different spirits cannot be seen as discrete from one another: the amount of vital heat spread by vital spirits is needed for nutrition to work; nutrition is needed for any living being to subsist; sense perception, internally apprehensive powers and intellectual activities can be affected and impaired by failures in both the sensitive and the digestive capacities. The strict interdependence among physiological processes is particularly evident in the case of animal spirits, defined by Albert as the highest result of a digestion-like process of the body's internal humors. This is a clear heritage of Arabic authors such as Avicenna and Costa Ben Luca. The mutual influence of these physiological structures emerges in many cases that have been observed in this book, viz. the diseases consequent to aging and caused by loss in digestive capacities, the embryonic malformations caused by defects in the formative power found in the semen and so on. While emphasizing the complexity of the anatomical and physiological structure based on the humoral system, Albert conceives the different physiological processes underlying the activity of the different soul powers as the gradually higher rungs of a hypothetical ladder with intellectual activities at its top. While in his early theology this is less self-evident and primarily relates to the zoological realm, in the Aristotelian paraphrases this strongly unifying perspective on the psychophysiology of all living beings (plants included) becomes the backbone of Albert's account. Albert builds his mature psychophysiology by pushing analogical comparisons between the world of plants and that of animals and humans on a number of concepts (age, radical moisture, formative power, the possession of a heart-like organ, of vein-like channels etc.), which highlights his 'universal' perspective on physiology. The close comparison between human, animal and plant physiology developed from originally medical or zoological notions is beyond a reasonable doubt the defining trait of Albert's mature psychophysiology. Like many of his contemporaries, we have seen how Albert employs a relevant amount of medical learning which he borrows from his Arabic sources.

There are two main differences between Albert and his contemporaries, though. In the *De homine* and his early theology, Albert pays greater attention to the physiological details of the organs of the different psychic powers and their related physiological processes. The only exception is represented by Peter of Spain. In the Aristotelian paraphrases, Albert typically enlarges the use of these notions to the world of plants and minerals whenever possible. At this time, there were no authors who were developing such a comprehensive philosophical vision, especially concerning botany. As opposed to this unifying pattern of studying natural phenomena in living beings, though, Albert emphasizes how humans are set apart from all other creatures because of their different essence and quiddity and not because of some behavioristic differences. Humans are the one and only *nexus Dei et mundi* and "intellect alone" (*homo inquantum homo solus est intellectus*), a difference that cannot be matched by any other creature and which is reflected in the peculiar structure of the brain. In humans, the middle brain ventricle hosts cogitative power, which is in itself no faculty of the soul. Rather, it designates a series of activities of the sensitive soul which require the assistance of the intellect, as in the case of recollection or the inchoative phase of intellectual activity, viz. the abstraction of the universal from a singular image. The structure of the brain shows how the different physiological arrangement of the brain parts is meant to host different psychological capacities, viz. those of the rational soul, which draw an essential line between humans and animals. Moreover, this difference is marked when considering sense perception: much as animals are better at certain forms of sense perception, they will never be able—like humans—to make more cognitively complex use of similar sense data. Ultimately, Albert roots this strong and essential superiority of all human faculties in their partaking of an immortal and rational soul.

2.2 *Spirit and the Body–Soul Relationship*

Albert's doctrine of the soul and his interpretation of the psychosomatic integrity of the living being in psychophysiological processes offers a clear case study of his strong Avicennian influence. On the one hand, Albert argues that the humoral and elemental complexion of spirits explains the different degrees in cognitive abilities among animals and among humans themselves, their different receptivity to sense stimulus, the decay in cognitive and rational abilities because of aging, the influence of endogenous and exogenous factors at play in birth, the formation of dream images, promptness in instinctual reactions, good memory and so on. Even more, according to Albert the material complexion of spirit impacts the effectiveness of the highest among psychic powers, viz. intellectual capacities. This is by all appearances a long-lasting *Nachleben* of Galen's idea that the faculties of the soul follow upon bodily mixture. Despite its strong relevance for material and efficient causation and its ability to influence the working of the soul powers, spirit is no active cause in itself. According to Albert, the source of its life-giving capacity derives from the emanation of the form of life from the soul (*vita influxu ab anima*) given a suitable arrangement of the body (*organizatio*).

Life becomes such when body, humors and spirit reach the suitable material organization required for the form of life (*forma vitae*) to be situated in the body. Thus, the immaterial form of life and, in the end, the soul, is the agent that enables spirit to spread life and the powers of the soul into the body. Many times across his works Albert explains the spread of the soul's virtues trough the body to luminous rays which transmit heavenly virtues to the sublunar world (*spiritus moventur in corpore sicut lumina in mundo*).

Still, the role of the soul and of the emanated form of life explains only that the material action of spirit cannot exhaust a causal explanation of psychophysiological process. By the conclusions drawn so far, it is not clear why the complexion of spirit and humors is such, viz. life-giving or suitable for any other activity of the soul. According to Albert, the presence of the soul, located in the heart (*sedes et domicilium animae*), is that for whose sake the body parts of the hylomorphic compound are arranged. In like manner, the different natures of the different faculties of the soul explains why the different kinds of fluids necessary for the exercise of the soul's powers, among which are spirits, have certain specific humoral and elemental properties. As has emerged through this long philosophical survey, the nature of spirits may be at variance between plants, animals and humans because of their different souls. On a different level, variation in elemental and humoral composition of spirits can explain on a more general level the differences between plants, animals and humans, a less general and interspecific level of differences among animals (e.g. perfect vs. imperfect animals) and a very restrict difference among individuals of the same species (animals of the same species, humans from different places/sexes/ages).

For instance, the clarity and brightness of certain spirits in animals such as eagles can be required for them to see from afar. By contrast, animals such as flies have no memory because of their excessively cold complexion of spirit. On another note, the particularly hot nature of natural spirits is required to carry out digestion in the stomach. By contrast, animal spirits must be subtle and clear to be able to vehiculate the most complex cognitive objects of the internally apprehensive faculties. These two examples, which apparently have nothing to do with each other, explain by contrast how spirits can differ either within animals (or among animals, plants and humans) and even within the kinds of spirits themselves (natural, vital, animal) in view of the activity for the sake of which they operate. Spirits are simply inert without the body's life-giving principle, viz. the soul. The nature of the soul (and its faculties) as a final cause of bodily organization in all living beings explains therefore the variance in the complexion of spirits both among themselves and among those found in different kinds of living beings. Certainly, the material endogenous and exogenous factors can influence the action of spirits and the soul's faculties. Yet, material and efficient causes alone fail to explain fully the active role of spirits, tied to their reception of the *forma vitae*, and the teleology of the material composition of spirits themselves, dependent on the nature of the soul existing in a certain body and its faculties, whose exercise requires a physiological structure of its own. The goal of the different activities of the soul faculties, together with the nature of the soul *qua* teleological principle of organization of a certain living being, is therefore the key to understand the teleology of the material compositions of spirit. Reconsidering Galen's famous words, the faculties of the soul do not only follow upon the bodily mixtures, but also determine the complexion of the material substances meant for the exercise of these faculties.

Albert's psychophysiology of life and perception rests therefore on the emanative structure of the soul and its nature as the final cause of the organizations of body parts and substances, among which are spirits. This heritage is doubtlessly Avicennian and is developed in the Aristotelian paraphrases in a manner that closely resembles the doctrine of the *De medicinis cordialibus*. Despite this strong Avicennian influence, Albert's approach is solidly grafted on an Aristotelian understanding of natural teleology. The remarkable element of difference between Albert and Avicenna is the use of this model (soul-spirit-humors etc.) of causality in all ensouled beings, not just humans and animals, but also plants.

The final picture emerging from Albert's doctrine of spirit is one in which life- and perception-related activities cannot be thought of as physiologically discrete from each other and are analogous in all living beings, plants included, only when considering the material substances (spirits, radical moisture, humors) involved. On the other hand, humans emerge as essentially different from all other living beings *qua* intellect alone and different even in the lowest faculties of the soul because of their partaking in the rational soul. In the second place, Albert's psychophysiological explanations of life- and perception-related activities appear to be incomplete when they do not combine material (spirits, humors, body) and immaterial (soul) causes, key to understanding the causal agency of spirits, the teleology of their material composition and of the actions of the soul powers carried out by spirits *qua* their vehicle. To complete the puzzle of Albert's psychophysiology, we need to turn to his connection between spirit, soul and the providential structure of the cosmos.

2.3 *Natural Teleology*

This last piece of the puzzle can be gathered by considering the origin of the human, animal and vegetative soul. Surely, spirit is defined as the vehicle of the form of life emanated by the soul. The reception of this form of life in the body through the mediation of the spirit grants that the soul imparts life to the body *qua* cause and principle of life. This idea, explicitly expounded in Albert's *De morte et vita* and *De spiritu et respiratione*, is only embryonic in Albert's early works. The soul, though, does not come to be on its own or in a manner that is abstract from external reality, which is ruled by divine providence. For this reason, to understand fully what the primary agent of life is, we need to go back to the origin of the soul.

In the case of plants and animals, their souls are educed (*eductio formae*) from matter. In Albert's psychophysiology, eduction from matter is possible if and only if the formative power possesses a suitable elemental (viz. thermic) composition that is able to turn the matter of the embryo or of the plant from life in potentiality into life in actuality. This suitable thermic composition, especially the amount of vital heat in the formative power, is derived from a combination of seminal and heavenly heat, the latter of which depends on the motion of heavenly bodies. According to Albert, there is little doubt that the motion of the heavens is regulated by God's providence. Thus, divine providence becomes the ruling cause of all natural processes in the living world, among which is the origin of the soul. By bringing Albert's providential structure of the cosmos to its most extreme consequences, even the spirit found in spontaneously generated animals and plants is an effect of divine providence regulating the motion of stars and heavens. This is because the existence of spontaneously generated living beings is produced by the cosmic component of formative power linked to the motion of heavenly bodies and their astral influences, all ruled by divine providence. By contrast, the case of the human soul is more intuitive. In humans, the action of spirits becomes life-giving as soon as the soul created by God transmits into the whole body the form of life (*forma vitae*) by the means of emanation from the soul in itself. This transmission, though, must be met by a suitable bodily organization in terms of primary qualities, humors and spirits. Spirit appears thus to be the instrument of God as the First Cause, since the life-giving capacity of the spirit is a direct effect of the soul's capacity to make a body alive. In turn, the soul, whatever its nature (human, animal, vegetative), is a more or less direct effect of divine providence and its creating and ruling action over the universe.

This combination between Neoplatonic emanative psychology and Galenic medicine is strongly influenced by Avicenna, and blended with insights from Averroes. Following in and enlarging the footsteps of his illustrious predecessors, Albert shaped a model of psychophysiology that granted the interdependence of life-related processes, the psychosomatic unity of the living being and the providential structure of the cosmos. At the end of this philosophical journey, we can now clearly observe how all of the above-listed, key aspects of Albert's spirit theory result from a development and original reworking of medico-philosophical doctrines inherited from authors of the Islamic world.

Bibliography

Sources and Literature

Unedited Sources

Adam of Buckfield, De sensu et sensato (London, The Wellcome Historical Medical Library, 3, fols. 53vb–60ra).

Averroes, Capitulum de corde (Paris, Bibliothèque National de France, 16110, fols. 99va–100rb).

Edited Sources

Al-Kindī, Liber de somno et visione quem edidit Jacobus Alchindus, magister vero Gerardus Cremonensis transtulit ex arabico in latinum, ed. Albino Nagy in *Die philosophischen Abhandlungen des Ja'qūb ben Isḥāq al-Kindī* (Beiträge zur Geschichte der Philosophie des Mittelalters 5/2), Münster 1897, 12–27.

Al-Kindī, Risāla fī Māhiyyat al-nawm wa-l-ru'yā, in: Rasā'il falsafiyya, ed. Muḥammad 'Abdu Abū Rīda, Cairo/Frankfurt a.M., 1950–1953/1999, 293–311.

Alanus de Insulis, De fide catholica contra haereticos sui temporis, praesertim Albigenses liber quatuor, PL 210, Paris 1855.

Albertus Magnus, De aetate sive de iuventute et senectute, ed. Auguste Borgnet, Paris 1890 (Ed. Paris. IX, 305–321).

Albertus Magnus, De anima, ed. Clemens Stroick, Münster 1968 (Ed. Colon. VII/1).

Albertus Magnus, De animalibus, ed. Hermann Stadler, Albertus Magnus: De animalibus libri XXVI, I–II (Beiträge zur Geschichte der Philosophie des Mittelalters 15–16), Münster 1916–1920.

Albertus Magnus, De bono, ed. Heinrich Kühle u. a., Münster 1951 (Ed. Colon. XXVIII).

Albertus Magnus, De causis et processu universitatis a prima causa, ed. Winfried Fauser, Münster 1993 (Ed. Colon. XVII/2).

Albertus Magnus, De divinatione. *Über die Weissagung, Latin-German, with a transl. and intr. by Silvia Donati* (Herders Bibliothek der Philosophie des Mittelalters 48), Freiburg im Breisgau 2020.

Albertus Magnus, De homine, ed. Henryk Anzulewicz-Joachim Söder, Münster 2008 (Ed. Colon. XXXIV/2).

Albertus Magnus, De memoria et reminiscentia, ed. Silvia Donati, Münster 2017 (Ed. Colon. VII/2a, 113–137).

Albertus Magnus, De morte et vita, ed. Auguste Borgnet, Paris 1890 (Ed. Paris. IX, 345–373).

Albertus Magnus, De motibus animalium, ed. Auguste Borgnet, Paris 1890 (Ed. Paris. IX, 257–303).

Albertus Magnus, De natura et origine animae, ed. Bernhard Geyer, Münster 1955 (Ed. Colon. XII, 3–44).

Albertus Magnus, De natura loci, ed. Paul Hossfeld, Münster 1980 (Ed. Colon. V/2, 1–46).

Albertus Magnus, De nutrimento et nutrito, ed. Silvia Donati, Münster 2017 (Ed. Colon. VII/2a, 1–18).

Albertus Magnus, De quattuor coaequaevis, ed. Auguste Borgnet, Paris 1895 (Ed. Paris. XXXIV, 307–789).

Albertus Magnus, De resurrectione, ed. Albert Ohlmeyer o.s.b, Münster 1958 (Ed. Colon. XXVI, 237–354).

Albertus Magnus, De sacramentis, ed. Albert Ohlmeyer o.s.b., Münster 1958 (Ed. Colon. XXVI, 1–170).

Albertus Magnus, De sensu et sensato, ed. Silvia Donati, Münster 2017 (Ed. Colon. VII/2a, 19–112).

Albertus Magnus, De sex principiis, ed. Ruth Meyer, Münster 2006 (Ed. Colon. I/2, 1–80).

Albertus Magnus, De somno et vigilia, ed. Auguste Borgnet, Paris 1890 (Ed. Paris. IX, 121–212).

Albertus Magnus, De spiritu et respiratione, ed. Auguste Borgnet, Paris 1890 (Ed. Paris. IX, 213–255).

Albertus Magnus, De unitate intellectus, ed. Alfons Hufnagel, Münster 1975 (Ed. Colon. XVII/1, 1–29)

Albertus Magnus, De vegetabililibus libri VII, ed. Ernst Meyer/Karl Jessen, Alberti Magni … De vegetabilibus libri VII, Berlin 1867/Frankfurt a.M. 1982.

Albertus Magnus, De vegetabilibus, *Buch VI, Traktat 2: lateinisch-deutsch. Transl. and comm. by Klaus Biewer. With an introd. by Rudolf Schmitz* (Quellen und Studien zur Geschichte der Pharmazie 62), Stuttgart 1992.

Albertus Magnus, Ethica, ed. Auguste Borgnet, Paris 1891 (Ed. Paris. VII).

Albertus Magnus, Metaphysica, ed. Bernhard Geyer, Münster 1960 und 1964 (Ed. Colon. XVI/1–2).

Albertus Magnus, Meteora, ed. Paul Hossfeld, Münster 2003 (Ed. Colon. VI/1).

Albertus Magnus, Mineralia, ed. Auguste Borgnet, Paris 1890 (Ed. Paris. V, 1–116).

Albertus Magnus, *On Animals: a Medieval* Summa Zoologica, *transl. and annotated by Kenneth F. Kitchell and Irven Resnick* (Foundations of Natural History) Baltimore 2020, 2 voll.

Albertus Magnus, Physica, ed. Paul Hossfeld, Münster 1987 und 1993 (Ed. Colon. IV/1–2).

Albertus Magnus, Quaestio de prophetia: *visione, immaginazione e dono profetico*, ed. Anna Rodolfi. Per Verba, 24. Firenze: Edizioni del Galluzzo per la Fondazione Ezio Franceschini, 2009.

Albertus Magnus, Quaestiones super De animalibus, ed. Ephrem Filthaut, Münster 1955 (Ed. Colon. XII, 77–321).

Albertus Magnus, Summa theologiae sive de mirabili scientia Dei libri I pars I quaestiones 1–50A, ed. Dionysius Siedler, Münster 1978 (Ed. Colon. XXXIV/1).

Albertus Magnus, Super Ethica, ed. Wilhelm Kübel, Münster 1968, 1972, 1987 (Ed. Colon. XIV/1–2).

Albertus Magnus, Super I Sententiarum (dist. 1–25), ed. Auguste Borgnet, Paris 1893 (Ed. Paris. XXV); Super I Sententiarum (dist. 26–48), ed. Auguste Borgnet, Paris 1893 (Ed. Paris. XXVI); Super II Sententiarum, ed. Auguste Borgnet, Paris 1894 (Ed. Paris. XXVII); Super III Sententiarum, ed. Auguste Borgnet, Paris 1894 (Ed. Paris. XXVIII); Super IV Sententiarum (dist. 1–22), ed. Auguste Borgnet, Paris 1893 (Ed. Paris. XXIX); Super IV Sententiarum (dist. 23–50), ed. Auguste Borgnet, Paris 1894 (Ed. Paris. XXX).

Albertus Magnus, Topica, ed. Auguste Borgnet, Paris 1890 (Ed. Paris II, 525–713).

Alfredus Anglicus, De motu cordis, ed. Clemens Baeumker, *Des Alfred von Sareshel* (Alfredus Anglicus) *Schrift* De motu cordis (Beiträge zur Geschichte der Philosophie des Mittelalters 23/1–2), Münster 1923.

Algazel (Abū Hāmid al-Ġazālī), Maqāṣid al-falāsifa: fi-l-manṭiq wa-l-ḥikma al-ilāhiyya wa-l-ḥikma al-ṭabī'iyya, Cairo 1936.

Algazel (Abū Hāmid al-Ġazālī), Metaphysica, ed. Joseph. T. Muckle, *Algazel's Metaphysics: A Mediaeval Translation*, Toronto 1933/1967.

Algazel (Abū Hāmid al-Ġazālī), Metaphysica [IV.1–5], ed. Eva St. Clair, "Algazel on the Soul: A Critical Edition," *Traditio* 2005 (60) 47–84.

Aristotle, De anima [Translatio vetus], ed. Decorte-Brams, Iacob Veneticus translator Aristotelis. De anima, textus editionis quae paratur a +J. Decorte et J. Brams [A.L. XII.1]. Text available online at the ALD: https:/clt.brepolis.net/ald.

Aristotle, De generatione animalium [Translatio Scoti], ed. Aafke M.I. van Oppenraaij. *Aristotle*, De animalibus. *Michael Scot's Arabic-Latin Translation, Part Three: Books XV–XIX: Generation of Animals* (Aristoteles Semitico-Latinus 5/2), Leiden-New York-Köln 1992.

Aristotle, De longitudine et brevitate vitae [Translatio vetus], ed. Manuel Alonso, in *Pedro Hispano: Obras filosóficas*, III (Instituto de Filosofía "Luis Vives", Serie A Núm. 4), Madrid 1952, 403–411.

Aristotle, De memoria et reminiscentia [Translatio vetus]. Textus praevius a S. Donati paratus [A.L. XIV.1]. Text available online at the at the ALD only: https:/clt.brepolis.net/ald.

Aristotle, Aristotle's De motu animalium: Symposium Aristotelicum, ed. Christof Rapp/Oliver Primavesi. *With a new critical edition of the Greek text by Oliver Primavesi and an English transl. by Benjamin Morison*, Oxford 2020.

Aristotle, Aristotle's De motu animalium. *Text with translation, commentary, and interpretive essays by Martha Craven Nussbaum*, Princeton 1978.

Aristotle, De sensu et sensato [Translatio vetus], textus qui a L. Peeters paratur [A.L. XIII.1]. Text available online at the ALD only: https:/clt.brepolis.net/ald.

Aristotle, Aristotelis De somno et vigilia liber adiectis veteribus translationibus et Theodori Metochitae commentario, ed. Hendryk J. Drossaart Lulofs, Leiden 1943.

Aristotle, De sompno et vigilia II–III [Vetus translatio–Nova translatio], ed. Hendrik J. Drossaart Lulofs, Aristotelis De insomniis et De divinatione per somnum. *A new Edition of the Greek Text with the Latin Translations*, Index verborum (Philosophia antiqua 2), Leiden 1947. Text available online at the ALD: https://clt.brepolis.net/ald.

Aristotle, *On sleep and dreams. Text and translation with introduction and notes by D. Gallop*, Warminster 1996.

Aristotle, *The Complete Works of Aristotle: the Revised Oxford Translation*, ed. Jonathan Barnes, 2 voll., Princeton 1984.

Arnaldus de Villanova, Tractatus de humido radicali, ed. Michael R. McVaugh, *with an intr. and comm. by Chiara Crisciani and Giovanna Ferrari* (Opera Medica Omnia, V.2), Barcelona 2010.

Averroes (Ibn Rušd), Al-Kulliyyāt fī al-tibb, ed. Mohamed ʿĀ. al-Ǧābirī, Beirut 1999.

Averroes (Ibn Rušd), Colliget (Aristotelis opera cum Averrois commentariis, Suppl. I), Venice 1562/Frankfurt a.M. 1962, fol. 1ra–172vb.

Averroes (Ibn Rušd), Commentarium magnum in Aristotelis de anima libros, ed. Frederick St. Crawford (Corpus commentariorum Averrois in Aristotelem, Versionum Latinarum VI/1), Cambridge (Mass.) 1953.

Averroes (Ibn Rušd), Compendium libri Aristotelis de memoria et reminiscentia, ed. Henricus Blumberg, Averrois Cordubensis, Compendia librorum Aristotelis qui Parva Naturalia vocantur (Corpus commentariorum Averrois in Aristotelem: Versionum Arabicorum VII), Cambridge (Mass.) 1972, 36–51.

Averroes (Ibn Rušd), Compendium libri Aristotelis de memoria et reminiscentia, ed. Aemilia L. Shields/Henricus Blumberg, Averrois Cordubensis Compendia librorum Aristotelis qui Parva naturalia vocantur (Corpus Commentariorum Averrois in Aristotelem, Versionum Latinarum VII), Cambridge (Mass.) 1949, 47–72.

Averroes (Ibn Rušd), Compendium libri Aristotelis de sensu et sensato, ed. Henricus Blumberg, Averrois Cordubensis, Compendia librorum Aristotelis qui Parva Naturalia vocantur (Corpus commentariorum Averrois in Aristotelem: Versionum Arabicorum VII), Cambridge (Mass.) 1972, 1–35.

Averroes (Ibn Rušd), Compendium libri Aristotelis de sensu et sensato, ed. Aemilia L. Shields/Henricus Blumberg, Averrois Cordubensis Compendia librorum Aristotelis qui Parva naturalia vocantur (Corpus Commentariorum Averrois in Aristotelem, Versionum Latinarum VII), Cambridge (Mass.) 1949, 3–44.

Averroes (Ibn Rušd), In Aristotelis Metaphysicorum libros XIII commentarius (Aristotelis Opera cum Averrois Commentariis, vol. 8), ed. Venezia 1562/Frankfurt a.M. 1962, fols. 1ra–355vb.

Averroes (Ibn Rušd), In Aristotelis Meteorologicorum libros IV commentarius (Aristotelis Opera cum Averrois Commentariis, vol. 5), ed. Venezia 1562/Frankfurt a.M. 1962, fols. 400ra–487vb.

Averroes (Ibn Rušd), Tafsir mā ba'da at-ṭabī'a. *Texte arabe inédit établi par Maurice Bouyges*. 2nd edition, Beyrouth 1973.

Avicenna (Ibn Sīnā), Abbreviatio de animalibus (Ed. Venezia 1508/Frankfurt a.M. 1961, 29–64).

Avicenna (Ibn Sīnā), Al-Šifā', al-Ṭabī'iyyāt VIII: al-Ḥayawān, ed. 'Abd al-Ḥalīm Muntaṣir Sa'īd Zāyid, 'Abdallāh Ismā'īl, Cairo 1970.

Avicenna (Ibn Sīnā), Al-Šifā', al-Ṭabī'iyyāt, Ma'ādin wa-Āṯār 'ulwiyya, ed. 'Abd al-Ḥalīm Muntaṣir Sa'īd Zāyid, 'Abdallāh Ismā'īl, Cairo 1965.

Avicenna (Ibn Sīnā), Al-Šifā', al-Ṭabī'iyyāt, al-Samā' al-ṭābī'ī, ed. Sa'īd Zāyid. With a preface by Ibrāhīm Madkūr, Cairo, 1983.

Avicenna (Ibn Sīnā), Canon (Ed. Venezia 1507/Hildesheim 1964).

Avicenna (Ibn Sīnā), De anima, ed. Fazlur Rahman, *Avicenna's* De anima (*Arabic text*): *being the Psychological Part of* Kitāb al-Shifā', London 1959.

Avicenna (Ibn Sīnā), De diluviis, ed. Manuel Alonso Alonso, "Homenaje a Avicena en su milenario. Las traducciones de Juan Gonzáles de Burgos y Salomón," *Al-Andalus* 14 (1949): 306–308.

Avicenna (Ibn Sīnā), De medicinis cordialibus, ed. Simone Van Riet, Liber de anima, Avicenna Latinus: Liber de anima seu sextus de naturalibus, IV–V, Louvain–Leiden 1968, 187–210.

Avicenna (Ibn Sīnā), *Les plantes; texte etab. et éd. par Abd el-Halim Montasir, Said Zayed, Abdullah Ismail; introd. by Ibrahim Madkour*, Cairo 1965.

Avicenna (Ibn Sīnā), Liber de anima, ed. Simone Van Riet, Avicenna Latinus: Liber de anima seu sextus de naturalibus, I–II–III, Louvain–Leiden 1972; IV–V, Louvain–Leiden 1968.

Avicenna (Ibn Sīnā), Liber de philosophia prima, ed. Simone Van Riet, Avicenna Latinus: Liber de philosophia prima sive scientia divina I–IV, Louvain–Leiden 1977; V–X, Louvain–Leiden 1980.

Avicenna (Ibn Sīnā), Sufficientia, ed. Simone Van Riet, Avicenna Latinus: Liber primus naturalium, Louvain-la-Neuve–Leiden 1992.

Avicenna (Ibn Sīnā), Qānūn fī al-ṭibb, ed. Institute of History of Medicine and Medical Research, vols. I, III.1, and IV. New Delhi 1402/1982.

Avicenna (Ibn Sīnā), Risāla fī-'adwiya al-qalbiyya, ed. Muhammad Zuhayr al-Bābā, Min mu'allafāt Ibn Sīnā al-ṭibbiyya, Aleppo 1404/1984, 221–294.

Avicenna (Ibn Sīnā), *The Metaphysics of The Healing* (al-Šifā': al-Ilāhiyyāt), *a parallel English-Arabic Text; translated, introduced and annotated by Michael E. Marmura*, Provo 2005.

Boezio di Dacia. *Sull'eternità del mondo. Sui sogni. Sul Sommo bene*, ed. Luca Bianchi (La coda di paglia 41), Milano 2017.

Bonaventura, In quatuor libros Sententiarum, ed. PP. Collegii a S. Bonaventura, Quaracchi 1882–1889 (Opera omnia I–IV).

Constantinus Africanus, Pantegni [Theorica IV.19], ed. Charles Burnett, "The Chapter on the Spirits in the *Pantegni* of Constantine the African," in *Constantine the African and ʿAlī Ibn al-ʿAbbās al-Mağūsī. The* Pantegni *and Related Texts*, ed. Charles Burnett/Danielle Jacquart (Studies in Ancient Medicine 10), Leiden-New York-Köln 1994, 99–120.

Costa Ben Luca (Qusṭā ibn Lūqā al-Baʿlabakkī), *Épître sur la différence entre l'esprit et l'âme, éd. et trad. par Gérard Troupeau avec la collaboration de Joseph Dagher*, Hommage d'André Miquel, Paris 2011.

Costa Ben Luca (Qusṭā ibn Lūqā al-Baʿlabakkī), De differentia spiritus et animae, ed. Wilcox, *On the Transmission and Influence of Qusta Ibn Luqa's "On the Difference Between Spirit and Soul,"* Unpublished Ph.D. Thesis, New York 1985, 143–191.

Costa Ben Luca (Qusṭā ibn Lūqā al-Baʿlabakkī), *Le livre des caractéres*, ed. Paul Sbath, "Le livre des caractères de Qosta ibn Louqa (1941) grand savant et célèbre médecin du IXe siècle," *Bulletin de l'Institut d'Égypte* 23 (1941), 103–169.

David de Dinanto, *I testi di David di Dinant: filosofia della natura e metafisica a confronto con il pensiero antico. Introduzione ed edizione dei testi di Elena Casadei* (Testi, Studi, Strumenti, 20) Spoleto 2008.

Galfridus de Aspall, *Questions on Aristotle's Physics, ed. Silvia Donati and Cecilia Trifogli, English Translation by E. Jennifer Ashworth and Cecilia Trifogli* (Auctores Britannici Medii Aevi, 26/1) Oxford 2017.

Haly Abbas (ʿAlī ibn ʿAbbās al-Mağūsī), Kāmil al-ṣināʿa al-ṭibbiyya, ed. Khālid Harbī, Alexandria, 2018.

Isaac ben Salomon Israeli (Isaac Iudaeus), Liber de definicionibus, ed. Joseph Th. Muckle, in: Archives d'Histoire Doctrinale et Littéraire du Moyen Âge 12/13 (1937/38) 299–340.

Johannes de Fonte, Auctoritates Aristotelis, Senecae, Boethii, Platonis, Appuleii Africani, Porphyrii et Gilberti Porretani, ed. Jacqueline Hamesse, *Les* Auctoritates Aristotelis: *Un florilège medieval. Étude historique et édition critique* (Philosophes médiévaux 17) Louvain–Paris 1974.

Johannes de Rupella, Summa de anima, ed. Jacques G. Bougerol (Textes philosophiques du Moyen Âge 19), Paris 1995.

Johannitius, Isagoge ad Techne Galieni, ed. Gregor Maurach, *Sudhoffs Archiv* 62 (1978/2): 148–174.

Nemesius, De natura hominis, ed. Gérard Verbeke/Josep R. Moncho (Corpus Latinum Commentariorum in Aristotelem Graecorum, Suppl. 1) Leiden 1975.

Nicolaus Damascenus, De plantis, ed. Hendrik J. Drossaart Lulofs/E.L.J. Poortman (Verhandelingen der Koninklijke nederlandse akademie van wetenschappen, Afd. letterkunde. Nieuwe reeks, 139), Amsterdam 1989.

Nicolaus Physicus, Anathomia, ed. Franz Redeker, *Die* "Anatomia magistri Nicolai phisici" *und ihr Verhältnis zur* Anatomia Cophonis *und* Richardi (Diss.), Leipzig 1917, 29–59.

Petrus de Hibernia, Expositio et quaestiones in Aristotelis librum de longitudine et brevitate vitae, ed. Michael Dunne (Philosophes médiévaux 30), Louvain-la-Neuve/Paris, 1993.

Peter of Spain, Questiones super libros De animalibus Aristotelis, ed. Francisca Navarro Sanchez, Farnham-Burlington 2015.

Peter of Spain, Tractatus de longitudine et brevitate vite, ed. Manuel Alonso, in *Pedro Hispano: Obras filosóficas*, III (Instituto de Filosofía 'Luis Vives.' Serie A Núm. 4), Madrid 1952, 413–490.

Philippus Cancellarius, Summa de bono, ed. Nikolaus Wicki (Corpus Philosophorum Medii Aevi. Opera Philosophica Mediae Aetatis Selecta, 2), vol. 1, Bern 1985.

Richardus (Anglicus), *Die Anatomie des* Richardus, ed. Victor Tarrasch, Berlin 1898.

Roger Bacon, Opus maius, ed. John H. Bridges, *The* 'Opus majus' *of Roger Bacon*, I–II und Suppl., Oxford 1879–1900/Frankfurt a.M. 1964.

Roger Bacon, De erroribus medicorum, ed. Andrew G. Little/Edward Withington, Roger Bacon: De retardatione accidentium senectutis cum aliis opusculis de rebus medicinalibus (Opera hactenus inedita Rogeri Baconi IX), Oxford 1928, 150–179.

Rolandus Cremonensis, Summae magistri Rolandi Cremonensis. Liber secundus, ed. Alessio Cortesi/Umberto Midali (Quaderni della Cortesiana 10), Bergamo 2017.

⟨Summa fratris Alexandri⟩ Doctoris Irrefragabilis Alexandri de Hales Ordinis Minorum Summa Theologica, I–IV, Quaracchi 1924–1948.

Thomas Aquinas, Sentencia libri de anima, ed. Fratres Praedicatores (Renée-Antoine Gauthier), Roma–Paris 1984 (Ed. Leon. XLV/1).

Thomas Aquinas, Sentencia libri de sensu et sensato, ed. Fratres Praedicatores (Renée-Antoine Gauthier), Roma–Paris 1985 (Ed. Leon. XLV/2, 1–101).

Thomas Aquinas, Super Boetium De trinitate, ed. Fratres Predicatores (Pierre M.J. Gils), Roma–Paris 1992 (Ed. Leon. L).

William of Auvergne, Tractatus de anima (Opera omnia, II, Supll.), Paris 1674/Frankfurt a.M. 1963, 65–228.

Literature

Albertus-Magnus-Institut. *Albertus Magnus und sein System der Wissenschaften. Schlüsseltexte in Übersetzung*. Lateinisch–Deutsch. Intr., transl. and prepared for print by Hannes Möhle/Henryk Anzulewicz/Maria Burger/Silvia Donati/Ruth Meyer/Martin Bredenbeck/Susana Bullido del Barrio, Münster 2011.

Alpina, Tommaso. "Al-Ǧūzǧānī's Insertion of On Cardiac Remedies in Avicenna's Book of the Soul: the Latin Translation as a Clue to his Editorial Activity on the Book of the Cure?" *Documenti e studi sulla tradizione filosofica medievale* 28 (2017): 365–400.

Alpina, Tommaso. "Exercising Impartiality to Favor Aristotle: Avicenna and 'the accomplished anatomists' (*aṣḥāb al-tašrīḥ al-muḥaṣṣilūna*)." *Arabic Sciences and Philosophy* 32 (2022/2): 137–178.

Alpina, Tommaso. "Is Nutrition a Sufficient Condition for Life? Avicenna's Position between Natural Philosophy and Medicine." In *Nutrition and Nutritive Soul in Aristotle and Aristotelianism*, ed. Roberto Lo Presti/Giouli Korobili, 221–258. Topics in Ancient Philosophy 9, Berlin 2020.

Alpina, Tommaso. "Retaining, Remembering, Recollecting. Avicenna's Account of Memory and Its Sources." In Memory and Recollection in the Aristotelian Tradition: Essays on the Reception of Aristotle's *De memoria et reminiscentia*, ed. Véronique Decaix/Christina Thomsen Thörnqvist (Studia Artistarum 47), Turnhout 2021, 67–92.

Alpina, Tommaso. *Subject, Definition, Activity. Framing Avicenna's Science of the Soul* (Scientia Graeco Arabica 28), Berlin 2021.

Alpina, Tommaso. "The Soul *of*, the Soul *in itself*, and the 'Flying Man' Experiment." *Arabic Sciences and Philosophy* 28 (2018): 187–224.

Alpina, Tommaso. "Translating Method: Inference from Behavior to Anatomy in Avicenna's Zoology." In *Premodern Experience of the Natural World in Translation*, ed. Katja Krause/Maria Auxent/Drora Weil, New York 2023, 136–152.

Altmann, Alexander–Stern, Samuel M. *Isaac Israeli A Neoplatonic Philosopher of the Early Tenth Century*, Oxford 1958.

Amerini, Fabrizio. *Aquinas on the Beginning and End of Human Life*, transl. by Mark Henninger, Cambridge-London 2013.

Anzulewicz, Henryk. "Anthropologie des Albertus Magnus als Ort des Dialogs zwischen den *sancti* und *philosophi*." In *Dialog. Auf dem Weg zur Wahrheit und zum Glauben*, ed. Frano Prcela, Zagreb-Mainz 1996, 47–53.

Anzulewicz, Henryk. "Die Aristotelische Biologie in den Frühwerken des Albertus Magnus." In *Aristotle's Animals in the Middle Ages and the Renaissance*, ed. Carlos Steel/Guy Guldentops/Pieter Beullens (Mediaevalia Lovaniensia [series 1] 27), Leuven 1999, 159–188.

Anzulewicz, Henryk. "Die Denkstruktur des Albertus Magnus. Ihre Dekodierung und ihre Relevanz für die Begrifflichkeit und Terminologie." In *L'élaboration du vocabulaire philosophique au Moyen Âge. Actes du Colloque international de Louvain-la-Neuve et Leuven 12–14 septembre 1998 organisé par la Société Internationale pour l'Étude de la Philosophie Médiévale*, ed. Jacqueline Hamesse/Carlos Steel (Rencontres de Philosophie Médiévale 8), Turnhout: 2000, 369–396.

Anzulewicz, Henryk. "Der Einfluß der Gestirne auf die sublunare Welt und die menschliche Willensfreiheit nach Albertus Magnus." In *Actes de la* v*ème* *Conference Annuelle de la* SEAC*, Gdansk 1997* (Swiatowit. Supplement [Series H: Anthropology] 2), Warszawa–Gdansk: 1999, 263–277.

Anzulewicz, Henryk. "Konzeptionen und Perspektiven der Sinneswahrnehmung im System Alberts des Großen." In *I cinque sensi* (Micrologus Library 10), Firenze 2002, 199–238.

Anzulewicz, Henryk. "*Memoria* und *reminiscentia* bei Albertus Magnus." In *La mémoire du temps au Moyen Age*, ed. Agostino Paravicini Bagliani (Micrologus Library 12), Firenze 2005, 163–200.

Anzulewicz, Henryk. "Person und Werk des David von Dinant im Literarischen Zeugnis Alberts des Grossen." *Mediaevalia Philosophica Polonorum* 34 (2001): 15–58.

Anzulewicz, Henryk. "Psychophysiology, Geographical Regions, and Climates: Albert the Great on the Natural Prerequisites of Scientific Aptitude in Human Beings." In *Medieval Science between Emergence and Inheritance: Albert the Great and His Arabic Sources* (Philosophy in the Abrahamitic Traditions, forthcoming 2023).

Anzulewicz, Henryk. "*Solus homo est nexus Dei et mundi.* Albertus Magnus über den Menschen." In Multifariam. *Homenaje a los profesores Anneliese Meis, Antonio Bentué y Sergio Silva*, ed. Samuel Fernández/Juan Noemi/ Rodrigo Polanco, (Anales de la Facultad de Teología. Supplementos a Teología y Vida 1), Santiago 2010, 321–335.

Anzulewicz, Henryk. "The Concept of Man in Albert the Great." In *A companion to Albert the Great: Theology, Philosophy and the Sciences*, ed. Irven Resnick (Brill's Companions to Christian Tradition 38), Leiden 2013, 325–346.

Anzulewicz, Henryk. "Zur Theorie des menschlichen Lebens nach Albertus Magnus. Theologische Grundlegungen und ihre bioethischen Implikationen." *Studia Mediewistyczne* 33 (1998): 35–49.

Anzulewicz, Henryk–Krause, Katja. "Albert der Große und sein holistisches Konzept menschlicher Erkenntnis." In Veritas et subtilitas. *Truth and Subtlety in the History of Philosophy. Essays in Memory of Burkhard Mojsisch (1944–2015)*, ed. Tengiz Iremadze/Udo Reinhold Jeck (Bochumer Studien zur Philosophie 59), Amsterdam 2018, 157–194.

Anzulewicz, Henryk–Rigo, Caterina. "*Reductio ad esse divinum.* Zur Vollendung des Menschen nach Albertus Magnus." In *Ende und Vollendung. Eschatologische Perspektiven im Mittelalter*, ed. Jan A. Artsen/Martin Pickavé (Miscellanea Medievalia 29), Berlin 2002, 388–416.

Ashley, Benedict. "Albert the Great on the Cogitative Power." In *A Companion to Albert the Great: Theology, Philosophy and the Sciences*, ed. Irven Resnick (Brill's Companions to Christian Tradition 38), Leiden-Boston 2013, 299–324.

Asùa, Miguel de. "Medicine and Philosophy in Peter of Spain's Commentary on the *De animalibus*." In *Aristotle's Animals in the Middle Ages and the Renaissance*, ed. Carlos Steel/Guy Guldentops/Pieter Beullens (Mediaevalia Lovaniensia [Series 1]. 27), Leuven 1999, 189–211.

Asùa, Miguel de. "Mineral, Plants and Animals from A to Z. The Inventory of the Natural World in Albert the Great's *philosophia naturalis*." In *Albertus Magnus. Zum Gedenken nach 800 Jahren. Neue Zugänge, Aspekte und Perspektiven*, ed. Walter Senner/Henryk Anzulewicz (Quellen und Forschungen zur Geschichte des Dominikanerordens, Neue Folge 10), Berlin 2001, 389–400.

Asùa, Miguel de. "Peter of Spain, Albert the Great and the *Quaestiones de animalibus*." *Physis* 34 (1997): 1–30.

Asùa, Miguel de. "War and Peace. Medicine and Natural Philosophy in Albert the Great." In *A Companion to Albert the Great: Theology, Philosophy and the Sciences*, ed. Irven Resnick (Brill's Companions to Christian Tradition 38), Leiden-Boston 2013, 269–297.

Baffioni, Carmela. "Further Notes on Averroes' Embryology and the Question of the 'Female Sperm'." In *Averroes and the Aristotelian Heritage*, ed. Carmela Baffioni, Napoli 2004, 159–172.

Bautier, Anne Marie. "'*Spiritus*' dans les textes antérieurs à 1200. Itinéraire lexicographique médiolatin: du souffle vital à l'au delà maléfique." In Spiritus. *IV Colloquio Internazionale del Lessico Intellettuale Europeo. Roma, 7–9 gennaio 1983*, ed. Marta Fattori/Massimo Bianchi, Roma 1984, 77–92.

Belo, Catarina. *Chance and Determinism in Avicenna and Averroes* (Islamic Philosophy, Theology and Science 69), Leiden–Boston 2017.

Bennett, David. "Avicenna's Dreaming in Context." In *Forms of Representation in the Aristotelian Tradition. Volume Two: Dreaming*, ed. Christina Thomsen Thörnqvist/Juhana Toivanen (Philosophia Antiqua 162), Leiden-Boston 2022, 88–109.

Berryman, Silvia. "Aristotle on *Pneuma* and Animal Self-Motion." *Oxford Studies in Ancient Philosophy* 23 (2002): 85–97.

Berti, Enrico. "Natura e generazione degli animali in Aristotele." Kriterion. *Revista de filosofia* 51 (2010): 489–504.

Bertolacci, Amos-Alpina, Tommaso. "Introduction." *Documenti e Studi sulla Tradizione Filosofica Medievale* 28 (2017): vii–xviii.

Bertolacci, Amos. "Averroes against Avicenna on Human Spontaneous Generation. The Starting-Point of a Lasting Debate." In *Renaissance Averroism and Its Aftermath: Arabic Philosophy in the Early Modern Europe, International Archives of the History of Ideas*, ed. Anna Akasoy/Guido Giglioni (Archives internationales d'histoire des idées 21), Dordrecht 2013, 37–54.

Bertolacci, Amos. "The Metaphysical Proof of Prophecy in Avicenna." In *Prophecy and Prophets in the Middle Ages*, ed. Alessandro Palazzo/Anna Rodolfi (Micrologus Library 103), Firenze 2020, 39–75.

Bianchi, Luca. "*Rusticus mendax*: Marcantonio Zimara e la fortuna di Alberto Magno nel Rinascimento italiano." In Studi sull'aristotelismo del Rinascimento, ed. Luca Bianchi (Subsidia Mediaevalia Patavina 5), Padova 2003, 209–223.

Bianchi, Luca. "*Ubi desinit physicus, ibi medicus incipit.*" In Summa doctrina et certa experientia. *Studi su medicina e filosofia per Chiara Crisciani*, ed. Gabriella Zuccolin (Micrologus 79), Firenze 2017, 5–27.

Bieniak-Nowak, Magdalena. *The Soul-Body Problem at Paris, ca. 1200–1250. Hugh of St. Cher and his Contemporaries* (Ancient and Medieval Philosophy [Series 1] 42), Leuven 2010.

Bieniak Nowak, Magdalena. "The Soul-Body Union in the *Summa Halensis*." In *The Legacy of Early Franciscan Thought*, ed. Lydia Schumacher (Veroffentlichungen des Grabmann-Institutes zur Erforschung der mittelalterlichen Theologie und Philosophie 67), Berlin 2021, 37–48.

Biesterfeldt, Hans Hinrich. "Zur medizinischen Terminologie des arabisch-islamischen Mittelalters." In *La formation du vocabulaire scientifique et intellectuel dans le monde arabe. Sous la direction de Danielle Jacquart* (Études sur le vocabulaire intellectuel du Moyen Âge/Civicima 7), Turnhout 1994, 66–90.

Black, Deborah. "Averroes on Imagination as A Cognitive Power." In *Contextualizing Premodern Philosophy Explorations of the Greek, Hebrew, Arabic, and Latin Traditions*, ed. Katja Krause/Luis Xavier López-Farjeat/Nicholas A. Oschman, New York 2022, 341–360.

Black, Deborah. "Estimation (*Wahm*) in Avicenna. The Logical and Psychological Dimensions." *Dialogue. Canadian Philosophical Association* 32 (1993/2): 219–258.

Black, Deborah. "Imagination and Estimation: Arabic Paradigms and Western Transformations." *Topoi* 19 (2000): 59–75.

Black, Deborah. "Memory, Individuals, and The Past in Averroes' Psychology." *Medieval Philosophy and Theology* 5 (1996): 161–187.

Bloch, David. *Aristotle on Memory and Recollection. Text, Translation, Interpretation, and Reception in Western Scholasticism* (Philosophia Antiqua 110), Leiden 2007.

Bonin, Therése. *Creation as Emanation: the Origin of Diversity in Albert the Great's On the causes and the procession of the universe*, Notre Dame 2001.

Bonin, Therése. "The Emanative Psychology of Albertus Magnus." *Topoi* 19 (2000): 45–57.

Bono, James J. "Medical Spirits and the Medieval Language of life." *Traditio* 40 (1984): 91–130.

Burnett, Charles. " "Magister Johannis Hispalensis" and Costa Ben Luca's *De differentia spiritus et animae*." In *Arabic into Latin in the Middle Ages. The translators and their Intellectual and Social Context*, ed. Charles Burnett (Collected Studied Series 939) London 2009, 221–267.

Burnett, Charles. "The Introduction of Aristotle's Natural Philosophy into Great Britain: a Preliminary Survey of the Manuscript Evidence." In *Aristotle in Britain during the Middle Ages: Proceedings of the International Conference at Cambridge, 8–11 April 1994, organized by the Société Internationale pour l'Étude de la Philosophie Médiévale*, ed. John Marenbon (Rencontres de Philosophie Médiévale 5) Turnhout 1996, 21–49.

Cadden, Joan. "Albertus Magnus' Universal Physiology: the Example of Nutrition." In *Albertus Magnus and the Sciences. Commemorative Essays*, ed. James A. Weisheipl (Studies and Texts 49), Toronto 1980, 231–248.

Cadden, Joan. *Meanings of Sex Difference in the Middle Ages. Medicine, Science and Culture* (Cambridge History of Medicine), Cambridge 1995.

Caparello, Adriana. *Senso e interiorità in Alberto Magno*, Rome 1993.

Carusi, Paola. "Vecchiaia avanzata e qualità della vita nel Canone di Avicenna." In Vita longa. *Vecchiaia e durata della vita nella tradizione medica e aristotelica antica e medievale*, ed. Chiara Crisciani/Luciana Repici/Pietro B. Rossi (Micrologus Library 33), Firenze 2009, 41–60.

Cerami, Cristina. *Génération et substance: Aristote et Averroès entre physique et métaphysique* (Scientia Graeco-Arabica 18), Berlin 2015.

Cerami, Cristina. "Function and Instrument Toward a New Criterion of the Scale of Being in Aristotle's Generation of Animals." In *Aristotle's Generation of Animals: A Critical Guide*, ed. Andrea Falcon and David Lefebvre (Cambridge Critical Guides), Cambridge 2018, 130–149.

Cerrito, Amalia. "Botany as Science and exegetical Tool in Albert the Great." *Aisthesis* 11 (2018/1): 97–107.

Cerrito, Amalia. "Disclosing the Hidden Life of Plants. Theories of the Vegetative Soul in Albert the Great's *De vegetabilibus et plantis*." In *Vegetative Powers. The roots of life in Ancient, Medieval and Early Modern Natural Philosophy*, ed. Fabrizio Baldassarri/Andreas Blank (International Archives of the History of Ideas/Archives internationales d'histoire des idées 234), Cham 2021, 105–122.

Charles, David. "Aristotle on Hypothetical Necessity and Irreducibility." *Pacific Philosophical Quarterly* 69 (1988): 1–53.

Chenu, Marie-Dominique. "*Spiritus*. Le vocabulaire de l'âme au XIIe siècle." *Revue des Sciences philosophiques et théologiques* 41 (1957/2): 217–232.

Connell, Sophia. "Aristotle's Explanation of Monstruous Births and Deformities." In *Aristotle's Generation of Animals. A Critical Guide*, ed. Andrea Falcon and David Lefébvre (Cambroidge Critical Guides), Cambridge 2018, 207–223.

Connell, Sophia. *Aristotle on Female Animals: A Study of the Generation of Animals* (Cambridge Classical Studies), Cambridge 2016.

Cova, Luciano. "Seme e generazione umana nelle opere teologiche di Alberto Magno." In Summa doctrina et certa experientia. *Studi su medicina e filosofia in onore di Chiara Crisciani*, ed. Gabriella Zuccolin (Micrologus 79), Firenze 2017, 237–256.

Crisciani, Chiara. "Aspetti del dibattito sull'umido radicale nella cultura del tardo medioevo (sec. XIII–XIV)." *Arxiu de textos catalans antics* 23/24 (2005): 333–380.

Crisciani, Chiara. "Medicine and Philosophy in the Middle Ages. Sisters, Companions, Rivals." In *Medical Ethics. Premodern Negotiations between Medicine and Philosophy*, ed. Marta Gadebusch Bondio, Stuttgart 2014, 29–46.

Crisciani, Chiara. "Premesse e promesse di lunga vita: tra teologia e pratica terapeutica (secolo XIII)." In Vita longa. *Vecchiaia e durata della vita nella tradizione medica e aristotelica antica e medievale*, ed. Chiara Crisciani/Luciana Repici/Pietro B. Rossi (Micrologus Library 33), Firenze 2009, 61–86.

D'Ancona, Cristina. "La doctrine de la création '*mediante intelligentia*' dans le *Liber de Causis* et dans ses sources." *Revue des Sciences philosophiques et théologiques* 76 (1992/2): 209–223.

De Leemans, Pieter. "*Secundum viam naturae et doctrinae*. Lire le *De motu animalium* et les *Parva Naturalia* d'Aristote au Moyen Âge." In *Les* Parva Naturalia *d'Aristote. Fortune antique et médiévale*, ed. Cristophe Grellard/Pierre-Marie Morel, Paris 2010, 197–220.

De Vaux, Richard. "La première entrée d'Averroës chez les Latins." *Revue des Sciences philosophiques et théologiques* 22 (1933) 193–245.

Decaix, Véronique. "What is memory of? Albert the Great on the proper object of memory." In *Memory and recollection in the Aristotelian tradition: essays on the reception of Aristotle's* De memoria et reminiscentia, ed. Véronique Decaix/Christina Thomsen Thörnqvist (Studia Artistarum 47), Turnhout 2021, 153–167.

Demaitre, Luke–Travill, Anthony A. "Human Embryology and Development in the Works of Albertus Magnus." In *Albertus Magnus and the Sciences. Commemorative Essays*, ed. James A. Weisheipl (Studies and Texts 49), Toronto 1980, 405–440.

Dewan, Lawrence. "St. Albert, the Sensibles, and Spiritual Being." In *Albertus Magnus and the Sciences, Commemorative Essays*, ed. James A. Weisheipl (Studies and Texts 49), Toronto 1980, 291–320.

Di Martino, Carla. Ratio particularis. *La doctrine des sens internes d'Avicenne à Thomas D'Aquin. Contribution a l'étude de la tradition arabo-latine de la psychologie d'Aristote* (Études de philosophie médiévale XCIV), Paris 2008.

Donati, Silvia. "Albert the Great as a Commentator of Aristotle's *De somno et vigilia*: The Influence of the Arabic Tradition." In Parva naturalia *in Greek, Arabic and Latin aristotelianism. Supplementing the Science of the Soul*, ed. Börje Bydén/Filip Radovanovic (Studies in the History of Philosophy of Mind 17), Cham, 2018, 169–209.

Donati, Silvia. "Alberts des Großen Konzept der *scientiae naturales*: Zur Konstitution einer peripatetischen Enzyklopädie der Naturwissenschaften." In *Albertus Magnus und der Ursprung der Universitätsidee: die Begegnung der Wissenschaftskulturen im 13. Jahrhundert und die Entdeckung des Konzepts der Bildung durch Wissenschaft*, ed. Ludger Honnefelder, Berlin 2011, 354–381.

Donati, Silvia. "Dreams and Divinatory Dreams in Albert the Great's *Liber De somno et vigilia*." In *Contemplation and Philosophy: Scholastic and Mystical Modes of Medieval Philosophical Thought. A Tribute to Kent Emery, Jr.*, ed. Roberto Hoffmeister Pich/Andreas Speer (Studien und Texte zur Geistesgeschichte des Mittelalters 125), Leiden-Boston, 2018, 178–215.

Draelants, Isabelle. "La reproduction imparfaite: les *gusanes* et l'état larvaire des insectes chez Albert le Grand." In *Fragmented Nature: Medieval Latinate Reasoning on the Natural World and Its Order*, ed. Mattia Cipriani/Nicola Polloni, London-New York 2022, 151–172.

Ebbesen, Sten-Thomsen-Thörnqvist, Christina-Décaix, Veronique. "Questions on *De sensu et sensato*, *De memoria* and *De somno et vigilia*." *Bulletin de Philosophie Médiévale* 57 (2015): 59–115.

Eckert, Willehad P. "Albert-Legenden." In *Albert der Grosse: seine Zeit, sein Werk, seine Wirkung*, ed. Albert Zimmermann and prepared for print by Gudrun Vuillemin-Diem (Miscellanea Medievalia 14), Berlin 1981, 1–23.

Fancy, Nahyan. "Generation in Medieval Islamic Medicine." In *Reproduction. Antiquity to the Present Day*, ed. Nick Hopwood/Rebecca Flemming/Lauren Kassell, Cambridge 2008, 129–140.

Fatigati, Michael. "Avicenna on Vegetative Faculties and the Life of Plants." In *Vegetative Powers. The Roots of Life in Ancient, Medieval and Early Modern Natural Philosophy*, ed. Fabrizio Baldassarri/Andreas Blank (International Archives of the History of Ideas Archives internationales d'histoire des idées 234), Cham 2021, 73–89.

Fattori, Marta. "Sogni e temperamenti." In *I sogni nel Medioevo: Seminario Internazionale, Roma 2–4 ottobre 1983*, ed. Tullio Gregory, Roma 1983, 87–109.

Fidora, Alexander. "Zum epistemologischen Status der Medizin in der *Summa Avicennae* und bei Thomas von Aquin." In *Handlung und Wissenschaft—Action and Science. Die Epistemologie der praktischen Wissenschaften im 13. und 14. Jahrhundert—The Epistemology of the Practical Sciences in the 13th and 14th Centuries*, ed. Matthias Lutz-Bachmann/Alexander Fidora (Wissenskultur und Gesellschaftlicher Wandel 29), Berlin 2008, 97–105.

Fioravanti, Gianfranco. "*Phantasia*: tra Aristotele e i Peripatetici." In *Immaginario e immaginazione nel Medioevo. Atti del convegno della Società italiana per lo studio del pensiero medievale (S.I.S.P.M.), Milano, 25–27 settembre 2008*, ed. Maria Bettetini/Francesco Paparella/Roberto Furlan (Textes et Études du Moyen Âge 51), Louvain-La-Neuve 2009, 163–178.

Fioravanti, Gianfranco. "*Politiae Orientalium et Aegyptiorum*. Alberto Magno e la Politica aristotelica." *Annali della Scuola Normale Superiore di Pisa*, 9 ([Series 3] 1979): 195–246.

Frede, Dorothea. "The Cognitive Role of Phantasia in Aristotle." In *Essays on Aristotle's* De anima, ed. Amelie O. Rorty/ Martha C. Nussbaum, Oxford 1995, 280–296.

Freudenthal, Gad. *Aristotle's Theory of Material Substance: Heat and Pneuma, Form and Soul* (Clarendon Paperbacks), Oxford 1998.

Freudenthal, Gad. "The Medieval Astrologization of Aristotle's Biology: Averroes on the Role of the Celestial Bodies in the Generation of Animate Beings." *Arabic Sciences and Philosophy* 12 (2002): 111–137.

Freudenthal, Gad. "The Astrologization of the Aristotelian Cosmos: Celestial Influences on the Sublunary World in Aristotle, Alexander of Aphrodisias, and Averroes." In *New Perspectives on Aristotle's* De caelo, ed. Alan C. Bowen/Cristopher Wilberding (Philosophia Antiqua 117), Leiden 2009, 239–281.

Gätje, Helmut. "Der *Liber de sensu et sensato* von Alfārābī bei Albertus Magnus." *Oriens Christianus* 48 (1964): 107–116.

Geiger, Louis-Bertrand. "La vie acte essentiel de l'âme—l'esse acte de l'essence d'après Albert le Grand." *Études d'histoire littéraire et doctrinale*, Montréal-Paris, 1962, 49–116.

Gelber, Jessica. "Females in Aristotle's Embryology." In *Aristotle's Generation of Animals. A Critical Guide* (Cambridge Critical Guides), ed. Andrea Falcon/David Lefébvre, Cambridge 2018, 171–187.

Giralt, Sebastià. "*Aristoteles imperfectus*: Natural Divination, Dream and Prophecy in the Latin Middle Ages (1210–1310)." In *Die mantischen Künste und die Epistemologie prognostischer Wissenschaften*, ed. Alexander Fidora, with the collaboration of Katrin Bauer (Beihefte zum Archiv für Kulturgeschichte 74), Köln-Weimar-Wien 2013, 23–59.

Gregoric, Pavel. *Aristotle on the Common Sense*, Oxford 2007.

Gregory, Tullio. "I sogni e gli astri." In *I sogni nel Medioevo, Seminario Internazionale. Roma, 2–4 ottobre 1983*, ed. Tullio Gregory, Rome 1985, 111–148.

Grellard, Cristophe. "La récéption médiévale du *De somno et vigilia*. Approche anthropologique et épistémologique du rêve, d'Albert le grand à Jean Buridan." *Les* Parva Naturalia *d'Aristote. Fortune antique et médiévale*, ed. Christophe Grellard/Pierre-Marie Morel, Paris 2010, 221–237.

Groisard, Jocelyn. "Hybridity and Sterility in Aristotle's Generation of Animals." In *Aristotle's Generation of Animals. A Critical Guide* (Cambridge Critical Guides), ed. Andrea Falcon/David Lefébvre, Cambridge 2018, 153–170.

Gutas, Dimitri. "Avicenna's Marginal Glosses on *De anima* and the Greek Commentatorial Tradition." In *Philosophy, Science and Exegesis in Greek, Arabic and Latin Commentaries*, 2 vols., ed. Peter. Adamson/Hans Baltussen, London 2004 (vol. 2), 77–88.

Gutas, Dimitri. "Imagination and Transcendental Knowledge in Avicenna." In *Arabic theology, Arabic philosophy: From the many to the one: Essays in celebration of Richard M. Frank*, ed. James E. Montgomery, Leuven 2006, 337–354.

Gutas, Dimitri. "Medical Theory and Scientific Method in the Age of Avicenna." In *Before and After Avicenna. Proceedings of the First Conference of the Avicenna Study Group*, ed. David C. Reisman, with the assistance of Ahmed H. Al-Rahim (Islamic Philosophy, Theology and Sciences: Texts and Sources 52), Leiden 2003, 145–162.

Hall, Thomas. "Life, Death and Radical Moisture." *Clio Medica* 6 (1971): 3–23.

Hansberger, Rotraud E. "Averroes on Divinatory Dreaming." In *Forms of Representation in the Aristotelian Tradition. Volume Two: Dreaming*, ed. Christina Thomsen Thörnqvist/Juhana Tovainen (Philosophia Antiqua 162) Leiden 2022, 110–149.

Hansberger, Rotraud E. "Averroes on the Internal Senses." In *Interpreting Averroes. Critical Essays*, ed. Peter Adamson/Matteo Di Giovanni, Cambridge 2019, 138–157.

Hansberger, Rotraud E. "How Aristotle Came to Believe in God-given Dreams: The Arabic Version of *De divinatione per somnum*." In *Dreaming Across Boundaries. The Interpretation of Dreams in Islamic Lands*, ed. Louise Marlow Cambridge, Mass./London 2008, 50–77.

Hansberger, Rotraud E. "*Kitāb al-Hiss wa-l-mahsūs*. Aristotle's *Parva naturalia* in Arabic Guise." In *Les* Parva naturalia *d'Aristote. Fortune antique et médiévale*, ed. Christophe Grellard/Pierre-Marie Morel, Paris 2010, 143–162.

Hansberger, Rotraud E. "Length and Shortness of Life between Philosophy and Medicine: the Arabic Aristotle and his Medical Readers." In *Philosophy and Medicine in the Formative Period of Islam*, ed. Peter Adamson/Peter Pormann (The Warburg Institute Colloquia 31), London 2018, 48–64.

Hansberger, Rotraud E. "Representation of Which Reality? 'Spiritual Forms' and '*maʿānī*' in the Arabic Adaptation of Aristotle's *Parva Naturalia*." In *The* Parva Naturalia *in Greek, Arabic and Latin Aristotelianism. Supplementing The Science of The Soul*, ed. Borje Bydén/Filip Radovic (Studies in the History of Philosophy of Mind 17), Cham 2020, 99–122.

Harvey, Ruth E. *The Inward Wits: Psychological Theory in the Middle Ages and the Renaissance* (The Warburg Institute Surveys 6), London 1975.

Hasse, Dag N. *Avicenna's* De Anima *in the Latin West. The Formation of a Peripatetic Philosophy of the Soul 1160–1300* (Warburg Institute Studies and Texts 1), London–Turin, 2000.

Hasse, Dag N. "Avicenna's 'Giver of Forms' in Latin Philosophy, especially in the Works of Albertus Magnus." In *The Arabic, Hebrew and Latin reception of Avicenna's Metaphysics*, ed. Dag Nikolaus Hasse/Amos Bertolacci (Scientia Graeco-Arabica 23), Berlin-Boston 2018, 225–249.

Hasse, Dag N. "*Plato arabico-latinus*: Philosophy—Wisdom Literature—Occult Sciences." In *The Platonic Tradition in the Middle Ages. A Doxographic Approach, a Doxographic Approach*, ed. Stephen Gersh/ Maarten J.F.M. Hoenen, with the assist. of Pieter Th. van Wingerden (Filosofia Platonica 184), Berlin-New York 2002, 31–65.

Hasse, Dag N. "Spontaneous Generation and the Ontology of Forms in Greek, Arabic, and Medieval Latin Sources." In *Classical Arabic Philosophy: Sources and Reception*, ed. Peter Adamson (Warburg Institute Colloquia 11), London–Turin, 2007, 150–175.

Hasse, Dag N. "The Early Albertus Magnus and his Arabic Sources on the Theory of the Soul." *Vivarium* 46 (2008): 232–252.

Hasse, Dag N. "The Soul's Faculties." In *The Cambridge History of Medieval Philosophy*, ed. Robert Pasnau, Cambridge 2010, 305–319.

Hödl, Ludwig. "*Opus naturae est opus intelligentiae*: Ein neuplatonisches Axiom im aristotelischen Verstandnis des Albertus Magnus." In *Averroismus im Mittelalter und Renaissance*, ed. Friedrich Niewöhner/Loris Sturlese, Zurich 1994, 132–148.

Hossfeld, Paul. "Albertus Magnus über die Frau." *Trierer Theologische Zeitschrift* 91 (1982): 235–237.

Jacquart, Danielle. "Avicenne et la nosologie galénique: l'exemple des maladie de cerveau." In *Perspectives arabes et médiévales sur la tradition scientifique et philosophique grecque: actes du colloque de la SIHSPAI (Société internationale d'histoire des sciences et de la philosophie arabes et islamiques), Paris, 31 mars–3 avril 1993*, ed. Ahmad Hasnawi/Abdelali Elamrani Jamal/Maroun Aouad; with a preface of Roshdi Rashed (Orientalia Lovaniensia Analecta 79), Leuven-Paris 1997, 217–226.

Jacquart, Danielle. "La médecine entre théorique et pratique: retour sur quelque définitions originelles." In *Handlung und Wissenschaft—Action and Science. Die Epistemologie der praktischen Wissenschaften im 13. und 14. Jahrhundert—The Epistemology of the Practical Sciences in the 13th and 14th Centuries*, ed. Matthias Lutz-Bachmann/Alexander Fidora (Wissenskultur und Gesellschaftlicher Wandel 29), Berlin 2008, 34–42.

Jacquart, Danielle. "La notion philosophico-médicale de *spiritus* dans l'*Avicenna Latinus*." In *Body and Spirit in the Middle Ages. Literature, Philosophy, Medicine*, ed. Gaia Gubbini, Berlin 2020, 15–33.

Jacquart, Danielle-Micheau, Françoise. *La medecine arabe et l'Occident médiéval* (Collection Islam-Occident 7), Paris 1990.

Jacquart, Danielle-Thomasset, Claude. "Albert le Grand et les problems de la sexualité." *History and Philosophy of the Life Sciences 3* (1981/3): 73–93.

Janssens, Julius. "The Notions of Wāhib Al-Suwar (Giver of Forms) and Wāhib Al-ʿAql (Bestower of intelligence) in Ibn Sīnā." In *Intellect et imagination dans la Philosophie Médiévale. Actes du 11. Congrès international de philosophie médiévale de la Société internationale pour l'étude de la philosophie médiévale (S.I.E.P.M.): Porto, du 26 au 31 août 2002*, ed. Maria Candido Pacheco/José Meirinhos, Turnhout 2006, vol. 1, 551–562.

Jordan, Mark. "Medicine and Natural Philosophy in Aquinas." In *Thomas von Aquin. Werk und Wirkung im Licht Neuer Forschung*, ed. Albert Zimmermann (Miscellanea Medievalia 19), Berlin 1988, 133–146.

Kahn, Charles. "Sensation and Consciousness in Aristotle's Psychology." In *Articles on Aristotle. Vol. IV. Psychology and Aesthetics*, ed. Jonathan Barnes/Malcholm Schofield/Richard Sorabji, London 1979, 1–31.

Kaukua, Jari. "Avicenna on the Soul's Activity in Perception." In *Active Perception in the History of Philosophy. From Plato to Modern Philosophy*, ed. José Filipe Silva/Mikko Yrjönsuuri (History of the Philosophy of the Mind 14), New York 2014, 99–116.

King, Richard A.H. *Aristotle on Life and Death*, London, 2001.

Kjeller-Johansen, Thomas. "Aristotle on the Sense of Smell." *Phronesis* 41 (1999/1): 1–19.

Kjeller-Johansen, Thomas. *The Powers of Aristotle's Soul* (Oxford Aristotle Studies), Oxford 2012.

Köhler, Theodor W. De quolibet modo hominis: *Alberts der Großen philosophischer Blick auf den Menschen* (Lectio Albertina 10), Münster 2009.

Köhler, Theodor W. Homo animal nobilissimum. *Konturen des spezifisch Menschlichen in der naturphilosophischen Aristoteleskommentierung des dreizehnten Jahrhunderts*, 3 voll. (Studien und Texten zur Geistesgeschichte des Mittelalters 114.2/1), Leiden-Boston 2014, vol. 2/1.

Korobili, Giouli. "Aristotle on the Role of Heat in Plant Life." In *Nutrition and Nutritive Soul in Aristotle and Aristotelianism*, ed. Roberto Lo Presti/Giouli Korobili (Topics in Ancient Philosophy 9), Berlin 2020, 153–169.

Krause, Katja. "Albert the Great on Animal and Human Origin in his Early Works." *Lo Sguardo (Rivista di filosofia)* 18 (2015/2): 205–232.

Krause, Katja. "Grenzen der Philosophie. Albert der Großen Kommentar zu *De animalibus* und die Medizin." *Documenti e Studi sulla Tradizione Filosofica Medievale* 30 (2019) 265–293.

Krause, Katja. "Medieval Philosophy of Nature popularized? Albert the Great's *De animalibus*." In *The Popularization of Philosophy in Medieval Islam, Judaism, and Christianity*, ed. Marieke Abram/Steven Harvey/Lukas Muehlethaler (Philosophy in the Abrahamic Traditions of the Middle Ages 3), Turnhout 2022, 185–201.

Krause, Katja-Anzulewicz, Henryk. "Albert the Great's *interpretatio*. Converting libraries into a scientific system." In *Premodern Translation: Comparative Approaches to Cross-Cultural Transformations*, ed. Sonja Brentjes/Alexander Fidora (Contact and Transmission; Intercultural Encounters from Late Antiquity to the Early Modern World 2), Turnhout 2021, 89–132.

Krause, Katja–Anzulewicz, Henryk. "From Content to Method: the *Liber de causis* in Albert the Great." In *Reading Proclus and the Book of Causes. Volume 1. Western Scholarly Networks and Debates*, ed. Dragos Calma, Leiden 2019, 180–208.

Kress, Emily. "Aristotle on Spontaneous Generation, Spontaneity, and Natural Processes." *Oxford Studies of Ancient Philosophy* 58 (2020): 157–204.

Kruger, Steven. *Dreaming in the Middle Ages* (Cambridge Studies in Medieval Literature 14), Cambridge 1992.

Kruk, Remke. "Ibn Sina on Animals. Between the First Teacher and the Physician." In *Avicenna and his heritage: Acts of the International Colloquium: Leuven-Louvain-La-Neuve, September 8-September 11, 1999*, ed. Jules Janssens/Daniel De Smet (Ancient and Medieval Philosophy [Series 1] 28), Leuven 2002, 325–341.

Kruk, Remke. "A Frothy Bubble: Spontaneous Generation in the Medieval Islamic Tradition." *Journal of Semitic Studies* 35 (1990): 265–310.

Lammer, Andreas. *The Elements of Avicenna's Physics. Greek Sources and Arabic Innovations* (Scientia Graeco-Arabica 20), Berlin 2018.

Lennox, James. "Teleology, Chance, and Aristotle's Theory of Spontaneous Generation." *Journal of the History of Philosophy* 20 (1982): 219–238.

Lennox, James. "Why Animals Must Keep Their Cool: Aristotle on the Need for Respiration (and Other Forms of Cooling)." In *Heat, Pneuma and Soul in Ancient Phi-*

losophy and Science, ed. Hynek Bartoš/Colin Guthrie-King, Cambridge, 2020, 217–242.

Lenzi, Massimiliano. "Alberto e Tommaso sullo statuto dell'anima umana." In *Archives d'histoire doctrinale et littéraire du Moyen Âge* 74 (2007/1) 27–58.

Lindberg, Charles. *Theories of Vision from al-Kindi to Kepler*, Chicago 1976.

Lizzini, Olga L. "Flusso, preparazione appropriata e *incohatio formae*: brevi osservazioni su Avicenna e Alberto Magno." In *Medioevo e filosofia: per Alfonso Maierù*, ed. Massimiliano Lenzi/Cesare A. Musatti/Luisa Valente (I Libri di Viella 150), Roma 2013, 129–150.

Lizzini, Olga L. "Representation and Reality: On the Definition of Imaginative Prophecy in Avicenna." In *The* Parva Naturalia *in Greek, Arabic and Latin Aristotelianism. Supplementing the Science of the Soul*, ed. Borje Bydén/Filip Radovic (Studies in the History of Philosophy of Mind 17), Cham 2020, 133–154.

Loconsole, Mario. "A Theory on the Formation of Minerals. Albert the Great and the Constitution of Scientific Mineralogy." *Quaestio* 20 (2020): 369–387.

Martorelli Vico, Romana. *Medicina e filosofia: per una storia dell'embriologia medievale nel 13. e 14. Secolo* (Hippocratica Civitas 4), Milano 2002.

McVaugh, Michael R. "The '*Humidum Radicale*' in 13th-century Medicine." *Traditio* 30 (1974): 259–283.

Meroni, Michele. "*Natura confortata per medicinam operatur per se*. The Role of Medicine in Albert's Early Theology and Aristotelian Paraphrases." *Quaestio. Yearbook of the History of Metaphysics. Special Issue Natural Philosophy in Albert the Great—A Dialogue of Disciplines*, ed. Mario. Loconsole/Evelina Miteva/Marilena Panarelli, 23 (2023): 109–136.

Miteva, Evelina. "*Iam patet igitur veritas eius quae dixit Aristoteles, et causa deceptionis Gallieni*. Philosophers vs. Medics in Albertus Magnus Account on Human Body." In *Irrtum-Error-Erreur*, ed. Maxime Mauriége/Andreas Speer (Miscellanea Medievalia 40), Berlin 2018, 107–122.

Miteva, Evelina. "Pygmies, Twins, Monsters: Human Nature on Its Borderline in Albert the Great." In *The Embodied Soul: Aristotelian Psychology and Physiology in Medieval Europe between 1200 and 1420*, ed. Marek Gensler/Monika Mansfeld/Monika Michalowska (Historical-Analytical Studies on Nature, Mind and Action 11), Cham 2022, 73–90.

Miteva, Evelina. "The Bodies of Adam and Eve: The (Missed?) Encounter of Natural Philosophy and Theology in the Account of Albert the Great." *Syzetesis* 7 (2019/2): 393–404.

Müller, Jörn. *Albertus Magnus über Gedächtnis, Erinnern und Wiederinnerung. Eine Philosophische Lekture von* De Memoria et Reminiscentia *mit Übersetzung* (Lectio Albertina 17), Münster 2017.

Musallam, Basil. "AVICENNA X. Medicine and Biology." In *Encyclopaedia Iranica*, III/1,

94–99, available online at http://www.iranicaonline.org/articles/avicenna-x (accessed on April 22, 2019).

Nardi, Bruno. *Studi di filosofia medievale*, Roma 1960.

Nelis, Tilke. "A Jumble of Writings: Commentaries on Aristotle's *De Longitudine et Brevitate Vitae* attributed to Adam of Buckfield." *Early Science and Medicine* 27 (2022/1): 1–56.

Niebyl, Paul H. "Old Age, Fever, and the Lamp Metaphor." *Journal of the History of Medicine* 26 (1971/4): 351–368.

Nocke, Franz J. *Sakrament und personaler Vollzug bei Albertus Magnus* (Beiträge zur Geschichte der Philosophie und Theologie des Mittelalters 41/4), Münster 1967.

Oelze, Anselm. *Animal Rationality. Later Medieval Theories 1250–1350* (Investigating Medieval Philosophy 12), Leiden-Boston 2018.

Oggins, Robin S. "Albert on Falcons and Hawks." In *Albertus Magnus and the Sciences. Commemorative Essays*, ed. James A. Weisheipl (Studies and Texts 49), Toronto 1980, 441–462.

Palanciuc, Ana. "Nemesius d'Émése." In *Dictionnaire des Philosophes Antiques*, ed. Richard Goulet, Paris 2005, vol. 4, 648–650.

Palazzo, Alessandro. "*Hi sunt natura servi.* Natural inequality and Biology in Albert the Great's *Politica*." In Homo—natura—mundus. *Human beings and their relationships. Proceedings of the XIV International Congress of the Société Internationale pour l'Étude de la Philosophie Médiévale, July 24–28, 2017, Porto Alegre, Brazil*, ed. Roberto Hofmeister Pich/Alfredo Carlos Storck/Alfredo Santiago Culleton (Rencontres de Philosophie Médiévale 22), Turnhout 2020, 613–631.

Palazzo, Alessandro. "*Philosophi aliter loquuntur de prophetia quam sancti.* Alberto il Grande e la profezia naturale." In *Immaginario e immaginazione nel Medioevo. Atti del convegno della Società italiana per lo studio del pensiero medievale* (*S.I.S.P.M.*), *Milano, 25–27 settembre 2008*, ed. Maria Bettetini/Francesco Paparella, with the collaboration of Roberto Furlan (Féderation Internationale des Instituts d'Études Médiévales, Textes et Études du Moyen Âge 51), Louvain-La-Neuve, 2009, 179–201.

Palazzo, Alessandro. "The Scientific Significance of Fate and Celestial Influences in Some Mature Works by Albert the Great: *De fato, De somno et vigilia, De intellectu et intelligibili, Mineralia*." In Per perscrutationem philosophicam. *Neue Perspektiven der mittelalterlichen Forschung*, ed. Alessandra Beccarisi/Ruedi Imbach/Pasquale Porro (Corpus Philosophorum Teutonicorum Medii Aevi, Beiheft 4), Hamburg 2008, 55–78.

Panarelli, Marilena. "Bodily Prerequisites of the Mind: The Spirit as the Highest Product of Digestion." In *The Embodied Soul: Aristotelian Psychology and Physiology in Medieval Europe between 1200 and 1420*, ed. Marek Gensler/Monika Mansfeld/Monika Michalowska (Historical-Analytical Studies on Nature, Mind and Action 11), Cham 2022, 115–132.

Panarelli, Marilena. "Converting Death into Life: Spontaneous Generation from Aristotle's Biology to Albert the Great's Analysis of Plants." *Quaestio* 22 (2022): 493–508.

Panarelli, Marilena. "How Do Plants Live and Grow? Radical Moisture and Digestion in Albert the Great's *De animalibus*." *Quaestio* 20 (2020): 347–367.

Panarelli, Marilena. "*Spiritus* and Plant Formation. Albert the Great's Restatement of the Pseudo-Aristotelian *De plantis*." in: Rencontres de Philosophie Médiévale (forthcoming).

Perfetti, Stefano. *Nature imperfette. Umano, subumano ed animale nel pensiero di Alberto Magno. Con un saggio di Amalia Cerrito* (Philosophica, serie viola 226), Pisa 2020.

Perfetti, Stefano. "Nelle gabbie del fissismo etologico complessità del desiderio umano e monotonia dell'appetito animale in Tommaso d'Aquino." In *Il Desiderio nel Medioevo*, ed. Alessandro Palazzo, Roma 2014, 111–129.

Perfetti, Stefano. "Rigenerazione degli animali? Alberto Magno tra *Parva Naturalia* e *De animalibus*." In Vita longa. *Vecchiaia e durata della vita nella tradizione medica e aristotelica antica e medievale*, ed. Chiara Crisciani/Luciana Repici/Pietro B. Rossi (Micrologus Library 33), Firenze 2009, 149–168.

Pormann, Peter E. "Avicenna on Medical Practice, Epistemology, and the Physiology of the Inner Senses." In *Interpreting Avicenna. Critical Essays*, ed. Peter Adamson, Cambridge 2013, 91–108.

Repici, Luciana. "Aristotele, l'anima e l'incorruttibilità: note su *De longitudine et brevitate vitae*, 1–3." *Antiquorum philosophia*: an international journal 1 (2007): 283–305.

Repici, Luciana. *Uomini capovolti. Le piante nel pensiero dei greci*, Bari 2000.

Repici, Luciana. "Tutto invecchia per opera del tempo. Senilità e senescenza in Aristotele." In *Vita longa*. Vecchiaia e durata della vita nella tradizione medica e aristotelica antica e medievale, ed. Chiara Crisciani/Luciana Repici/Pietro B. Rossi (Micrologus Library 33), Firenze 2009, 19–40.

Reynolds, Philip L. *Food and the Body. Some Peculiar Questions in High Medieval Theology* (Studien und Texte zur Geistesgeschichte des Mittelalters 69), Leiden 1999.

Richardson, Kara. "Two Arguments for Natural Teleology from Avicenna's *Sifā'*." *History of Philosophy Quarterly* 32 (2015/2): 123–140.

Ricklin, Thomas. "Albert le Grand, commentateur: L'exemple du *De somno et vigilia* III, 1." *Freiburger Zeitschrift für Philosophie und Theologie* 45 (1998): 31–55.

Riddle, John M.-Mulholland, James A. "Albert on Stones and Minerals." In *Albertus Magnus and the Sciences*, ed. James A. Weisheipl, Toronto 1980, 203–234.

Rodolfi, Anna. Cognitio obumbrata: *lo statuto epistemologico della profezia nel secolo 13* (Micrologus Library 74), Firenze 2016.

Rodolfi, Anna. *Il concetto di materia nell'opera di Alberto Magno* (Corpus philosophorum Medii Aevi. Testi e studi, 18), Firenze 2004.

Rodolfi, Anna. "Sogno e profezia in Alberto Magno." In Scientia, Fides, Theologia. *Studi di filosofia medievale in onore di Gianfranco Fioravanti*, ed. Stefano Perfetti, Pisa 2011, 193–215.

Roest, Bert. "Divination, Visions and Prophecy according to Albert the Great." In Media Latinitas. *A Collection of Essays to Mark the Occasion of the Retirement of L.J. Engels*, ed. Renée I.A. Nip et al./H. van Dijk/E.M.C. van Houts/ C.H.J.M. Kneepkens/ G.A.A. Kortekaas (Instrumenta Patristica et Mediaevalia 28), Turnhout 1996, 323–328.

Rigo, Caterina. "Zur Redaktionsfrage der Frühschriften des Albertus Magnus." In *Albertus Magnus und die Anfänge der Aristoteles-Rezeption im lateinischen Mittelalter: vom Richardus Rufus bis zu Franciscus de Mayronis*, ed. Ludger Honnefelder/Rega Wood/Mechthild Dreyer, Marc-Aeilko Aris (Subsidia Albertina 1), Münster 2005, 325–374.

Rigo, Caterina. "Zur Rezeption des Moses Maimonides im Werk des Albertus Magnus." In *Albertus Magnus. Zum Gedenken nach 800 Jahren*, ed. Walter Senner/Henryk Anzulewicz (Quellen und Forschungen zur Geschichte des Dominikanerordens, NS 10), Berlin 2001, 29–66.

Rubio Hernandez, Paloma. "The Role of the *Intentio Individualis* in Albert the Great's Sense Perception Theory." In *The Embodied Soul: Aristotelian Psychology and Physiology in Medieval Europe between 1200 and 1420*, ed. Marek Gensler/Monika Mansfeld/Monika Michalowska (Historical-Analytical Studies on Nature, Mind and Action 11), Cham 2022, 133–161.

Runggaldier, Edmund. *Die Menschliche Seele bei Albertus Magnus: Ein nicht-reduktionistischer Beitrag zum Leib-Seele-Problem* (Lectio Albertina 11), Münster 2010.

Rutkin, Darrel. "Astrology and Magic." In *A Companion to Albert the Great: Theology, Philosophy and the Sciences*, ed. Irven Resnick (Brill's Companions to Christian Tradition 38), Leiden-Boston 2013, 459–476.

Schipperges, Heinrich. "Das medizinische Denken bei Albertus Magnus." In *Albertus Magnus. Doctor Universalis 1280/1980*, ed. Gerbert Meyer, OP/Albert Zimmermann (Walberberger Studien, Philosophische Reihe 6), Mainz 1980, 279–294.

Schlosser, Marianne. Lucerna in caliginoso loco. *Aspekte des Prophetie-Begriffes in der scholastischen Theologie*, (Veröffentlichungen des Grabmann-Institutes zur Erforschung der mittelalterlichen Theologie und Philosophie. Neue Folge 43), Paderborn 2000.

Seidl, Horst. "Zur Geistseele im menschlichen Embryo nach Aristoteles, Albert dem Großen und Thomas von Aquin. Ein Diskussionsbeitrag." Salzburger Jahrbuch für Philosophie 31 (1986): 37–63.

Signori, Marco. "Prophecy and the Authority of the Prophet in al-Ġazālī's *Maqāṣid al-falāsifa*." In *Prophecy and Prophets in the Middle Ages*, ed. Alessandro Palazzo/Anna Rodolfi (Micrologus Library 103), Firenze 2020, 77–100.

Siraisi, Nancy. "The Medical Learning of Albertus Magnus." In *Albertus Magnus and the Sciences. Commemorative essays*, ed. James A. Weisheipl (Studies and Texts 49), Toronto 1980, 379–404.

Smith, A. Mark. "Perception." In *The Cambridge History of Medieval Philosophy*, ed. Robert Pasnau/Christina van Dyke, Cambridge 2010, vol. 1, 334–345.

Snyder, Steven C. "Albert the Great, *Incohatio Formae*, and the Pure Potentiality of Matter." *American Catholic Philosophical Quarterly* 70 (1996): 63–82.

Söder, Joachim. "Albert der Große über Sinne und Träume. Beobachtungen am Traumtraktat vom *De homine*." In *I cinque sensi* (Micrologus Library 10), Firenze 2002, 239–250.

Sorabji, Richard. "From Aristotle to Brentano: the Development of the Concept of Intentionality." *Oxford Studies in Ancient Philosophy*, Supplementary Volume (1991) 227–259.

Sorabji, Richard. "Body and Soul in Aristotle." In *Articles on Aristotle (IV). Psychology and Aesthetics*, ed. Jonathan Barnes/Malcolm Schofield/Richard Sorabji, London 1979, 42–64.

Steel, Carlos. "A Philological Diet for Philosophers. Aristippus' Translation of Book IV of Aristoteles' *Meteorology* and Albert the Great." In *L'Antichità classica nel pensiero medievale. Atti del convegno della Società Italiana per lo Studio del Pensiero Medievale (SISPM). Trento, 27–29 settembre 2010*, ed. Alessandro Palazzo (Textes et études du Moyen âge 61), Porto 2011, 79–106.

Steneck, Nicholas H. "Albert on the Classification and Localization of Internal Senses." *Isis* 65 (1974/2): 193–211.

Steneck, Nicholas H. "Albert on the Psychology of Sense Perception." In *Albertus Magnus and the Sciences, Commemorative Essays*, ed. James A. Weisheipl (Studies and Texts 49), Toronto 1980, 263–290.

Struve, Thieleman. "Die Anthropologie des Alfredus Anglicus in ihrer Stellung zwischen Platonismus und Aristotelismus." In *Staat und Gesellschaft im Mittelalter*, ed. Thieleman Struve (Historische Forschungen 80), Berlin 2004, 29–52.

Sturlese, Loris. "Il razionalismo filosofico e scientifico di Alberto il Grande." In *Storia della filosofia tedesca nel Medioevo: Il secolo XIII* (Accademia toscana di scienze e lettere La Colombaria 149), Firenze 1990, 69–125.

Taylor, Richard. "Averroes and the Philosophical Account of Prophecy." *Studia Graeco-Arabica* 8 (2018): 287–304.

Takahashi, Adam. "Nature, Formative Power and Intellect in the Natural Philosophy of Albert the Great." *Early Science and Medicine* 13 (2008): 451–481.

Takahashi, Adam. *Interpreting Aristotle's Cosmos: Albert the Great as a Reader of Averroes*, Ph.D. Dissertation: Radboud University Nijmegen, 2017.

Tellkamp, Jörg-Alejandro. "Albert the Great on the Structure and Function of Inner Senses." In *The Judeo-Christian-Islamic Heritage: Philosophical and theological per-*

spectives, ed. Richard Taylor/Irfan A. Omar (Marquette Studies in Philosophy 75), Milwaukee 2012, 305–324.

Temkin, Owsei. "On Galen's Pneumatology." *Gesnerus* 8 (1951): 180–189.

Thomsen Thörnqvist, Christina: "How Dreams Are Made: Some Latin Medieval Commentators on Dream Formation in Aristotle's *De insomniis*." In *Forms of Representation in the Aristotelian Tradition. Volume Two: Dreaming*, ed. Christina Thomsen Thörnqvist/Juhana Tovainen (Philosophia Antiqua 162), Leiden-Boston: 2022, 150–177.

Thijssen, Jos M. "Twins as Monsters: Albertus Magnus's Theory of the Generation of Twins and its Philosophical Context." *Bulletin of the History of Medicine* 61 (1987/2): 237–246.

Van der Eijk, Philip. *Medicine and Philosophy in Classical Antiquity: Doctors and Philosophers on Nature, Soul, Health and Disease*. Cambridge 2005.

Van der Luugt, Maaike. "L'animation de l'embryon humain et le statut de l'enfant à naître dans la pensée médiévale." In *Formation et animation de l'embryon dans l'Antiquité et au Moyen Âge*, ed. Luc Brisson/Marie-Heléne Congourdeau/Jean-Luc Solère (Histoire de doctrines de l'Antiquité Classique 38), Paris 2008, 234–254.

Van der Luugt, Maaike. *Le ver, le démon et la vierge. Les théories médiévales de la génération extraordinaire: une étude sur les rapports entre théologie, philosophie naturelle et médecine* (L'ane d'or) Paris 2004.

Weisheipl, James A. "The Axiom '*Opus naturae est opus intelligentiae*' and its Origins." In *Albertus Magnus—Doctor Universalis 1280/1980*, ed. Gerbert Meyer OP/Albert Zimmermann, Mainz 1980, 441–463.

Wieland, Georg. *Zwischen Natur und Vernunft: Alberts des Großen Begriff von Menschen* (Lectio Albertina 2), Münster 1999.

Wilcox, Judith Carol. "Qusta ibn Luqa's *De differentia spiritus et animae* in Medieval Considerations of the Internal Senses." In *Corpo e anima, sensi interni e intelletto dai secoli 13.–14. ai post-cartesiani e spinoziani: atti del Convegno internazionale, Firenze, Dipartimento di scienze dell'educazione e dei processi culturali e formativi, 18–20 settembre 2003*, ed. Graziella Federici Vescovini/Valeria Sorge/Carlo Vinti (Textes et études du Moyen âge 30), Turnhout 2005, 55–77.

Wilcox, Judith Carol. *On the Transmission and Influence of Qusıa Ibn Luqa's "On the Difference Between Spirit and Soul."* Unpublished Ph.D. Thesis, New York 1985.

Wilson, Leonard J. "Erasistratus, Galen and the *pneuma*." *Bulletine of History of Medicine* 33 (1959): 293–314.

Wisnovsky, Robert. *Avicenna's Metaphysics in Context*, London 2003.

Wisnovsky, Robert. "Final and Efficient Causation in Avicenna's Cosmology and Theology." *Quaestio* 2 (2002) 97–123.

Wöllmer, Gila. "Albert the Great and His Botany." In *A Companion to Albert the Great: Theology, Philosophy and the Sciences*, ed. Irven Resnick (Brill's Companions to Christian Tradition 38), Leiden 2013, 221–268.

Wood, Rega. "Spirituality and Perception in Medieval Aristotelian Natural Philosophy." In *Medieval Perceptual Puzzles: Theories of Sense Perception in the 13. and 14. Centuries*, ed. Elena Baltuta (Investigating Medieval Philosophy 13), Leiden 2020, 153–178.

Ziegler, Joseph. "Medicine and Immortality in Terrestrial Paradise." In *Religion and Medicine in the Middle Ages*, ed. Peter Biller/Joseph Ziegler (York Studies in Medieval Theology 3), York 2001, 201–242.

Zuccolin, Gabriella. *I gemelli nel Medioevo. Questioni filosofiche, mediche e teologiche*, Como-Pavia 2019.

Zuccolin, Gabriella. "Il corpo sanabile e l'embrione. Il *De animalibus* come punto di intersezione tra filosofia naturale e medicina." In *La Medicina nel Basso Medioevo. Tradizioni e conflitti. Atti del 55. Convegno storico internazionale, Todi, 14–16 ottobre 2018* (Atti dei convegni del Centro italiano di studi sul basso Medioevo, Nuova Serie, 32) Spoleto 2019, 113–134.

Zuccolin, Gabriella. "L'umidità come principio di vita. Trasformazioni di un tema classico tra medicina e filosofia." In *Fonti, flussi, onde: l'acqua tra realtà e metafora nel pensiero antico, medievale e moderno*, ed. Massimiliano Lenzi/ Olga L. Lizzini/Pina Totaro/Luisa Valente (Micrologus Library 111), Firenze 2022, 27–52.

Zuccolin, Gabriella. "Monkeys, Pygmies, and Human Beings. *Sensus disciplinales* and the Hierarchy of Living Beings in Albert the Great." In Aristotle's *De sensu* in the Latin Tradition, 1250–1650, ed. Mattia Mantovani/Roberto Zambiasi/Gabriella. Zuccolin (Micrologus Library, Special Issue 31), Firenze 2023, 85–104.

Zuccolin, Gabriella. "Seed Matters. Albert the Great on Human Generation as a Disciplinary Conflict." *Quaestio. Yearbook of the History of Metaphysics. Special Issue Natural Philosophy in Albert the Great—A Dialogue of Disciplines*, ed. Mario Loconsole, Evelina Miteva, Marilena Panarelli, 23 (2023): 55–84.

Index of Modern Authors

Index of Ancient Sources

Index of Names and Subjects